# Spring Boot
## 源码解读与原理分析

LinkedBear ◎著

人民邮电出版社

北　京

图书在版编目（CIP）数据

Spring Boot源码解读与原理分析 / LinkedBear著
． -- 北京：人民邮电出版社，2023.2
ISBN 978-7-115-60137-7

Ⅰ．①S… Ⅱ．①L… Ⅲ．①JAVA语言—程序设计
Ⅳ．①TP312.8

中国版本图书馆CIP数据核字(2022)第185763号

## 内 容 提 要

Spring Boot 是目前 Java EE 开发中颇受欢迎的框架之一。依托于底层 Spring Framework 的基础支撑，以及完善强大的特性设计，Spring Boot 已成为业界流行的应用和微服务开发基础框架。

本书共 14 章，分为 4 个部分。第 1 部分介绍 Spring Boot 底层依赖的核心容器，以及底层 Spring Framework 的两大核心特性 IOC 和 AOP；第 2 部分从底层源码角度深入剖析 Spring Boot 的全方位生命周期，包括 SpringApplication、IOC 容器、嵌入式 Web 容器和 AOP 模块的生命周期；第 3 部分针对项目开发中整合的主流场景，介绍场景模块中的核心装配和关键机制原理，如 JDBC 中的事务、Web 中的核心控制器等；第 4 部分与 Spring Boot 的运行部署相关，针对不同运行场景讲解 Spring Boot 的启动引导方式。

阅读本书之前，读者需要先对 Spring Framework 和 Spring Boot 有基本的理解与简单的框架使用经验或项目开发经验。本书的重点是 Spring Boot 的设计、思想和原理，无论是对于已经有一定基础的开发者还是已熟练使用 Spring Boot 并希望进一步提升技能和水平的开发者，本书都是他们透彻研究 Spring Boot 源码和原理的理想选择。

◆ 著　　LinkedBear
　　责任编辑　傅道坤
　　责任印制　王　郁　胡　南
◆ 人民邮电出版社出版发行　北京市丰台区成寿寺路 11 号
　　邮编　100164　电子邮件　315@ptpress.com.cn
　　网址　https://www.ptpress.com.cn
　　北京七彩京通数码快印有限公司印刷
◆ 开本：787×1092　1/16
　　印张：27.25　　　　　　　　　2023 年 2 月第 1 版
　　字数：676 千字　　　　　　　2025 年 1 月北京第 6 次印刷

定价：129.80 元

读者服务热线：（010）81055410　印装质量热线：（010）81055316
反盗版热线：（010）81055315
广告经营许可证：京东市监广登字 20170147 号

# 作者简介

LinkedBear，Java 开发工程师、底层技术研究者与分享者，倾心研究 Spring 技术体系多年，对 Spring、Spring Boot 等框架有独到的见解，拥有丰富的框架体系实践经验和架构封装经验。

# 致谢

创作本书的过程漫长而艰辛，首先特别感谢自己能狠下心来从头开始著书，得益于自己耕耘于专业擅长领域，以及渴望将自己的见解和经验分享给更多软件开发行业的朋友们，本书才能顺利完成并与各位见面。在整个创作过程中，我的家人一直在背后默默地支持着我，这使我能全身心投入到著书当中，借此机会祝他们健康平安！

本书得以顺利出版，离不开人民邮电出版社的编辑老师们，尤其是与我对接的傅道坤老师，傅老师从专业图书编辑的角度，提供了非常宝贵的经验以及专业的指导和支持，没有编辑老师们的辛勤付出就不会有本书的顺利出版，谢谢你们！

还要感谢在本书创作中帮助过我的人，包括稀土掘金的运营老师、前期一直支持我的小册的读者。本书的前身以掘金小册的形式发布，在宣传推广和阅读反馈中得到了非常多的帮助，正是得益于你们的反馈，才使得在编写本书时能够做到讲解清晰透彻、内容主次分明。

最后要感谢的是正在阅读本书的您，感谢您选择本书作为陪伴您学习 Spring Boot 源码和底层原理的"搭档"，希望本书能帮助您在学习 Spring Boot 和 Spring Framework 时汲取到尽可能多的原理、思想、方法等知识。衷心希望本书能给正在阅读的您带来帮助。祝您阅读愉快，前程似锦！

# 前言

Spring Boot 从诞生至今已有数个年头，依托其简单易用、覆盖场景广泛、满足分布式应用快速开发等特性，迅速成为互联网软件开发的首选基础框架。Spring Boot 本身设计强大、巧妙，内部蕴含着令许多开发者和使用者争相学习的设计思想。无论读者目前的开发水平如何，通过研究 Spring Boot 与 Spring Framework 的源码与底层设计，都会使能力得到不同程度的提升。但是，能够做到深挖源码和理解原理设计的人实在少之又少，总结关键原因主要包含以下几点。

- 阅读源码是一件难度极大且费时费力的工作，对开发者而言，单枪匹马深入底层研究的投入产出比太低。
- 框架源码的底层过于复杂，尤其是对于经历了近 20 年迭代的经典框架，其内部设计之精炼、覆盖之全面、结构之庞大可想而知，这也为想要研究源码的开发者提升了门槛。
- 借助网络可以找到与源码解读相关的资料和博客，但由于大多不成体系、没有来龙去脉等，引发读者"读不懂""没听说过"等尴尬体验，长期出现这种情况会引起深入学习的负反馈，最终导致退缩甚至放弃。

基于以上原因，开发者对于学习 Spring Boot 与 Spring Framework 的原理与设计是有意愿的，但由于门槛高、难度大、难成体系等障碍因素导致望而却步。为了给各位开发者提供一段相对合理、平滑、系统的源码阅读与学习轨迹，我前后花了两年半的时间编写本书，总结了自己对 Spring Boot 与 Spring Framework 的研究和理解，希望能对正在探究和准备开始学习 Spring Boot 原理的开发者提供一些帮助。

## 本书组织结构

本书分为 4 个部分，包括 Spring Boot 底层依赖的核心容器、Spring Boot 的生命周期原理分析、Spring Boot 整合常用开发场景，以及 Spring Boot 应用的运行。具体内容如下。

- 第 1 章，"Spring Boot 整体概述"：从整体层面回顾 Spring Boot 与底层依赖的 Spring Framework，以及 Spring Boot 的核心特性、Spring Boot 原生支持的技术场景整合等。
- 第 2 章，"Spring Boot 的自动装配"：从组件装配开始，讲解 Spring Boot 自动装配的技术基础支撑，随后拆解分析 Spring Boot 核心主启动类注解@SpringBootApplication，配合常见的 WebMvc 场景进行自动装配的实例分析。
- 第 3 章，"Spring Boot 的 IOC 容器"：全方面讲解 Spring Boot 底层依赖的 IOC 容器模型，以及 IOC 容器中的关键组件与设计，包括 Environment、BeanDefinition 和后置处理器等。
- 第 4 章，"Spring Boot 的核心引导：SpringApplication"：从宏观角度理解 Spring Boot 的核心引导启动类 SpringApplication，并初步了解内部的关键设计和总体生命周期。
- 第 5 章，"Spring Boot 的 AOP 支持"：回顾 Spring Framework 的 AOP 模块，以及 Spring Boot 整合 AOP 后的关键开关和装配底层。

- 第 6 章，"Spring Boot 准备容器与环境"：完整地讲解一个 Spring Boot 应用的启动流程中的 SpringApplication 部分，包含创建 SpringApplication、启动 SpringApplication 的关键逻辑，以及不同 Spring Boot 版本间设计的对比与扩展。
- 第 7 章，"IOC 容器的刷新"：全方位讲解 ApplicationContext 的容器初始化全流程，该部分对原生 Spring Framework 的 IOC 容器初始化逻辑进行深度剖析和原理解读。
- 第 8 章，"Spring Boot 容器刷新扩展：嵌入式 Web 容器"：针对 Spring Boot 的关键特性"嵌入式 Web 容器"展开讲解，以嵌入式 Tomcat 为例讲解嵌入式 Web 容器的模型设计、初始化机制和启动流程等。
- 第 9 章，"AOP 模块的生命周期"：从 AOP 模块的核心后置处理器切入，讲解 AOP 机制在收集增强器、代理 bean 对象的创建过程以及调用代理对象时内部的执行机制。
- 第 10 章，"Spring Boot 整合 JDBC"：讲解 Spring Boot 整合 JDBC 场景的自动装配，以及整合事务模块后的生效原理、控制原理和事务传播行为原理。
- 第 11 章，"Spring Boot 整合 MyBatis"：讲解 Spring Boot 整合 MyBatis 操控数据库的自动装配，以及 MyBatis 的核心组件 SqlSessionFactory 的初始化流程。
- 第 12 章，"Spring Boot 整合 WebMvc"：讲解 Spring Boot 整合 WebMvc 场景的自动装配，以及 WebMvc 中的核心组件，随后通过具体的场景讲解核心控制器 DispatcherServlet 的工作全流程。
- 第 13 章，"Spring Boot 整合 WebFlux"：从响应式编程开始讲解 WebFlux 的基础和快速使用，随后讲解 Spring Boot 整合 WebFlux 场景的自动装配，以及核心控制器 DispatcherHandler 的工作全流程。
- 第 14 章，"运行 Spring Boot 应用"：通过不同的 Spring Boot 应用打包方式分别讲解应用的引导启动流程，并讲解 2.3 版本引入的优雅停机特性。

## 目标读者

本书并不是一本 Spring Boot 的入门书，因此需要读者至少了解 Spring Boot 和 Spring Framework，并有基本的使用经验。除此之外，还希望读者对 Java SE、Java EE 的相关基础知识有一定的掌握。因此本书更适合以下读者阅读：

- 会使用 Spring Boot、Spring Framework；
- 有实际项目的开发经验，但不满足于浅层次使用现状；
- 能熟练使用 Spring Boot，但没有深入挖掘深层次特性和高层级使用；
- 职业规划目标为技术总监、架构师等高级技术岗位；
- 技术广度足够，但深度有限；
- 被 Spring Boot、Spring Framework 问题困扰的求职者；
- 有意向对 Spring 生态深入探究的研究者。

## 表达约定

本书中出现的部分名词可能会出现多种不同的称呼，以下是部分专有名词的映射关系。

- Spring Framework：指 Spring 框架，简称为 Spring。

- Bean：Spring Framework 中管理的组件对象（概念）。
- bean 对象：容器中真实存在的组件对象实例。
- IOC 容器：泛指 ApplicationContext，当上下文讲解 BeanFactory 时则指代 BeanFactory。
- Web 容器：Servlet 容器与 NIO 容器的统称，而不仅限于 Tomcat、Jetty 等 Servlet 容器。

在源码分析的过程中，考虑到框架底层的源码实现可能比较复杂，为了使读者把控源码的整体逻辑，不在细节上消耗过多精力，本书中出现的框架源码部分会有适当省略与删减。以下是几个源码片段的省略示例。

```java
protected List<String> getCandidateConfigurations(AnnotationMetadata metadata,
        AnnotationAttributes attributes){
    List<String> configurations = SpringFactoriesLoader.loadFactoryNames (getSpringFactories
LoaderFactoryClass(), getBeanClassLoader());
    // assert ......
    return configurations;
}

protected Class<?> getSpringFactoriesLoaderFactoryClass() {
    return EnableAutoConfiguration.class;
}
```

```java
@Bean(name = DEFAULT_DISPATCHER_SERVLET_BEAN_NAME)
public DispatcherServlet dispatcherServlet(WebMvcProperties webMvcProperties) {
    DispatcherServlet dispatcherServlet = new DispatcherServlet();
    // DispatcherServlet 的定制化 ......
    return dispatcherServlet;
}
```

```java
public static String getProperty(String key) {
    checkKey(key);
    // JMX ......
    // 从 Properties 中取值
    return props.getProperty(key);
}
```

## 源码版本

在编写本书时，Spring Boot 的最新正式版本是 2.5.x，而至本书定稿时 Spring Boot 已更新到 2.6.x，且 2.7.x 与 3.0.0 也正在 SNAPSHOT 版。考虑到主流的 Spring Boot 及底层依赖的 Spring Framework 版本分别为 2.x 与 5.x，而同样的大版本下的子版本中大部分是特性的添加和优化，不会使内部骨架和核心产生很大变动。基于以上几种情况，本书最终定稿时采用的 Spring Boot 源码解读基准版本是 2.3.11.RELEASE 与 2.5.3，在没有特别说明版本时，本书中引用的源码均基于 Spring Boot 2.3.11.RELEASE。

## 纠错、源码、课件

Spring Boot 和 Spring Framework 在当下的应用范围广泛，版本迭代比较频繁，加之作者本人的技术水平有限，所以书中出现错误在所难免。虽然作者在编写本书时已经对每个知识点反复研究和推敲，力求讲解得尽可能正确、精准、易懂，但仍然不敢保证内容没有错误。读者如果发现本书中有任何错误，或者想给本书及作者提供任何建议，欢迎通过以下方式与作者取得联系，以便于及时修正本书的错误和疏漏。

- 邮箱：LinkedBear@163.com。
- 微信公众号：老熊说 Spring。
- GitHub 博客：https://linkedbear.github.io。

本书的勘误将发布在微信公众号与 GitHub 博客中，请各位读者及时关注，以便获取最新的更新信息。

本书附带的所有测试代码及课件已托管至 GitHub 平台，欢迎各位读者下载参考。GitHub 仓库地址：https://github.com/LinkedBear/spring-boot-source-analysis-epubit。

LinkedBear（苏振志）

2022 年 6 月

# 资源与支持

本书由异步社区出品,社区(https://www.epubit.com/)为您提供相关资源和后续服务。

## 提交勘误

作者和编辑尽最大努力来确保书中内容的准确性,但难免会存在疏漏。欢迎您将发现的问题反馈给我们,帮助我们提升图书的质量。

当您发现错误时,请登录异步社区,按书名搜索,进入本书页面,单击"提交勘误",输入勘误信息,单击"提交"按钮即可。本书的作者和编辑会对您提交的勘误进行审核,确认并接受后,您将获赠异步社区的100积分。积分可用于在异步社区兑换优惠券、样书或奖品。

## 扫码关注本书

扫描下方二维码,您将会在异步社区微信服务号中看到本书信息及相关的服务提示。

## 与我们联系

我们的联系邮箱是 contact@epubit.com.cn。

如果您对本书有任何疑问或建议,请您发邮件给我们,并请在邮件标题中注明本书书名,以便我们更高效地做出反馈。

如果您有兴趣出版图书、录制教学视频,或者参与图书技术审校等工作,可以发邮件给本书的责任编辑(fudaokun@ptpress.com.cn)。

如果您来自学校、培训机构或企业,想批量购买本书或异步社区出版的其他图书,也可以发邮件给我们。

如果您在网上发现有针对异步社区出品图书的各种形式的盗版行为,包括对图书全部或部分内容的非授权传播,请您将怀疑有侵权行为的链接通过邮件发给我们。您的这一举动是对作者权益的保护,也是我们持续为您提供有价值的内容的动力之源。

## 关于异步社区和异步图书

"**异步社区**"是人民邮电出版社旗下IT专业图书社区,致力于出版精品IT技术图书和相关

# 资源与支持

学习产品,为作译者提供优质出版服务。异步社区创办于 2015 年 8 月,提供大量精品 IT 技术图书和电子书,以及高品质技术文章和视频课程。更多详情请访问异步社区官网 https://www.epubit.com。

"**异步图书**"是由异步社区编辑团队策划出版的精品 IT 专业图书的品牌,依托于人民邮电出版社的计算机图书出版积累和专业编辑团队,相关图书在封面上印有异步图书的 LOGO。异步图书的出版领域包括软件开发、大数据、AI、测试、前端、网络技术等。

异步社区

微信服务号

# 目录

## 第 1 部分  Spring Boot 底层依赖的核心容器

第 1 章　Spring Boot 整体概述 ········ 3
1.1　Spring Framework ············· 3
　　1.1.1　Spring Framework 的历史 ····· 4
　　1.1.2　IOC 与 AOP ··············· 4
1.2　Spring Boot 与 Spring Framework ··· 4
1.3　Spring Boot 的核心特性 ·········· 5
1.4　Spring Boot 的体系 ············· 5
1.5　开发第一个 Spring Boot 应用 ······ 6
　　1.5.1　创建项目 ················· 6
　　1.5.2　编写简单代码 ············ 10
1.6　小结 ························ 11

第 2 章　Spring Boot 的自动装配 ······ 12
2.1　组件装配 ···················· 12
　　2.1.1　组件 ··················· 12
　　2.1.2　手动装配 ················ 13
　　2.1.3　自动装配 ················ 13
2.2　Spring Framework 的模块装配 ···· 14
　　2.2.1　模块 ··················· 14
　　2.2.2　快速体会模块装配 ········ 15
　　2.2.3　导入配置类 ·············· 17
　　2.2.4　导入 ImportSelector ······· 19
　　2.2.5　导入 ImportBeanDefinitionRegistrar ················ 21
　　2.2.6　扩展：DeferredImportSelector ··· 22
2.3　Spring Framework 的条件装配 ···· 24
　　2.3.1　基于 Profile 的装配 ········ 24
　　2.3.2　基于 Conditional 的装配 ···· 26
2.4　SPI 机制 ···················· 28
　　2.4.1　JDK 原生的 SPI ··········· 29
　　2.4.2　Spring Framework 3.2 的 SPI ··· 30
2.5　Spring Boot 的装配机制 ········· 32
　　2.5.1　@ComponentScan ········ 33
　　2.5.2　@SpringBootConfiguration ··· 34
　　2.5.3　@EnableAutoConfiguration ··· 35

2.6　WebMvc 场景下的自动装配原理 ·················· 42
　　2.6.1　Servlet 容器的装配 ········ 43
　　2.6.2　DispatcherServlet 的装配 ···· 46
　　2.6.3　SpringWebMvc 的装配 ····· 48
2.7　小结 ························ 53

第 3 章　Spring Boot 的 IOC 容器 ····· 54
3.1　Spring Framework 的 IOC 容器 ··· 54
　　3.1.1　BeanFactory ·············· 55
　　3.1.2　ApplicationContext ········ 66
　　3.1.3　选择 ApplicationContext 而不是 BeanFactory ········· 74
3.2　Spring Boot 对 IOC 容器的扩展 ··· 75
　　3.2.1　WebServerApplicationContext ··· 75
　　3.2.2　AnnotationConfigServletWebServerApplicationContext ····· 75
　　3.2.3　ReactiveWebApplicationContext ················· 76
3.3　选用注解驱动 IOC 容器的原因 ···· 76
　　3.3.1　配置方式的对比 ·········· 76
　　3.3.2　约定大于配置下的选择 ···· 77
3.4　Environment ·················· 77
　　3.4.1　Environment 概述 ········· 77
　　3.4.2　Environment 的结构与设计 ··· 78
　　3.4.3　Environment 与 IOC 容器的关系 ···················· 80
3.5　BeanDefinition ················ 81
　　3.5.1　理解元信息 ·············· 81
　　3.5.2　BeanDefinition 概述 ······· 81
　　3.5.3　BeanDefinition 的结构与设计 ··· 82
　　3.5.4　体会 BeanDefinition ······· 85
　　3.5.5　BeanDefinitionRegistry ····· 88
　　3.5.6　设计 BeanDefinition 的意义 ··· 89
3.6　后置处理器 ·················· 89
　　3.6.1　理解后置处理器 ·········· 89
　　3.6.2　BeanPostProcessor ········ 90

3.6.3 BeanPostProcessor 的扩展……91
3.6.4 BeanFactoryPostProcessor……93
3.6.5 BeanDefinitionRegistryPost
Processor……94
3.6.6 后置处理器对比……95
3.7 IOC 容器的启动流程……96
3.8 小结……100

## 第 4 章 Spring Boot 的核心引导：SpringApplication……101
4.1 总体设计……101
4.1.1 启动失败的错误报告……101
4.1.2 Bean 的延迟初始化……103
4.1.3 SpringApplication 的定制……103
4.1.4 Web 类型推断……104
4.1.5 监听与回调……104
4.1.6 应用退出……106

4.2 生命周期概述……107
4.2.1 创建 SpringApplication……107
4.2.2 启动 SpringApplication……107
4.2.3 应用退出……108
4.3 小结……108

## 第 5 章 Spring Boot 的 AOP 支持……109
5.1 Spring Framework 的 AOP 回顾……109
5.1.1 AOP 术语……109
5.1.2 通知类型……110
5.2 Spring Boot 使用 AOP……110
5.3 AOP 的开关：@EnableAspectJAutoProxy……111
5.3.1 AspectJAutoProxyRegistrar……112
5.3.2 AnnotationAwareAspectJAutoProxyCreator……114
5.4 小结……117

# 第 2 部分 Spring Boot 的生命周期原理分析

## 第 6 章 Spring Boot 准备容器与环境……121
6.1 创建 SpringApplication……122
6.1.1 推断 Web 环境……122
6.1.2 设置初始化器……123
6.1.3 设置监听器……125
6.1.4 确定主启动类……126
6.1.5 与 Spring Boot 1.x 的区别……127
6.1.6 与 Spring Boot 2.4.x 的区别……128
6.2 启动 SpringApplication……129
6.2.1 前置准备……130
6.2.2 获取 SpringApplicationRunListeners……133
6.2.3 准备运行时环境……135
6.3 IOC 容器的创建与初始化……137
6.3.1 打印 Banner……137
6.3.2 创建 IOC 容器……140
6.3.3 初始化 IOC 容器……142
6.3.4 刷新 IOC 容器……145
6.3.5 Spring Boot 2.4.x 的新特性……145
6.4 IOC 容器刷新后的回调……148

6.5 小结……149

## 第 7 章 IOC 容器的刷新……150
7.1 初始化前的预处理……152
7.1.1 初始化属性配置……152
7.1.2 初始化早期事件的集合……154
7.2 obtainFreshBeanFactory——初始化 BeanFactory……154
7.2.1 注解驱动的 refreshBeanFactory……155
7.2.2 XML 驱动的 refreshBeanFactory……155
7.3 prepareBeanFactory——BeanFactory 的预处理动作……156
7.3.1 ApplicationContextAwareProcessor……157
7.3.2 自动注入的支持……158
7.3.3 ApplicationListenerDetector……159
7.4 postProcessBeanFactory——BeanFactory 的后置处理……160
7.4.1 回调父类方法……161
7.4.2 组件扫描&解析手动传入的配置类……164

7.5 invokeBeanFactoryPost
    Processors——执行 BeanFactory
    PostProcessor ·················· 164
    7.5.1 现有的后置处理器分类 ········· 165
    7.5.2 执行最高优先级的 BeanDefinition
          RegistryPostProcessor ·········· 165
    7.5.3 执行其他 BeanDefinition
          RegistryPostProcessor ·········· 166
    7.5.4 回调 postProcessBeanFactory
          方法 ························ 167
    7.5.5 BeanFactoryPostProcessor
          的分类 ······················ 168
    7.5.6 执行 BeanFactoryPostProcessor ····· 168
    7.5.7 重要的后置处理器：Configuration
          ClassPostProcessor ············ 169
7.6 registerBeanPostProcessors——
    初始化 BeanPostProcessor ······· 185
    7.6.1 BeanPostProcessorChecker ······· 186
    7.6.2 MergedBeanDefinitionPost
          Processor 被重复注册 ············ 187
    7.6.3 PriorityOrdered 类型的后置
          处理器 ······················ 188
7.7 initMessageSource——初始化
    国际化组件 ····················· 188
7.8 initApplicationEventMulticaster——
    初始化事件广播器 ··············· 190
7.9 onRefresh——子类扩展的
    刷新动作 ······················ 191
7.10 registerListeners——注册
     监听器 ······················· 191
7.11 finishBeanFactoryInitialization——
     初始化剩余的单实例 bean 对象 ····· 192
     7.11.1 beanFactory.preInstantiate
            Singletons ················· 193
     7.11.2 getBean ·················· 193
     7.11.3 createBean ················ 199
     7.11.4 doCreateBean ·············· 201
     7.11.5 SmartInitializingSingleton ····· 215
7.12 finishRefresh——刷新后的
     动作 ························ 216
     7.12.1 LifecycleProcessor ·········· 216
     7.12.2 getLifecycleProcessor().
            onRefresh() ··············· 217
7.13 resetCommonCaches——
     清除缓存 ···················· 217

7.14 ApplicationContext 初始化中
     的扩展点 ···················· 218
     7.14.1 invokeBeanFactoryPost
            Processors ················ 218
     7.14.2 finishBeanFactoryInitialization··· 219
7.15 循环依赖的解决方案 ············ 221
     7.15.1 循环依赖的产生 ············ 221
     7.15.2 循环依赖的解决模型 ········· 222
     7.15.3 基于 setter/@Autowired 的
            循环依赖 ·················· 222
     7.15.4 基于构造方法的循环依赖 ····· 230
     7.15.5 基于原型 Bean 的循环依赖 ···· 230
     7.15.6 引入 AOP 的额外设计 ········ 231
7.16 小结 ······················· 232

## 第 8 章 Spring Boot 容器刷新扩展：
        嵌入式 Web 容器 ············ 233

8.1 嵌入式 Tomcat 简介 ············ 233
    8.1.1 嵌入式 Tomcat 与普通 Tomcat ··· 234
    8.1.2 Tomcat 整体架构 ············· 234
    8.1.3 Tomcat 的核心工作流程 ········ 235
8.2 Spring Boot 中嵌入式容器
    的模型 ······················ 236
    8.2.1 WebServer ················· 236
    8.2.2 WebServerFactory ············ 236
    8.2.3 ServletWebServerFactory 和
          ReactiveWebServerFactory ······ 237
    8.2.4 ConfigurableServletWebServer
          Factory ···················· 237
8.3 嵌入式 Web 容器的初始化
    时机 ························ 237
    8.3.1 创建 WebServer ············· 238
    8.3.2 Web 容器关闭相关的回调 ······ 241
8.4 嵌入式 Tomcat 的初始化 ········ 242
    8.4.1 获取 Context ··············· 243
    8.4.2 阻止 Connector 初始化 ········ 244
    8.4.3 启动 Tomcat ··············· 244
    8.4.4 阻止 Tomcat 结束 ············ 246
8.5 嵌入式 Tomcat 的启动 ·········· 248
8.6 小结 ······················· 249

## 第 9 章 AOP 模块的生命周期 ······ 250

9.1 @EnableAspectJAutoProxy ······ 250
9.2 AnnotationAwareAspectJAuto
    ProxyCreator ················· 252
    9.2.1 类继承结构 ················· 253

9.2.2 初始化时机 ················· 253
9.2.3 作用时机 ··················· 254
9.3 Advisor 与切面类的收集 ········ 257
　9.3.1 收集增强器的逻辑 ······· 257
　9.3.2 收集原生增强器 ·········· 258
　9.3.3 解析 AspectJ 切面封装增强器 ··· 259
9.4 TargetSource 的设计 ············· 266
　9.4.1 TargetSource 的设计 ······ 267
　9.4.2 TargetSource 的好处 ······ 267
　9.4.3 TargetSource 的结构 ······ 267
　9.4.4 Spring Framework 中提供的 TargetSource ··············· 268
9.5 代理对象生成的核心：wrapIfNecessary ················ 268
9.5.1 getAdvicesAndAdvisorsFor
　　　Bean ························· 269
9.5.2 createProxy ················· 274
9.6 代理对象的底层执行逻辑 ······ 277
　9.6.1 DemoService#save ········ 277
　9.6.2 获取增强器链 ·············· 278
　9.6.3 执行增强器 ················· 281
　9.6.4 JDK 动态代理的执行底层 ··· 285
　9.6.5 AspectJ 中通知的底层实现 ··· 287
9.7 AOP 通知的执行顺序对比 ······ 289
　9.7.1 测试代码编写 ·············· 289
　9.7.2 Spring Framework 5.x 的顺序 ··· 290
　9.7.3 Spring Framework 4.x 的顺序 ··· 291
9.8 小结 ······························ 292

# 第 3 部分　Spring Boot 整合常用开发场景

## 第 10 章　Spring Boot 整合 JDBC ········· 295
10.1 Spring Boot 整合 JDBC
　　　项目搭建 ······················· 295
　10.1.1 初始化数据库 ············ 295
　10.1.2 整合项目 ·················· 296
　10.1.3 编写测试代码 ············ 296
10.2 整合 JDBC 后的自动装配 ······ 297
　10.2.1 配置数据源 ··············· 298
　10.2.2 创建 JdbcTemplate ······ 302
　10.2.3 配置事务管理器 ········· 303
10.3 声明式事务的生效原理 ········ 303
　10.3.1 TransactionAutoConfiguration ················ 303
　10.3.2 TransactionManagementConfigurationSelector ····· 305
　10.3.3 AutoProxyRegistrar ······ 305
　10.3.4 ProxyTransactionManagementConfiguration ··············· 307
10.4 声明式事务的控制全流程 ····· 309
　10.4.1 CglibAopProxy#intercept ··· 309
　10.4.2 TransactionInterceptor ··· 310
10.5 声明式事务的传播行为控制 ··· 319
　10.5.1 修改测试代码 ············ 320
　10.5.2 PROPAGATION_REQUIRED ··· 321
　10.5.3 PROPAGATION_REQUIRES_NEW ························· 327

10.6 小结 ···························· 330

## 第 11 章　Spring Boot 整合 MyBatis ····· 332
11.1 MyBatis 框架概述 ············· 332
11.2 Spring Boot 整合 MyBatis
　　　项目搭建 ······················· 333
11.3 自动装配核心 ··················· 334
　11.3.1 场景启动器的秘密 ······ 334
　11.3.2 MybatisLanguageDriverAutoConfiguration ·············· 335
　11.3.3 MybatisAutoConfiguration ··· 335
11.4 小结 ···························· 342

## 第 12 章　Spring Boot 整合 WebMvc ····· 343
12.1 整合 WebMvc 的核心
　　　自动装配 ······················· 343
12.2 WebMvc 的核心组件 ·········· 344
　12.2.1 DispatcherServlet ········ 344
　12.2.2 Handler ···················· 345
　12.2.3 HandlerMapping ········· 345
　12.2.4 HandlerAdapter ·········· 347
　12.2.5 ViewResolver ············ 348
12.3 @Controller 控制器装配原理 ··· 349
　12.3.1 初始化 RequestMapping
　　　　 的入口 ···················· 349
　12.3.2 processCandidateBean ··· 350
　12.3.3 detectHandlerMethods ··· 350

12.4　DispatcherServlet 的工作
　　　全流程解析 ·················· 352
　　12.4.1　DispatcherServlet#service ······ 352
　　12.4.2　processRequest ············ 353
　　12.4.3　doService ················ 354
　　12.4.4　doDispatch ··············· 356
　　12.4.5　DispatcherServlet 工作
　　　　　　全流程小结 ·············· 372
12.5　小结 ······················· 372

## 第 13 章　Spring Boot 整合 WebFlux ····· 374
13.1　快速了解响应式编程与
　　　Reactor ···················· 374
　　13.1.1　命令式与响应式 ············ 374
　　13.1.2　概念和思想的回顾与引入 ······ 375
　　13.1.3　快速体会 Reactor 框架 ········ 377
13.2　快速使用 WebFlux ············ 380
　　13.2.1　WebMvc 的开发风格 ········ 380
　　13.2.2　逐步过渡到 WebFlux ········ 381
　　13.2.3　WebFlux 的函数式开发 ······· 382
　　13.2.4　WebMvc 与 WebFlux 的
　　　　　　对比 ···················· 383

13.3　WebFlux 的自动装配 ·········· 384
　　13.3.1　ReactiveWebServerFactory
　　　　　　AutoConfiguration ········· 384
　　13.3.2　WebFluxAutoConfiguration ···· 385
　　13.3.3　WebFluxConfig ············· 386
　　13.3.4　EnableWebFluxConfiguration ···· 387
　　13.3.5　WebFluxConfigurationSupport ··· 388
13.4　DispatcherHandler 的传统
　　　方式工作原理 ··············· 390
　　13.4.1　handle 方法概览 ············ 390
　　13.4.2　筛选 HandlerMapping ········ 391
　　13.4.3　搜寻 HandlerAdapter 并执行 ··· 393
　　13.4.4　返回值处理 ················ 394
　　13.4.5　工作流程小结 ·············· 395
13.5　DispatcherHandler 的函数式
　　　端点工作原理 ··············· 396
　　13.5.1　HandlerMapping 的不同 ······ 396
　　13.5.2　HandlerAdapter 的不同 ······· 397
　　13.5.3　返回值处理的不同 ·········· 398
　　13.5.4　工作流程小结 ·············· 399
13.6　小结 ······················· 399

# 第 4 部分　Spring Boot 应用的运行

## 第 14 章　运行 Spring Boot 应用 ·········· 403
14.1　部署打包的两种方式 ·········· 403
　　14.1.1　以可独立运行 jar 包的
　　　　　　方式 ···················· 403
　　14.1.2　以 war 包的方式 ············ 404
14.2　基于 jar 包的独立运行机制 ···· 405
　　14.2.1　可运行 jar 包的前置知识 ····· 405
　　14.2.2　Spring Boot 的可运行 jar
　　　　　　包结构 ·················· 405
　　14.2.3　JarLauncher 的设计及
　　　　　　工作原理 ················ 407

14.3　基于 war 包的外部 Web 容器
　　　运行机制 ··················· 412
　　14.3.1　Servlet 3.0 规范中引导应用
　　　　　　启动的说明 ·············· 413
　　14.3.2　Spring BootServletInitializer
　　　　　　的作用和原理 ············ 413
14.4　Spring Boot 2.3 新特性：
　　　优雅停机 ··················· 415
　　14.4.1　测试优雅停机场景 ·········· 416
　　14.4.2　优雅停机的实现原理 ········ 417
14.5　小结 ······················· 419

# 第 1 部分
# Spring Boot 底层依赖的核心容器

▶ 第 1 章  Spring Boot 整体概述

▶ 第 2 章  Spring Boot 的自动装配

▶ 第 3 章  Spring Boot 的 IOC 容器

▶ 第 4 章  Spring Boot 的核心引导：SpringApplication

▶ 第 5 章  Spring Boot 的 AOP 支持

# 第1章 Spring Boot 整体概述

**本章主要内容：**
◇ Spring Framework 与 Spring Boot 概述；
◇ Spring Boot 与 Spring Framework 的关系；
◇ Spring Boot 的核心特性；
◇ Spring Boot 的体系；
◇ 构建和开发第一个基于 Spring Boot 的应用。

自 2013 年开始，Pivotal 公司的 Spring Framework 开发团队就在着手准备 Spring Boot 项目的开发，他们的想法是打造一个"支持不依赖外部容器的 Web 应用程序体系结构"。2014 年 3 月 28 日，Spring Boot 1.0.0 版正式对外发布，这意味着基于 Spring Framework 的 Web 应用开发进入了一个新的模式：**开发者不必再纠结烦琐的配置、环境的部署等琐碎问题，而只需专注于业务的开发。**

就其本身而言，Spring Boot 不是一个全新的框架，而是基于 Spring Framework 的"二次封装"，因此首先需要对 Spring Framework 有一个全面、清晰的认识。

## 1.1 Spring Framework

Spring Framework 是由 Rod Johnson 与 Juergen Hoeller 为首发起的一个开源的、松耦合的、分层的、可配置的一站式企业级 Java 开发框架。它的核心是 IOC 与 AOP，可以更容易地构建企业级 Java 应用，并且可以根据应用开发组件的需要，整合对应的技术。为了使读者更容易理解，下面对其中几个关键词分别进行阐述。

- **IOC 与 AOP**：Spring Framework 的两大核心特性，即**控制反转**（Inverse of Control，IOC）、**面向切面编程**（Aspect Oriented Programming，AOP）。
- **松耦合**：IOC 和 AOP 两大特性可以尽可能地将对象之间的关系解耦。
- **可配置**：提供外部化配置的方式，可以灵活地配置容器及容器中的 Bean。
- **一站式**：覆盖企业级开发中的所有领域（包括 JavaWeb、分布式、微服务，甚至 Java SE、GUI 项目等）。
- **第三方整合**：Spring Framework 可以很方便地整合第三方技术（如持久层框架 MyBatis 和 Hibernate、表现层框架 SpringWebMvc 和 Struts2、权限校验框架 Spring Security 和 Shiro 等）。

总的来看，Spring Framework 是一个功能极其强大的基础框架，其设计考虑之周全，使得几乎任何 Java 应用都可以从中受益。

### 1.1.1　Spring Framework 的历史

Spring Framework 的诞生是为了替代 J2EE 时期的 EJB 规范体系。EJB 作为初期流行的 J2EE 企业级项目开发技术解决方案，设计过于复杂和笨重，导致当时的 Java 开发者对 EJB 不满，并造成了一种矛盾的现象：大家怀疑 EJB 难用，但又不敢说 EJB 难用。

在这些开发者中，有一小部分优秀且大胆的开发者敢于发声、敢于质疑，他们认为 EJB 的确复杂且笨重，于是 Rod Johnson（也就是 Spring Framework 的创始人之一）在 2002 年写了一本书，书名为 *Expert One-on-One J2EE Design and Development*，其中对当时 J2EE 应用的架构和框架存在的臃肿、低效等问题提出了质疑，并且积极寻找和探索解决方案。2004 年 Spring Framework 1.0.0 横空出世。随后 Rod Johnson 又写了一本书，这本书在当时的 J2EE 开发界引起了巨大轰动，它就是著名的 *Expert One-on-One J2EE Development without EJB*。Rod Johnson 在这本书中直接指明完全可以不使用 EJB 开发 J2EE 应用，而是可以用一种更轻量级、更简单的框架来代替，这就是 Spring Framework。

时间证明，使用了 Spring Framework 后的项目，在开发阶段的效率的确比 EJB 高，而且 Spring Framework 提供的一些特性也比 EJB 强大。于是开发者开始慢慢转用 Spring Framework，并逐步淘汰了 EJB。随着后续 Spring Framework 版本的迭代，加上越来越多的开发者使用和认可 Spring Framework，如今它已经成为现代 Java EE 开发的标杆。

### 1.1.2　IOC 与 AOP

Spring Framework 的两大核心特性，分别是**控制反转**（Inverse of Control，IOC）、**面向切面编程**（Aspect Oriented Programming，AOP）。

IOC 的最直接体现，就是作为 Spring Framework 的**核心容器**，这个核心容器又被称为 IOC 容器，它在内部管理了基于 Spring Framework 的应用中会用到的所有组件（即 Bean）。在实际的项目开发中，可以通过一些模式注解（如`@Component` 等）标注在指定的类上，配合组件扫描，可以实现组件装配到核心容器。如果使用带有特定意义的模式注解，则 Spring Framework 会认定其有特殊功能（如对于`@Controller` 注解标注的类，SpringWebMvc 会认定其为一个 Web Controller，具有请求处理、视图跳转等功能）。

容器除了可以管理 bean 对象，还可以对这些对象进行增强，使其具有一些其他的特性（如对于`@Transactional` 注解标注的类，在引入和开启事务管理期间，其方法执行时会自动应用事务），而增强的核心就是依赖 AOP 技术。使用 AOP 技术可以通过预编译/运行时动态代理的方式，对目标对象动态添加功能特性。AOP 的应用可以使核心业务逻辑与系统级服务（如事务控制、日志审计、权限校验等）分离，从而实现组件功能的"可插拔"。

## 1.2　Spring Boot 与 Spring Framework

简单了解 Spring Framework 后，下面聚焦 Spring Boot 框架。Spring Boot 本身不是一个新的框架，而是基于 Spring Framework 之上进行的"二次封装"，因此 Spring Boot 的底层还是 Spring

Framework，基于 Spring Boot 的应用仍然是 Spring Framework 的应用，Spring Boot 做的只是帮助开发者整合不同场景下的依赖，以及提供默认的配置等。可以这样理解，Spring Boot 是开发者与 Spring Framework 之间的一道中间层，它帮助开发者完成部分基于 Spring Framework 的项目的配置、管理、部署等工作，目的是为开发者"减负"，让开发者专注于项目中的业务开发，而无须把一部分精力浪费在项目环境搭建和琐碎的配置上。

另外注意一点，WebMvc 对于 Spring Boot 而言只是生态中的一个模块，通过引入 WebMvc 的依赖，可以使项目支持 SpringWebMvc 的开发，同时也会引入对应的嵌入式 Web 容器来支撑项目的运行。

## 1.3 Spring Boot 的核心特性

Spring Boot 设计之初的目的是简化基于 Spring Framework 的项目搭建和应用开发，而不是替代 Spring Framework，因此 Spring Boot 提供了以下几个核心特性来帮助开发者省略/简化配置，以及构建企业级应用。

- **约定大于配置**（convention over configuration）：Spring Boot 对日常开发中比较常见的场景都提供了约定的默认配置，并基于自动装配机制，将场景中通常必需的组件都注册好，以此来达到少配置、甚至不配置就能正常启动项目的效果。
- **场景启动器 starter**：Spring Boot 对常用的场景都进行了整合，将这些场景中所需的依赖都收集整理到一个依赖中，并在其中添加默认的配置，使项目开发中只需要导入一个依赖，即可实现场景技术的整合。
- **自动装配**：Spring Boot 基于 Spring Framework 的模块装配+条件装配，可以在具体场景下自动引入所需的配置类并解析执行，而且可以根据项目代码中已经配置的内容，动态注册缺少/必要的组件，以此实现约定大于配置的效果。
- **嵌入式 Web 容器**：Spring Boot 在运行时可以不依赖外部的 Web 容器，而是使用内部嵌入式 Web 容器来支撑应用的运行。也正因如此，基于 Spring Boot 的应用可以直接以一个单体应用的 jar 包运行。
- **生产级的特性**：Spring Boot 提供了一些很有用的生产运维型的功能特性，比如健康检查、监控指标、外部化配置等。

## 1.4 Spring Boot 的体系

截至编写本书时，Spring Boot 的版本已经发展到 2.5.x 和 2.6.x（2.7.x 和 3.x 也发布了里程碑版本），已经整合了非常多的技术场景。以下内容列举了 Spring Boot 支持的常见场景整合。

- SpringWebMvc & SpringWebFlux——Web 应用开发。
- Thymeleaf & Freemarker——Web 视图渲染。
- Spring Security——安全控制。
- Spring Data Access——数据访问（SQL & NoSQL）。
- Spring Cache——缓存实现。
- Spring Message——消息中间件（JMS & AMQP）。
- Spring Quartz——定时任务。

- Spring Distribution Transaction——分布式事务（JTA）。
- Spring Session——分布式 Session。
- Container Image——容器镜像构建支持。

……

可以发现，Spring Boot 可以整合的技术场景非常多，项目需要用到特定的场景或技术时，只需导入对应的启动器依赖，之后编写少量配置甚至不配置，就可以达到场景整合的效果。另外，基于 starter 场景启动器的整合不需要开发者考虑版本问题，Spring Boot 早已帮助开发者考虑并适配好，开发者只需导入后使用即可。

> 提示：本书使用 SpringWebMvc 指代基于 Servlet 的 Web 开发，读者可能对 SpringWebMvc 的更熟悉的叫法是 SpringMVC，这两者指代的是同一项技术。为了更好、更清楚地区分 WebMvc 与 WebFlux，本书后续提到的所有 SpringMVC 统称为 SpringWebMvc。

## 1.5 开发第一个 Spring Boot 应用

在本章的最后一节中，我们来构建一个最基本的 Spring Boot 应用。

> 提示：在开始进行第一个 Spring Boot 应用开发之前，先强调一下使用的 Spring Boot 版本。本书使用的 Spring Boot 版本主要有两个：2.3.11 与 2.5.3。之所以选择这两个版本，是考虑到两个重要的原因。其一，Spring Boot 2.3.11 是基于 Spring Framework 5.2.x 的较新版本，而 Spring Boot 2.5.3 是基于 Spring Framework 5.3.x 的现行较新版本，两个 Spring Framework 版本之间会有一些差异，因此本书通过两个 Spring Boot 版本来顺势区分 Spring Framework 的版本。其二，Spring Boot 2.4.x 推出了一些新的特性，这些特性与 Spring Boot 2.3.x 及之前版本的开发方式、底层实现有所不同，因此读者也需要分别研究这两个 Spring Boot 大版本的底层设计。综合以上两个因素，加上本书编写的时间，最终选择 2.3.11 和 2.5.3 这两个版本作为研究对象。

### 1.5.1 创建项目

创建基于 Spring Boot 项目的方式有很多种，本书主要回顾两种常用的方式。

#### 1. 基于 Spring Initializer

在 Spring Boot 的官方网站中，OVERVIEW 选项卡的底部有一个醒目的板块（见图 1-1），它告诉我们可以通过 Spring Initializer 来初始化基于 Spring Boot 的项目。

图 1-1　Spring Boot 的官方网站下方有跳转至 Spring Initializer 的入口

单击 Spring Initializr 即可跳转到在线的 Spring 项目初始化向导网站，网站的界面是如图 1-2 所示的表单。

1.5 开发第一个 Spring Boot 应用

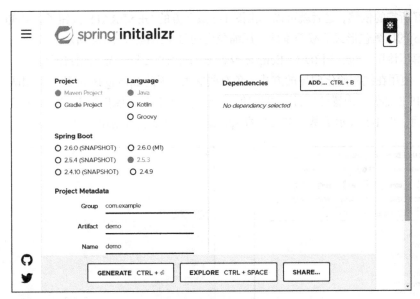

图 1-2 Spring Initializer 的初始界面

表单中的可选项较多，简单说明如下。

- Project：项目创建工具，可选 Maven 或 Gradle。
- Language：项目编码所使用的开发语言，可选 Java、Kotlin、Groovy。
- Spring Boot：选择所使用的 Spring Boot 版本（注意此处只能选择最新的 3 个大版本的最新 RELEASE 版和紧跟着要"转正"的 SNAPSHOT 版）。
- Project Metadata：项目的基本信息，包含 Group、Artifact、Spring Boot 启动类所在的包、打包方式、Java 语言版本等。
- Dependencies：项目所使用的依赖，这个位置可以搜索并选择 Spring Boot 预先准备好的一些 starter 场景启动器（见图 1-3）。

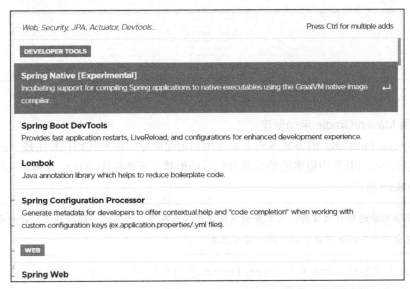

图 1-3 搜索、选择项目所需的依赖

选择好所需的依赖，之后就可以单击图 1-2 最下方的 GENERATE 按钮了，生成的代码会以 [工程名.zip]的压缩包形式下载至本地。压缩包内的目录结构如图 1-4 所示。

之后解压该压缩包，用 IDE（Eclipse 或者 Intellij IDEA 等）导入项目即可。

当然，使用在线初始化向导的方式未免有些麻烦，因此 Spring Boot 的开发团队给目前市场上主流的 IDE 都做了内置的插件，可以使用 Spring Initializer 插件来更方便地创建基于 Spring Boot 的项目。图 1-5 展示了基于 IDEA 的 Spring Initializer 项目创建入口。

图 1-4　生成的代码目录结构

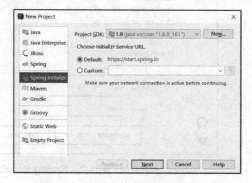

图 1-5　IDEA 中整合的 Spring Initializer

后续的操作与在线 Spring Initializer 基本一致，包括填写的项目信息与依赖的选项也非常相似（见图 1-6），这里不再赘述。

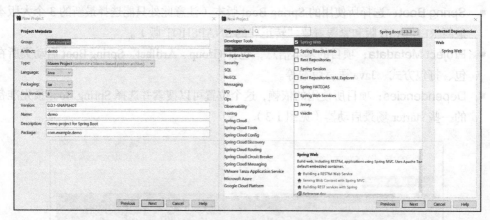

图 1-6　项目的基本信息填写与在线 Spring Initializer 一致

### 2. 使用 Maven/Gradle 手动创建

使用 Spring Initializer 的方式本质上是使用初始化向导的固定代码模板生成 Maven/Gradle 项目。在实际的项目开发中更多的情况是自行手动创建。下面使用 IDEA 开发工具新建一个简单的 Spring Boot 项目。

> 💡 提示：本书后续的所有项目都使用本节创建的目录作为基准。为了统一维护项目，此处会创建两个项目（包含一个 parent 项目，用于统一管理版本）。

在 IDEA 中先创建一个空项目 Empty Project，再依次创建一个 parent 模块和 quickstart 模块。创建完成后的项目结构如图 1-7 所示。

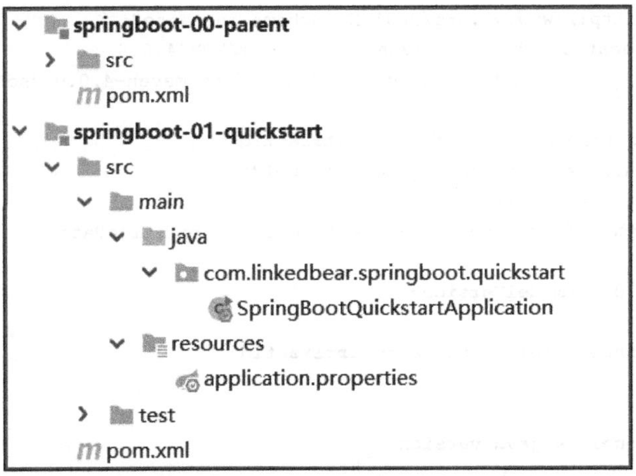

## 1.5 开发第一个 Spring Boot 应用

```
▼ 📁 springboot-00-parent
    ▶ 📁 src
      📄 pom.xml
▼ 📁 springboot-01-quickstart
    ▼ 📁 src
        ▼ 📁 main
            ▼ 📁 java
                ▼ 📁 com.linkedbear.springboot.quickstart
                    © SpringBootQuickstartApplication
            ▼ 📁 resources
                📄 application.properties
        ▶ 📁 test
      📄 pom.xml
```

图 1-7 最终创建完成的项目结构

其中 parent 模块的 `pom.xml` 只需要继承 `spring-boot-starter-parent` 项目,定义模块所使用的 Java 版本,如代码清单 1-1 所示。

**│代码清单 1-1　springboot-00-parent 的 pom.xml 定义**

```xml
<?xml version="1.0"encoding="UTF-8"?>
<project xmlns="http://maven.apache.org/POM/4.0.0"
    xmlns:xsi="http://www.w3.org/2001/XMLSchema-instance"
    xsi:schemaLocation="http://maven.apache.org/POM/4.0.0
                        http://maven.apache.org/xsd/maven-4.0.0.xsd">
    <modelVersion>4.0.0</modelVersion>
    <parent>
        <groupId>org.springframework.boot</groupId>
        <artifactId>spring-boot-starter-parent</artifactId>
        <version>2.3.11.RELEASE</version>
    </parent>
    <groupId>com.linkedbear.springboot</groupId>
    <artifactId>springboot-00-parent</artifactId>
    <version>1.0-RELEASE</version>
    <packaging>pom</packaging>
    
    <modules>
        <module>../springboot-01-quickstart</module>
    </modules>
    
    <properties>
        <java.version>1.8</java.version>
    </properties>
</project>
```

代码清单 1-2 的 quickstart 模块继承自 parent 模块,并添加了 `spring-boot-starter-web` 依赖,即可完成基于 WebMvc 的场景整合,如代码清单 1-2 所示。

**│代码清单 1-2　springboot-01-quickstart 的 pom.xml 定义**

```xml
<?xml version="1.0"encoding="UTF-8"?>
<project xmlns="http://maven.apache.org/POM/4.0.0"
```

```xml
        xmlns:xsi="http://www.w3.org/2001/XMLSchema-instance"
        xsi:schemaLocation="http://maven.apache.org/POM/4.0.0
                            http://maven.apache.org/xsd/maven-4.0.0.xsd">
    <parent>
        <artifactId>springboot-00-parent</artifactId>
        <groupId>com.linkedbear.springboot</groupId>
        <version>1.0-RELEASE</version>
        <relativePath>../springboot-00-parent/pom.xml</relativePath>
    </parent>
    <modelVersion>4.0.0</modelVersion>

    <artifactId>springboot-01-quickstart</artifactId>

    <properties>
        <java.version>1.8</java.version>
    </properties>

    <dependencies>
        <dependency>
            <groupId>org.springframework.boot</groupId>
            <artifactId>spring-boot-starter-web</artifactId>
        </dependency>
    </dependencies>

    <build>
        <plugins>
            <plugin>
                <groupId>org.springframework.boot</groupId>
                <artifactId>spring-boot-maven-plugin</artifactId>
            </plugin>
        </plugins>
    </build>
</project>
```

Spring Boot 主启动类非常简单,不再详细介绍,如代码清单 1-3 所示。

### 代码清单 1-3　Spring Boot 主启动类

```java
@SpringBootApplication
public class SpringBootQuickstartApplication {

    public static void main(String[]args) {
        SpringApplication.run(SpringBootQuickstartApplication.class, args);
    }
}
```

经过以上步骤后,一个最简单的 Spring Boot 项目创建完成。

## 1.5.2　编写简单代码

下面编写一个简单的 Controller 来测试 WebMvc 场景的效果。在 `com.linkedbear.springboot.quickstart` 包下新建一个 `controller` 包,并添加 `HelloController` 类,定义 `hello` 方法用于接受请求并响应,如代码清单 1-4 所示。

代码清单 1-4　HelloController.java

```
@RestController
public class HelloController {

    @GetMapping("/hello")
    public String hello() {
        return "hello springboot";
    }
}
```

> 提示：使用@RestController 后 Controller 中所有被@RequestMapping 标注的方法相当于添加了@ResponseBody 注解，会将方法的返回值序列化为 JSON 响应给客户端。

代码编写完成后，程序运行的预期是启动项目，访问/hello 请求，可以响应"hello springboot"的字符串内容。

运行 SpringBootQuickstartApplication 主启动类，控制台会打印 Tomcat started on port(s):8080 (http) with context path''，说明项目已经成功启动。在浏览器中访问 http://localhost:8080/hello，可以成功响应（见图 1-8），说明 springboot-01-quickstart 项目已经正确创建完成。

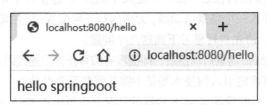

图 1-8　hello 请求可以成功响应

## 1.6　小结

本章主要回顾了 Spring Framework 的基本知识与核心特性、Spring Boot 的体系与核心特性，以及使用 IDEA 创建简单的 Spring Boot 项目。Spring Boot 本身是基于 Spring Framework 之上封装的更易于开发的"脚手架"，借助 Spring Boot 约定大于配置的原则，在创建项目时，只需要极少量的配置甚至不需要任何配置，就可以成功整合所需的开发场景。

# 第 2 章 Spring Boot 的自动装配

**本章主要内容：**
- ◇ 理解组件装配的概念和设计；
- ◇ Spring Framework 的模块装配；
- ◇ Spring Framework 的条件装配；
- ◇ Spring Framework 3.2 特性——SPI 机制；
- ◇ Spring Boot 的装配机制与 @EnableAutoConfiguration 核心注解分析；
- ◇ WebMvc 场景下的自动装配原理分析。

Spring Boot 中最重要的特性之一是"约定大于配置"，这在第 1 章的简单示例项目中已经展现得淋漓尽致。仅需导入一个场景启动器，无须编写任何配置，应用即可运行在 8080 端口上。在进入本章的介绍之前，请先自行思考下面这几个问题。

- 为什么导入 WebMvc 场景启动器后，即使没有编写任何配置代码，应用也可以正常启动？
- Spring Boot 如何确定引入的技术场景中需要哪些重要组件？
- 为什么项目在没有配置任何 Web 容器的情况下也可以正常启动 Web 服务？

> 💡 **提示：** 上述提到的 Web 容器指的是包括 Tomcat、Jetty、Undertow 等 Servlet 容器以及 Netty 等非阻塞 Web 容器在内的所有能部署 Web 项目的应用服务器。由于 Spring Boot 2.x 不仅支持基于 Servlet 的 Web 开发，还引入了 SpringWebFlux 的异步非阻塞式 Web 开发，因此此处不单指 Servlet 容器或者 NIO 容器，而是以 Web 容器统称。
> 如果没有特殊说明，本书后续出现的所有 Web 容器均指代 Servlet 容器和 NIO 容器。

想要了解 Spring Boot 底层完成的工作，最重要的是掌握 Spring Boot 的**自动装配机制**。自动装配本身是一个相对新的概念，它基于 Spring Framework 原生组件装配进行了延伸。下面首先了解组件装配及相关概念。

## 2.1 组件装配

Spring Framework 本身有一个 IOC 容器，该容器中会统一管理其中的 bean 对象，bean 对象可以理解为**组件**。要理解组件装配，首先需要理解组件的概念。

### 2.1.1 组件

基于 Spring Framework 的应用在整合第三方技术时，要把第三方框架中的核心 API 配置到

Spring Framework 的配置文件或注解配置类中,以供 Spring Framework 统一管理。此处的配置是关键,通过编写 XML 配置文件或注解配置类,将第三方框架中的核心 API 以对象的形式注册到 IOC 容器中。这些核心 API 对象会在适当的位置发挥其作用,以支撑项目的正常运行。IOC 容器中的核心 API 对象本身就是一个个的 bean 对象,即**组件**;将核心 API 配置到 XML 配置文件或注解配置类的行为称为**组件装配**。

Spring Framework 本身只有一种组件装配方式,即**手动装配**,而 Spring Boot 基于原生的手动装配,通过**模块装配+条件装配+SPI** 机制,可以完美实现组件的自动装配。下面分别就手动装配和自动装配的概念简单展开解释。

### 2.1.2 手动装配

所谓手动装配,指的是开发者在项目中通过编写 XML 配置文件、注解配置类、配合特定注解等方式,将所需的组件注册到 IOC 容器(即 `ApplicationContext`)中。代码清单 2-1 中的三种方式都是手动装配的体现。

**代码清单 2-1　三种手动装配方式**

```
<!-- 基于 XML 配置文件的手动配置 -->
<bean id="person" class="com.linkedbear.springboot.component.Person"/>
// 基于注解配置类的手动装配
@Configuration
public class ExampleConfiguration {
    
    @Bean
    public Person person() {
        return new Person();
    }
}
// 基于组件扫描的手动装配
@Component
public class DemoService {}

@Configuration
@ComponentScan("com.linkedbear.springboot")
public class ExampleConfiguration {}
```

从上述代码中可以提取出一个共性:手动装配都需要**亲自编写配置信息,将组件注册到 IOC 容器中**。

### 2.1.3 自动装配

Spring Boot 的核心特性之一是组件的**自动装配**。自动装配的核心是,本应该由开发者编写的配置,转为框架自动根据项目中整合的场景依赖,合理地做出判断并装配合适的 Bean 到 IOC 容器中。相较于手动装配,自动装配关注的重点是整合的场景,而不是每个具体的场景中所需的组件。鉴于关注的重点和粒度,自动装配更应该考虑应用全局的组件配置。Spring Boot 利用模块装配+条件装配的机制,可以在开发者不进行任何干预的情况下注册默认所需的组件,也可以基于开发者自定义注册的组件装配其他必要的组件,并合理地替换默认的组件注册(即覆盖

## 第 2 章  Spring Boot 的自动装配

默认配置)。由此可以概括一点：Spring Boot 的自动装配具有非侵入性。

> 例如，当整合 `spring-jdbc` 时，如果项目中已经注册了 `JdbcTemplate`，则 Spring Boot 提供的默认的 `JdbcTemplate` 就不会再创建。

同时，Spring Boot 的自动装配可以实现配置禁用，通过在 `@SpringBootApplication` 或者 `@EnableAutoConfiguration` 注解上标注 `exclude/excludeName` 属性，可以禁用默认的自动配置类。这种禁用方式在 Spring Boot 的全局配置文件中声明 `spring.autoconfigure.exclude` 属性时同样适用。

Spring Boot 的自动装配所用的底层机制全部来自 Spring Framework。下面先了解和回顾 Spring Framework 的基础知识。

> 💡 提示：由于 Spring Boot 本身已不推荐使用 XML 配置文件的方式构建应用，因此后续内容中如果没有特殊说明，所使用的所有配置均基于注解驱动。

## 2.2  Spring Framework 的模块装配

模块装配是自动装配的核心，它可以把一个模块所需的核心功能组件都装配到 IOC 容器中，并保证装配的方式尽可能简单。Spring Framework 中引入模块装配的表现形式是在 3.1 版本后引入大量 `@EnableXXX` 注解，通过标注 `@EnableXXX` 系列注解，可以实现快速激活和装配对应的模块。

当下主流的 Spring Boot 版本是 2.x，对应的 Spring Framework 版本是 5.x，但是从 5.x 的官方文档中已经很难找到 `@EnableXXX` 的介绍，读者可以在 Spring Framework 3.1.0 的官方文档的 3.1.5 节中找到介绍 `@EnableXXX` 注解的使用，该文档中提供了一些示例，示例中不乏有读者可能熟悉的以下注解。

- `@EnableTransactionManagement`：开启注解事务驱动。
- `@EnableWebMvc`：激活 SpringWebMvc 整合 Web 开发。
- `@EnableAspectJAutoProxy`：开启注解 AOP 编程。
- `@EnableScheduling`：开启调度功能（定时任务）。

……

上述的部分注解会在后续章节中出现，目前暂不介绍。

模块装配围绕的核心是"模块"。下面先理解模块的概念。

### 2.2.1  模块

模块可以理解成一个个可以分解、组合、更换的独立单元。模块与模块之间可能存在一定的依赖，模块的内部通常是高内聚的，一个模块通常用于解决一个独立的问题（比如引入事务模块是为了处理数据库操作的 ACID 特性）。按照上述理解，可以将项目开发中编写的功能代码看作一个个模块，项目底层封装的一个个组件也可以看作模块。

简单总结起来，模块通常具有以下几个特性：
- 独立的；
- 功能高内聚；

- 可相互依赖；
- 目标明确。

### 2.2.2 快速体会模块装配

理解模块的概念后，下面通过几个示例快速体会模块装配的设计。在开始之前请读者记住使用模块装配的核心原则：**自定义注解+`@Import`导入组件**。

#### 1. 模块装配场景

为了更好地让读者体会模块装配的设计，这里假设一个场景：使用代码模拟构建一个**酒馆**，酒馆里有**吧台**、**调酒师**、**服务员**和**老板** 4 种不同的实体元素。在该场景中，酒馆可看作 `ApplicationContext`，酒馆里的吧台、调酒师、服务员、老板等元素可看作一个个**组件**。使用代码模拟实现的最终目的是，可以**通过一个注解，把以上元素全部填充到酒馆中**。

> 提示：假设的场景仅配合代码完成演示，对于其中具体的结构设计，本书不过多深入，感兴趣的读者可以在练习时自行发挥。

目标明确后，下面开始动手实操。首先实现最简单的装配方式：普通 Bean 的装配。

#### 2. 声明自定义注解

示例场景的目标是构建一个酒馆，根据 Spring Framework 对于模块装配的注解命名风格，此处定义一个自定义注解`@EnableTavern`，如代码清单 2-2 所示。

**代码清单 2-2　@EnableTavern 注解的声明**

```
@Documented
@Retention(RetentionPolicy.RUNTIME)
@Target(ElementType.TYPE)  // 该注解只能标注在类上
public @interface EnableTavern {

}
```

> 提示：自定义注解上应标注必要的元注解，代表该注解在运行时起效，并且只能标注在类上。

要使`@EnableTavern`注解发挥作用，需要配合模块装配中的最核心注解`@Import`，该注解要标注在`@EnableTavern`上，且注解中需要传入 `value` 属性的值。借助 IDE 可以了解`@Import`注解的使用方式，如代码清单 2-3 所示。`value` 属性的文档注释中已经写明，`@Import`注解可以导入配置类、`ImportSelector` 的实现类、`ImportBeanDefinitionRegistrar` 的实现类，以及普通类。

**代码清单 2-3　@Import 注解**

```
@Target(ElementType.TYPE)
@Retention(RetentionPolicy.RUNTIME)
@Documented
public @interface Import {

    /**
```

```
 *{@link Configuration @Configuration},{@link ImportSelector},
 *{@link ImportBeanDefinitionRegistrar}, or regular component classes to import.
 */
Class<?>[]value();
}
```

### 3. 声明老板类

本节先演示普通类的导入。酒馆必须由老板经营,下面定义一个"老板"类,该类中不需要声明任何属性和方法。

```
public class Boss {}
```

> 提示:注意 Boss 类中不需要标注@Component 注解,因为@Component 是需要配合@ComponentScan 使用的,而本节内容不会涉及组件扫描机制。

接下来在@EnableTavern 的@Import 注解中填入 Boss 类,如代码清单 2-4 所示,这就意味着如果一个配置类上标注了**@EnableTavern** 注解,就会触发**@Import** 的效果,向容器中导入一个 **Boss** 类的 **Bean**。

**代码清单 2-4** 将 Boss 类标注到@EnableTavern 注解中,代表装配一个 Boss 类对象

```
@Documented
@Retention(RetentionPolicy.RUNTIME)
@Target(ElementType.TYPE)
@Import(Boss.class)
public @interface EnableTavern {

}
```

### 4. 创建配置类

注解驱动的测试离不开配置类。接下来声明一个 TavernConfiguration 配置类,并标注@Configuration 和@EnableTavern 注解。TavernConfiguration 中不需要定义其他内容,也无须注册其他 bean 对象,如代码清单 2-5 所示。

**代码清单 2-5** 添加新的注解配置类 TavernConfiguration

```
@Configuration
@EnableTavern
public class TavernConfiguration { }
```

### 5. 编写启动类测试

核心代码编写完毕,最后编写一个测试启动类,以检验组件装配的效果。测试启动类如代码清单 2-6 所示。

**代码清单 2-6** 测试装配普通类

```
public class TavernApplication {

    public static void main(String[] args) {
        ApplicationContext ctx = new AnnotationConfigApplicationContext(TavernConfiguration.class);
```

```
        Boss boss = ctx.getBean(Boss.class);
        System.out.println(boss);
    }
}
```

通过运行 main 方法，可以发现使用 getBean 能够正常提取 Boss 对象，说明 Boss 类已经被注册到 IOC 容器中，并创建了一个对象，以此就完成了最简单的模块装配。

```
com.linkedbear.springboot.assemble.a_module.component.Boss@b9afc07
```

### 2.2.3 导入配置类

到这里可能有读者会产生疑惑：原本通过 @Configuration + @Bean 注解就能完成的工作，换用 @Import 注解后代码量却增加了，这不是"徒增功耗"吗？如果你也有这种疑问，请不要着急，仔细观察 @Import 的 value 属性允许传入的类，可以发现普通类是最简单的方式，而其余几种类型更为重要。

如果需要直接导入项目中现有的一些配置类，使用 @Import 也可以直接加载进来。本节会编写一个有关调酒师的独立配置类，并通过 @Import 注解导入。

#### 1. 声明调酒师类

考虑到在真实场景中，酒吧的调酒师通常不止一位，因此调酒师的模型类中需要定义一个 **name** 属性，以对不同的调酒师加以区分，如代码清单 2-7 所示。

**代码清单 2-7　调酒师的模型类**

```java
public class Bartender {
    
    private String name;
    
    public Bartender(String name) {
        this.name = name;
    }
    
    // getter ......
}
```

#### 2. 注册调酒师对象

通过注解配置类的方式，可以一次性注册多个相同类的 bean 对象。下面编写一个配置类 BartenderConfiguration，并用 @Bean 注册两个不同的 Bartender 类，如代码清单 2-8 所示。

**代码清单 2-8　注册调酒师的配置类 BartenderConfiguration**

```java
@Configuration
public class BartenderConfiguration {
    
    @Bean
    public Bartender zhangxiaosan() {
        return new Bartender("张小三");
    }
```

```
    @Bean
    public Bartender zhangdasan() {
        return new Bartender("张大三");
    }
}
```

此处注意,如果读者用 IDEA 编码开发,此时这个类会有黄色警告,提示配置类 Bartender Configuration 还没有被使用(事实也确实如此,目前的代码中并没有用到它)。想让 BartenderConfiguration 起作用,只需在 @EnableTavern 的 @Import 中把这个配置类一并导入,如代码清单 2-9 所示。

**代码清单 2-9　在 @EnableTavern 注解中添加 BartenderConfiguration 配置类**

```
@Import({Boss.class, BartenderConfiguration.class})
public @interface EnableTavern {}
```

> 提示:注意此处有一个细节,如果读者在自行练习时编写的启动类或者配置类上使用了**组件扫描(包扫描)**,恰好把这个配置类扫描到了,就会导致即使没有使用 @Import 导入配置类 BartenderConfiguration,Bartender 调酒师也会被注册到 IOC 容器中。这里一定要细心,组件扫描本身会扫描配置类,并且会使其生效。如果既想用包扫描又不想扫描到 BartenderConfiguration,只需把 BartenderConfiguration 移至其他包中,从而使组件扫描时找不到它。

### 3. 测试运行

修改启动类,使用 ApplicationContext 的 getBeansOfType 方法可以一次性提取出 IOC 容器中指定类的所有 bean 对象,如代码清单 2-10 所示。

**代码清单 2-10　测试装配配置类**

```
public static void main(String[] args) {
    AnnotationConfigApplicationContext ctx = new AnnotationConfigApplicationContext(TavernConfiguration.class);
    Stream.of(ctx.getBeanDefinitionNames()).forEach(System.out::println);
    System.out.println("--------------------------");
    Map<String, Bartender> bartenders = ctx.getBeansOfType(Bartender.class);
    bartenders.forEach((name, bartender) -> System.out.println(bartender));
}
```

通过运行 main 方法,可以发现控制台成功打印出两个调酒师对象,说明注解配置类的装配正确完成。

```
// IOC 内部的组件已省略打印
tavernConfiguration
com.linkedbear.springboot.assemble.a_module.component.Boss
com.linkedbear.springboot.assemble.a_module.config.BartenderConfiguration
zhangxiaosan
zhangdasan
--------------------------
com.linkedbear.springboot.assemble.a_module.component.Bartender@23bb8443
com.linkedbear.springboot.assemble.a_module.component.Bartender@1176dcec
```

> 提示：注意一个细节，BartenderConfiguration 配置类也被注册到 IOC 容器中，并成为一个 Bean。

## 2.2.4 导入 ImportSelector

借助 IDE 打开 ImportSelector，会发现它是一个接口，它的功能可以从 javadoc 中读到一些信息：

> Interface to be implemented by types that determine which @Configuration class(es) should be imported based on a given selection criteria, usually one or more annotation attributes.
>
> ImportSelector 是一个接口，它的实现类可以根据指定的筛选标准（通常是一个或者多个注解）决定导入哪些配置类。

文档注释阐述的关键是 ImportSelector 可以导入配置类，但其实它也可以导入普通类。被 ImportSelector 导入的类，最终会在 IOC 容器中以单实例 Bean 的形式创建并保存。下面演示 ImportSelector 如何使用。

### 1. 声明吧台类+配置类

吧台的模型类设计也是定义一个最简单的类即可，无须过度设计。

```java
public class Bar {}
```

ImportSelector 不仅可以导入配置类，也可以导入普通类，代码清单 2-11 中是一个新的配置类，演示两种类皆可。

**代码清单 2-11　BarConfiguration 配置类中注册 Bar**

```java
@Configuration
public class BarConfiguration {

    @Bean
    public Bar bbbar() {
        return new Bar();
    }
}
```

### 2. 编写 ImportSelector 的实现类

接下来是 ImportSelector 的实现类，新定义一个 BarImportSelector，以实现 ImportSelector 接口并实现对应的 selectImports 方法，如代码清单 2-12 所示。

**代码清单 2-12　BarImportSelector 实现 ImportSelector 接口**

```java
public class BarImportSelector implements ImportSelector {

    @Override
    public String[] selectImports(AnnotationMetadata importingClassMetadata) {
        return new String[0];
    }
}
```

注意，selectImports 方法的返回值是一个 String 类型的数组，要获知这样设计的目的，需要参考 selectImports 方法的文档注释：

> Select and return the names of which class(es) should be imported based on the AnnotationMetadata of the importing @Configuration class.
>
> 根据导入的 `@Configuration` 类的 `AnnotationMetadata` 选择并返回要导入的类的类名。

这里要强调的重点是：返回**一组类名**（一定是**全限定类名**，因为如果没有全限定类名就无法定位具体的类）。由此就在这里把上面定义的 `Bar` 和 `BarConfiguration` 的类名写入，如代码清单 2-13 所示。

**代码清单 2-13　selectImports 方法返回要注册的 Bean 的全限定类名**

```java
public String[] selectImports(AnnotationMetadata importingClassMetadata) {
    return new String[] {Bar.class.getName(), BarConfiguration.class.getName()};
}
```

最后在 `@EnableTavern` 的 `@Import` 注解中将 `BarImportSelector` 导入即可，如代码清单 2-14 所示。

**代码清单 2-14　在 @EnableTavern 注解中添加 BarImportSelector 导入**

```java
@Import({Boss.class, BartenderConfiguration.class, BarImportSelector.class})
public @interface EnableTavern {}
```

### 3. 测试运行

修改启动类的 `main` 方法。为了更明显地体现出容器中的 bean 对象，这次只打印 IOC 容器中所有 bean 对象的名称。运行 main 方法，控制台会打印出两个 Bar（下列代码中倒数第一行和第三行），说明 `ImportSelector` 可以导入普通类和配置类。

```
// IOC 内部的组件已省略打印 ……
tavernConfiguration
com.linkedbear.springboot.assemble.a_module.component.Boss
com.linkedbear.springboot.assemble.a_module.config.BartenderConfiguration
zhangxiaosan
zhangdasan
com.linkedbear.springboot.assemble.a_module.component.Bar
com.linkedbear.springboot.assemble.a_module.config.BarConfiguration
bbbar
```

> 💡 提示：注意一个细节，`BarImportSelector` 本身没有注册到 IOC 容器中。

### 4. ImportSelector 的灵活性

到这里读者可能会觉得很奇怪：上面示例中是直接取现有类的全限定名，这种设计似乎使复杂度变高了！但是请读者明白一点：`ImportSelector` 的核心是可以使开发者采用**更灵活的声明式向 IOC 容器注册 Bean**，其重点是可以灵活地指定要注册的 Bean 的类。由于是传入全限定名的字符串，那么如果这些**全限定名以配置文件的形式存放在项目可以读取的位置**，是不是可以**避免组件导入的硬编码问题**？因此 `ImportSelector` 的作用非常大，在 Spring Boot 的自动装配中，底层就是利用了 `ImportSelector`，实现从 `spring.factories` 文件中读取自动配置类，2.5 节中将会讲到。

## 2.2.5 导入 ImportBeanDefinitionRegistrar

如果说 `ImportSelector` 是以声明式导入组件,那么 `ImportBeanDefinitionRegistrar` 可以解释为以编程式向 IOC 容器中注册 bean 对象,不过它实际导入的是 `BeanDefinition` (Bean 的定义信息)。有关 `BeanDefinition` 的详细讲解可以参考 3.5 节,此处不展开讲解,读者在这里先对 `ImportBeanDefinitionRegistrar` 有一些快速的了解即可,后面在第 7 章中会详细讲解 `ImportBeanDefinitionRegistrar` 的获取和引导回调的原理。

### 1. 声明服务员类

离最后的酒馆建成只剩下一组服务员类,同样以最简单的模型类定义即可。

```java
public class Waiter {}
```

> 提示:这里没有把服务员的模型类设计得很复杂,因为本节的目的是使读者了解和学会模块装配,而不是仔细研究 `BeanDefinition` 的复杂 Bean 定制。

### 2. 编写 ImportBeanDefinitionRegistrar 的实现类

接下来编写 `WaiterRegistrar`,使它实现 `ImportBeanDefinitionRegistrar` 接口,如代码清单 2-15 所示。

**代码清单 2-15　WaiterRegistrar 实现 ImportBeanDefinitionRegistrar**

```java
public class WaiterRegistrar implements ImportBeanDefinitionRegistrar {
    @Override
    public void registerBeanDefinitions(AnnotationMetadata metadata, BeanDefinitionRegistry registry) {
        registry.registerBeanDefinition("waiter", new RootBeanDefinition(Waiter.class));
    }
}
```

对于这里的写法读者先不必仔细研究,可以先跟着示例代码写一遍。这里简单解释一下 `registerBeanDefinition` 方法传入的两个参数,第一个参数是 Bean 的名称(即 id),第二个参数中传入的 `RootBeanDefinition` 要指定 Bean 的字节码(`.class`),这种方式相当于向 IOC 容器中注册了一个普通的单实例 bean(最终效果与组件扫描、`@Bean` 注解注册 Bean 的效果相同)。

> 提示:有关 `BeanDefinition` 的相关讲解,可参考 3.5 节。

最后把 `WaiterRegistrar` 标注在 `@EnableTavern` 的 `@Import` 注解中,即完成了 `ImportBeanDefinitionRegistrar` 的导入,如代码清单 2-16 所示。

**代码清单 2-16　在 @EnableTavern 注解中添加 ImportBeanDefinitionRegistrar 导入**

```java
@Import({Boss.class, BartenderConfiguration.class, BarImportSelector.class, WaiterRegistrar.class})
public @interface EnableTavern {}
```

### 3. 测试运行

直接重新运行 `main` 方法,控制台可以打印出服务员对象(下列代码中的最后一行),证明

## 第 2 章　Spring Boot 的自动装配

使用 `ImportBeanDefinitionRegistrar` 的组件装配也成功了。

```
// IOC 内部的组件已省略打印……
tavernConfiguration
com.linkedbear.springboot.assemble.a_module.component.Boss
com.linkedbear.springboot.assemble.a_module.config.BartenderConfiguration
zhangxiaosan
zhangdasan
com.linkedbear.springboot.assemble.a_module.component.Bar
com.linkedbear.springboot.assemble.a_module.config.BarConfiguration
bbbar
waiter
```

> 提示：`WaiterRegistrar` 也没有注册到 IOC 容器中。

### 2.2.6　扩展：DeferredImportSelector

本节的最后扩展是 `ImportSelector` 的一个子接口 `DeferredImportSelector`，这个接口来自 Spring Framework 4.0，它提供了类似于 `ImportSelector` 的组件装配机制，但执行时机比普通的 `ImportSelector` 晚。这里先解释一下执行时机，`ImportSelector` 接口的处理时机是在注解配置类的解析期间，此时配置类中的 `@Bean` 方法等还没有被解析，而 `DeferredImportSelector` 的处理时机是注解配置类完全解析后，此时配置类的解析工作已全部完成，这样做的目的主要是为了**配合下面要提到的条件装配**（条件装配也来自 Spring Framework 4.0，所以可以理解为它是配合工作的）。

下面通过一个简单的测试示例，体会 `DeferredImportSelector` 的执行时机。

**1. DeferredImportSelector 的执行时机**

在上述测试代码中编写一个新的 `WaiterDeferredImportSelector`，使它实现 `DeferredImportSelector`，其作用是导入新的服务员 bean 对象，如代码清单 2-17 所示。

**代码清单 2-17　新增 WaiterDeferredImportSelector 导入服务员**

```java
public class WaiterDeferredImportSelector implements DeferredImportSelector {
    
    @Override
    public String[] selectImports(AnnotationMetadata importingClassMetadata) {
        System.out.println("WaiterDeferredImportSelector invoke ......");
        return new String[] {Waiter.class.getName()};
    }
}
```

> 提示：注意代码中添加了一行控制台打印，这样可以便于观察到 `DeferredImportSelector` 的执行时机。

同样，为其余的两种组件 `ImportSelector` 和 `ImportBeanDefinitionRegistrar` 也添加上控制台打印，如代码清单 2-18 所示。

**代码清单 2-18　补充其他组件的控制台打印**

```java
public class BarImportSelector implements ImportSelector {

    @Override
    public String[] selectImports(AnnotationMetadata importingClassMetadata) {
        System.out.println("BarImportSelector invoke ......");
        return new String[] {Bar.class.getName(), BarConfiguration.class.getName()};
    }
}

public class WaiterRegistrar implements ImportBeanDefinitionRegistrar {

    @Override
    public void registerBeanDefinitions(AnnotationMetadata metadata, BeanDefinitionRegistry registry) {
        System.out.println("WaiterRegistrar invoke ......");
        registry.registerBeanDefinition("waiter", new RootBeanDefinition(Waiter.class));
    }
}
```

测试之前，不要忘记给 @EnableTavern 注解的 @Import 上补充 WaiterDeferredImportSelector 的导入。

如此编写完成后，重新运行 TavernApplication，观察控制台的打印：

```
BarImportSelector invoke ......
WaiterDeferredImportSelector invoke ......
WaiterRegistrar invoke ......
```

可以发现，**DeferredImportSelector** 的执行时机比 **ImportSelector** 的确晚，但比 **ImportBeanDefinitionRegistrar** 早。至于为什么要这样设计，下面讲到基于 Conditional 的条件装配时再来说明。

**2. 扩展：DeferredImportSelector 的分组概念**

Spring Framework 5.0.5 中对 DeferredImportSelector 加入了新的概念：分组。简单地理解，引入了分组的概念后可以对不同的 DeferredImportSelector 加以区分。上面在编写代码时读者可能没有感知到，实际上 DeferredImportSelector 有一个默认的 getImportGroup 方法，如代码清单 2-19 所示。

**代码清单 2-19　DeferredImportSelector 中添加了分组的概念**

```java
default Class<? extends Group> getImportGroup() {
    return null;
}
```

这个 getImportGroup 方法可以指定一个实现了 DeferredImportSelector.Group 接口的类型，其可以对 DeferredImportSelector 加以区分。不过读者不必对它过于在意，在 Spring Framework 和 Spring Boot 中使用它的地方非常少，因此只需要了解 DeferredImportSelector 可以分组。

> 💡 提示：Spring Boot 的自动装配部分有一个 DeferredImportSelector 分组特性的使用，2.5.3 节会特别提到。

## 2.3 Spring Framework 的条件装配

模块装配可以一次性导入一个场景中所需的组件，但如果只靠模块装配的内容，还不足以实现完整的组件装配。仍以酒馆为例，如果将这套代码模拟的环境放到**一片荒野**，此时吧台还在，老板还在，但是调酒师会因为环境恶劣而跑掉（荒郊野外不会有闲情逸致的人去喝酒），所以在这种场景下，调酒师就不应该注册到 IOC 容器了。在这种模拟的场景中，如果只使用模块装配是无法实现的，因为只要配置类中声明了 @Bean 注解的方法，这个方法的返回值就一定会被注册到 IOC 容器中，并最终成为一个 bean 对象。

因此，为了解决在不同场景/条件下满足不同组件的装配，Spring Framework 提供了两种条件装配的方式：基于 Profile 和基于 Conditional。

### 2.3.1 基于 Profile 的装配

Spring Framework 3.1 中就已经引入了 Profile 的概念，下面先了解一下 Profile 的定义。

#### 1. 理解 Profile

Spring Framework 的官方文档中并没有对 Profile 进行过多描述，而是借助一篇官网的博客文章来详细介绍 Profile 的使用，此外在 @Profile 注解的 javadoc 上也有一些简短的描述：@Profile 注解可以标注在组件上，当一个配置属性（并不是文件）激活时，它才会起作用，而激活这个属性的方式有很多种（启动参数、环境变量、web.xml 配置等）。

简单概括一下，Profile 提供了一种"基于环境的配置"：根据当前项目的不同运行时环境，可以动态地注册与当前运行环境匹配的组件。

#### 2. 使用@Profile 注解

下面来实际使用一下 Profile 机制，以满足上面提到的新需求：城市与荒野。

（1）Bartender 添加@Profile

需求描述中提到，荒郊野外下调酒师不会再工作，在这种假设下调酒师就不会在荒郊野外的环境下存在，而只会在城市存在。用代码来表达，就是在注册调酒师的配置类上标注 @Profile，如代码清单 2-20 所示。

**代码清单 2-20　为注册调酒师的配置类添加@Profile 注解**

```
@Configuration
@Profile("city")
public class BartenderConfiguration {
    @Bean
    public Bartender zhangxiaosan() {
        return new Bartender("张小三");
    }

    @Bean
    public Bartender zhangdasan() {
        return new Bartender("张大三");
    }
}
```

## 2.3 Spring Framework 的条件装配

（2）编程式设置运行时环境

如果现在直接运行 `TavernProfileApplication` 的 `main` 方法，控制台中不会打印 `zhangxiaosan` 和 `zhangdasan`（已省略一些内部的组件打印）：

```
tavernConfiguration
com.linkedbear.springboot.assemble.b_profile.component.Boss
com.linkedbear.springboot.assemble.b_profile.component.Bar
com.linkedbear.springboot.assemble.b_profile.config.BarConfiguration
bbbar
waiter
```

为什么会出现这种情况呢？默认情况下，`ApplicationContext` 中的 Profile 为 "default"，与上面 `@Profile("city")` 不匹配，`BartenderConfiguration` 就不会生效，这两个调酒师也不会被注册到 IOC 容器中。要想让调酒师注册到 IOC 容器中，就需要给 `ApplicationContext` 中设置一下激活的 Profile，如代码清单 2-21 所示。

**代码清单 2-21　为 ApplicationContext 设置 Profile**

```java
public static void main(String[] args) {
    AnnotationConfigApplicationContext ctx = new AnnotationConfigApplicationContext();
    // 为 ApplicationContext 的环境设置正在激活的 Profile
    ctx.getEnvironment().setActiveProfiles("city");
    ctx.register(TavernConfiguration.class);
    ctx.refresh();
    Stream.of(ctx.getBeanDefinitionNames()).forEach(System.out::println);
}
```

注意，代码清单 2-21 中初始化 `ApplicationContext` 的逻辑与之前不同。`AnnotationConfigApplicationContext` 在创建对象时，如果直接传入了配置类，则会立即初始化 IOC 容器，在不传入配置类的情况下，内部不会执行初始化逻辑，而是要等到手动调用其 `refresh` 方法后才会初始化 IOC 容器，在初始化 IOC 容器的过程中，会顺便将环境配置一并处理，因此，为了避免不必要的麻烦，这里使用手动初始化 IOC 容器的方式。

修改完成后，重新运行 `main` 方法，控制台可以成功打印 `zhangxiaosan` 和 `zhangdasan`。

```
tavernConfiguration
com.linkedbear.springboot.assemble.b_profile.component.Boss
com.linkedbear.springboot.assemble.b_profile.config.BartenderConfiguration
zhangxiaosan
zhangdasan
com.linkedbear.springboot.assemble.b_profile.component.Bar
com.linkedbear.springboot.assemble.b_profile.config.BarConfiguration
bbbar
waiter
```

（3）命令行参数设置运行时环境

上面编程式配置虽然已经可以使用，但这种方式并不实用。将 Profile 硬编码在 Java 代码中本身就是一种 "坏味道"，如果需要切换 Profile，则需要修改 Java 代码后重新编译。Spring Framework 考虑到了这种情况，所以它提供了很多灵活的 Profile 配置方式。下面演示最容易实现的一种：命令行参数配置。

要测试命令行参数的环境变量，需要在 IDEA 中配置启动选项，如图 2-1 所示。

25

按照图 2-1 的方式配置好之后，在 main 方法中改回原来的构造方法传入配置类的形式并运行，控制台仍然会打印 zhangxiaosan 和 zhangdasan。

修改传入的 JVM 参数，将 city 改成 wilderness，重新运行 main 方法，发现控制台不再打印 zhangxiaosan 和 zhangdasan，说明使用 JVM 命令行参数也可以控制 Profile。

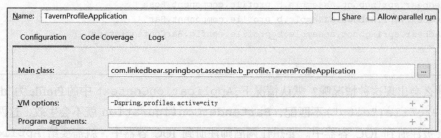

图 2-1　IDEA 设置命令行启动参数

**3. @Profile 运用于实际开发**

Profile 机制在 Spring Boot 中使用得非常经典，使用 `spring.profiles.active` 属性可以激活指定的环境配置，`application.properties` 文件都可以通过加 profile 后缀来区分不同环境下的配置文件（`application-dev.properties`、`application-prod.properties`）。

举一个简单的例子，如果需要根据不同的环境配置不同的嵌入式 Web 容器端口，则可以声明几个不同的配置文件，以分别定义不同的端口配置，如代码清单 2-22 所示。当全局 `application.properties` 中激活 `dev` 环境时，`application-dev.properties` 中的配置生效，项目就会运行在 8787 端口。

**代码清单 2-22　Spring Boot 中 Profile 的应用**

```
# application-dev.properties
server.port=8787

# application-prod.properties
server.port=8989

# application.properties
spring.profiles.active=dev # 激活 dev 的配置
```

**4. Profile 的不足**

Profile 固然很强大，但它仍有一些无法控制的地方。下面将场景进一步复杂化：吧台应由老板安置，如果酒馆中连老板都没有，那么吧台也不应该存在。在这种情况下，只使用 Profile 机制便无法实现，因为 Profile 控制的是**整个项目的运行环境**，无法根据单个 Bean 的因素决定是否装配。基于这种情况，出现了第二种条件装配的方式：基于 `@Conditional` 注解。

## 2.3.2　基于 Conditional 的装配

Conditional，意为**条件**，这个概念比 Profile 更直接明了。按照惯例，首先对 Conditional 有个清楚的认识。

**1. 理解 Conditional**

`@Conditional` 是在 Spring Framework 4.0 版本正式推出的，它可以使 Bean 的装配基于一

些指定的条件。换句话说，被标注@Conditional 注解的 Bean 要注册到 IOC 容器时，必须满足@Conditional 上指定的所有条件才允许注册。

在 Spring Framework 的官方文档中没有对@Conditional 的介绍，而是引导读者直接参考 javadoc，而 javadoc 中描述的内容大致可以总结为：@Conditional 注解可以指定匹配条件，而被@Conditional 注解标注的"组件类/配置类/组件工厂方法"必须满足@Conditional 中指定的所有条件，才会被创建/解析。

### 2. @Conditional 的使用

继续实现上面的需求：吧台依赖老板的存在。在 BarConfiguration 的 Bar 注册中，要指定 Bar 的创建需要 Boss 的存在，反映到代码上就是在 bbbar 方法上标注@Conditional，如代码清单 2-23 所示。

**代码清单 2-23　在 BarConfiguration 中为 bbbar 添加装配条件**

```java
@Bean
@Conditional(???)
public Bar bbbar(){
    return new Bar();
}
```

注意，@Conditional 注解中需要传入一个 Condition 接口的实现类数组，说明使用原生条件装配还需要编写条件判断类作为匹配依据。下面声明一个 ExistBossCondition 条件判断类，用来判断 IOC 容器中是否存在 Boss 对象，如代码清单 2-24 所示。编写完成后将该条件判断类放入@Conditional 注解中。

**代码清单 2-24　判断 Boss 是否存在的条件判断类**

```java
public class ExistBossCondition implements Condition {
    
    @Override
    public boolean matches(ConditionContext context, AnnotatedTypeMetadata metadata) {
        return context.getBeanFactory().containsBeanDefinition(Boss.class.getName());
    }
}
```

> 💡 提示：matches 方法中使用 **BeanDefinition** 而不是 Bean 做判断，这是因为考虑的是当条件匹配时 Boss 对象可能尚未创建，导致条件匹配出现偏差。

下面重新运行测试启动类的 main 方法，发现吧台被成功创建：

```
tavernConfiguration
com.linkedbear.springboot.assemble.c_conditional.component.Boss
com.linkedbear.springboot.assemble.c_conditional.component.Bar
com.linkedbear.springboot.assemble.c_conditional.config.BarConfiguration
bbbar
waiter
```

为了检验上面的@Conditional 的确起了作用，可以将@EnableTavern 注解中导入的 Boss 类去掉，重新运行测试启动类，会发现 Boss 和 bbbar 均不会被打印，说明@Conditional 注解生效。

### 3. ConditionalOn×××系列注解

Spring Boot 中针对@Conditional 注解扩展了一系列的条件注解，下面是几个常用的条件装配注解。

- @ConditionalOnClass & @ConditionalOnMissingClass：检查当前项目的类路径下是否包含/缺少指定类。
- @ConditionalOnBean & @ConditionalOnMissingBean：检查当前容器中是否注册/缺少指定 Bean。
- @ConditionalOnProperty：检查当前应用的属性配置。
- @ConditionalOnWebApplication & @ConditionalOnNotWebApplication：检查当前应用是否为 Web 应用。
- @ConditionalOnExpression：根据指定的 SpEL 表达式确定条件是否满足。

> 提示：注意，在 Spring Boot 的官方文档中针对@ConditionalOnBean 注解的特别说明是，这些@ConditionalOn×××注解通常都用在自动配置类中，对于普通的配置类最好避免使用，以免出现判断偏差。

以上注解在 2.6 节研究 Spring Boot 的自动装配场景实例中会遇到，读者一定要熟记于心。

## 2.4 SPI 机制

下面继续扩展与自动装配相关的服务提供方接口（Service Provider Interface，SPI）机制。关于 SPI 的来源，需要读者熟悉设计模式的**依赖倒转原则**。依赖倒转原则中提到，**应该依赖接口而不是实现类**，但接口最终要有实现类落地。如果程序因业务调整，需要替换某个接口的实现类，就不得不改动实现类的创建，也就是修改源码。SPI 机制的出现解决了这个问题，它通过一种"服务寻找"的机制，**动态地加载接口/抽象类对应的具体实现类**。简单了解 SPI 的概念后，读者是否联想到了 IOC？的确，SPI 的确有 IOC 的"味道"，它把接口的具体实现类的定义和声明权交给了外部化的配置文件。

通过图 2-2 来更简单地理解 SPI 机制，图 2-2 中的接口可以有多个实现类，通过 SPI 机制，可以将一个接口需要创建的实现类的对象都罗列在一个特殊的文件中，SPI 机制会依次将这些实现类的对象进行创建并返回。

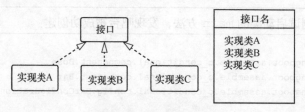

图 2-2 SPI 设计示意

原生 JDK 的 SPI 机制可以通过一个指定的接口/抽象类找到预先配置好的实现类（并创建实现类的对象）。JDK 1.6 中新增了 SPI 的实现，Spring Framework 3.2 也引入了 SPI 的实现，而且比 JDK 的实现更加强大。下面分别讲解两种 SPI 的实现。

### 2.4.1　JDK 原生的 SPI

对于 JDK 原生的 SPI，读者只需简单了解，因为它的使用范围相对有限，只能通过接口或抽象类来加载具体的实现类，所以很多框架没有直接采用 JDK 的 SPI，而是自定义了一套更强大的实现（Spring Framework、Dubbo 等框架都有自定义的 SPI 机制）。

下面通过一个简单的示例讲解使用 JDK 原生的 SPI 机制。

#### 1. 定义接口+实现类

简单示例不侧重于代码的复杂度，只需要定义一个接口+两个实现类，并模拟一套 Dao 接口的不同数据库访问支持，如代码清单 2-25 所示。无须定义接口的方法，只从类名上对实现类加以区分即可。

**代码清单 2-25　定义 SPI 机制测试的接口+实现类**

```
public interface DemoDao {}

public class DemoMySQLDaoImpl implements DemoDao {}

public class DemoOracleDaoImpl implements DemoDao {}
```

#### 2. 声明 SPI 文件

JDK 的 SPI 需要遵循以下规范：所有定义的 SPI 文件都必须放在项目的 **META-INF/services** 目录下，且文件名必须命名为接口或抽象类的全限定名，文件内容为接口或抽象类的具体实现类的全限定名；如果出现多个具体实现类，则每行声明一个类的全限定名，多个类之间没有分隔符。

如果要给上面定义的 DemoDao 接口声明 SPI 文件，则文件名应该是全限定名：com.linkedbear.springboot.assemble.d_spi.bean.DemoDao（注意没有后缀名），而文件的内容是这些具体实现类的全限定名。

```
com.linkedbear.springboot.assemble.d_spi.bean.DemoMySQLDaoImpl
com.linkedbear.springboot.assemble.d_spi.bean.DemoOracleDaoImpl
```

#### 3. 测试获取

接下来编写测试启动类，使用 JDK 提供的一个 `ServiceLoader` 类来加载 SPI 文件中定义的实现类，并直接打印到控制台，如代码清单 2-26 所示。

**代码清单 2-26　测试 JDK 原生的 SPI 机制**

```
public class JdkSpiApplication {

    public static void main(String[] args) throws Exception {
        ServiceLoader<DemoDao> serviceLoader = ServiceLoader.load(DemoDao.class);
        serviceLoader.iterator().forEachRemaining(dao -> {
            System.out.println(dao);
        });
    }
}
```

运行测试启动类的 `main` 方法，控制台可以成功打印出 DemoDao 的两个实现类对象，这说明 JDK 原生的 SPI 机制已成功使用。

```
com.linkedbear.springboot.assemble.d_spi.bean.DemoMySQLDaoImpl@50040f0c
com.linkedbear.springboot.assemble.d_spi.bean.DemoOracleDaoImpl@2dda6444
```

### 2.4.2 Spring Framework 3.2 的 SPI

Spring Framework 中的 SPI 相较于 JDK 原生的 SPI 更加高级实用,因为它不仅限于接口或抽象类,而可以是**任何一个类、接口或注解**。也正是因为 Spring Framework 的 SPI 可以支持注解作为索引,所以在 Spring Boot 中大量用到 SPI 机制加载自动配置类和特殊组件等(即 2.5 节中即将讲到的大名鼎鼎的 **@EnableAutoConfiguration**)。因此 Spring Framework 的 SPI 更值得读者认真学习。

#### 1. 声明 SPI 文件

Spring Framework 的 SPI 文件也有相应的规范,需要将 SPI 文件放在项目的 **META-INF** 目录下,且文件名必须为 **spring.factories**。该文件其实是一个 properties 文件,如代码清单 2-27 所示。

**代码清单 2-27　spring.factories 文件**

```
com.linkedbear.springboot.assemble.d_spi.bean.DemoDao=\
com.linkedbear.springboot.assemble.d_spi.bean.DemoMySQLDaoImpl,\
com.linkedbear.springboot.assemble.d_spi.bean.DemoOracleDaoImpl
```

可以发现,`spring.factories` 文件定义的规则是:被检索的类/接口/注解的全限定名作为 properties 的 key,具体要检索的类(注意不是实现类)的全限定名作为 value,多个类之间英文逗号分隔。

> 提示:注意,`spring.factories` 文件中定义的 key 和 value 可以毫无关联,仅凭这一点就比 JDK 的 SPI 要灵活得多。

#### 2. 测试获取

使用 Spring Framework 的 SPI 机制时,加载 `spring.factories` 文件的 API 是 **SpringFactoriesLoader**,它不仅可以加载声明的类的对象,而且可以直接把预先定义好的全限定名都提取出来,如代码清单 2-28 所示。

**代码清单 2-28　使用 Spring Framework SPI 加载**

```
public class SpringSpiApplication {
    
    public static void main(String[] args) {
        List<DemoDao> demoDaos = SpringFactoriesLoader
            .loadFactories(DemoDao.class, SpringSpiApplication.class.getClassLoader());
        demoDaos.forEach(dao -> {
            System.out.println(dao);
        });
        
        System.out.println("-------------------------------------------------");
        
        List<String> daoClassNames = SpringFactoriesLoader
```

```
        .loadFactoryNames(DemoDao.class, SpringSpiApplication.class.getClassLoader());
    daoClassNames.forEach(className -> {
        System.out.println(className);
    });
    }
}
```

运行测试运行类的 main 方法,控制台也可以正常打印对象与全限定名,说明 Spring Framework 的 SPI 机制已正确使用。

```
com.linkedbear.springboot.assemble.d_spi.bean.DemoMySQLDaoImpl@763d9750
com.linkedbear.springboot.assemble.d_spi.bean.DemoOracleDaoImpl@5c0369c4
------------------------------------------------
com.linkedbear.springboot.assemble.d_spi.bean.DemoMySQLDaoImpl
com.linkedbear.springboot.assemble.d_spi.bean.DemoOracleDaoImpl
```

### 3. Spring SPI 机制的实现原理

下面深入底层源码,以了解 Spring Framework 的 SPI 机制是如何实现的,这部分有助于读者更好地理解 SPI 以及它的利用方式。

SPI 的核心使用方法是 `SpringFactoriesLoader.loadFactoryNames` 方法,通过这个方法可以获得指定全限定名对应配置的所有类的全限定类名,`loadFactoryNames` 方法实现的底层逻辑如代码清单 2-29 所示(关键注释已在其中标注)。

**代码清单 2-29　SpringFactoriesLoader.loadFactoryNames 底层实现**

```java
// 存储 SPI 机制加载的类及其映射
private static final Map<ClassLoader, MultiValueMap<String, String>> cache
        = new ConcurrentReferenceHashMap<>();
// SPI 的文件名必须为 spring.factories
public static final String FACTORIES_RESOURCE_LOCATION = "META-INF/spring.factories";

public static List<String> loadFactoryNames(Class<?> factoryType,
        @Nullable ClassLoader classLoader){
    String factoryTypeName = factoryType.getName();
    // 利用缓存机制提高加载速度
    return loadSpringFactories(classLoader).getOrDefault(factoryTypeName,Collections.emptyList());
}

private static Map<String, List<String>> loadSpringFactories(@Nullable ClassLoader classLoader) {
    // 解析之前先检查缓存,有则直接返回
    MultiValueMap<String, String> result = cache.get(classLoader);
    if (result != null) {
        return result;
    }

    try {
        // 真正的加载动作,利用类加载器加载所有的 spring.factories(多个),并逐个配置解析
        Enumeration<URL> urls = (classLoader != null?
                classLoader.getResources(FACTORIES_RESOURCE_LOCATION):
                classLoader.getSystemResources(FACTORIES_RESOURCE_LOCATION));
        result = new LinkedMultiValueMap<>();
        while (urls.hasMoreElements()) {
```

```
            // 提取出每个spring.factories文件
            URL url = urls.nextElement();
            UrlResource resource = new UrlResource(url);
            // 以Properties的方式读取spring.factories
            Properties properties = PropertiesLoaderUtils.loadProperties(resource);
            for(Map.Entry<?, ?> entry : properties.entrySet()) {
                // 逐个收集配置项映射
                String factoryTypeName = ((String) entry.getKey()).trim();
                // 如果一个key配置了多个value,则用英文逗号隔开,此处会做分隔
                for (String factoryImplementationName : StringUtils.commaDelimitedListToString
Array((String) entry.getValue())) {
                    result.add(factoryTypeName, factoryImplementationName.trim());
                }
            }
        }
        cache.put(classLoader, result);
        return result;
    } // catch ......
}
```

下面简单介绍 loadFactoryNames 方法的逻辑。SpringFactoriesLoader 中有一块缓存区,这块缓存区会在 SPI 机制第一次被利用时将项目类路径下所有的 spring.factories 文件都加载并解析、存入缓存区(空间换时间的思想)。解析的逻辑也不难,它会将每一个 spring.factories 文件当作 properties 文件解析,并提取每一对映射关系,保存到 Map 中,最终存入全局缓存。

整体逻辑不算很难,读者可以对照源码注释,配合 IDE 的 Debug 运行几次加以体会。

## 2.5　Spring Boot 的装配机制

了解了 Spring Framework 的组件装配和 SPI 机制,下面就可以正式进入 Spring Boot 的装配机制阶段。Spring Boot 的自动装配实际就是**模块装配+条件装配+SPI 机制**的组合使用,而这一切都凝聚在 Spring Boot 的主启动类的 @SpringBootApplication 注解上。

@SpringBootApplication 注解是由三个注解组合而来的复合注解,它来源于 Spring Boot 1.2.0,在此之前,代码清单 2-30 中的这三个注解需要开发者自己手动声明。

**代码清单 2-30　@SpringBootApplication 注解的组成**

```
@SpringBootConfiguration
@EnableAutoConfiguration
@ComponentScan(excludeFilters = {
    @Filter(type = FilterType.CUSTOM, classes = TypeExcludeFilter.class),
    @Filter(type = FilterType.CUSTOM, classes = AutoConfigurationExcludeFilter.class)
})
public @interface SpringBootApplication
```

只要在 Spring Boot 的主启动类上标注 @SpringBootApplication 注解,就会触发组件的自动装配(@EnableAutoConfiguration)和组件扫描(@ComponentScan)。下面分别讲解这三个注解的作用。

## 2.5.1 @ComponentScan

@ComponentScan 组件扫描,其本身来自 Spring Framework,放在@SpringBootApplication 注解上组合的意图是扫描主启动类所在包及其子包下的所有组件,这也解释了为什么 Spring Boot 的启动类要放到所有类所在包的最外层。不过跟平时项目开发中使用的方式有点不同的是,在 @SpringBootApplication 注解组合时@ComponentScan 额外添加了两个过滤条件:

```
@ComponentScan(excludeFilters = {
    @Filter(type = FilterType.CUSTOM, classes = TypeExcludeFilter.class),
    @Filter(type = FilterType.CUSTOM, classes = AutoConfigurationExcludeFilter.class)
})
```

既然 Spring Boot 有意设置了排除过滤器,那么它一定有特殊的作用,值得探究。

### 1. TypeExcludeFilter

由类名理解 TypeExcludeFilter 是一个类型排除的过滤器,它提供了一种扩展机制,可以让开发者向 IOC 容器中注册一些自定义的组件过滤器,以便在组件扫描的过程中过滤它们。文档注释中特意说明了 TypeExcludeFilter 的使用方式:只需要在应用上下文中注册 TypeExcludeFilter 类的子类,并重写 match 方法,Spring Boot 会自动找到这些 TypeExcludeFilter 的子类并调用它们。有关效果的测试读者可以自行编写测试代码体会,本书不再演示。下面重点研究底层的运行原理和机制。

具体来看,TypeExcludeFilter 的底层实现是从 BeanFactory,即 IOC 容器中找出所有类为 TypeExcludeFilter 的 Bean,并依次调用它们匹配,如代码清单 2-31 所示。

**代码清单 2-31　TypeExcludeFilter 的核心逻辑实现**

```java
public boolean match(MetadataReader metadataReader,
        MetadataReaderFactory metadataReaderFactory) throws IOException {
    if (this.beanFactory instanceof ListableBeanFactory && getClass() == TypeExcludeFilter.class) {
        for (TypeExcludeFilter delegate : getDelegates()) {
            // 提取出所有 TypeExcludeFilter 并依次调用其 match 方法检查是否匹配
            if (delegate.match(metadataReader, metadataReaderFactory)) {
                return true;
            }
        }
    }
    return false;
}

private Collection<TypeExcludeFilter> getDelegates() {
    Collection<TypeExcludeFilter> delegates = this.delegates;
    if (delegates == null) {
        // 此处会从 BeanFactory 中提取出所有类为 TypeExcludeFilter 的 Bean
        delegates = ((ListableBeanFactory) this.beanFactory).getBeansOfType(TypeExcludeFilter.class).values();
        this.delegates = delegates;
    }
    return delegates;
}
```

整个处理逻辑比较简单，它会从 BeanFactory 中提取出所有类为 TypeExcludeFilter 的 Bean，并缓存到本地（空间换时间的设计体现），后续在进行组件扫描时会依次调用这些 TypeExcludeFilter 对象，检查被扫描的类是否符合匹配规则。

> 提示：由于 Spring Boot 中设置的 TypeExcludeFilter 属于排除性质的过滤器，因此被匹配到的 Bean 不会被注册到 IOC 容器中。

### 2. AutoConfigurationExcludeFilter

AutoConfigurationExcludeFilter 的类名已经将其作用体现得很明确，它是用来排除自动配置类的过滤器。注意，它是在组件扫描阶段排除自动配置类，而不是放弃自动配置类不管。它的底层过滤逻辑中要判断的规则有两条：一个类是配置类；且是自动配置类。

检查的规则在源码中设计得比较简单，如代码清单 2-32 所示。被 @Configuration 注解标注的类都是注解配置类；同时，被定义在 spring.factories 中 @EnableAutoConfiguration 注解的配置类为自动配置类。可能读者看到这里有点费解，2.5.3 节中将进行解释，这里只需要知道 AutoConfigurationExcludeFilter 这个类型过滤器会在组件扫描阶段过滤掉自动配置类。

**代码清单 2-32　AutoConfigurationExcludeFilter 的核心逻辑实现**

```java
public boolean match(MetadataReader metadataReader,
        MetadataReaderFactory metadataReaderFactory) throws IOException {
    return isConfiguration(metadataReader) && isAutoConfiguration(metadataReader);
}

private boolean isConfiguration(MetadataReader metadataReader) {
    // 检查是配置类的原则：是否被@Configuration注解修饰
    return metadataReader.getAnnotationMetadata().isAnnotated(Configuration.class.getName());
}

private boolean isAutoConfiguration(MetadataReader metadataReader) {
    // 检查是自动配置类的原则：是否被定义在 spring.factories 中的 EnableAutoConfiguration 中
    return getAutoConfigurations().contains(metadataReader.getClassMetadata().getClassName());
}

protected List<String> getAutoConfigurations() {
    if (this.autoConfigurations == null) {
        this.autoConfigurations = SpringFactoriesLoader
                .loadFactoryNames(EnableAutoConfiguration.class, this.beanClassLoader);
    }
    return this.autoConfigurations;
}
```

## 2.5.2　@SpringBootConfiguration

@SpringBootConfiguration 注解并不是一个神秘的扩展，它本身仅组合了一个 Spring Framework 的 @Configuration 注解而已。

```java
@Configuration
public @interface SpringBootConfiguration
```

## 2.5 Spring Boot 的装配机制

所以简单地理解，一个类标注了 `@SpringBootConfiguration` 注解也就是标注了 `@Configuration` 注解，代表这个类是一个注解配置类。

> 如果读者要探究 `@SpringBootConfiguration` 有什么额外的作用，本书可以提供一个线索以供参考。在 Spring Boot 的官方文档 Spring Boot Features 部分中有一节"features.testing.spring-boot-applications.detecting-configuration"提及了 `@SpringBootConfiguration` 注解。文档中提到，Spring Boot 运行测试时会寻找被 `@SpringBootApplication` 或者 `@SpringBootConfiguration` 注解标注的类，之后开始测试工作。而在 Spring Boot 的源码中，如果读者引入了 `spring-boot-starter-test` 的依赖，可以发现这个注解在 `SpringBootTestContextBootstrapper` 的 `getOrFindConfigurationClasses` 方法中被调用，而这个方法的名称恰好就叫"寻找配置类"，由此推断与官方文档描述的内容相吻合。

### 2.5.3 @EnableAutoConfiguration

最后一个注解是重头戏，`@EnableAutoConfiguration` 注解承载了 Spring Boot 自动装配的"灵魂"。不必着急阅读源码的结构，先请读者仔细解读 Spring Boot 的开发者是如何描述自动装配的。

**1. javadoc 的描述**

为了方便读者能更方便、更容易地阅读和理解 `@EnableAutoConfiguration` 注解的 javadoc，本节会将注释拆分为多段解读。

*Enable auto-configuration of the Spring Application Context, attempting to guess and configure beans that you are likely to need. Auto-configuration classes are usually applied based on your classpath and what beans you have defined. For example, if you have* `tomcat-embedded.jar` *on your classpath you are likely to want a* `TomcatServletWebServerFactory` *(unless you have defined your own* `ServletWebServerFactory` *bean). When using* `SpringBootApplication`*, the auto-configuration of the context is automatically enabled and adding this annotation has therefore no additional effect.*

第一段主要讲解的是注解本身的作用。标注 `@EnableAutoConfiguration` 注解后，会启用 Spring 应用上下文（即 `ApplicationContext`）的自动配置，并且 Spring Boot 会尝试猜测和配置当前项目中可能需要的 Bean。通常根据项目的类路径和定义的 Bean 就可以合理地应用自动配置类。例如，如果项目的类路径下引用了 `tomcat-embedded.jar` 包，则当前项目很有可能需要配置一个 `TomcatServletWebServerFactory`（除非项目中已经手动注册了一个 `ServletWebServerFactory` 类型的 Bean）。另外，当使用 `@SpringBootApplication` 注解时，将自动启用上下文的自动装配，因此也就不需要再声明 `@EnableAutoConfiguration` 了（本身 `@SpringBootApplication` 注解已经组合了 `@EnableAutoConfiguration`）。

*Auto-configuration tries to be as intelligent as possible and will back-away as you define more of your own configuration. You can always manually* `exclude()` *any configuration that you never want to apply (use* `excludeName()` *if you don't have access to them). You can also exclude them via the* `spring.autoconfigure.exclude` *property. Auto-configuration is always applied after user-defined beans have been registered.*

紧接着的一段主要解释了 Spring Boot 自动配置的机制和禁用方法。Spring Boot 的自动配置会尽可能地智能化，并会在项目中注册更多自定义配置的时候自动退出（即被覆盖）。开发者也可以直接禁用某些不需要/不想用的自动配置（使用 `exclude` 属性，或者在无法访问时使用 `excludeName` 属性），或者通过全局配置文件的 `spring.autoconfigure.exclude` 属性排除这些自动配置类。最后一点，Spring Boot 的自动配置类的触发时机是在项目中自定义的配置加载完毕后应用。

*The package of the class that is annotated with `@EnableAutoConfiguration`, usually via `@SpringBootApplication`, has specific significance and is often used as a 'default'. For example, it will be used when scanning for `@Entity` classes. It is generally recommended that you place `@EnableAutoConfiguration` (if you're not using `@SpringBootApplication`) in a root package so that all sub-packages and classes can be searched.*

接下来的一段阐述了组件扫描的相关事宜：被 `@EnableAutoConfiguration` 或 `@SpringBootApplication` 注解标注的类，其所在的包有特殊含义，通常被定义为"默认值"。这个默认值体现在组件扫描、JPA 实体类扫描、Mapper 接口扫描等。Spring Boot 默认扫描被 `@EnableAutoConfiguration` 注解标注的类所在包及其子包的所有组件。

*Auto-configuration classes are regular Spring `Configuration` beans. They are located using the `SpringFactoriesLoader` mechanism (keyed against this class). Generally auto-configuration beans are `@Conditional` beans (most often using `@ConditionalOnClass` and `@ConditionalOnMissingBean` annotations).*

最后一段提到了自动配置类与 Spring Framework SPI 机制的关系。Spring Boot 的自动配置类本身也是一些普通的配置类，只不过它们的加载是通过 Spring Framework 的 SPI 机制（即 `SpringFactoriesLoader`）。另外，Spring Boot 的自动配置类也是条件配置类（被 `@Conditional` 系列注解标注，其最常见的如使用 `@ConditionalOnClass` 和 `@ConditionalOnMissingBean` 标注）。

通读下来，读者是否可以体会到一点：Spring Boot 的文档注释写得相当好，非常清晰。最后简单总结 `@EnableAutoConfiguration` 注解的作用：标注该注解后，会启用 Spring Boot 的自动装配，根据导入的依赖、上下文配置合理加载默认的自动配置。

`@EnableAutoConfiguration` 本身是一个组合注解，它还包含了两个注解，下面逐一来解释。

```
@AutoConfigurationPackage
@Import(AutoConfigurationImportSelector.class)
```

### 2. @AutoConfigurationPackage

`@AutoConfigurationPackage` 注解本身组合了一个 `@Import` 注解，它导入了一个内部类 `AutoConfigurationPackages.Registrar`。

```
@Import(AutoConfigurationPackages.Registrar.class)
public @interface AutoConfigurationPackage
```

`@AutoConfigurationPackage` 注解所做的是将主启动类所在的包记录下来，注册到 `AutoConfigurationPackages` 中。由于 `@SpringBootApplication` 注解中已经组合了

## 2.5 Spring Boot 的装配机制

这个@AutoConfigurationPackage，因此在此处也可以提取到主启动类所在的包。

> 💡 提示：在 Spring Boot 2.3.0 版本以前，@AutoConfigurationPackage 注解没有任何属性，标注了该注解即确定了主启动类所在包（约定）。在 Spring Boot 2.3.0 版本及以后，注解中多了两个属性：basePackages 和 basePackageClasses。它们可以手动指定应用的根包/根路径（配置）。如果没有手动指定，则仍然采用默认的主启动类所在包（约定大于配置）。为什么 Spring Boot 要执着于记录应用的根包呢？因为这个根包路径在整合第三方 starter 时有特殊作用，下面会讲到。

简单介绍了@AutoConfigurationPackage 注解本身，下面把目标转移到导入的内部类 Registrar 上。Registrar 本身是一个 ImportBeanDefinitionRegistrar，它的作用是以编程式向 IOC 容器中注册 bean 对象，而 Registrar 要注册的对象实际上是默认主启动类所在的包路径（也就是@AutoConfigurationPackage 注解要记录的根包），如代码清单 2-33 所示。

**代码清单 2-33　AutoConfigurationPackages.Registrar**

```java
static class Registrar implements ImportBeanDefinitionRegistrar, DeterminableImports {

    @Override
    public void registerBeanDefinitions(AnnotationMetadata metadata,
            BeanDefinitionRegistry registry) {
        register(registry, new PackageImports(metadata).getPackageNames().toArray(new String[0]));
    }

    @Override
    public Set<Object> determineImports(AnnotationMetadata metadata) {
        return Collections.singleton(new PackageImports(metadata));
    }
}
```

观察代码清单 2-33 中的两个方法，其中 determineImports 方法暂且不提，主要关注 ImportBeanDefinitionRegistrar 的核心 registerBeanDefinitions 方法。这里有两点要注意，分别是 register 方法本身，以及传入的一个包名数组。

（1）PackageImports

注意 register 方法的第二个参数，它利用 PackageImports 导出了一组包名，而包名的来源是一个 AnnotationMetadata，这个 AnnotationMetadata 本质上可以理解为@AutoConfigurationPackage 注解本身。换句话说，这个 PackageImports 类提取出@AutoConfigurationPackage 注解中定义的两个属性（basePackages 与 basePackageClasses）。实际上源码也是如此，它做了一个约定大于配置的设计，PackageImports 的构造方法中会先提取注解中的这两个属性，如果两个属性都没有定义则会提取主启动类所在的包名，具体逻辑如代码清单 2-34 所示。

**代码清单 2-34　PackageImports 中提取应用的根包名，体现了约定大于配置**

```java
private static final class PackageImports {

    private final List<String> packageNames;

    PackageImports(AnnotationMetadata metadata) {
```

```java
        AnnotationAttributes attributes = AnnotationAttributes.fromMap(metadata
                .getAnnotationAttributes(AutoConfigurationPackage.class.getName(),false));
        List<String> packageNames = new ArrayList<>();
        // 先提取两个属性
        for (String basePackage : attributes.getStringArray("basePackages")) {
            packageNames.add(basePackage);
        }
        for (Class<?> basePackageClass : attributes.getClassArray("basePackageClasses")) {
            packageNames.add(basePackageClass.getPackage().getName());
        }
        // 如未提取到,则提取被注解标注的类的所在包,即Spring Boot的主启动类所在的包
        if (packageNames.isEmpty()) {
            packageNames.add(ClassUtils.getPackageName(metadata.getClassName()));
        }
        this.packageNames = Collections.unmodifiableList(packageNames);
    }
```

提取出根包路径后,下面是调用外层的 `register` 方法。

**(2) register 方法**

请读者注意一点,这个 `register` 方法并没有定义在 `Registrar` 内部类上,而是外部 `AutoConfigurationPackages` 的一个静态方法,如代码清单 2-35 所示。

**代码清单 2-35　register 方法注册根包路径**

```java
private static final String BEAN = AutoConfigurationPackages.class.getName();

public static void register(BeanDefinitionRegistry registry, String... packageNames) {
    // 检查IOC容器中是否包含指定的Bean,即AutoConfigurationPackages
    if (registry.containsBeanDefinition(BEAN)){
        BeanDefinition beanDefinition = registry.getBeanDefinition(BEAN);
        ConstructorArgumentValues constructorArguments = beanDefinition.getConstructorArgumentValues();
        // 添加构造方法的参数:应用的根包路径
        constructorArguments.addIndexedArgumentValue(0,addBasePackages(constructorArguments, packageNames));
    } else {
        // 如果没有Bean,则此处构造定义信息,并注册到IOC容器中
        GenericBeanDefinition beanDefinition = new GenericBeanDefinition();
        beanDefinition.setBeanClass(BasePackages.class);
        beanDefinition.getConstructorArgumentValues().addIndexedArgumentValue(0, packageNames);
        beanDefinition.setRole(BeanDefinition.ROLE_INFRASTRUCTURE);
        registry.registerBeanDefinition(BEAN, beanDefinition);
    }
}
```

注意观察这部分逻辑:`register` 方法要先判断 IOC 容器中是否包含一个名为 `AutoConfigurationPackages` 的 Bean,如果有则给这个 Bean 添加构造方法的参数,如果没有则创建一个全新的 Bean 注册到 IOC 容器中。这里暂且不提 Bean 注册的逻辑,将重点放在这两个分支逻辑中的共性:添加构造方法的参数。这个构造方法的参数就是上面在 `PackageImports` 中刚提取出的 `packageNames`,即应用的根包路径。

至于最终这些根包路径放到哪里了，可以关注一下 else 分支中的 BeanClass：BasePackages.class，它的结构相当简单，内部只维护了一个 packages 集合存放这些根包路径，如代码清单 2-36 所示。

**代码清单 2-36　BasePackages 的结构相当简单**

```
static final class BasePackages {

    private final List<String> packages;

    BasePackages(String... names) {
        List<String> packages = new ArrayList<>();
        for (String name : names) {
            if (StringUtils.hasText(name)) {
                packages.add(name);
            }
        }
        this.packages = packages;
    }
```

如果保存当前 Spring Boot 应用的根包路径仅仅是为了提供给 Spring Boot 内部使用，那么这个设计似乎有一点多余。回想一下，Spring Boot 的强大之处中非常重要的一点是更方便地整合第三方技术。以读者非常熟悉的 MyBatis 为例，当项目中引入 mybatis-spring-boot-starter 依赖后，可以在 IDE 中打开 MyBatisAutoConfiguration 类。在这个配置类中可以找到 AutoConfiguredMapperScannerRegistrar 这样一个组件，如代码清单 2-37 所示。

**代码清单 2-37　AutoConfiguredMapperScannerRegistrar 中使用了 basePackages（节选）**

```
public static class AutoConfiguredMapperScannerRegistrar
        implements BeanFactoryAware, ImportBeanDefinitionRegistrar {

    private BeanFactory beanFactory;

    @Override
    public void registerBeanDefinitions(AnnotationMetadata importingClassMetadata,
            BeanDefinitionRegistry registry) {
        // ......
        List<String> packages = AutoConfigurationPackages.get(this.beanFactory);
        // ......

        BeanDefinitionBuilder builder = BeanDefinitionBuilder
                .genericBeanDefinition(MapperScannerConfigurer.class);
        builder.addPropertyValue("processPropertyPlaceHolders", true);
        builder.addPropertyValue("annotationClass", Mapper.class);
        builder.addPropertyValue("basePackage", StringUtils.collectionToCommaDelimitedString(packages));
        // ......
        registry.registerBeanDefinition(MapperScannerConfigurer.class.getName(), builder.getBeanDefinition());
    }
```

注意，AutoConfiguredMapperScannerRegistrar 是一个 ImportBeanDefinition

Registrar，它会向 IOC 容器中注册一个 `MapperScannerConfigurer`，熟悉 MyBatis 的读者肯定对它不陌生，它是用于整合 Spring Framework 进行 Mapper 接口扫描的组件。在 `AutoConfiguredMapperScannerRegistrar` 中会将 `basePackages` 提取出并设置到 `MapperScannerConfigurer` 中，以备后续 `MapperScannerConfigurer` 的 Mapper 接口扫描工作，由此也就体现出 `basePackages` 的作用了。

有了对`@AutoConfigurationPackage` 和`@ComponentScan` 注解的了解，也就理解了为什么 Spring Boot 的主启动类要在所有类的最外层了。

### 3. AutoConfigurationImportSelector

`@EnableAutoConfiguration` 的另一个导入组件是 `AutoConfigurationImportSelector`，它本身是一个 `DeferredImportSelector`，除了具备 `ImportSelector` 的作用，2.2.6 节中曾讲到，实现了 `DeferredImportSelector` 接口的 `ImportSelector` 执行时机更晚，所以更适合做一些补充性工作，正好自动装配的设计就是约定大于配置，项目中已经有的配置不会重复注册，项目中没有配置的部分会予以补充，而负责补充的任务就交给了自动配置类，`AutoConfigurationImportSelector` 的作用就是加载这些自动配置类，所以这就形成了一条很清晰的逻辑链条。

下面聚焦 `AutoConfigurationImportSelector` 本身，在 2.2.6 节中提到过高版本的 Spring Framework 中给 `DeferredImportSelector` 添加了分组的概念，所以真正起作用的方法不是 `selectImports`,而是 `DeferredImportSelector` 接口的另一个方法 `getImportGroup` 返回的类，如代码清单 2-38 所示（这个返回的 `AutoConfigurationGroup` 才是真正起作用的）。

**代码清单 2-38　AutoConfigurationImportSelector 中声明了导入所属组**

```
public Class<? extends Group> getImportGroup() {
    return AutoConfigurationGroup.class;
}
```

`AutoConfigurationGroup` 本身实现了 `DeferredImportSelector.Group` 接口，这个接口中又定义了一个 `process` 方法，这个方法才是真正负责加载所有自动配置类的入口。`AutoConfigurationGroup` 中的实现逻辑如代码清单 2-39 所示。

**代码清单 2-39　AutoConfigurationGroup 中包含加载自动配置类的方法实现**

```
public void process(AnnotationMetadata annotationMetadata,
        DeferredImportSelector deferredImportSelector) {
    // 断言……
    AutoConfigurationEntry autoConfigurationEntry = ((AutoConfigurationImportSelector) deferredImportSelector).getAutoConfigurationEntry(annotationMetadata);
    this.autoConfigurationEntries.add(autoConfigurationEntry);
    for (String importClassName : autoConfigurationEntry.getConfigurations()) {
        this.entries.putIfAbsent(importClassName, annotationMetadata);
    }
}
```

这个方法首先会检查传入的 `deferredImportSelector` 是否由 `AutoConfigurationImportSelector` 导入而来，该细节通常不会受影响，所以这里必然会通过。下面的

## 2.5 Spring Boot 的装配机制

getAutoConfigurationEntry 方法是加载自动配置类的核心逻辑（见代码清单 2-40），其源码步骤稍多，不过主干逻辑只有三步：加载自动配置类→移除被去掉的自动配置类→封装 Entry 返回。这里加载自动配置类的动作就是利用 Spring Framework 的 SPI 机制，从 spring.factories 中提取出所有 @EnableAutoConfiguration 对应的配置值（见代码清单 2-41）。可以通过三种途径移除被去掉的自动配置类：@SpringBootApplication 或 @EnableAutoConfiguration 注解的 exclude、excludeName 属性，以及全局配置文件的 spring.autoconfigure.exclude 属性配置。底层源码会提取出这三个位置配置的自动配置类并移除（见代码清单 2-42）。

**代码清单 2-40　getAutoConfigurationEntry 获取要加载的自动配置类**

```java
protected AutoConfigurationEntry getAutoConfigurationEntry(AnnotationMetadata annotation
Metadata){
    if (!isEnabled(annotationMetadata)) {
        return EMPTY_ENTRY;
    }
    // 加载注解属性配置
    AnnotationAttributes attributes = getAttributes(annotationMetadata);
    // 真正加载自动配置类的动作
    List<String> configurations = getCandidateConfigurations(annotationMetadata, attributes);
    // 自动配置类去重
    configurations = removeDuplicates(configurations);
    // 获取显式配置了要移除的自动配置类
    Set<String> exclusions = getExclusions(annotationMetadata, attributes);
    checkExcludedClasses(configurations, exclusions);
    // 移除动作
    configurations.removeAll(exclusions);
    configurations = getConfigurationClassFilter().filter(configurations);
    // 广播 AutoConfigurationImportEvent 事件
    fireAutoConfigurationImportEvents(configurations, exclusions);
    return new AutoConfigurationEntry(configurations, exclusions);
}
```

**代码清单 2-41　利用 SPI 机制加载所有配置了 @EnableAutoConfiguration 的自动配置类**

```java
protected List<String> getCandidateConfigurations(AnnotationMetadata metadata,
        AnnotationAttributes attributes){
    List<String> configurations = SpringFactoriesLoader
            .loadFactoryNames(getSpringFactoriesLoaderFactoryClass(), getBeanClassLoader());
    // 断言 ......
    return configurations;
}

protected Class<?> getSpringFactoriesLoaderFactoryClass() {
    return EnableAutoConfiguration.class;
}
```

**代码清单 2-42　getExclusions 获取被显式移除的自动配置类**

```java
protected Set<String> getExclusions(AnnotationMetadata metadata,
        AnnotationAttributes attributes){
```

```
Set<String> excluded = new LinkedHashSet<>();
    excluded.addAll(asList(attributes,"exclude"));
    excluded.addAll(Arrays.asList(attributes.getStringArray("excludeName")));
    excluded.addAll(getExcludeAutoConfigurationsProperty());
    return excluded;
}
```

经过如上所述的流程后，需要被加载的自动配置类会全部收集完毕，并在最后返回这些自动配置类的全限定名，存入 `AutoConfigurationGroup` 的一个缓存中，后续 IOC 容器会提取出这些自动配置类并解析，完成自动装配的加载。

> 提示：Spring Boot 官方支持的场景启动器中，对应的自动配置类通常都放到 `spring-boot-autoconfigure` 依赖的 `spring.factories` 文件中统一管理。读者在自行探究场景启动器时，可以尝试以该文件中 `EnableAutoConfiguration` 配置的自动配置类为切入点学习。

**4．小结**

简单总结一下 Spring Boot 的核心 `@SpringBootApplication` 和自动装配机制。

- `@SpringBootApplication` 包含 `@ComponentScan` 注解，可以默认扫描当前包及子包下的所有组件。
- `@EnableAutoConfiguration` 中包含 `@AutoConfigurationPackage` 注解，可以记录下最外层根包的位置，以便第三方框架整合使用。
- `@EnableAutoConfiguration` 导入的 `AutoConfigurationImportSelector` 可以利用 Spring Framework 的 SPI 机制加载所有自动配置类。

## 2.6　WebMvc 场景下的自动装配原理

了解了 Spring Boot 的自动装配机制后，下面研究一个实际且常见的场景：当项目整合 SpringWebMvc 后 Spring Boot 的自动装配都做了什么。

在引入 `spring-boot-starter-web` 的依赖后，Spring Boot 会进行 WebMvc 环境的自动装配，处理的核心是一个叫作 `WebMvcAutoConfiguration` 的类，首先观察这个类做了什么，如代码清单 2-43 所示。

**代码清单 2-43　WebMvcAutoConfiguration 的类头定义**

```
@Configuration(proxyBeanMethods = false)
// 当前环境必须是 WebMvc（Servlet）环境
@ConditionalOnWebApplication(type = Type.SERVLET)
// 当前运行环境的 classpath 中必须有 Servlet 类、DispatcherServlet 类和 WebMvcConfigurer 类
@ConditionalOnClass({ Servlet.class, DispatcherServlet.class, WebMvcConfigurer.class})
// 如果没有自定义 WebMvc 的配置类，则使用本自动配置
@ConditionalOnMissingBean(WebMvcConfigurationSupport.class)
@AutoConfigureOrder(Ordered.HIGHEST_PRECEDENCE + 10)
// 当前自动配置会在以下几个配置类的解析后再处理
@AutoConfigureAfter({ DispatcherServletAutoConfiguration.class,
                TaskExecutionAutoConfiguration.class,
                ValidationAutoConfiguration.class})
public class WebMvcAutoConfiguration
```

由代码清单 2-43 可以发现，要想使 `WebMvcAutoConfiguration` 配置类生效，需要满足以下几个条件：

- 当前环境必须是 WebMvc（即 Servlet 环境），导入了 WebMvc 的依赖后，该条件默认生效；
- 当前类路径下必须有 `Servlet`、`DispatcherServlet`、`WebMvcConfigurer` 这几个类，如果缺少其中任何一个类，WebMvc 的自动装配就无效（即上述几个类是构造 WebMvc 环境的重要成员）；
- 只有项目中未自定义注册 `WebMvcConfigurationSupport` 的类或者子类，自动装配才会生效。

针对第一个条件，这里先解释一下，当同时导入 WebMvc 和 WebFlux 时，则 WebMvc 生效，WebFlux 无效，有关这部分内容会在 6.1.1 节中讲解。

关于最后一个条件，这里有必要展开解释。`WebMvcAutoConfiguration` 之所以要检查容器中是否未包含 `WebMvcConfigurationSupport`，是为了避免与 `@EnableWebMvc` 注解冲突，因为查看源码时可以发现，`@EnableWebMvc` 注解本身导入了一个 `WebMvcConfigurationSupport` 的子类 `DelegatingWebMvcConfiguration`，它可以支持使用 `WebMvcConfigurer` 配合完成 WebMvc 的组件定制化和扩展配置，而不需要手动注册 WebMvc 的核心组件。`WebMvcAutoConfiguration` 本身就已经支持了这种方式，不需要再额外标注 `@EnableWebMvc` 注解，因此这里就做了一个约定大于配置的设计：如果项目的配置类中没有标注 `@EnableWebMvc` 注解，则 Spring Boot 认为采用约定的配置；而标注了 `@EnableWebMvc` 注解后，Spring Boot 认为项目中需要覆盖约定的自动装配，转而使用自定义的配置，所以此处不再使 `WebMvcAutoConfiguration` 生效。

`WebMvcAutoConfiguration` 在解析之前，注解中提到了另一个非常重要的配置类 `DispatcherServletAutoConfiguration`，显然它是 `DispatcherServlet` 相关的自动配置类，而 `DispatcherServletAutoConfiguration` 又不是排序最靠前的，它排在另一个自动配置类 `ServletWebServerFactoryAutoConfiguration` 之后，如代码清单 2-44 所示。

**代码清单 2-44　ServletWebServerFactoryAutoConfiguration 处理排序更靠前**

```
@AutoConfigureOrder(Ordered.HIGHEST_PRECEDENCE)
@Configuration(proxyBeanMethods = false)
@ConditionalOnWebApplication(type = Type.SERVLET)
@ConditionalOnClass(DispatcherServlet.class)
// ServletWebServerFactoryAutoConfiguration 解析完后才能轮到它
@AutoConfigureAfter(ServletWebServerFactoryAutoConfiguration.class)
public class DispatcherServletAutoConfiguration
```

所以大体上可以梳理出 WebMvc 场景的自动装配环节：Servlet 容器的装配→DispatcherServlet 的装配→WebMvc 核心组件的装配。下面就这三个环节逐一展开讲解。

### 2.6.1　Servlet 容器的装配

首先来看嵌入式 Servlet 容器的装配，这里又导入了几个组件，分别是一个 `BeanPostProcessorsRegistrar` 和几个 Embedded 容器内部类，如代码清单 2-45 所示。

### 代码清单 2-45　ServletWebServerFactoryAutoConfiguration 的类头定义

```
@Configuration(proxyBeanMethods = false)
@AutoConfigureOrder(Ordered.HIGHEST_PRECEDENCE)
@ConditionalOnClass(ServletRequest.class)
@ConditionalOnWebApplication(type = Type.SERVLET)
@EnableConfigurationProperties(ServerProperties.class)
@Import({ServletWebServerFactoryAutoConfiguration.BeanPostProcessorsRegistrar.class,
        ServletWebServerFactoryConfiguration.EmbeddedTomcat.class,
        ServletWebServerFactoryConfiguration.EmbeddedJetty.class,
        ServletWebServerFactoryConfiguration.EmbeddedUndertow.class})
public class ServletWebServerFactoryAutoConfiguration
```

**1. EmbeddedTomcat、EmbeddedJetty 和 EmbeddedUndertow**

这三个容器其实是一码事。学过 Spring Boot 整合 WebMvc 开发的读者都清楚，默认情况下 Spring Boot 会整合嵌入式 Tomcat (EmbeddedTomcat) 作为可独立运行 jar 文件的嵌入式 Web 容器。如果需要切换嵌入式 Web 容器，只需要把原本的嵌入式 Tomcat 依赖移除掉，再添加新的嵌入式 Servlet 容器依赖，如代码清单 2-46 所示。

### 代码清单 2-46　切换嵌入式 Servlet 容器为 Jetty 的方法

```xml
<dependency>
    <groupId>org.springframework.boot</groupId>
    <artifactId>spring-boot-starter-web</artifactId>
    <exclusions>
        <exclusion>
            <groupId>org.springframework.boot</groupId>
            <artifactId>spring-boot-starter-tomcat</artifactId>
        </exclusion>
    </exclusions>
</dependency>
<dependency>
    <groupId>org.springframework.boot</groupId>
    <artifactId>spring-boot-starter-jetty</artifactId>
    <scope>provided</scope>
</dependency>
```

底层真正确定应该实例化哪个嵌入式 Web 容器，这在本质上是由这几个嵌入式内部类决定的。下面以 EmbeddedTomcat 为例，以查看它的内部设计，如代码清单 2-47 所示。

### 代码清单 2-47　EmbeddedTomcat 处理的组件

```
@Configuration(proxyBeanMethods = false)
@ConditionalOnClass({ Servlet.class, Tomcat.class, UpgradeProtocol.class })
@ConditionalOnMissingBean(value = ServletWebServerFactory.class,
                search = SearchStrategy.CURRENT)
static class EmbeddedTomcat {

    @Bean
    TomcatServletWebServerFactory tomcatServletWebServerFactory(
            ObjectProvider<TomcatConnectorCustomizer> connectorCustomizers,
```

```
            ObjectProvider<TomcatContextCustomizer> contextCustomizers,
            ObjectProvider<TomcatProtocolHandlerCustomizer<?>> protocolHandlerCustomizers) {
        TomcatServletWebServerFactory factory = new TomcatServletWebServerFactory();
        factory.getTomcatConnectorCustomizers().addAll(connectorCustomizers.orderedStream()
.collect(Collectors.toList()));
        factory.getTomcatContextCustomizers().addAll(contextCustomizers.orderedStream().
collect(Collectors.toList()));
        factory.getTomcatProtocolHandlerCustomizers().addAll(protocolHandlerCustomizers.
orderedStream().collect(Collectors.toList()));
        return factory;
    }
}
```

注意 `EmbeddedTomcat` 上面标注的注解,它需要只有在当前项目的类路径下包含 Tomcat 类时,当前配置类才会生效,而配置类中注册的是一个 `TomcatServletWebServerFactory`,它负责创建嵌入式 Tomcat 容器。具体嵌入式 Tomcat 是如何创建的,以及嵌入式 Tomcat 的初始化流程,在第 8 章中将详细展开讲解。

> 扩展一个小细节,代码清单 2-47 的 `TomcatServletWebServerFactory` 的定制器中,最下面的 `TomcatProtocolHandlerCustomizer` 是在 Spring Boot 2.2.0 版本后才有的,低版本的 Spring Boot 中并没有这个扩展点,其余两种定制器是自 2.0.0 版本出现的。在之前的版本中,只能通过定义 `WebServerFactoryCustomizer` 的实现类来传入这些特殊的定制器,而 Spring Boot 2.2.0 版本之后可以直接定义这些定制器,并标注 `@Component` 注册到 IOC 容器中,Spring Boot 会自动应用这些定制器。

### 2. BeanPostProcessorsRegistrar

`ServletWebServerFactoryAutoConfiguration` 导入的另一个组件是一个后置处理器的注册器,读者可能还不了解后置处理器的概念和使用,不过这不妨碍理解 `BeanPostProcessorsRegistrar` 本身的作用,如代码清单 2-48 所示。

**代码清单 2-48　BeanPostProcessorsRegistrar**

```
public static class BeanPostProcessorsRegistrar implements ImportBeanDefinitionRegistrar,
                                                            BeanFactoryAware {

    private ConfigurableListableBeanFactory beanFactory;
    // setBeanFactory ......

    @Override
    public void registerBeanDefinitions(AnnotationMetadata importingClassMetadata,
            BeanDefinitionRegistry registry) {
        if (this.beanFactory == null) {
            return;
        }
        registerSyntheticBeanIfMissing(registry, "webServerFactoryCustomizerBeanPostProcessor",
                WebServerFactoryCustomizerBeanPostProcessor.class);
        registerSyntheticBeanIfMissing(registry, "errorPageRegistrarBeanPostProcessor",
                ErrorPageRegistrarBeanPostProcessor.class);
    }

    private void registerSyntheticBeanIfMissing(BeanDefinitionRegistry registry,
            String name, Class<?> beanClass) {
```

```
    if (ObjectUtils.isEmpty(this.beanFactory.getBeanNamesForType(beanClass, true, false))) {
        RootBeanDefinition beanDefinition = new RootBeanDefinition(beanClass);
        beanDefinition.setSynthetic(true);
        registry.registerBeanDefinition(name, beanDefinition);
    }
}
```

注意观察代码清单 2-48 的结构与实现，`BeanPostProcessorsRegistrar` 本身是一个 `ImportBeanDefinitionRegistrar`，它注册的组件是两个后置处理器，简单解释它们的作用：`WebServerFactoryCustomizerBeanPostProcessor` 负责执行所有 `WebServerFactoryCustomizer`（嵌入式 Web 容器的定制器），`ErrorPageRegistrarBeanPostProcessor` 负责向嵌入式 Web 容器中注册默认的错误提示页面。有关这两个组件的底层设计，读者可以结合 IDE 自行查看，这里不再展开。

**3. 两个定制器的注册**

除了 `ServletWebServerFactoryAutoConfiguration` 上面的 `@Import` 注解导入的组件，这个自动配置类中还注册了两个定制器，如代码清单 2-49 所示。

**代码清单 2-49　`ServletWebServerFactoryAutoConfiguration` 注册的两个定制器**

```
//方法名、变量名有精简
@Bean
public ServletWebServerFactoryCustomizer WebServerCustomizer(ServerProperties prop) {
    return new ServletWebServerFactoryCustomizer(prop);
}

@Bean
@ConditionalOnClass(name = "org.apache.catalina.startup.Tomcat")
public TomcatServletWebServerFactoryCustomizer tomcatCustomizer(ServerProperties prop) {
    return new TomcatServletWebServerFactoryCustomizer(prop);
}
```

注意观察这两个定制器，它们都把 `ServerProperties` 作为构造方法的参数传入到定制器中，而 `ServerProperties` 本身定义了有关 Web 容器的一些配置（如端口、上下文路径、开启 SSL 等），这些配置对应 Spring Boot 全局配置文件的 `server.*` 部分。实质上，此处注册的两个定制器是在底层将全局配置文件中定义的配置属性实际应用在嵌入式 Web 容器中，达到 "外部配置内部生效" 的效果。有关定制器在底层是如何运行的，在 8.3.1 节讲解嵌入式 Tomcat 中会提到。

至此，有关 `ServletWebServerFactoryAutoConfiguration` 的装配内容全部完成。

### 2.6.2　DispatcherServlet 的装配

接下来是有关 `DispatcherServlet` 的装配。在讲解 `DispatcherServlet` 相关的组件装配之前，需要读者回顾 Spring Boot 如何注册 Servlet 原生三大组件。

**1. Spring Boot 注册 Servlet 原生组件**

基于 Spring Boot 的项目，其底层都会采用 Servlet 3.0 及以上的规范。Servlet 3.0 规范中不再使用 `web.xml`，而是使用注解的方式配合 Servlet 容器扫描完成原生组件的注册。Spring Boot

本身并不默认支持扫描 Servlet 三大组件，所以它提供了另外两种注册方式。

- **Servlet 原生组件扫描 `@ServletComponentScan`**
  原生组件扫描适用于**项目中存在自定义的 Servlet 原生组件**。它的使用方式是，在 Spring Boot 的主启动类上标注 `@ServletComponentScan` 注解后，即可像 `@ComponentScan` 那样自动扫描主启动类所在包及其子包下的所有 Servlet 原生组件。这种扫描方式需要配合 Servlet 3.0 规范中引入的 `@WebServlet`、`@WebFilter`、`@WebListener` 注解才能完成原生组件的扫描与装配（就像 `@ComponentScan` 配合 `@Component` 注解一样）。
- **借助辅助注册器 RegistrationBean**
  辅助注册器的方式适用于**引入项目依赖的 jar 包中存在 Servlet 原生组件**。由于所依赖的第三方库中的代码不可修改，因此靠 Servlet 组件扫描的方式是不现实的。由此 Spring Boot 引入了一种注册器机制，针对 Servlet 原生三大组件提供了三个对应的 RegistrationBean 辅助注册。

代码清单 2-50 中提供了一个辅助注册器的简单示例。

**代码清单 2-50　RegistrationBean 的简单使用示例**

```java
public class DemoServlet extends HttpServlet {
    @Override
    protected void doGet(HttpServletRequest req, HttpServletResponse resp)
            throws ServletException, IOException {
        resp.getWriter().println("demo servlet");
    }
}

public class DemoServletRegistryBean extends ServletRegistrationBean<DemoServlet> {
    public DemoServletRegistryBean(DemoServlet servlet, String...urlMappings) {
        super(servlet, urlMappings);
    }
}

@Configuration
public class ServletConfiguration {
    @Bean
    public DemoServletRegistryBean demoServletRegistryBean() {
        return new DemoServletRegistryBean(new DemoServlet(), "/demo/servlet");
    }
}
```

### 2. DispatcherServlet 的注册

回顾了 Servlet 原生组件的注册方式，下面紧接着了解 `DispatcherServlet` 的注册。代码清单 2-51 展示了 `DispatcherServletAutoConfiguration` 中的核心源码，其中包含上述提到的 RegistrationBean 辅助注册器。

**代码清单 2-51　注册 DispatcherServlet**

```java
@EnableConfigurationProperties(WebMvcProperties.class)
protected static class DispatcherServletConfiguration {

    @Bean(name = DEFAULT_DISPATCHER_SERVLET_BEAN_NAME)
```

```java
    public DispatcherServlet dispatcherServlet(WebMvcProperties webMvcProperties) {
        DispatcherServlet dispatcherServlet = new DispatcherServlet();
        // DispatcherServlet 的定制化 ......
        return dispatcherServlet;
    }

    @Bean
    @ConditionalOnBean(MultipartResolver.class)
    @ConditionalOnMissingBean(name = DispatcherServlet.MULTIPART_RESOLVER_BEAN_NAME)
    public MultipartResolver multipartResolver(MultipartResolver resolver) {
        return resolver;
    }
}

@EnableConfigurationProperties(WebMvcProperties.class)
@Import(DispatcherServletConfiguration.class)
protected static class DispatcherServletRegistrationConfiguration {

    @Bean(name = DEFAULT_DISPATCHER_SERVLET_REGISTRATION_BEAN_NAME)
    @ConditionalOnBean(value = DispatcherServlet.class,
                       name = DEFAULT_DISPATCHER_SERVLET_BEAN_NAME)
    public DispatcherServletRegistrationBean registration(DispatcherServlet dispatcherServlet,
    WebMvcProperties webMvcProperties, ObjectProvider<MultipartConfigElement> multipartConfig) {
        DispatcherServletRegistrationBean registration = new DispatcherServletRegistrationBean
                    (dispatcherServlet, webMvcProperties.getServlet().getPath());
        registration.setName(DEFAULT_DISPATCHER_SERVLET_BEAN_NAME);
        registration.setLoadOnStartup(webMvcProperties.getServlet().getLoadOnStartup());
        multipartConfig.ifAvailable(registration::setMultipartConfig);
        return registration;
    }
}
```

整体来看，注册 DispatcherServlet 的核心源码包含两部分：DispatcherServlet Configuration 负责注册 DispatcherServlet 本身，其中有实例化 Dispatcher Servlet 的动作，还有对其进行的一些定制化操作；DispatcherServletRegistrationConfiguration 负责将 DispatcherServlet 注册到 Web 容器中，它利用 ServletRegistrationBean 的派生类 DispatcherServletRegistrationBean 将 DispatcherServlet 注册到 Web 容器中，以完成 DispatcherServlet 的装配工作。

有关源码中具体的实现细节，感兴趣的读者可以自行研究，本书不展开讨论。

### 2.6.3 SpringWebMvc 的装配

嵌入式 Web 容器和 DispatcherServlet 都装配完成后，最后是 WebMvc 的装配，这部分的主要装配内容都集中在 WebMvcAutoConfiguration 的两个内部类 WebMvcAutoConfigurationAdapter 与 EnableWebMvcConfiguration 中，所以研究重点是这两个内部类。

### 1. WebMvcAutoConfigurationAdapter

首先来看 `WebMvcAutoConfigurationAdapter`，这个类本身是一个 `WebMvcConfigurer`。`WebMvcAutoConfigurationAdapter` 整体可以理解为一个以 WebMvc 配置为主的配置器。`WebMvcAutoConfigurationAdapter` 重写了 `WebMvcConfigurer` 的大量方法，也注册了一些新的 Bean，拆分来看各组件。

（1）配置 HttpMessageConverter

重写 `configureMessageConverters` 方法，目的是配置默认的 `HttpMessageConverter`，如代码清单 2-52 所示。`HttpMessageConverter` 是一个消息转换器，它的作用对象是 `@RequestBody` 和 `@ResponseBody` 注解标注的 Controller 方法，分别完成请求体到参数对象的转换以及响应对象到响应体的转换。默认情况下 Spring Boot 整合 WebMvc 时，底层会自动依赖 Jackson 作为 JSON 支持，所以这里会配置一个 `MappingJackson2HttpMessageConverter` 作为消息转换器的实现。

**代码清单 2-52　配置 HttpMessageConverter**

```
@Override
public void configureMessageConverters(List<HttpMessageConverter<?>> converters) {
    this.messageConvertersProvider.ifAvailable((customConverters) ->
            converters.addAll(customConverters.getConverters()));
}
```

（2）异步支持

`configureAsyncSupport` 的作用是配置异步请求的支持。Spring Boot 在底层已经默认准备好了一个异步线程池，支持 Controller 层使用异步处理的方式接收请求。线程池在另一个自动配置类 `TaskExecutionAutoConfiguration` 中已经创建完成，而且 Bean 的名称刚好是 `applicationTaskExecutor`，所以这里可以顺利地获取并设置，如代码清单 2-53 所示。异步支持的最终落地会在 `RequestMappingHandlerAdapter` 中得以体现，感兴趣的读者可以结合源码自行探究，这里不作为重点展开讲解。

**代码清单 2-53　configureAsyncSupport 配置异步支持**

```
public static final String APPLICATION_TASK_EXECUTOR_BEAN_NAME = "applicationTaskExecutor";

@Override
public void configureAsyncSupport(AsyncSupportConfigurer configurer){
    if (this.beanFactory.containsBean(TaskExecutionAutoConfiguration.APPLICATION_TASK_EXECUTOR_
BEAN_NAME)) {
        Object taskExecutor = this.beanFactory
                .getBean(TaskExecutionAutoConfiguration.APPLICATION_TASK_EXECUTOR_BEAN_NAME);
        if (taskExecutor instanceof AsyncTaskExecutor) {
            configurer.setTaskExecutor(((AsyncTaskExecutor) taskExecutor));
        }
    }
    Duration timeout = this.mvcProperties.getAsync().getRequestTimeout();
    if (timeout != null) {
        configurer.setDefaultTimeout(timeout.toMillis());
    }
}
```

> 提示：SpringWebMvc 在 4.0 及以后的版本支持异步请求，使用方式是在 Controller 方法中不再直接返回响应对象，而是返回 Callable 或者 DeferredResult 对象，具体可以参照官方文档，本书不再展开。

（3）注册视图解析器

接下来是视图解析器的注册。源码中一共注册了 3 个视图解析器，之所以没有全部展示出来，是因为 BeanNameViewResolver 已几乎不再使用（一个 Bean 只能处理一个页面，放在当下的大环境下非常不实用）。代码清单 2-54 中列举的两个视图解析器中，对于 InternalResourceViewResolver，想必读者都不陌生，它通过路径前后缀拼接的方式解析逻辑视图名称，常见于原生 Spring Framework+SpringWebMvc+MyBatis 的项目技术栈中配置，用于处理 JSP 页面配置，不过 Spring Boot 默认已经不支持 JSP，所以不必再研究这个视图解析器。另一个是 ContentNegotiatingViewResolver，它是顶层级的视图解析器，负责将视图解析的工作交由不同的代理 ViewResolver 来实现，以处理不同的逻辑视图。换言之，ContentNegotiatingViewResolver 做的核心工作是中心转发。

**代码清单 2-54　注册默认的视图解析器**

```
@Bean
@ConditionalOnMissingBean
public InternalResourceViewResolver defaultViewResolver() {
    InternalResourceViewResolver resolver = new InternalResourceViewResolver();
    resolver.setPrefix(this.mvcProperties.getView().getPrefix());
    resolver.setSuffix(this.mvcProperties.getView().getSuffix());
    return resolver;
}

@Bean
@ConditionalOnBean(ViewResolver.class)
@ConditionalOnMissingBean(name = "viewResolver",
                    value = ContentNegotiatingViewResolver.class)
public ContentNegotiatingViewResolver viewResolver(BeanFactory beanFactory) {
    ContentNegotiatingViewResolver resolver = new ContentNegotiatingViewResolver();
    resolver.setContentNegotiationManager(beanFactory.getBean(ContentNegotiationManager.class));
    resolver.setOrder(Ordered.HIGHEST_PRECEDENCE);
    return resolver;
}
```

> 提示：如果项目中整合 Thymeleaf 或 FreeMarker 作为页面模板引擎，默认的页面路径也不会在 InternalResourceViewResolver 配置，而是在具体的页面模板引擎对应的自动配置类中体现（如 Thymeleaf 对应的 ThymeleafAutoConfiguration 中注册了一个 SpringResourceTemplateResolver 专门负责解析逻辑视图）。

（4）国际化的支持

LocaleResolver 是 SpringWebMvc 针对国际化支持的核心接口，它主要的作用是解析请求中的语言标志参数或者请求头中的 Accept-Language 参数，并将解析的参数存放到指定的位置中，它通常配合 LocaleChangeInterceptor 使用，如代码清单 2-55 所示。注意，只有配置了 spring.mvc.locale 配置项后，LocaleResolver 才会被创建。

**代码清单 2-55 注册国际化支持组件**

```
@Bean
@ConditionalOnMissingBean
@ConditionalOnProperty(prefix = "spring.mvc", name = "locale")
public LocaleResolver localeResolver() {
    if (this.mvcProperties.getLocaleResolver() == WebMvcProperties.LocaleResolver.FIXED) {
        return new FixedLocaleResolver(this.mvcProperties.getLocale());
    }
    AcceptHeaderLocaleResolver localeResolver = new AcceptHeaderLocaleResolver();
    localeResolver.setDefaultLocale(this.mvcProperties.getLocale());
    return localeResolver;
}
```

（5）RequestContextHolder 的支持

最后一个比较值得关注的组件是 RequestContextFilter，如代码清单 2-56 所示。乍一看会感觉比较眼熟，因为读者可能或多或少地用到过一个可以全局获取到 HttpServletRequest 和 HttpServletResponse 的 API：RequestContextHolder。而支撑 RequestContextHolder 获取的组件就是 RequestContextFilter。

**代码清单 2-56　RequestContextHolder 的支持**

```
@Bean
@ConditionalOnMissingBean({ RequestContextListener.class, RequestContextFilter.class })
@ConditionalOnMissingFilterBean(RequestContextFilter.class)
public static RequestContextFilter requestContextFilter() {
    return new OrderedRequestContextFilter();
}
```

代码清单 2-57 中展示了 RequestContextFilter 的核心过滤逻辑，doFilterInternal 方法是 RequestContextFilter 的核心过滤方法，其中调用了下方的 initContextHolders 方法，将 requestAttributes 放入 RequestContextHolder 中，这样后续在 Controller 层的任意位置都可以获取 HttpServletRequest 和 HttpServletResponse 对象（异步线程例外）。

**代码清单 2-57　RequestContextFilter 的核心工作逻辑**

```
@Override
protected void doFilterInternal(HttpServletRequest request, HttpServletResponse response,
        FilterChain filterChain) throws ServletException, IOException {
    ServletRequestAttributes attributes = new ServletRequestAttributes(request, response);
    // 初始化 ContextHolder
    initContextHolders(request, attributes);

    try {
        filterChain.doFilter(request, response);
    }// finally ......
}

private void initContextHolders(HttpServletRequest request,
        ServletRequestAttributes requestAttributes){
```

```
        LocaleContextHolder.setLocale(request.getLocale(), this.threadContextInheritable);
        RequestContextHolder.setRequestAttributes(requestAttributes, this.threadContextInheritable);
        // logger ......
}
```

## 2. EnableWebMvcConfiguration

WebMvcAutoConfigurationAdapter 的类头上使用@Import 注解导入了 EnableWebMvcConfiguration，所以 EnableWebMvcConfiguration 也会随之生效，这个类中注册了很多 WebMvc 会用到的核心组件，同样值得关注。

（1）注册 HandlerMapping

HandlerMapping 处理器映射器的作用是根据请求的 URI 去匹配查找能处理的 Handler，如代码清单 2-58 所示。目前主流的使用 WebMvc 的方式都是使用@RequestMapping 注解定义的 Handler 请求处理器，所以这里默认直接注册了一个 RequestMappingHandlerMapping。

**代码清单 2-58  注册 HandlerMapping**

```
// 参数名有精简
@Bean
@Primary
@Override
public RequestMappingHandlerMapping requestMappingHandlerMapping(
        @Qualifier("mvcContentNegotiationManager") ContentNegotiationManager manager,
        @Qualifier("mvcConversionService") FormattingConversionService conversionService,
        @Qualifier("mvcResourceUrlProvider") ResourceUrlProvider resourceUrlProvider) {
    // Must be @Primary for MvcUriComponentsBuilder to work
    return super.requestMappingHandlerMapping(manager, conversionService, resourceUrlProvider);
}
```

（2）注册 HandlerAdapter

处理器适配器 HandlerAdapter 会拿到 HandlerMapping 匹配成功的 Handler，并用合适的方式执行 Handler 的逻辑，如代码清单 2-59 所示。使用@RequestMapping 注解定义的 Handler，其底层负责执行的适配器就是 RequestMappingHandlerAdapter。

**代码清单 2-59  注册 HandlerAdapter**

```
// 参数名有精简
@Bean
@Override
public RequestMappingHandlerAdapter requestMappingHandlerAdapter(
        @Qualifier("mvcContentNegotiationManager") ContentNegotiationManager manager,
        @Qualifier("mvcConversionService") FormattingConversionService conversionService,
        @Qualifier("mvcValidator") Validator validator) {
    RequestMappingHandlerAdapter adapter = super.requestMappingHandlerAdapter(manager,
            conversionService, validator);
    adapter.setIgnoreDefaultModelOnRedirect(
            this.mvcProperties == null || this.mvcProperties.isIgnoreDefaultModelOnRedirect());
    return adapter;
}
```

### （3）静态资源加载配置

WebMvc整合页面时静态资源必不可少，addResourceHandlers方法中会默认配置几个常用的约定好的静态文件的存放位置，其中不乏读者熟悉的/resources、/static，以及依赖了webjars的/webjars/**，如代码清单2-60所示。配置好这些静态文件的存放路径后，页面文件就可以直接引用这些路径下的静态文件了。

**代码清单2-60　静态资源加载配置**

```
private static final String[] CLASSPATH_RESOURCE_LOCATIONS = { "classpath:/META-INF/resources/",
    "classpath:/resources/", "classpath:/static/", "classpath:/public/" };
private String[] staticLocations = CLASSPATH_RESOURCE_LOCATIONS;

@Override
protected void addResourceHandlers(ResourceHandlerRegistry registry) {
    super.addResourceHandlers(registry);
    if (!this.resourceProperties.isAddMappings()) {
        logger.debug("Default resource handling disabled");
        return;
    }
    addResourceHandler(registry, "/webjars/**", "classpath:/META-INF/resources/webjars/");
    addResourceHandler(registry, this.mvcProperties.getStaticPathPattern(),
            this.resourceProperties.getStaticLocations());
}
```

关于其他组件和配置，感兴趣的读者可以结合源码自行查看，本书不过多展开。

## 2.7　小结

本章主要讲解了有关Spring Boot的核心特性：自动装配的机制和原理。模块装配、条件装配、SPI机制是自动装配的实现基础，掌握基础的装配方式才能更好地理解自动装配的设计和内涵。后续通过WebMvc场景的自动装配实例，进一步体会自动装配在具体场景中发挥的作用。

# 第 3 章 Spring Boot 的 IOC 容器

**本章主要内容：**
- 理解、对比 Spring Framework 中的 IOC 容器；
- 了解 Spring Boot 对 IOC 容器模型的扩展；
- 理解 Environment 的设计以及与 IOC 容器的关系；
- 理解 BeanDefinition 的设计；
- 理解各种后置处理器的设计；
- 了解 ApplicationContext 的容器核心启动流程。

第 2 章中系统地讲解了 Spring Boot 的自动装配机制，无论是原生的装配，还是自动装配，最终组件都会装配到 IOC 容器中。Spring Framework 的 IOC 容器设计复杂且强大，如果要在整体层面把控 Spring Boot，理解 IOC 容器的设计是必不可少的一环。本章会系统地剖析 Spring Framework 中的 IOC 容器设计，以及 Spring Boot 对其的利用与扩展。

## 3.1 Spring Framework 的 IOC 容器

有关 Spring Framework 中的 IOC 容器模型，读者都知道有一个 ApplicationContext 接口，不过它不是顶层接口，它还有一个父接口 BeanFactory。在 Spring Framework 的官方文档中，有一个片段解释了这两者的关系：

> The org.springframework.beans and org.springframework.context packages are the basis for Spring Framework's IoC container. The BeanFactory interface provides an advanced configuration mechanism capable of managing any type of object. ApplicationContext is a subinterface of BeanFactory. It adds:
> - Easier integration with Spring's AOP features
> - Message resource handling (for use in internationalization)
> - Event publication
> - Application-layer specific contexts such as the WebApplicationContext for use in web applications.

org.springframework.beans 和 org.springframework.context 包是 Spring Framework 的 IOC 容器的基础。BeanFactory 接口提供了一种高级配置机制，能够管理任何类型的对象。ApplicationContext 是 BeanFactory 的子接口，它增加了以下几个特性：

- 与 Spring Framework 的 AOP 功能轻松集成；
- 消息资源处理（用于国际化）；
- 事件发布；
- 应用层特定的上下文，例如 Web 应用程序中使用的 `WebApplicationContext`。

由这段文档的描述可以初步得到一个结论：`BeanFactory` 是 IOC 容器的基础抽象，`ApplicationContext` 包含 `BeanFactory` 的所有功能，且扩展了更多实用特性。下面分别来了解 `BeanFactory` 与 `ApplicationContext` 的抽象设计。

> 提示：下面的部分会大量引用 Spring Framework 中的 javadoc，Spring Framework 的 javadoc 质量非常高，建议各位读者对该部分深入学习时，一定要结合 javadoc 来学习。

### 3.1.1 BeanFactory

**BeanFactory** 是 IOC 容器的顶层抽象，它仅定义最基础的 bean 对象的管理。借助 IDEA 可以生成原生 Spring Framework 中 `BeanFactory` 及其派生接口的继承结构，如图 3-1 所示。

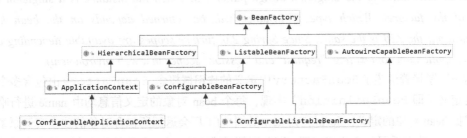

图 3-1　BeanFactory 和它的子接口

注意观察图 3-1 中的继承关系，`BeanFactory` 的扩展接口中有部分 `ApplicationContext` 相关的子接口，有关这部分内容放在 3.1.2 节讲解。为了能让读者更清楚地分辨和学习 `BeanFactory` 与 `ApplicationContext` 的体系，下面重点来看只与 `BeanFactory` 相关的接口与实现类，如图 3-2 所示。

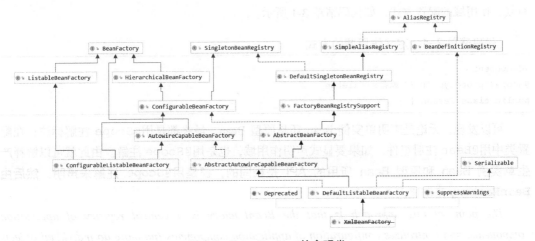

图 3-2　BeanFactory 的实现类

由图 3-2 可以发现 `BeanFactory` 的结构体系非常庞大。下面就其中最核心的几个接口和实现类做重点介绍。

#### 1. BeanFactory 根接口

作为 Spring Framework 中顶层的容器接口，`BeanFactory` 的作用一定是最简单、最核心的。以下内容会结合 javadoc 分段讲解。

*The root interface for accessing a Spring bean container. This is the basic client view of a bean container; further interfaces such as* `ListableBeanFactory` *and* `ConfigurableBeanFactory` *are available for specific purposes.*

本段内容来自 `BeanFactory` 文档注释的开始，解释了 `BeanFactory` 是 Spring Framework 中用于访问 Bean 容器的最基本的根接口，下面的扩展都是为了实现某些额外的特性（**层次性、可搜索性、可配置性**等）。

*This interface is implemented by objects that hold a number of bean definitions, each uniquely identified by a String name. Depending on the bean definition, the factory will return either an independent instance of a contained object (the Prototype design pattern), or a single shared instance (a superior alternative to the Singleton design pattern, in which the instance is a singleton in the scope of the factory). Which type of instance will be returned depends on the bean factory configuration: the API is the same. Since Spring 2.0, further scopes are available depending on the concrete application context (e.g. "request" and "session" scopes in a web environment).*

这一段紧接着讲述了 `BeanFactory` 中定义的**作用域概念**。`BeanFactory` 由包含多个 bean 对象的定义（即 `BeanDefinition`）实现，每个 bean 对象的定义信息均由 name 进行唯一标识。根据 bean 对象的定义，Spring Framework 中的工厂会返回所包含对象的独立实例原型模式（prototype），或者返回单个共享实例，即单实例模式（singleton）的替代方案，其中实例是工厂作用域中的单实例。返回 bean 对象的实例类型取决于 bean 对象工厂的配置：API 是相同的。从 Spring Framework 2.0 开始，根据具体的 `ApplicationContext` 落地实现，可以使用更多作用域（例如 Web 环境中的 request 和 session 作用域）。

请注意，文档注释中的一句话比较难理解"返回 bean 的实例类型取决于 bean 对象工厂的配置：API 是相同的"。如何理解后半句"API 是相同的"？这需要读者回顾一下定义 Bean 的时候，作用域的配置方式，如代码清单 3-1 所示。

**代码清单 3-1 定义作用域的方式**

```
@Component
@Scope("prototype") // 或者 singleton
public class Person { }
```

可以发现，无论是声明单实例 Bean 还是原型 Bean，最终都是用 @Scope 注解标注；在配置类中用 @Bean 注册组件，如果要显式声明作用域，也是用 @Scope 注解。由此就可以解释产生单实例 Bean 和原型 Bean 所用的 API 是相同的，都是用 **@Scope** 注解来声明，然后由 **BeanFactory** 来创建。

*The point of this approach is that the BeanFactory is a central registry of application components, and centralizes configuration of application components (no more do individual objects need to read properties files, for example). See chapters 4 and 11 of "Expert One-on-One J2EE Design and Development" for a discussion of the benefits of this approach.*

这一段内容重点强调了 `BeanFactory` **与环境配置的继承**。`BeanFactory`本身是所有bean对象的注册中心，所有的 bean 最终都在 `BeanFactory` 中创建和保存。另外 `BeanFactory` 中还集成了配置信息，它可以通过加载外部的 properties 文件，将配置文件的属性值设置到 bean 对象中。

不过请读者注意，这里有关集成配置的概念其实是相对陈旧的。自 Spring Framework 3.1 之后出现了一个新的概念 `Environment`（3.4 节会展开讲解），它才是真正实现环境和配置存储的容器。

*Note that it is generally better to rely on Dependency Injection ("push" configuration) to configure application objects through setters or constructors, rather than use any form of "pull" configuration like a BeanFactory lookup. Spring's Dependency Injection functionality is implemented using this BeanFactory interface and its subinterfaces.*

这部分内容跟`BeanFactory`的关系不是特别大，它阐述的是 Spring Framework 官方在IOC 的两种实现上的权衡：**推荐使用依赖注入（DI）、尽可能避免依赖查找（DL）**。文档中提示我们，通常最好使用依赖注入（"推"的配置），通过 setter 方法或构造方法注入的方式配置应用程序对象，而不是使用任何"拉"形式的配置（如借助 `BeanFactory` 进行依赖查找）。这里的一对概念特别经典：**依赖注入的思想是"推"**，它主张把组件所需的依赖"推"到组件的成员上；**依赖查找的思想是"拉"**，组件需要哪些依赖需要组件自己去 IOC 容器中"拉取"，这对概念有助于我们理解和对比 IOC 的两种实现方式——**依赖注入和依赖查找**。

这段注释的最后补充了一句，Spring Framework 的依赖注入特性是使用 `BeanFactory` 接口及其子接口实现的，这个在后面的 `AbstractAutowireCapableBeanFactory` 中会提到。

*Normally a BeanFactory will load bean definitions stored in a configuration source (such as an XML document), and use the org.springframework.beans package to configure the beans. However, an implementation could simply return Java objects it creates as necessary directly in Java code. There are no constraints on how the definitions could be stored: LDAP, RDBMS, XML, properties file, etc. Implementations are encouraged to support references amongst beans (Dependency Injection).*

这一段主要说明的是 `BeanFactory` **支持多种类型的 Bean 配置源**。通常情况下 `BeanFactory` 会加载存储在配置源（例如 XML 配置文件）中 Bean 的定义，并使用 `org.springframework.beans` 包的 API 来配置 Bean。然而 `BeanFactory` 的实现可以根据需要直接在 Java 代码中返回它创建的 Java 对象。Bean 定义的存储方式没有任何限制，它可以是 LDAP（轻型目录访问协议）、RDBMS（关系型数据库系统）、XML、properties 文件等。对于这些存储方式读者只需要知道最常用的方式：XML 配置文件、注解配置类、模式注解+组件扫描。

*In contrast to the methods in ListableBeanFactory, all of the operations in this interface will also check parent factories if this is a HierarchicalBeanFactory. If a bean is not found in this factory instance, the immediate parent factory will be asked. Beans in this factory instance are supposed to override beans of the same name in any parent factory.*

这一段注释说明的是 `BeanFactory` 可实现**层次性**。与 `ListableBeanFactory` 中的方法相比，`BeanFactory` 中的所有操作还将检查父级工厂（`BeanFactory` 本身可以支持父子结构，这个父子结构的概念和实现由 HierarchicalBeanFactory 完成）。如果在 BeanFactory 实例中没有找到指定的 bean 对象，则会在父工厂中搜索查找。BeanFactory 实例中的 Bean 应该覆盖任何父工厂中的同名 Bean。

*Bean factory implementations should support the standard bean lifecycle interfaces as far as possible. The full set of initialization methods and their standard order is:*
- *BeanNameAware's* setBeanName
- *BeanClassLoaderAware's* setBeanClassLoader

……

*On shutdown of a bean factory, the following lifecycle methods apply:* ……

最后一段注释粗略说明了一件事情：BeanFactory 中设有**完整的生命周期控制机制**。BeanFactory 接口的内部实现了尽可能支持标准 Bean 的生命周期接口（Aware 系列接口、用于初始化的 InitializingBean 接口、用于容器初始化完成后启动的 Lifecycle 接口等）。

最后简单总结一下 BeanFactory 的基础特性：
- 基础的容器；
- 定义了作用域的概念；
- 集成环境配置；
- 支持多种类型的配置源；
- 层次性的设计；
- 完整的生命周期控制机制。

对于上述的部分特性，读者可能在现阶段理解起来比较吃力，随着后面学习的不断深入，这些特性会逐一得到解释。

**2. HierarchicalBeanFactory**

从 HierarchicalBeanFactory 这个类的名称上可以很容易地理解，它是一个体现了**层次性**的 BeanFactory。有了层次性的特性，BeanFactory 就有了**父子结构**。它的 javadoc 比较简单，我们简单介绍一下。

*Sub-interface implemented by bean factories that can be part of a hierarchy.*

*The corresponding setParentBeanFactory method for bean factories that allow setting the parent in a configurable fashion can be found in the ConfigurableBeanFactory interface.*

javadoc 解释得非常清楚，HierarchicalBeanFactory 是 BeanFactory 的子接口，实现了 HierarchicalBeanFactory 接口的 IOC 容器具有层次性。它本身有一个 getParentBeanFactory 方法可以获取父级容器，用 ConfigurableBeanFactory 的 setParentBeanFactory 方法设置父级容器。

另外请读者注意一点：当调用 BeanFactory 的 getBean 方法时，如果这个 BeanFactory 是具有层次性的 Bean，getBean 方法会从当前 BeanFactory 开始查找是否存在指定的 Bean，如果找不到就依次向上查找父级 BeanFactory，直到找到为止返回，或者最终找不到而抛出 NoSuchBeanDefinitionException。

到这里，请读者思考一个问题：如果当前 BeanFactory 中有指定的 Bean，父级 BeanFactory 中可能还有 Bean 吗？

答案是可能有，因为**即便存在父子关系，它们在本质上也是不同的容器**。因此有可能找到多个相同的 Bean。换言之，`@Scope` 中声明的 `singleton` 只是在单独某一个容器中是单实例的，但有了层次性结构后，对于整体的多个容器就不是单实例的了。

### 3. ListableBeanFactory

`ListableBeanFactory` 这个类名称的前缀已经体现出它的特性：**可列举**。实现了 `ListableBeanFactory` 接口的 IOC 容器具有的最关键能力是，可以将容器中的所有 bean 对象都列举出来。下面分段解读 javadoc。

*Extension of the BeanFactory interface to be implemented by bean factories that can enumerate all their bean instances, rather than attempting bean lookup by name one by one as requested by clients. BeanFactory implementations that preload all their bean definitions (such as XML-based factories) may implement this interface.*

第一段注释比较好理解，`ListableBeanFactory` 本身也是 `BeanFactory` 接口的扩展实现，它的扩展功能是能在获得 `BeanFactory` 时直接把容器中的 **Bean 全部取出**（相当于提供了**可迭代的特性**），而不是逐一用 name 获取（逐一获取很麻烦，而且可能取不全）。

*If this is a HierarchicalBeanFactory, the return values will not take any BeanFactory hierarchy into account, but will relate only to the beans defined in the current factory. Use the BeanFactoryUtils helper class to consider beans in ancestor factories too.*

紧接着的这一段给我们提供了一个关键信息：如果当前 `BeanFactory` 同时也是 `HierarchicalBeanFactory`，则返回值**会忽略** `BeanFactory` 的**层次性结构**，而仅与当前 `BeanFactory` 中定义的 Bean 有关。换言之，`ListableBeanFactory` **只会列举当前容器的** bean 对象，上文中提到 `BeanFactory` 具有层次性，在这种前提下，在列举所有 bean 对象时，就需要斟酌到底是获取包括父级容器在内的所有 bean 对象，还是只获取当前容器中的 bean 对象。Spring Framework 在斟酌之后选择了只获取当前容器中的 bean 对象。如果的确想获取所有 bean 对象，可以借助 `BeanFactoryUtils` 工具类来实现（工具类中有不少名称以 "`IncludingAncestors`" 结尾的方法，表示可以同时获取父级容器）。

*The methods in this interface will just respect bean definitions of this factory. They will ignore any singleton beans that have been registered by other means like org.spring framework.beans.factory.config.ConfigurableBeanFactory's registerSingleton method, with the exception of getBeanNamesForType and getBeansOfType which will check such manually registered singletons too. Of course, BeanFactory's getBean does allow transparent access to such special beans as well. However, in typical scenarios, all beans will be defined by external bean definitions anyway, so most applications don't need to worry about this differentiation.*

本段注释强调了一点：`ListableBeanFactory` **会有选择性地列举**。`ListableBeanFactory` 中的方法将仅遵循当前工厂的 Bean 定义，它们将忽略通过其他方式（例如 `ConfigurableBeanFactory` 的 `registerSingleton` 方法）注册的任何单实例 Bean（但 `getBeanNamesForType` 和 `getBeansOfType` 除外，它们也会检查这种手动注册的单实例 Bean）。

文档注释比较难理解，我们换一种说法：作为一个"可迭代"的 `BeanFactory`，按理来讲至少应该把当前容器中的所有 bean 对象都列举出来，但是**有些 Bean 在迭代期间会被忽略掉而不列举**。这似乎有点让人摸不着头脑，明明是可以列举出容器中的所有 bean 对象，但又不能全部列举，这样的设计是否存在问题？还是这样的设计蕴含着特殊的设计？下面通过一个小测试来解释 Spring Framework 的这一设计。

（1）无法迭代的 Bean？

我们可以定义两个简单类：猫和狗。

```
public class Cat { }
public class Dog { }
```

之后编写 XML 配置文件，只将 Cat 注册到 IOC 容器中，如代码清单 3-2 所示。

**代码清单 3-2　测试只将 Cat 注册到 IOC 容器中**

```xml
<?xml version="1.0" encoding="UTF-8"?>
<beans xmlns="http://www.springframework.org/schema/beans"
    xmlns:xsi="http://www.w3.org/2001/XMLSchema-instance"
    xsi:schemaLocation="http://www.springframework.org/schema/beans
                        https://www.springframework.org/schema/beans/spring-beans.xsd">
    
    <bean class="com.linkedbear.springboot.container.beanfactory.bean.Cat"/>
</beans>
```

下面的代码清单 3-3 是使用 XML 配置文件加载 BeanFactory，注意此处不能使用 ApplicationContext，而只能选择 BeanFactory 的最终实现 DefaultListableBeanFactory（虽然超出了目前介绍的范围，但是不要担心，下面会讲解），使用这种方式，也可以加载 XML 配置文件，以完成 BeanFactory 的构建。

**代码清单 3-3　测试打印容器中的所有 Bean**

```java
public class ListableBeanFactoryApplication {
    
    public static void main(String[] args) throws Exception{
        ClassPathResource resource = new ClassPathResource ("container/listable-container.xml");
        DefaultListableBeanFactory beanFactory = new DefaultListableBeanFactory();
        XmlBeanDefinitionReader beanDefinitionReader = new XmlBeanDefinitionReader(beanFactory);
        beanDefinitionReader.loadBeanDefinitions(resource);
        // 直接打印容器中的所有 Bean
        System.out.println("加载 XML 文件后容器中的 Bean 如下：");
        Stream.of(beanFactory.getBeanDefinitionNames()).forEach(System.out::println);
    }
}
```

此时如果直接打印 BeanFactory 中的 Bean，可以发现只有一个 Cat：

```
加载 XML 文件后容器中的 Bean 如下：
com.linkedbear.springboot.container.beanfactory.bean.Cat#0
```

目前为止这些操作都是我们极为熟悉的，接下来我们在 main 方法中添加几行代码，以完成 Dog 的注册，如代码清单 3-4 所示。

**代码清单 3-4　手动注册 Dog 后测试打印所有 Bean**

```java
public static void main(String[] args) throws Exception {
    // ...
```

```
// 手动注册一个单实例 Bean
beanFactory.registerSingleton("doggg", new Dog());
// 再打印容器中的所有 Bean
System.out.println("手动注册单实例 Bean 后容器中的所有 Bean 如下: ");
Stream.of(beanFactory.getBeanDefinitionNames()).forEach(System.out::println);
}
```

再次运行 `ListableBeanFactoryApplication` 的 `main` 方法,会发现控制台仍然只打印了一个 `Cat`:

```
加载 XML 文件后容器中的 Bean 如下:
com.linkedbear.springboot.container.beanfactory.bean.Cat#0
```

```
手动注册单实例 Bean 后容器中的所有 Bean 如下:
com.linkedbear.springboot.container.beanfactory.bean.Cat#0
```

这个运行结果似乎与我们预期的不符,代码清单 3-4 中显式地构建了一个 `Dog` 对象,并注册到 `BeanFactory` 中,但是迭代打印容器中的 bean 对象时却没有打印 `Dog` 对象。这就印证了上面文档注释中的那句话:`ListableBeanFactory` 在获取容器内的所有 bean 对象时,不会把手动注册(即使用 `registerSingleton` 方法)的 bean 对象取出,也就是文档注释中所说的"忽略通过其他方式"。

不过这个时候可能会有读者提出疑问:容器中真的有 `Dog` 吗?答案是肯定的,我们可以通过 `BeanFactory` 的 `getBean` 方法成功获取(读者可以自行测试),说明 `Dog` 对象本身是存在的。

(2)设计选择性列举的目的

无法迭代的现象出现了,Spring Framework 为什么要如此设计呢?借助 IDE,查看 `ConfigurableBeanFactory` 的 `registerSingleton` 方法调用,可以发现在 `AbstractApplicationContext` 的 `prepareBeanFactory` 方法中有一些使用,如代码清单 3-5 所示。

**代码清单 3-5　底层注册特殊类型的 Bean**

```
//注册默认的环境 Bean
if (!beanFactory.containsLocalBean(ENVIRONMENT_BEAN_NAME)) {
    beanFactory.registerSingleton(ENVIRONMENT_BEAN_NAME, getEnvironment());
}
if (!beanFactory.containsLocalBean(SYSTEM_PROPERTIES_BEAN_NAME)) {
    beanFactory.registerSingleton(SYSTEM_PROPERTIES_BEAN_NAME,
            getEnvironment().getSystemProperties());
}
if (!beanFactory.containsLocalBean(SYSTEM_ENVIRONMENT_BEAN_NAME)) {
    beanFactory.registerSingleton(SYSTEM_ENVIRONMENT_BEAN_NAME,
            getEnvironment().getSystemEnvironment());
}
```

可以发现,源码在此处使用 `beanFactory.registerSingleton` 的方式注册了几个组件,而这些组件仅供 Spring Framework 内部使用,这样做的目的是 Spring Framework 不希望开发者直接操控它们,于是就使用了这种设计方式来隐藏这些内部组件。

如果这种设计不是很好理解,我们可以举另外一个例子,在 Windows 操作系统中,系统不希望用户随意改动系统内部使用的一些文件,因此会在文件资源管理器中设置一个选项,以隐藏受保护的操作系统文件（在控制面板→文件资源管理器选项中），如图 3-3 所示。

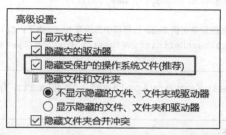

图 3-3　Windows 默认隐藏受保护的系统文件

默认情况下这个选项是勾选的（意思就是 Windows 默认会隐藏这些文件，由 Windows 和软件自行管理，不希望用户干涉），当然我们可以取消勾选它，这样文件资源管理器中也能显示那些操作系统文件，但倘若我们不懂操作系统，要是真的改动了它们，则计算机可能会出现一些意外的错误或故障。

#### 4. AutowireCapableBeanFactory

`AutowireCapableBeanFactory` 这个类的名称中有一个熟悉的概念：autowire（自动注入）。可见 `AutowireCapableBeanFactory` 是一个支持自动注入的 `BeanFactory`，同时也意味着 `AutowireCapableBeanFactory` 可以支持依赖注入。文档注释中下了很大的功夫解释 `AutowireCapableBeanFactory` 的作用，分段来看。

*Extension of the BeanFactory interface to be implemented by bean factories that are capable of autowiring, provided that they want to expose this functionality for existing bean instances.*

*This subinterface of BeanFactory is not meant to be used in normal application code: stick to BeanFactory or ListableBeanFactory for typical use cases.*

*Integration code for other frameworks can leverage this interface to wire and populate existing bean instances that Spring does not control the lifecycle of. This is particularly useful for WebWork Actions and Tapestry Page objects, for example.*

前两段注释最为重要，不过直接理解可能会有些困难，这里换一种方式来理解：`AutowireCapableBeanFactory` 可以支持组件的自动注入，而且**可以为一些现有的、没有被 Spring Framework 统一管理的 Bean 提供组件自动注入的支持**。紧接着文档解释了 `AutowireCapableBeanFactory` 如何使用，一般情况下，在正常的项目开发中不会接触到 `AutowireCapableBeanFactory`，而是在第三方框架与 Spring Framework 整合时才可能会用到。注意文档注释的描述：**利用此接口来连接和注入那些 Spring Framework 无法控制其生命周期的现有 bean 实例**。这其实已经把 `AutowireCapableBeanFactory` 的作用完整地描述出来了：`AutowireCapableBeanFactory` **通常用于与其他框架集成的场景**，如果其他框架的某些 bean 实例无法让 Spring Framework 接管，但又需要注入一些由 Spring Framework 管理的 bean 对象，就可以使用 `AutowireCapableBeanFactory` 辅助完成注入。

可能有读者不理解，在什么样的具体场景中会用到 `AutowireCapableBeanFactory` 的这个特性呢？试想如果我们要编写一个自定义的 `Servlet`，在这个 `Servlet` 中需要注入 Spring Framework 中管理的 Bean，应该如何实现呢？根据 IOC 的两种实现方式，Spring Framework 均可以提供解决方案。

- 依赖查找：由 `Servlet` 获取到 `HttpServletRequest` 后取得 `ServletContext`，借助 `WebAppliCationContextUtils` 可以获取。
- 依赖注入：给需要注入的组件上标注 `@Autowired` 注解，借助 `AutowireCapableBeanFactory` 辅助注入。

依赖查找的方式比较容易理解，在 Spring Framework 整合 Web 开发的场景中都有涉及；依赖注入的方式在下面一段 javadoc 中有解释。

*Note that this interface is not implemented by ApplicationContext facades, as it is hardly ever used by application code. That said, it is available from an application context too, accessible through ApplicationContext's getAutowireCapableBeanFactory() method.*

文档注释请我们注意，`ApplicationContext` 本身并没有实现 `AutowireCapableBeanFactory` 接口，需要我们通过 `ApplicationContext` 的方法间接获取 `AutowireCapableBeanFactory`。如果需要给自定义的 `Servlet` 等没有被 Spring Framework 统一管理的 `Bean` 注入组件，可以在获取 `AutowireCapableBeanFactory` 后调用其 API 实现依赖注入。

最后简单总结一下，`AutowireCapableBeanFactory` 这个 API 一般不需要我们去操纵，因为正常的项目开发中不会使用，但如果需要获取 `AutowireCapableBeanFactory`，可以通过 `ApplicationContext` 间接获取。

### 5. ConfigurableBeanFactory

这个接口需要引起重视，Spring Framework 中对于核心 API 的命名有非常强的规律性。当我们看到类名带有 **Configurable** 前缀时，意味着这个接口的行为有"写"的动作，而去掉 Configurable 前缀的接口只有"读"的动作。这里要提到一对概念：**可写与可读**。回想面向对象编程中，一个类的属性设置为 **private** 后，提供 **get** 方法意味着该属性**可读**，提供 **set** 方法意味着该属性**可写**。同样，Spring Framework 的这些 `BeanFactory`，包括后面的 `ApplicationContext` 中，都会有这样的设计：普通的 `BeanFactory` 只有 **get** 相关的操作，而 **Configurable** 前缀的 `BeanFactory` 或者 `ApplicationContext` 就具有了 **set** 相关的操作，如代码清单 3-6 所示。

**代码清单 3-6　ConfigurableBeanFactory 中的部分 set 方法**

```
void setBeanClassLoader(@Nullable ClassLoader beanClassLoader);
void setTypeConverter(TypeConverter typeConverter);
void addBeanPostProcessor(BeanPostProcessor beanPostProcessor);
```

读者务必先理解清楚这一对概念，之后我们就可以往下阅读了。

*Configuration interface to be implemented by most bean factories. Provides facilities to configure a bean factory, in addition to the bean factory client methods in the BeanFactory interface.*

*This bean factory interface is not meant to be used in normal application code: Stick to BeanFactory or org.springframework.beans.factory.ListableBeanFactory for typical needs. This extended interface is just meant to allow for framework-internal plug'n'play and for special access to bean factory configuration methods.*

文档注释并没有占很大篇幅，不过已经讲得很清楚了。文档中提到 Configurable

`BeanFactory` 已经提供了**带配置的功能**,可以调用它定义的方法对 `BeanFactory` 进行修改、扩展等操作。但是紧接着文档又提到 SpringFramework 不希望开发者用 `Configurable BeanFactory`,而是坚持使用最基础的 `BeanFactory`,原因是**原则上程序在运行期间不应该对 `BeanFactory` 再进行频繁的变动**,此时只应该有读的动作,而不应该出现写的动作(除非是确定要有对 `BeanFactory` 进行写的操作,且有把握)。

### 6. AbstractBeanFactory

上面介绍的都是有关 `BeanFactory` 接口部分的内容,接下来我们来看两个抽象实现和一个落地实现。首先是 `BeanFactory` 的最基础的抽象实现 `AbstractBeanFactory`,它只具有部分功能,并不是完整的实现。下面分段阅读文档注释。

*Abstract base class for BeanFactory implementations, providing the full capabilities of the ConfigurableBeanFactory SPI. Does not assume a listable bean factory: can therefore also be used as base class for bean factory implementations which obtain bean definitions from some backend resource (where bean definition access is an expensive operation).*

第一段注释中解释了 `AbstractBeanFactory` 作为 `BeanFactory` 接口下面第一个抽象的实现类,具有最基础的功能且可以从配置源(XML、LDAP、RDBMS 等)获取 Bean 的定义信息(即 `BeanDefinition`,要查看有关 `BeanDefinition` 的讲解请参阅 3.5 节,这里不作展开)。

*This class provides a singleton cache (through its base class DefaultSingletonBeanRegistry), singleton/prototype determination, FactoryBean handling, aliases, bean definition merging for child bean definitions, and bean destruction (org.springframework.beans.factory.DisposableBean interface, custom destroy methods). Furthermore, it can manage a bean factory hierarchy (delegating to the parent in case of an unknown bean), through implementing the org.springframework.beans.factory.HierarchicalBeanFactory interface.*

紧接着的一段中解释了 `AbstractBeanFactory` 对 **Bean 的支持**。`AbstractBeanFactory` 可以提供单实例 Bean 的缓存(通过其父类 `DefaultSingletonBeanRegistry`)、单实例/原型 Bean 的裁定 `FactoryBean` 处理、Bean 对象的别名存储、用于子 Bean 定义的 `BeanDefinition` 合并,以及 Bean 销毁(`DisposableBean` 接口,自定义 `destroy` 方法)。此外,`AbstractBeanFactory` 可以通过实现 `HierarchicalBeanFactory` 接口来管理 `BeanFactory` 的层次性结构(在未知 Bean 的情况下委托给父级工厂)。

*The main template methods to be implemented by subclasses are getBeanDefinition and createBean, retrieving a bean definition for a given bean name and creating a bean instance for a given bean definition, respectively. Default implementations of those operations can be found in DefaultListableBeanFactory and AbstractAutowireCapableBeanFactory.*

最后一段提到了一个关键信息:Spring Framework 中大量使用**模板方法模式**来设计核心组件。它的设计思路是:**父类提供逻辑规范,子类提供具体步骤的实现**。在文档注释中,我们可以看到 `AbstractBeanFactory` 中对 `getBeanDefinition` 和 `createBean` 这两个方法进行了规范上的定义,分别代表获取 Bean 的定义信息,以及创建 bean 对象,这两个方法在 Spring Framework 的 IOC 容器初始化阶段起着至关重要的作用。

> 💡 提示:`createBean` 是 Spring Framework 能管控的所有 Bean 的创建入口。

## 7. AbstractAutowireCapableBeanFactory

从 `AbstractAutowireCapableBeanFactory` 的类名可以看出，它是 `AutowireCapableBeanFactory` 接口的具体实现，也意味着它的内部实现组件自动装配的逻辑。其实 `AbstractAutowireCapableBeanFactory` 的作用不仅是这些，我们仔细阅读下面的 javadoc。

*Abstract bean factory superclass that implements default bean creation, with the full capabilities specified by the RootBeanDefinition class. Implements the AutowireCapableBeanFactory interface in addition to AbstractBeanFactory's createBean method.*

第一段注释提到 `AbstractAutowireCapableBeanFactory` 是实现了默认 bean 对象创建逻辑的 `BeanFactory` 接口的抽象类，它除了实现 `AbstractBeanFactory` 的 `createBean` 方法，还额外实现了 `AutowireCapableBeanFactory` 接口的方法，这就意味着 bean 对象真正的创建动作都在 `AbstractAutowireCapableBeanFactory` 中完成。

> 提示：其实 `createBean` 方法也不是最终实现 Bean 创建的方法，而是由另一个叫作 `doCreateBean` 的方法实现的，它同样在 `AbstractAutowireCapableBeanFactory` 中定义，而且是 protected 方法，没有子类重写它。有关 `createBean` 与 `doCreateBean` 方法的底层实现，在第 7 章会有详细讲解，此处不作展开。

*Provides bean creation (with constructor resolution), property population, wiring (including autowiring), and initialization. Handles runtime bean references, resolves managed collections, calls initialization methods, etc. Supports autowiring constructors, properties by name, and properties by type.*

紧接着的一段列出了 `AbstractAutowireCapableBeanFactory` 中实现的核心功能：bean 对象的创建、属性的赋值和依赖注入和 Bean 的初始化逻辑执行。这几个步骤即是创建 Bean 的最核心的三个步骤。

*The main template method to be implemented by subclasses is resolveDependency (DependencyDescriptor, String, Set, TypeConverter), used for autowiring by type. In case of a factory which is capable of searching its bean definitions, matching beans will typically be implemented through such a search. For other factory styles, simplified matching algorithms can be implemented.*

这一段内容告诉我们 `AbstractAutowireCapableBeanFactory` 并没有把所有的模板方法都实现，它保留了文档注释中提到的 `resolveDependency` 方法，这个方法的作用是**解析 Bean 的成员中定义的属性依赖关系**，会在下面提到的 `DefaultListableBeanFactory` 中真正实现。

*Note that this class does not assume or implement bean definition registry capabilities. See DefaultListableBeanFactory for an implementation of the ListableBeanFactory and BeanDefinition Registry interfaces, which represent the API and SPI view of such a factory, respectively.*

最后一段阐述的是 `AbstractAutowireCapableBeanFactory` 实现了对 Bean 的创建、赋值、注入和初始化的逻辑，但对于 Bean 的定义信息是如何进入 `BeanFactory` 的部分没有涉及。这其中包含两个流程：**Bean 的创建和 Bean 定义的注册**。这在后面涉及 `BeanDefinition` 的章节以及本书第 2 部分会详细讲解。

### 8. DefaultListableBeanFactory

最后要介绍的 `DefaultListableBeanFactory` 是**唯一目前底层正在使用的** `BeanFactory` **的落地实现**。它的设计相当重要，读者一定要对它足够重视。

*Spring's default implementation of the ConfigurableListableBeanFactory and BeanDefinitionRegistry interfaces: a full-fledged bean factory based on bean definition metadata, extensible through post-processors.*

第一段内容提到，`DefaultListableBeanFactory` 是 `BeanFactory` 最终的默认实现，通过源码可以了解，`DefaultListableBeanFactory` 已经没有 **abstract** 关键字的标注，说明 `DefaultListableBeanFactory` 是一个**成熟的落地实现类**。另外，`DefaultListableBeanFactory` 还实现了 `BeanDefinitionRegistry` 接口，具有 Bean 定义信息（即 Bean Definition）的统一管理能力。

> 💡 提示：这里我们多注意一点：`BeanDefinitionRegistry` 从字面意思理解是一个"**Bean 定义的注册器**"，而且我们能强烈地感受到它与 `BeanDefinition` 的关系极为紧密，在 3.5 节中会再次提到。

*Typical usage is registering all bean definitions first (possibly read from a bean definition file), before accessing beans. Bean lookup by name is therefore an inexpensive operation in a local bean definition table, operating on pre-resolved bean definition metadata objects.*

注意理解这一段内容，当 `DefaultListableBeanFactory` 要获取指定的 Bean 之前，应该先把 Bean 的定义信息注册进去，由此可知，`DefaultListableBeanFactory` 在 `AbstractAutowireCapableBeanFactory` 的基础上完成了**注册** Bean 定义信息的动作，而这个动作就是通过 `BeanDefinitionRegistry` 实现的。由此可以总结出一点：完整的 `BeanFactory` 对 Bean 的管理应该是，**先注册 Bean 的定义信息，再完成 Bean 的创建和初始化动作**。

*Note that readers for specific bean definition formats are typically implemented separately rather than as bean factory subclasses: see for example PropertiesBeanDefinitionReader and org.springframework.beans.factory.xml.XmlBeanDefinitionReader.*

最后一段内容是一个小提示：特定的 Bean 定义信息格式的解析器通常是单独实现的，而不是作为 `BeanFactory` 的子类实现的，有关这部分的内容参见 `PropertiesBeanDefinitionReader` 和 `XmlBeanDefinitionReader`。由此可以体现出，Spring Framework 对于**组件的单一职责把控得非常好**。`BeanFactory` 作为一个统一管理 Bean 组件的容器，它的核心工作就是控制 Bean 在创建阶段的生命周期与 bean 对象的统一管理，而对于 Bean 从哪里来、如何被创建、有哪些依赖要被注入，这些都与 `BeanFactory` 无关，而是有专门的组件来处理（包括上面提到的 `BeanDefinitionReader` 在内的一些其他组件）。

至此，有关 `BeanFactory` 相关的重要接口和实现类全部讲解完毕。了解 `BeanFactory` 的相关扩展和实现后，接下来是更为复杂的 `ApplicationContext` 体系。

### 3.1.2 ApplicationContext

`ApplicationContext` 是基于 `BeanFactory` 的扩展，它提供了更为强大的特性。我们

照例先看 `ApplicationContext` 接口和它的父接口、扩展接口之间的关系，如图 3-4 所示。

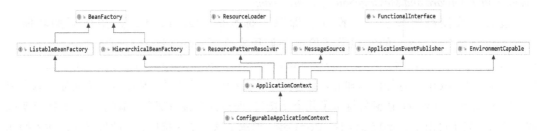

图 3-4　ApplicationContext 和它的子接口

从图 3-4 中可以清楚地看到，`ApplicationContext` 除了继承 `BeanFactory` 接口，还额外继承了几个功能性接口，这些接口共同组成了 `ApplicationContext` 扩展的几个核心特性。下面逐一展开讲解。

#### 1. ApplicationContext 的根接口

`ApplicationContext` 作为日常操作 Spring Framework 的核心 API，它必然是主角，然而 javadoc 的篇幅并不比 `BeanFactory` 长，下面分段讲解。

*Central interface to provide configuration for an application. This is read-only while the application is running, but may be reloaded if the implementation supports this.*

第一段话言简意赅，`ApplicationContext` 是 Spring Framework 的核心中央接口。在应用程序运行时，它是只读的，但是如果支持的话，它可以重新加载。注意**重新加载**这个概念，在下面的 `AbstractRefreshableApplicationContext` 中会体现。

*An ApplicationContext provides:*

- *Bean factory methods for accessing application components. Inherited from Listable BeanFactory.*
- *The ability to load file resources in a generic fashion. Inherited from the ResourceLoader interface.*
- *The ability to publish events to registered listeners. Inherited from the ApplicationEvent Publisher interface.*
- *The ability to resolve messages, supporting internationalization. Inherited from the Message Source interface.*
- *Inheritance from a parent context. Definitions in a descendant context will always take priority. This means, for example, that a single parent context can be used by an entire web application, while each servlet has its own child context that is independent of that of any other servlet.*

紧跟着的一段内容列出了 `ApplicationContext` 的核心功能，大致包含访问 Bean 的能力、加载文件资源、事件发布、国际化支持、层级关系支持。注意"层级关系支持"的概念，`ApplicationContext` 也是支持层级结构的，但这里文档注释的描述是父子上下文，容器与上下文要区分理解。上下文中包含容器，但又不仅仅是容器。容器只负责管理 Bean，但上下文中还包括动态增强、资源加载、事件监听机制等多方面扩展功能。

*In addition to standard BeanFactory lifecycle capabilities, ApplicationContext implementations* detect and invoke ApplicationContextAware beans as well as ResourceLoaderAware, Application EventPublisherAware and MessageSourceAware beans.

最后一段注释解释了一个设计，即除了标准的 `BeanFactory` 生命周期功能，`ApplicationContext` 的实现类还检测并调用实现了 `ApplicationContextAware`、`ResourceLoaderAware`、`ApplicationEventPublisherAware` 以及 `MessageSourceAware` 接口的 `Bean`。这句话的理解要结合图 3-4 的接口继承关系，从图 3-4 中明显可以发现 `ApplicationContext` 还额外继承了几个父接口，而这几个接口都有对应的 `Aware` 接口实现回调注入，只不过由于 `ApplicationContext` 继承了这几个父接口，因此最终回调注入的都是 `ApplicationContext` 本身。

### 2. ConfigurableApplicationContext

与上面提到的 `BeanFactory` 与 `ConfigurableBeanFactory` 相似，`ConfigurableApplicationContext` 也为 `ApplicationContext` 提供了"可写"的功能，实现了该接口的实现类可以由客户端代码修改其内部的某些配置。文档注释不算长，分段来看。

*SPI interface to be implemented by most if not all application contexts. Provides facilities to configure an application context in addition to the application context client methods in the ApplicationContext interface.*

第一段注释提到 `ConfigurableApplicationContext` 会被绝大多数 `ApplicationContext` 的最终实现类实现。除了 `ApplicationContext` 接口中的应用程序上下文客户端方法，还提供了用于配置应用程序上下文的功能。这可以从接口方法中得以体现，`ConfigurableApplicationContext` 接口中扩展了 `setParent`、`setEnvironment`、`addBeanFactoryPostProcessor`、`addApplicationListener` 等方法，这都是可以改变 `ApplicationContext` 本身的方法。

*Configuration and lifecycle methods are encapsulated here to avoid making them obvious to ApplicationContext client code. The present methods should only be used by startup and shutdown code.*

这段注释告诉我们，配置和与生命周期相关的方法都封装在 `ConfigurableApplicationContext` 中，目的是避免暴露给 `ApplicationContext` 的调用者。`ConfigurableApplicationContext` 的方法仅应该由启动和关闭代码使用。由这段话可以看出，虽然 `ConfigurableApplicationContext` 本身扩展了一些方法，但是一般情况下它不希望让开发者调用，而是只允许调用刷新容器（refresh）和关闭容器（close）方法。注意，这个"一般情况"是在程序运行期间的业务代码中，如果是为了定制化 `ApplicationContext` 或者对其进行扩展，`ConfigurableApplicationContext` 的扩展则会成为切入的主目标。

### 3. ApplicationContext 的父接口

`ApplicationContext` 的子接口只有 `ConfigurableApplicationContext`，但它还实现了几个父接口，这些接口共同组成了 `ApplicationContext` 扩展的几个核心特性，我们一一来看。

（1）EnvironmentCapable

capable 意为"有能力的"，在这里解释为"携带/组合"更为合适。这里有一个有助于理解 Spring Framework 源码的规律：在 Spring Framework 的底层源码中，如果一个接口的名称以

Capable 结尾，通常意味着可以通过这个接口的某个特定的方法（通常是 getxxx()）获取特定的组件。根据这个概念，`EnvironmentCapable` 接口中就应该通过 `getEnvironment()` 方法获取 `Environment`，事实上也确实如此。

`EnvironmentCapable` 接口与 `Environment` 强关联。`Environment` 本身也是 Spring Framework 中的一个很重要的设计，这个组件的设计在 3.4 节会讲到，这里先简单提一下。`Environment` 是 Spring Framework 中抽象出来的类似于**运行环境**的**独立抽象**，它内部存放着应用程序运行所需的一些配置。基于 Spring Framework 的项目在运行时包含两部分：**应用程序本身和应用程序的运行时环境**。

`EnvironmentCapable` 与 `ApplicationContext` 的关系是：只要获取到 `ApplicationContext` 的实例，就可以借助 `EnvironmentCapable` 接口获取到 `Environment`。如果是 `ConfigurableApplicationContext`，则可以获取到 `ConfigurableEnvironment`。

（2）MessageSource

`MessageSource` 是国际化支持的组件。国际化是针对不同地区、不同国家的访问，可以提供对应的符合用户阅读习惯（语言）的页面和数据。对于不同地区、使用不同语言的用户，需要分别提供对应语言环境的描述。`ApplicationContext` 使用**委托机制**对国际化予以支持。在 `ApplicationContext` 实现类的内部整合了一个 `MessageSource` 的真正实现，当我们调用 `ApplicationContext` 的国际化相关方法时，内部直接将调用转发到 `MessageSource` 的落地实现类对象中，以此完成国际化支持。

（3）ApplicationEventPublisher

从类名上理解，`ApplicationEventPublisher` 是一个**事件发布器**。Spring Framework 内部支持强大的事件监听机制，而 `ApplicationContext` 作为容器的顶层，自然也要实现观察者模式中广播器的角色。这里我们简单提一下 Spring Framework 中设计的事件驱动机制。

Spring Framework 中，体现观察者模式的特性就是事件驱动和监听器。**监听器充当订阅者，监听特定的事件；事件源充当被观察的主题，用来发布事件；IOC 容器本身也是事件广播器，可以理解成观察者**。Spring Framework 的事件驱动核心可以划分为 4 部分：**事件源、事件、广播器和监听器**。

- 事件源：发布事件的对象。
- 事件：事件源发布的信息/做出的动作。
- 广播器：事件真正广播给监听器的对象。
- 监听器：监听事件的对象。

在这个模型中，`ApplicationContext` 的角色是广播器：`ApplicationContext` 实现了 `ApplicationEventPublisher` 接口，具有**事件广播器发布事件的能力**；`ApplicationContext` 的内部组合了 `ApplicationEventMulticaster`，它组合了所有的监听器，具有**事件广播器广播事件的能力**。

（4）ResourcePatternResolver

`ResourcePatternResolver` 接口是 `ApplicationContext` 继承的功能性接口中最复杂的一个，从类名理解可以解释为"**资源模式解析器**"，实际上 `ResourcePatternResolver` 的作用是**根据特定的路径解析资源文件**。`ResourcePatternResolver` 本身是 `ResourceLoader` 的扩展，它可以支持 Ant 形式的带星号（*）的路径解析。另外，`ResourcePattern`

Resolver 本身是**基于路径匹配的解析器**，这种扩展实现的特点是可以**根据特殊的路径返回多个匹配到的资源文件**。借助文件匹配的机制，ApplicationContext 可以将所需要的资源文件加载到应用程序中，从而完成功能配置、属性赋值等工作。

> 💡 提示：补充关于 Ant 形式的路径匹配写法。
> - /WEB-INF/*.xml：匹配/WEB-INF 目录下的任意 XML 文件。
> - /WEB-INF/**/beans-*.xml：匹配/WEB-INF 目录下任意层级目录的 beans-开头的 XML 文件。
> - /**/*.xml：匹配任意 XML 文件。
>
> 以上是三种常用的 Ant 形式的路径匹配写法，感兴趣的读者可以从网上搜索学习更多关于 Ant 风格的写法。

#### 4. AbstractApplicationContext

介绍完接口，接下来是 ApplicationContext 的几个重要的实现类。我们先将 Spring Framework 中 ApplicationContext 的重要实现类都罗列出来，如图 3-5 所示。

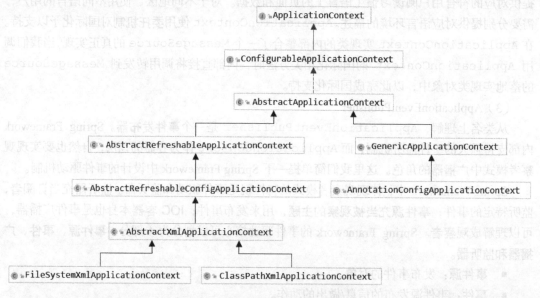

图 3-5 ApplicationContext 的实现类

从图 3-5 中可以发现，ApplicationContext 接口的众多实现类中，顶层的抽象类是 AbstractApplicationContext，这个类极为重要。AbstractApplicationContext 中定义和实现了**绝大部分应用上下文的特性和功能**，一定要重视。下面是文档注释的内容，分段来解读。

*Abstract implementation of the ApplicationContext interface. Doesn't mandate the type of storage used for configuration; simply implements common context functionality. Uses the Template Method design pattern, requiring concrete subclasses to implement abstract methods.*

第一段文档提到 AbstractApplicationContext 是 ApplicationContext 接口的抽象实现，它只是简单地实现了应用上下文的基本功能，并不强制约束配置的承载形式（XML、注解驱动等）。AbstractApplicationContext 的内部会使用大量模板方法规范整体功能逻辑，具体的实现由子类负责。总结一下，`AbstractApplicationContext` **只构建功能抽象**。

*In contrast to a plain BeanFactory, an ApplicationContext is supposed to detect special beans defined in its internal bean factory: Therefore, this class automatically registers BeanFactory PostProcessors, BeanPostProcessors, and ApplicationListeners which are defined as beans in the context.*

紧跟着的一段解释了 `ApplicationContext` 相较于 `BeanFactory` 对于**特殊类型 Bean 处理的扩展**。与普通的 `BeanFactory` 相比，`ApplicationContext` 应当能够检测在其内部 bean 对象工厂中定义的特殊 Bean，这类自动注册在上下文中定义为 Bean 的包括 `BeanFactoryPostProcessors`、`BeanPostProcessors` 和 `ApplicationListeners`。其实这几种 Bean 本身也是一些组件对象，只不过在 `ApplicationContext` 中它们会发挥更重要的作用，所以 `ApplicationContext` 将它们区分出来，并且给予它们发挥特殊功能的机会。

*A MessageSource may also be supplied as a bean in the context, with the name "messageSource"; otherwise, message resolution is delegated to the parent context. Furthermore, a multicaster for application events can be supplied as an "applicationEventMulticaster" bean of type ApplicationEventMulticaster in the context; otherwise, a default multicaster of type SimpleApplicationEventMulticaster will be used.*

这一段内容描述了对功能性接口扩展的支持，`ApplicationContext` 实现了国际化接口 `MessageSource`、事件广播器接口 `ApplicationEventMulticaster`，作为容器，它也会**把自己看成一个 Bean**，以支持不同类型的组件注入需要。

*Implements resource loading by extending DefaultResourceLoader. Consequently treats non-URL resource paths as class path resources (supporting full class path resource names that include the package path, e.g. "mypackage/myresource.dat"), unless the getResourceByPath method is overridden in a subclass.*

最后一段文档注释中解释了 `AbstractApplicationContext` 提供默认的加载资源文件策略。默认情况下，`AbstractApplicationContext` 加载资源文件的策略是直接继承 `DefaultResourceLoader` 的策略，从类路径下加载；但在 Web 项目中加载策略可能会发生改变，`AbstractApplicationContext` 的 Web 场景实现类可以从 `ServletContext` 中加载（扩展的子类 `ServletContextResourceLoader` 等）。

> 💡 提示：有关 `AbstractApplicationContext`，这里再多提一句。`AbstractApplicationContext` 中定义了一个特别重要的方法，它是控制 `ApplicationContext` 生命周期的核心方法：`refresh`。这个方法分为 13 步，包含一个 Spring Framework 应用上下文的所有重要步骤处理。关于这部分内容，我们先在 3.6 节简单接触，第 7 章 IOC 容器的生命周期中会详细展开讲解。

### 5. GenericApplicationContext

Spring Framework 的 IOC 容器最终落地有基于 **XML 配置文件**和**注解驱动**两种实现方式。Spring Boot 已经放弃了基于 XML 配置文件的实现方式，所以我们着重研究与注解驱动相关的 `ApplicationContext` 实现。`GenericApplicationContext` 是注解驱动 IOC 容器的第一个非抽象实现类，它已经具备 `ApplicationContext` 所有的基本能力了。对于它的文档注释我们也要重点关注。

*Generic ApplicationContext implementation that holds a single internal DefaultListableBean Factory instance and does not assume a specific bean definition format. Implements the BeanDefinitionRegistry interface in order to allow for applying any bean definition readers to it.*

第一段注释中提到 `GenericApplicationContext` 是一个通用 `ApplicationContext` 的实现，该实现拥有一个内部 `DefaultListableBeanFactory` 实例，并且不采用特定的 Bean 定义格式。另外它还实现了 `BeanDefinitionRegistry` 接口，以便允许将任何 Bean 定义读取器应用于该容器中。

这段注释的信息量很大，我们要把握两个重点：`GenericApplicationContext` 中组合了一个 `DefaultListableBeanFactory`，由此可以得到一个重要信息，即 `ApplicationContext` 并不是继承自 `BeanFactory` 的容器，而是组合了 `BeanFactory`；`GenericApplicationContext` 实现了 `BeanDefinitionRegistry`，所以 Bean 的定义信息可以由 `GenericApplicationContext` 注册到容器中。

*Typical usage is to register a variety of bean definitions via the BeanDefinitionRegistry interface and then call refresh() to initialize those beans with application context semantics (handling org.springframework.context.ApplicationContextAware, auto-detecting BeanFactoryPostProcessors, etc).*

第二段注释中又提到了 `BeanDefinitionRegistry`，`GenericApplicationContext` 实现了 `BeanDefinitionRegistry` 接口，可以自定义注册一些 Bean。然而在 `GenericApplicationContext` 中，它实现的定义注册方法 `registerBeanDefinition`，在底层还是调用的 `DefaultListableBeanFactory` 执行 `registerBeanDefinition` 方法，说明它没有对此做任何扩展，仅仅是**委托机制的体现**。

*In contrast to other ApplicationContext implementations that create a new internal BeanFactory instance for each refresh, the internal BeanFactory of this context is available right from the start, to be able to register bean definitions on it. refresh() may only be called once.*

这段注释中解释了重复刷新的设计。由于 `GenericApplicationContext` 中组合了一个 `DefaultListableBeanFactory`，而这个 `BeanFactory` 在 `GenericApplicationContext` 的构造方法中就已经初始化完毕，而初始化完毕的 `BeanFactory` **不允许在运行期间被重复刷新**。具体源码实现如代码清单 3-7 所示。

**代码清单 3-7　`GenericApplicationContext` 不允许重复刷新的底层逻辑**

```
public GenericApplicationContext() {
    // 内置的 beanFactory 在 GenericApplicationContext 创建时就已经初始化完毕
    this.beanFactory = new DefaultListableBeanFactory();
}

protected final void refreshBeanFactory() throws IllegalStateException {
    if (!this.refreshed.compareAndSet(false, true)) {
        // 利用 CAS，保证只能设置一次 true，如果出现第二次，就抛出重复刷新异常
        throw new IllegalStateException("GenericApplicationContext does not support multiple refresh attempts: just call 'refresh' once");
    }
    this.beanFactory.setSerializationId(getId());
}
```

如果在文档注释中没有直接说不允许重复刷新，而是在段落开头加一个"与……相反"的前缀，那么是因为有另一类 `ApplicationContext` 的设计不是这样的，它就是 `AbstractRefreshableApplicationContext`，下面会简单提一下。

*For the typical case of XML bean definitions, simply use ClassPathXmlApplicationContext or FileSystemXmlApplicationContext, which are easier to set up - but less flexible, since you can just use standard resource locations for XML bean definitions, rather than mixing arbitrary bean definition formats. The equivalent in a web environment is org.springframework.web.context.support.XmlWebApplicationContext.*

最后一段内容提到了 Web 环境相关的配置。对于 XML Bean 定义的典型情况，只需使用 `ClassPathXmlApplicationContext` 或 `FileSystemXmlApplicationContext`，因为它们更易于设置（但灵活性较差，只能从标准的资源配置文件中读取 XML Bean 定义，而不能混合使用任意 Bean 定义的格式）。在 Web 环境中，`GenericApplicationContext` 的替代方案是 `XmlWebApplicationContext`。

### 6. AnnotationConfigApplicationContext

`AnnotationConfigApplicationContext` 是 Spring Framework 中最常使用的注解驱动 IOC 容器，它本身继承自 `GenericApplicationContext`，那么自然它也只能刷新一次。文档注释中并没有花太多篇幅去描述 `AnnotationConfigApplicationContext`，我们快速了解即可。

*Standalone application context, accepting component classes as input - in particular @Configuration-annotated classes, but also plain @Component types and JSR-330 compliant classes using javax.inject annotations.*

第一段注释中说明 `AnnotationConfigApplicationContext` 是一个独立的注解驱动的 `ApplicationContext`，它接受组件类作为输入（特别是使用`@Configuration` 注解的类），还可以使用普通的`@Component` 类和符合 JSR-330 规范（使用 `javax.inject` 包的注解）的类。注意文档注释的措辞，它更重视的是被`@Configuration` 注解标注的类，因为一个被`@Configuration` 标注的类相当于一个 XML 文件，对一个注解驱动的 IOC 容器来讲，注解配置类就是驱动加载源。

*Allows for registering classes one by one using register(Class...) as well as for classpath scanning using scan(String...).*

*In case of multiple @Configuration classes, @Bean methods defined in later classes will override those defined in earlier classes. This can be leveraged to deliberately override certain bean definitions via an extra @Configuration class.*

这段文档注释告诉我们，`AnnotationConfigApplicationContext` 允许使用 `register(Class ...)` 方法直接传入指定的配置类，以及使用 `scan(String ...)` 方法进行类路径的包扫描。如果有多个`@Configuration` 类，则在后面的类中定义的`@Bean` 方法将覆盖在先前的类中定义的方法。这可以通过一个额外的`@Configuration` 类来故意覆盖某些 `BeanDefinition`。

### 7. AbstractRefreshableApplicationContext

最后简单了解一下基于 XML 配置文件驱动的 IOC 容器。之所以是简单了解，是因为 Spring

Boot 扩展出的 IOC 容器全都是基于注解驱动的，Spring Boot 已经不再将 XML 配置文件作为配置源承载的首选。

基于 XML 配置文件的 IOC 容器有一个特殊的特性：**可重复刷新**。得益于内部的设计，基于 XML 配置文件的 IOC 容器会在加载配置文件时才初始化内部的 `BeanFactory`，然后是解析配置文件、注册 bean 对象等动作。基于 XML 配置文件的 IOC 容器一般运用在传统的 Spring Framework + SpringWebMvc + MyBatis 技术栈的项目中，它们通常都使用 XML 作为配置源驱动项目运行。

基于 XML 配置文件的 IOC 容器的最终落地实现有我们熟悉的 `ClassPathXmlApplicationContext` 和 `FileSystemXmlApplicationContext`，还有支持 Web 环境的 `XmlWebApplicationContext`，只是在 Spring Boot 中这些落地实现都不再使用，所以本书不对这些实现类展开讨论，感兴趣的读者可以自行翻阅源码学习。

### 3.1.3 选择 ApplicationContext 而不是 BeanFactory

了解 `BeanFactory` 与 `ApplicationContext` 的介绍与对比之后，有关实际项目开发中应该选择使用 `BeanFactory` 还是 `ApplicationContext` 的问题，官方文档中提供了一个参考观点。

> You should use an `ApplicationContext` unless you have a good reason for not doing so, with `GenericApplicationContext` and its subclass `AnnotationConfigApplicationContext` as the common implementations for custom bootstrapping. These are the primary entry points to Spring's core container for all common purposes: loading of configuration files, triggering a classpath scan, programmatically registering bean definitions and annotated classes, and (as of 5.0) registering functional bean definitions.
>
> 你应该使用 `ApplicationContext`，除非有充分的理由能解释不使用它的原因。一般情况下，我们推荐将 `GenericApplicationContext` 及其子类 `AnnotationConfigApplicationContext` 作为自定义引导的常见实现。这些实现类是用于所有常见目的的 Spring Framework 核心容器的主要入口点：加载配置文件，触发类路径扫描，编程式注册 Bean 定义和带注解的类，以及（从 5.0 版本开始）注册功能性 Bean 的定义。

这段话的下面还提供了一张表，对比了 `BeanFactory` 与 `ApplicationContext` 的不同特性，如表 3-1 所示。

表 3-1　BeanFactory 与 ApplicationContext 的特性对比

| 特性 | BeanFactory | ApplicationContext |
| --- | --- | --- |
| Bean instantiation/wiring——Bean 的实例化和属性注入 | 是 | 是 |
| Integrated lifecycle management——生命周期管理 | 否 | 是 |
| Automatic `BeanPostProcessor` registration——Bean 后置处理器的支持 | 否 | 是 |
| Automatic `BeanFactoryPostProcessor` registration——BeanFactory 后置处理器的支持 | 否 | 是 |
| Convenient `MessageSource` access (for internalization)——消息转换服务（国际化） | 否 | 是 |
| Built-in `ApplicationEvent` publication mechanism——事件发布机制（事件驱动） | 否 | 是 |

由表 3-1 可见，`ApplicationContext` 相较于 `BeanFactory` 功能更强大，这也解释了为什么日常开发中都是使用 `ApplicationContext` 而很少接触 `BeanFactory`。

## 3.2 Spring Boot 对 IOC 容器的扩展

Spring Boot 本身并没有直接利用 Spring Framework 现有的 IOC 容器，而是针对嵌入式 Web 容器的这一核心特性扩展了更强大的 IOC 容器。另外，在 Spring Boot 2.x 中，底层依赖了 Spring Framework 5.x，这个大版本下一个典型的特性是新增了 WebFlux 模块，其与 WebMvc 平级，用于构建异步非阻塞式 Web 应用。Spring Boot 针对 WebMvc 与 WebFlux 这两种 Web 应用的搭建基础分别扩展了不同的 IOC 容器。本节我们简单了解一下 Spring Boot 针对 IOC 容器是如何扩展的。

> 提示：本节我们只简单了解，不再像上面那样展开讲解和讨论，因为 IOC 容器的根本设计还是 Spring Framework 的模型，Spring Boot 仅进行了扩展，读者一定要把握好重点。

### 3.2.1 WebServerApplicationContext

首先请读者了解一个概念：`WebServer`，这个概念来自 Spring Boot 2.x（在 Spring Boot 1.x 中称为 `EmbeddedServletContainer`，因为 Spring Boot 1.x 基于 Spring Framework4.x，当时还没有 WebFlux）。WebServer 可以简单理解为**嵌入式 Web 容器**（即便类名中没有体现嵌入式，依照 Spring Boot 1.x 的历史以及 javadoc 的描述也可以理解为有嵌入式的意思）。Spring Boot 为了设计和支持嵌入式 Web 容器，在 `ApplicationContext` 的基础上又扩展了一个 `WebServerApplicationContext` 子接口，如代码清单3-8 所示。

**代码清单 3-8　WebServerApplicationContext 扩展的核心方法**

```
public interface WebServerApplicationContext extends ApplicationContext {
    // 获取嵌入式 Web 容器
    WebServer getWebServer();
}
```

`WebServerApplicationContext` 的接口源码中有一个重要方法：`getWebServer`。这个方法可以获取当前应用的嵌入式 Web 容器实例，换言之，Spring Boot 中的 `ApplicationContext` 可以获取正在运行的 Web 容器。至于内部的 WebServer 是如何创建的，我们放到第 8 章专门来讲解 Spring Boot 嵌入式容器的生命周期。

### 3.2.2 AnnotationConfigServletWebServerApplicationContext

`WebServerApplicationContext` 的下面有几个扩展和落地实现，本书选择最常见的 Spring Boot 整合 WebMvc 场景的 IOC 容器最终落地实现 `AnnotationConfigServletWebServerApplicationContext` 来简单讲解。该类名虽然很长，但结构很完整也容易理解：注解驱动的、基于 Servlet 环境的、支持嵌入式 Web 容器的应用上下文。Spring Boot 默认使用注解驱动，所以落地实现都是用 `AnnotationConfig` 前缀的 `ApplicationContext` 实现类作为 IOC 容器支撑（具体的实现和引用见第 4 章）。既然是注解驱动，那么实现类中一定有类似于

`AnnotationConfigApplicationContext` 的特性：注解配置类的注册、模式注解的扫描等。而这些特性的底层支撑还是依靠 Spring Framework 的现有组件辅助完成。

另外，`AnnotationConfigServletWebServerApplicationContext` 继承的父类 `ServletWebServerApplicationContext` 中组合了嵌入式容器 `WebServer` 对象以及 `ServletConfig` 配置对象，并且也有创建嵌入式容器的逻辑。代码清单 3-9 展示了内部创建嵌入式 Web 容器的方法。

**代码清单 3-9　ServletWebServerApplicationContext 中组合了嵌入式容器**

```java
public class ServletWebServerApplicationContext
        extends GenericWebApplicationContext
        implements ConfigurableWebServerApplicationContext {

    private volatile WebServer webServer;
    private ServletConfig servletConfig;

    // 内部有定义创建嵌入式 Web 容器的方法
    private void createWebServer() { ... }
```

由此可以体现出，Spring 团队对于每个类的职责划分层次相当清晰，充分体现了单一职责原则和迪米特法则（又称为最少知识原则）。

### 3.2.3　ReactiveWebApplicationContext

Spring Boot 2.x 基于 Spring Framework 5.x，而 5.x 版本中引入了新的 WebFlux 模块，所以 IOC 容器就不只有基于 Servlet 环境的最终实现，Spring Boot 同样为 ReactiveWeb 扩展了 `ApplicationContext` 的支持：`ReactiveWebServerApplicationContext`。同样，它也有一套类似的"接口-实现类"体系，包括中间的 `GenericReactiveWebApplicationContext`（类比 `GenericWebApplicationContext`）、最终的落地实现 `AnnotationConfigReactiveWebServerApplicationContext`，与基于 Servlet 环境的命名方式如出一辙。

对于这几种基于 Reactive 环境的 IOC 容器，读者只需大概了解。对于 IOC 容器本身的设计，两种环境下的差距不大，如果充分掌握了 Servlet 环境下的 IOC 容器实现，Reactive 下环境的 IOC 容器也不难。

## 3.3　选用注解驱动 IOC 容器的原因

下面我们讨论一个话题：Spring Boot 为什么选用注解驱动的 IOC 容器，而放弃了曾经主推的 XML 配置？

要讨论这个话题，首先要明白注解驱动配置（即 JavaConfig）与 XML 配置文件的方式分别有什么特点，以及各自的优劣势。然后再根据 Spring Boot 的设计理念，体会两者的差异，揣度 Spring 团队当时选择注解驱动 IOC 容器的想法和动机。

### 3.3.1　配置方式的对比

要想搞明白 XML 配置文件与注解配置类的区别，最好先了解一下历史背景。Spring

Framework 刚出现的时候，只能使用 XML 配置文件的方式驱动 IOC 容器，直到 Spring Framework 3.x 后才支持注解驱动配置，所以首要的一点是：XML 配置文件是早期采用的配置方式，注解配置类则相对晚一些。

其次，XML 配置文件的方式本身修改起来灵活，无须重新编译，而且基于 XML 配置文件驱动的 IOC 容器可以反复刷新加载配置文件；而注解配置类在修改上就没那么灵活了，每次修改后都需要重新编译、打包运行，而且通过上面了解 `ApplicationContext` 的实现类我们知道，注解驱动的 IOC 容器只能加载一次，无法反复刷新。因此，从修改灵活性的角度上讲，XML 配置文件更为灵活。

最后要对比的是配置内容的编写，XML 配置文件的内容编写要符合相关的规范，对于规范以外的内容 Spring Framework 是不认可的。即便要扩展 XML 配置文件声明的内容，也需要配合 XML 解析逻辑支持才能运行。而注解配置类的方式就灵活多了，Spring Framework 3.1 提供的模块装配、4.0 提供的条件装配可以使配置的编写非常灵活，且可以根据不同场景的要求注册不同类型的组件。因此，从配置内容的编写上讲，注解配置类更胜一筹。

### 3.3.2　约定大于配置下的选择

简单对比 XML 配置文件与注解配置类之后，我们回到 Spring Boot 上。考虑一个问题：一开始设计 Spring Boot 的时候，它的核心设计之一是**约定大于配置**，这种设计更加强调项目环境与自定义配置间的配合协作（已经配置的不重复注册，没有配置的自动补充）。基于这种设计，我们再考虑 XML 配置文件与注解配置类是否可以满足约定大于配置的实现条件。

虽然基于 XML 配置文件的方式对于配置的编写更灵活、易修改，但整合具体场景制作为启动器时，这种优势会荡然无存，毕竟在实际项目开发中，我们不会轻易改动现有 jar 包中的文件（何况它还是配置文件），而且 XML 文件中定义了组件，会实实在在地注册到应用上下文中，运行时的灵活性较差；而基于注解配置类的方式支持**模块装配**、**条件装配**、**SPI 机制**等高级特性，依据这些高级特性，在整合具体场景制作启动器时，完全可以实现与项目中自定义配置相互配合，即实现了约定大于配置。经过这样的分析后，我们就更能理解为什么 Spring Boot 推荐使用注解驱动的配置方式。

## 3.4　Environment

接下来要讲解的内容是与 `ApplicationContext` 紧密相关的概念：**Environment**（环境）。

### 3.4.1　Environment 概述

首先整体了解一下 `Environment`，它是从 Spring Framework 3.1 开始引入的一个抽象模型，包含 Profile 与 Property 配置的信息，可以实现统一的配置存储和注入、配置属性的解析等行为。其中 Profile 实现了基于模式的环境配置（即条件装配），Property 则多用于外部化配置。

下面我们拆解出其中的要素分别解释。

#### 1. Profile

在第 2 章的条件装配中，我们已经讲过 Profile 的使用，当时推荐读者理解 Profile 是"**基于环境的配置**"，到了 Environment 这里，想必读者更能理解"环境"这个概念了。`Environment`

本身就组合了 Profile 的特性,用于区分不同的**环境模式**。根据不同的环境模式,Spring Framework 可以装配特定的 bean 对象(以及配置属性的注入)。

#### 2. Property

Property 作为键值对形式的数据结构,非常适合存储配置信息。在学习 Spring Framework 时我们就接触过,在 Property 文件中存储 JDBC 连接参数,并通过@PropertySource 注解或 <context:property-placeholder>标签引入这些 Property 文件到 IOC 容器中,以实现配置项分离的目的。在这种设计中,利用 @PropertySource 注解或 <property-placeholder>标签,将 Property 文件导入目标位置,其实就是 Environment 中,而导入配置后,最终解析配置项、给 Bean 组件注入属性值的行为都由 Environment 负责。

### 3.4.2 Environment 的结构与设计

简单了解 Environment 的概念和设计思想后,下面结合 API,了解一下 Environment 的上下级继承关系。借助 IDEA 可以形成 Environment 接口的上下级继承和派生关系,如图 3-6 所示。

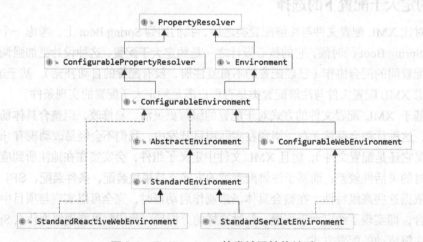

图 3-6 Environment 的类继承结构关系

从图 3-6 中可以得到两个简单的结论:Environment 不是顶层的根接口;Environment 的落地实现有 3 种,分别适应不同的场景。既然 Environment 不是顶层接口,中间也有一些派生和落地实现,我们选择其中几个重要的接口讲解。

#### 1. PropertyResolver

从类名上看,PropertyResolver 是一个属性解析器,它可以处理 XML 配置文件、注解配置类、普通 Bean 中用到的${}属性配置占位符。Environment 继承自 PropertySource,并且 Environment 中存放了 Profile 和 Property 属性配置,这就意味着 PropertyResolver 可以解析${}占位符,并提取出 Environment 中的 Property 配置完成赋值操作。

借助 IDE,可以大致浏览 PropertyResolver 接口的部分核心方法,如代码清单 3-10 所示。PropertyResolver 接口完成的工作主要包含两部分:配置属性的获取;占位符的解析。由此也就可以得出另一个结论:Environment **可以获取属性配置元信息,同时也可以解析占位符的信息。**

### 代码清单 3-10　PropertyResolver 接口的部分方法

```java
public interface PropertyResolver {

    // 检查所有配置属性中是否包含指定 key
    boolean containsProperty(String key);

    // 以 String 形式返回指定的配置属性的值
    String getProperty(String key);

    // 带默认值的获取
    String getProperty(String key, String defaultValue);

    // 指定返回类型的配置属性值获取
    <T> T getProperty(String key, Class<T> targetType);

    // 解析占位符
    String resolvePlaceholders(String text);
}
```

#### 2. ConfigurableEnvironment

经过对上面 BeanFactory 和 ApplicationContext 接口的分析和了解，看到 ConfigurableEnvironment 后，想必读者可以快速意识到它的特性：具备可配置的能力，接口中大概率会有 set 之类的方法。借助 IDE 看一下它的几个核心方法，如代码清单 3-11 所示。

### 代码清单 3-11　ConfigurableEnvironment 接口的部分方法

```java
public interface ConfigurableEnvironment extends Environment, ConfigurablePropertyResolver {

    // 激活指定 profiles
    void setActiveProfiles(String... profiles);

    // 追加激活指定 profile
    void addActiveProfile(String profile);

    // 设置默认情况下激活的 profiles
    void setDefaultProfiles(String... profiles);

    // 获取 PropertySource 的组合体
    MutablePropertySources getPropertySources();
}
```

前三个方法可以很容易地理解，它可以编程式设置 activeProfiles 和 defaultProfiles；最后一个方法 getPropertySources 比较特别，它用来获取所有的 PropertySource 对象，但返回值的类型是 MutablePropertySources，这个类型内部是一个 List 封装。

```java
public class MutablePropertySources implements PropertySources {
    private final List<PropertySource<?>> propertySourceList = new CopyOnWriteArrayList<>();
```

由此可以总结出一个小结论：以 Mutable 开头的类名，通常可能是一个类型的 List 组合封装。

> 提示：可以简单理解为一个 `PropertySource` 对象对应了一个配置源，这个配置源可能来自 Property 文件，可能来自系统环境变量、项目信息，也可能是自定义的加载方式（如数据库），但不管来自哪里，最终都会封装为键值对的形式存储在一个 `PropertySource` 对象中，统一由 `Environment` 管理。

### 3. AbstractEnvironment

最后了解一下所有 `Environment` 落地实现的父类 `AbstractEnvironment`。之所以看抽象而不看实现，是因为实现类中的代码量相当少，而且没有很重要的逻辑，大多数的实现都在 `AbstractEnvironment` 中体现。对于了解 `AbstractEnvironment` 的方式，读者不必过于深入，了解其中的核心设计即可，如代码清单 3-12 所示。

**代码清单 3-12　AbstractEnvironment 的部分核心成员**

```java
public abstract class AbstractEnvironment implements ConfigurableEnvironment {

    private final Set<String> activeProfiles = new LinkedHashSet<>();

    private final Set<String> defaultProfiles = new LinkedHashSet<>(getReservedDefaultProfiles());

    private final MutablePropertySources propertySources = new MutablePropertySources();

    private final ConfigurablePropertyResolver propertyResolver =
            new PropertySourcesPropertyResolver(this.propertySources);
```

代码清单 3-12 列出的是 `AbstractEnvironment` 的部分成员属性，它会存储默认的 Profile，以及声明激活的所有 Profile，所有的 `PropertySource` 也在这里存储。除此之外，请读者注意还有一个很重要的成员：`PropertySourcesPropertyResolver`。它也是一个 `PropertyResolver`，`AbstractEnvironment` 在此处组合了一个 `PropertyResolver` 的实现类，意图也很明显：只要是 `PropertyResolver` 接口下的方法要做的工作，全部交予 `PropertySourcesPropertyResolver` 完成，我们称这种设计为"委派"，它与代理、装饰者不同：委派仅仅是将方法的执行工作移交给另一个对象，而代理则可能会在此做额外的处理，装饰者会在方法执行前后做增强。

## 3.4.3　Environment 与 IOC 容器的关系

本节的最后，我们探讨一下 `Environment` 与 IOC 容器的关系。我个人倾向于如下的结构和关系理解：

- `Environment` 中包含 Profile 和 Property，这些配置信息会影响 IOC 容器中的 Bean 的注册与创建；
- `Environment` 是在 `ApplicationContext` 创建后才创建的，所以 `Environment` 应该伴随着 `ApplicationContext` 的存在而存在；
- `ApplicationContext` 中同时包含 `Environment` 和普通的 Bean 组件对象，从 `BeanFactory` 的视角来看，`Environment` 也是一个 Bean，只不过它的地位比较特殊而已。

基于这三点，Environment 与 IOC 容器以及内部的组件之间的结构关系如图 3-7 所示。

图 3-7　Environment 与 ApplicationContext 的结构关系

图 3-7 中 Environment 覆盖的范围比较大，这是因为考虑到 Environment 还要辅助处理配置属性值的解析和注入工作，需要它配合 ApplicationContext 协同完成，所有的组件、Bean 都可能需要 Environment 参与处理，所以这里将组件和 Bean 放入 Environment 所在的框中。

## 3.5　BeanDefinition

下面我们继续了解一个更为重要的类：BeanDefinition。这个类的设计相当重要，理解 BeanDefinition 对后续学习 Spring Boot 的扩展点、生命周期等部分有很大帮助。

### 3.5.1　理解元信息

> 提示：本节提及的元信息仅针对 Spring Framework 和 Spring Boot，不会涉及数据分析、数据挖掘等相关概念。

**元信息**，又可以理解为**元定义**，简单地说，它就是**定义的定义**。其概念比较难理解，我们可以举一个例子说明。

- 张三，男，18 岁
  - 它的元信息就是它的属性：Person {name, age, sex}
- 咪咪，美国短毛，黑白毛，主人是张三
  - 它的元信息就是它的属性：Cat {name, type, color, master}

在这个简单的示例中，可以体现出对象和类的关系：对象的属性由类定义，**类中包含对象的元信息**。

再往深层次想一步，类有元信息吗？当然有，Class 这个类中就包含一个类的所有定义（属性、方法、继承实现、注解等），所以我们可以说：Class **中包含类的元信息**。

既然对象、类都有自己的元信息，IOC 容器中的 bean 对象是否也有呢？当然也是有的，描述 Bean 的元信息的模型就是 BeanDefinition。

### 3.5.2　BeanDefinition 概述

BeanDefinition 描述了 Spring Framework 中 Bean 的元信息，包含 Bean 的类信息、属性、行为、依赖关系、配置信息等，并且可以在 IOC 容器的初始化阶段被拦截处理。

这句话很精炼，读者理解起来可能稍有困难，我们可以借助 IDE 大体浏览一下 BeanDefinition 读者的接口定义，从接口的方法定义上加以理解。BeanDefinition 接口的方法定义整体上包含以下几个部分。

- Bean 的类信息：全限定名（beanClassName）。
- Bean 的属性：作用域（scope）、是否默认 Bean（primary）、描述信息（description）等。
- Bean 的行为特性：是否延迟加载（lazy）、是否自动注入（autowireCandidate）、初始化/销毁方法（initMethod / destroyMethod）等。
- Bean 与其他 Bean 的关系：父 Bean 名称（parentName）、依赖的 Bean（dependsOn）等。
- Bean 的配置属性：构造方法参数（constructorArgumentValues）、属性变量值（propertyValues）等。

由此可见，BeanDefinition 几乎能把 Bean 的所有信息收集并封装起来，非常全面。

### 3.5.3 BeanDefinition 的结构与设计

大体理解 BeanDefinition 的基本概念后，下面我们来研究 BeanDefinition 接口的上下级继承和派生关系，借助 IDEA 可以形成如图 3-8 所示的继承关系图。

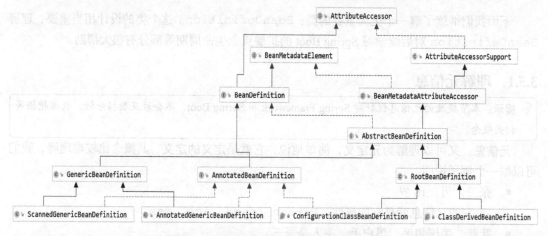

图 3-8 BeanDefinition 的类继承结构关系

可以发现 BeanDefinition 并不是顶层接口，它的上层还有其他的父接口，下层也有一些扩展实现，我们选择其中比较重要的几个接口和类简单了解一下。

#### 1. AttributeAccessor

AttributeAccessor 类名直译为属性访问器。了解 Java 内省机制的读者可能会对此产生一个猜测：这其中会不会定义了一些类似于 bean 对象中 get/set 方法的接口方法？确实如此，只不过 AttributeAccessor 设计得更全面，它不仅有获取和写入的操作，还能移除 Bean 的属性。从整体上看，它的设计与 Map 有些相似（都有 get、set、remove、contains 等操作），如代码清单 3-13 所示。

**代码清单 3-13　AttributeAccessor 接口的方法定义**

```
public interface AttributeAccessor {
    // 设置 bean 对象中属性的值
    void setAttribute(String name, @Nullable Object value);
```

```
    // 获取bean对象中指定属性的值
    Object getAttribute(String name);

    // 移除bean对象中的属性
    Object removeAttribute(String name);

    // 判断bean对象中是否存在指定的属性
    boolean hasAttribute(String name);

    // 获取bean对象的所有属性
    String[] attributeNames();
}
```

由 `AttributeAccessor` 接口，我们可以总结出 `BeanDefinition` 的第一个特性：`BeanDefinition` 继承自 `AttributeAccessor` 接口，具有配置 Bean 属性的功能。

> 提示：补充一点，在配置 Bean 的属性时，会涉及访问、修改、移除等操作。

### 2. BeanMetadataElement

`BeanMetadataElement` 作为 `BeanDefinition` 的另一个父接口，它的类名中包含 metadata 的概念，类名直译为**存放 Bean 的元信息的元素**。这个接口只有一个方法，其作用是获取 Bean 的资源来源，如代码清单 3-14 所示。

**代码清单 3-14　BeanMetadataElement**

```java
public interface BeanMetadataElement {

    default Object getSource(){
        return null;
    }
}
```

所谓的资源来源，就是 Bean 的文件/url 路径。一般情况下项目中所编写的所有注册到 IOC 容器中的 Bean，都是从本地磁盘上的 .class 文件加载进来的，所以此处获取的其实是一个 `Resource` 对象（或者可以理解为一个 `File` 对象）。

### 3. AbstractBeanDefinition

`AbstractBeanDefinition` 作为 `BeanDefinition` 的基本抽象实现，与前面的 `BeanFactory`、`ApplicationContext`、`Environment` 等核心组件相似，都是基本抽象实现中已经完成了相当一部分属性组成和逻辑实现。

借助 IDE 翻看 `AbstractBeanDefinition` 的成员，可以发现跟上面概述部分提到的核心组件基本吻合（类信息、作用域、属性、构造方法、生命周期等），如代码清单 3-15 所示。

**代码清单 3-15　AbstractBeanDefinition 的部分核心成员**

```java
// Bean 的全限定类名
private volatile Object beanClass;

// 默认的作用域为单实例
private String scope = SCOPE_DEFAULT;
```

```
// 默认Bean都不是抽象的
private boolean abstractFlag = false;

// 是否延迟初始化
private Boolean lazyInit;

// 同类型的首选Bean
private boolean primary = false;

// Bean的构造方法参数和参数值列表
private ConstructorArgumentValues constructorArgumentValues;

// Bean的属性和属性值集合
private MutablePropertyValues propertyValues;

// Bean的初始化方法、销毁方法
private String initMethodName;
private String destroyMethodName;

// Bean的资源来源
private Resource resource;
```

可能有读者会对此产生困惑：既然 AbstractBeanDefinition 中的成员已经齐全了，为什么还要单独抽取为一个抽象类呢？AbstractBeanDefinition 的 javadoc 中给我们提供了一个解释：

*Base class for concrete, full-fledged BeanDefinition classes, factoring out common properties of GenericBeanDefinition, RootBeanDefinition, and ChildBeanDefinition.*

如文档注释所述，它是 BeanDefinition 接口的抽象实现类，它剔除了 GenericBeanDefinition、RootBeanDefinition 和 ChildBeanDefinition 的共有属性。换言之，不同的 BeanDefinition 落地实现，其内部的属性还是有差异的，这也就是 AbstractBeanDefinition 存在的意义。

### 4. 几个落地实现类

下面我们看几个 BeanDefinition 的落地实现类。首先来看 GenericBeanDefinition，它的前缀 Generic 代表"通用的""一般的"，所以我们可以理解为 GenericBeanDefinition 具有一般性。GenericBeanDefinition 的源码实现非常简单，仅比 AbstractBeanDefinition 多了一个 parentName 属性。恰恰是由于这个设计，我们可以得出以下几个结论：

- AbstractBeanDefinition 已经完全可以构成 BeanDefinition 的实现；
- GenericBeanDefinition 就是 AbstractBeanDefinition 的非抽象扩展；
- GenericBeanDefinition 具有层次性（parentName 已经告诉了我们一切）。

与 GenericBeanDefinition 类似的一个设计是 ChildBeanDefinition，从类名上看 ChildBeanDefinition 一定是一个子定义信息，所以 ChildBeanDefinition 也有一个 parentName 的内部成员。不过 ChildBeanDefinition 与 GenericBeanDefinition 之间有一个小区别，因为 ChildBeanDefinition 已经在类名上体现出"子定义"的概念，所以它只有一个带 parentName 参数的构造方法（GenericBeanDefinition 有两个

构造方法）。

最后一个是 `RootBeanDefinition`，类名中有"根"的概念，这就意味着 `RootBeanDefinition` 只能作为单独的 `BeanDefinition` 或者父 `BeanDefinition` 出现（不能继承其他 `BeanDefinition`）。`RootBeanDefinition` 中的设计相对复杂，从源码的篇幅上就能看出来（接近 500 行，而 `GenericBeanDefinition` 只有 100 多行），这里我们关注一些重要的内部成员即可，如代码清单 3-16 所示。

**代码清单 3-16　RootBeanDefinition 的部分核心成员**

```java
// BeanDefinition 的引用持有，存放了 Bean 的别名
private BeanDefinitionHolder decoratedDefinition;

// Bean 上面的注解信息
private AnnotatedElement qualifiedElement;

// Bean 中的泛型
volatile ResolvableType targetType;

// BeanDefinition 对应的真实 Bean
volatile Class<?> resolvedTargetType;

// 是否是 FactoryBean
volatile Boolean isFactoryBean;
// 工厂 Bean 方法返回的类型
volatile ResolvableType factoryMethodReturnType;
// 工厂 Bean 对应的方法引用
volatile Method factoryMethodToIntrospect;
```

从代码清单 3-16 中可以发现，`RootBeanDefinition` 在 `AbstractBeanDefinition` 的基础上又扩展了一些 Bean 的其他信息：id 和别名、注解信息、工厂相关信息（是否为工厂 Bean、工厂类、工厂方法等）。而且这其中把一些反射相关的元素直接组合进来，可见 `BeanDefinition` 在底层做的事情更多。至于 `RootBeanDefinition` 都发挥了哪些更强大的作用，在本书第 2 部分会有更深入的讲解。

### 3.5.4　体会 BeanDefinition

下面我们结合一些简单的 Demo，帮助读者体会 `BeanDefinition` 的设计，以及其中封装 Bean 的相关内容。

#### 1. 基于组件扫描的 BeanDefinition

使用模式注解 + 组件扫描的方式，每扫描到一个类，就相当于构建了一个 `BeanDefinition`。为了方便演示扫描的效果，我们来构建一个简单的 `Person` 类，并标注 `@Component` 注解，如代码清单 3-17 所示。

**代码清单 3-17　定义简单类 Person 并标注模式注解**

```java
@Component
public class Person {
```

```java
    private String name;

    public String getName() {
        return name;
    }

    public void setName(String name) {
        this.name = name;
    }
}
```

使用 AnnotationConfigApplicationContext 可以在不编写注解配置类的情况下，直接扫描所有标注了 @Component 及派生注解的组件，扫描完成后直接从 ApplicationContext 中提取出 person 的 BeanDefinition，并打印出来查看其信息以及 BeanDefinition 的类型，如代码清单 3-18 所示。

**代码清单 3-18　测试获取 Person 的 BeanDefinition 信息**

```java
public class ComponentScanBeanDefinitionApplication {

    public static void main(String[] args){
        AnnotationConfigApplicationContext ctx = new AnnotationConfigApplicationContext(
                "com.linkedbear.springboot.beandefinition.bean");
        BeanDefinition personBeanDefinition = ctx.getBeanDefinition("person");
        System.out.println(personBeanDefinition);
        System.out.println(personBeanDefinition.getClass().getName());
    }
}
```

运行测试程序的 main 方法，控制台打印出来的是一个 `GenericBeanDefinition` 的派生类（用来标注它是一个被扫描到的 Bean），并且 person 对象的基本信息都已经收集并封装完成。

```
Generic bean: class [com.linkedbear.springboot.beandefinition.bean.Person]; scope= singleton; abstract=
false; lazyInit=null; autowireMode=0; dependencyCheck=0; autowire Candidate=true; primary=
false; factoryBeanName=null; factoryMethodName=null; initMethodName= null; destroyMethodName=
null; defined in file [E:\IDEA\spring-boot-source-analysis-epubit\springboot-03-ioccontainer\
target\classes\com\linkedbear\springboot\beandefinition\bean\Person.class]
org.springframework.context.annotation.ScannedGenericBeanDefinition
```

> 提示：留意一个细节，基于 @Component 注解解析出的 Bean，定义来源在类的 .class 文件中。

**2. 基于 @Bean 的 BeanDefinition**

要演示 @Bean 注解的 BeanDefinition 构造，就必须要编写注解配置类，我们可以声明一个最简单的配置类，并注册一个 Person 对象，如代码清单 3-19 所示。

**代码清单 3-19　编写注册了 Person 的注解配置类**

```java
@Configuration
public class BeanDefinitionConfiguration {

    @Bean
    public Person person(){
```

```
        return new Person();
    }
}
```

随后，用这个注解配置类驱动 IOC 容器，从中获取 person 的 BeanDefinition 信息并打印，如代码清单 3-20 所示。

**代码清单 3-20　测试获取 Person 的 BeanDefinition 信息**

```
public class AnnotationConfigBeanDefinitionConfiguration {

    public static void main(String[] args) {
        AnnotationConfigApplicationContext ctx = new AnnotationConfigApplicationContext(
                BeanDefinitionConfiguration.class);
        BeanDefinition personBeanDefinition = ctx.getBeanDefinition("person");
        System.out.println(personBeanDefinition);
        System.out.println(personBeanDefinition.getClass().getName());
    }
}
```

编写完成后运行测试程序的 `main` 方法，可以发现控制台打印的内容与前面基于组件扫描的 `BeanDefinition` 有很大区别：

```
Root bean: class [null]; scope=; abstract=false; lazyInit=null; autowireMode=3; dependencyCheck=0;
autowireCandidate=true; primary=false; factoryBeanName=beanDefinitionConfiguration;
factoryMethodName=person; initMethodName=null; destroyMethodName=(inferred); defined in
com.linkedbear.springboot.beandefinition.config.BeanDefinitionConfiguration
org.springframework.context.annotation.ConfigurationClassBeanDefinitionReader$Configuration
ClassBeanDefinition
```

具体区别如下。

- `BeanDefinition` 的类型是 Root Bean（`ConfigurationClassBeanDefinition` 继承自 `RootBeanDefinition`）。
- bean 对象的 `className` 不见了。
- 自动注入模式为 `AUTOWIRE_CONSTRUCTOR`（构造方法自动注入）。
- 有 factoryBean 属性：person 由 `BeanDefinitionConfiguration` 的 `person` 方法创建。

为什么两种不同的 Bean 定义方式在实际运行时会有如此大的差别呢？这里先简单提一下底层逻辑，具体的内容会在第二部分生命周期中展开探讨。

- 通过模式注解+组件扫描的方式构造的 `BeanDefinition`，它的扫描工具是 `ClassPathBeanDefinitionScanner`，扫描器会扫描指定包路径下包含特定模式注解的类，扫描器的核心工作方法是 `doScan`，它会调用父类 `ClassPathScanningCandidateComponentProvider` 的 `findCandidateComponents` 方法，创建 `ScannedGenericBeanDefinition` 并返回；
- 通过配置类 + `@Bean` 注解的方式构造的 `BeanDefinition` 最复杂，它涉及配置类的解析。配置类的解析要追踪到 `ConfigurationClassPostProcessor` 的 `processConfigBeanDefinitions` 方法，该方法会处理配置类，并交给 Config

urationClassParser 来解析配置类，提取出所有标注了 @Bean 的方法。随后这些方法又被 ConfigurationClassBeanDefinitionReader 解析，最终在底层创建 ConfigurationClassBeanDefinition 并返回。

> 💡 提示：目前读者只需对这些内容有一个最简单的了解和认知，不需要关心其内部的具体实现，切勿本末倒置。

### 3.5.5　BeanDefinitionRegistry

BeanDefinition 解析完成后，最终转化为 bean 对象，这之间有一个 BeanDefinition 存储（注册）、随后解析 BeanDefinition 生成 bean 对象的过程，而统一管理 BeanDefinition 的核心就是本节要讲解的 BeanDefinitionRegistry。BeanDefinitionRegistry 是**维护 BeanDefinition 的注册中心**，它内部存放了 IOC 容器中 Bean 的定义信息，同时 BeanDefinitionRegistry 也是支撑其他组件和动态注册 Bean 的重要组件。

首先，Registry 有注册表的意思，联想一下 Windows 中的注册表，它存放了 Windows 系统中应用程序和配置的信息。从底层源码的设计来看，BeanDefinitionRegistry **本质上是一个存放 BeanDefinition 的容器**。

```
// 源自 DefaultListableBeanFactory
private final Map<String, BeanDefinition> beanDefinitionMap = new ConcurrentHashMap<>(256);
```

其次，Registry 还有注册器的意思。既然 Map 有增删改查，那么作为 BeanDefinition 的注册器，BeanDefinitionRegistry 自然也有 BeanDefinition **的注册功能**。BeanDefinitionRegistry 中有 3 个方法，刚好对应了 BeanDefinition 的增、删、查：

```
void registerBeanDefinition(String beanName, BeanDefinition beanDefinition)
        throws BeanDefinitionStoreException;
void removeBeanDefinition(String beanName) throws NoSuchBeanDefinitionException;
BeanDefinition getBeanDefinition(String beanName) throws NoSuchBeanDefinitionException;
```

另外，请读者回顾一下 ImportBeanDefinitionRegistrar 的核心 registerBeanDefinitions 方法，它的入参就有一个 BeanDefinitionRegistry，我们取得 BeanDefinitionRegistry 后，就可以在方法体内自行构造 BeanDefinition 后注册到 BeanDefinitionRegistry。由此又说明了一点：BeanDefinitionRegistry **在底层会支撑其他组件运行**。

```
void registerBeanDefinitions(AnnotationMetadata metadata, BeanDefinitionRegistry registry);
```

最后我们要留意一点，BeanDefinitionRegistry 的主要实现是 DefaultListableBeanFactory，它也是 BeanFactory 的最核心落地实现，所以我们要意识到一点：Default ListableBeanFactory **不仅是 bean 对象的统一管理容器，而且是** BeanDefinition **的统一管理容器**。

> 提示：对于 `BeanDefinitionRegistry`，读者目前无须掌握太多有关它的具体使用，只需了解它是什么就够了，随着后面深入到 Spring Boot 的生命周期部分，我们还会频繁地与它打交道。

### 3.5.6 设计 BeanDefinition 的意义

最后我们讨论一个小问题：Spring Framework 为什么会设计 `BeanDefinition`？为什么没有选择直接注册 Bean 的方式？

的确，如果直接创建出 bean 对象，放入 IOC 容器，这样的设计简单，但同时对应了另一个问题：Bean 的管理机制过于简单，无法应对各种复杂的场景，而且无法针对某些特殊的 Bean 进行附加的处理（如 AOP 代理、事务增强支持等）。设计 `BeanDefinition`，其本质上比较接近面向对象开发中先编写 Class 类，再 new 出对象。Spring Framework 面对一个应用程序，也需要对其中的 Bean 进行定义抽取，只有抽取成可以统一类型/格式的模型，才能在后续的 bean 对象管理时进行统一管理，或者是对特定的 Bean 进行特殊化处理。而这一切最终落地到统一类型上就是 `BeanDefinition` 这个抽象化的模型。换一种更简单的说法，有了定义信息，按照既定的规则，就可以任意解析生成 bean 对象，也可以根据实际需求对解析和生成对象的过程任意扩展。

## 3.6 后置处理器

介绍完 `BeanDefinition` 后，接下来是本章最后一个非常重要的知识点：**后置处理器**。要学习后置处理器，首先要对这个概念有一个基础的认识。

### 3.6.1 理解后置处理器

Spring Framework 中设计的后置处理器主要分为两种类型：针对 bean 对象的后置处理器 `BeanPostProcessor`；针对 `BeanDefinition` 的后置处理器 `BeanFactoryPostProcessor`。这两种都是主要针对 IOC 容器中的 Bean，在其生命周期中进行一些切入处理和干预。`BeanPostProcessor` 切入的时机是在 bean 对象的初始化阶段前后添加自定义处理逻辑。而 `BeanFactoryPostProcessor` 切入的时机是在 IOC 容器的生命周期中，所有 `BeanDefinition` 都注册到 `BeanDefinitionRegistry` 后切入回调，它的主要工作是访问/修改已经存在的 `BeanDefinition`。

另外，`BeanPostProcessor` 与 `BeanFactoryPostProcessor` 都有一些扩展的子接口，它们切入的时机也不尽相同，图 3-9 从整体上描述了 IOC 容器与 bean 对象的初始化过程中后置处理器的切入扩展时机。

从图 3-9 中可以看出，后置处理器在整个 IOC 容器以及 bean 对象的生命周期中有非常多可以切入的部分，尽管具体的后置处理器名称对于读者可能还很陌生，但仅从切入点上看，我们可以直观地感受到一点：IOC 容器的可扩展性非常强，而扩展的手段中很大一部分就是借助后置处理器。

# 第 3 章 Spring Boot 的 IOC 容器

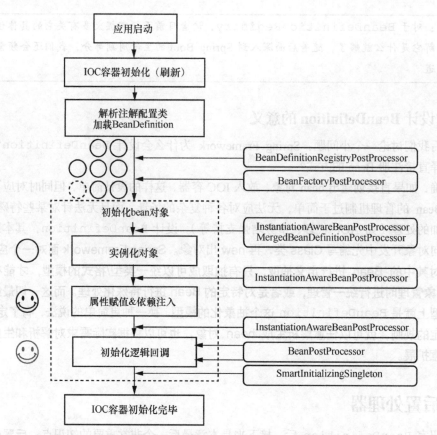

图 3-9 后置处理器在 IOC 容器与 bean 对象生命周期的切入时机

## 3.6.2 BeanPostProcessor

BeanPostProcessor 是针对 bean 对象的后置处理器,它本身的设计是一个包含两个方法的接口,如代码清单 3-21 所示。

**代码清单 3-21 BeanPostProcessor**

```java
public interface BeanPostProcessor {
    default Object postProcessBeforeInitialization(Object bean, String beanName)
            throws BeansException {
        return bean;
    }
    default Object postProcessAfterInitialization(Object bean, String beanName)
            throws BeansException {
        return bean;
    }
}
```

> 提示:Spring Framework 5.x 版本之前由于最低支持 Java 6,此处并没有默认的方法实现。

这两个方法的文档注释写得非常完善。postProcessBeforeInitialization 方法会在任何 bean 对象的初始化回调逻辑（例如 InitializingBean 的 afterPropertiesSet 或自定义 init-method）之前执行，而 postProcessAfterInitialization 方法会在任何 bean 对象的初始化回调逻辑之后执行。整个 bean 对象的生命周期中 BeanPostProcessor 的切入时机如图 3-10 所示。

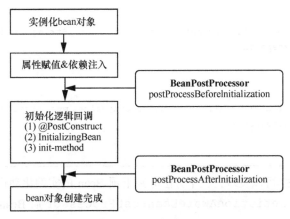

图 3-10　BeanPostProcessor 的切入时机

此外，对于 postProcessAfterInitialization 方法，还可以对那些 FactoryBean 创建出的真实对象进行后置处理。

### 3.6.3　BeanPostProcessor 的扩展

借助 IDEA，可以发现 BeanPostProcessor 只是一个根接口，它的下面还有几个派生的子接口，具体的派生关系如图 3-11 所示。

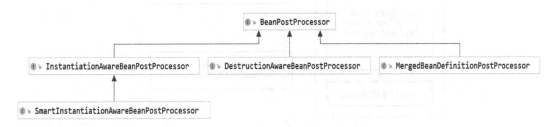

图 3-11　BeanPostProcessor 的扩展接口

下面我们逐一介绍其中重要的派生接口。

#### 1. InstantiationAwareBeanPostProcessor

InstantiationAwareBeanPostProcessor 的类名中带有一个 instantiation，这意味着 InstantiationAwareBeanPostProcessor 会干预对象的实例化阶段。参照 javadoc 的描述，我们可以大体了解到，InstantiationAwareBeanPostProcessor 会拦截并替换 bean 对象的默认实例化动作，也会拦截 bean 对象的属性注入和自动装配，并在此控制流程。从接口的方法设计上看，它在 BeanPostProcessor 的基础上扩展了 3 个新的方法，如代码清单 3-22 所示。

### 代码清单 3-22　InstantiationAwareBeanPostProcessor

```
default Object postProcessBeforeInstantiation(Class<?> beanClass, String beanName)
        throws BeansException {
    return null;
}

default boolean postProcessAfterInstantiation(Object bean, String beanName)
        throws BeansException {
    return true;
}

default PropertyValues postProcessProperties(PropertyValues pvs, Object bean, String beanName)
        throws BeansException {
    return null;
}
```

通过代码清单 3-22 可以发现，三个方法分别干预 Bean 的实例化动作前后以及切入属性赋值的动作。加入 InstantiationAwareBeanPostProcessor 后 Bean 的初始化阶段就变为图 3-12 所示的流程。

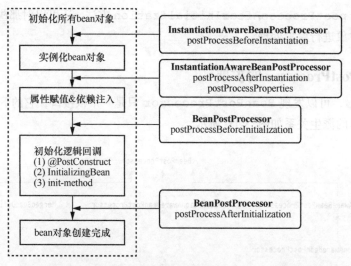

图 3-12　InstantiationAwareBeanPostProcessor 的切入时机

对于这些切入的方法具体会做什么，3.7 节会先简单提一下，详细的内容会统一放在第二部分生命周期中展开讲解。

#### 2. DestructionAwareBeanPostProcessor

与初始化相反，DestructionAwareBeanPostProcessor 切入的时机是 bean 对象的销毁阶段。当 IOC 容器关闭时，会先销毁容器中的所有单实例 Bean，而销毁的过程中，除了回调 bean 对象本身定义的 @PreDestroy 注解标注的方法、destory-method 等方法，还会回调所有 DestructionAwareBeanPostProcessor 类型的后置处理器来干预。简言之，DestructionAwareBeanPostProcessor 会在 ApplicationContext 的 close 方法调用时执行。

Spring Framework 中有一个 `DestructionAwareBeanPostProcessor` 的内置实现：监听器的引用释放回调。由于 `ApplicationContext` 中会注册一些 `ApplicationListener`，而这些 `ApplicationListener` 与 `ApplicationContext` 互相依赖（引用），因此在 IOC 容器销毁之前，就需要将这些引用断开，这样才可以进行对象的销毁和回收。

> 提示：这个实现相对来讲不是很重要，读者了解一下即可，不必耗费太多精力。

### 3. MergedBeanDefinitionPostProcessor

`MergedBeanDefinitionPostProcessor` 的类名中带有 "合并" 的概念，这里要解释一下 `BeanDefinition` 的合并。如果我们向 IOC 容器中注册的 Bean 是一个有父类的派生类（子类），那么 Spring Framework 在收集 Bean 中的信息时，不仅要收集当前类，还应该收集它的父类，而负责父类收集的工作，就交给 `MergedBeanDefinitionPostProcessor` 来完成。

`MergedBeanDefinitionPostProcessor` 接口中只额外扩展了一个方法，用于在 bean 对象的生命周期阶段中属性赋值和自动注入之前提前收集好 bean 对象需要注入的属性。只有提前收集好要注入的属性，IOC 容器底层才会把需要注入的属性都处理好。

```
void postProcessMergedBeanDefinition(RootBeanDefinition beanDefinition, Class<?> beanType,
        String beanName);
```

> 提示：注意 `postProcessMergedBeanDefinition` 方法只传入了 `BeanDefinition`，没有传入实例化后的空 bean 对象，这体现了迪米特法则（最少知识原则），即处理 `BeanDefinition` 的时候不应该知道 bean 对象的存在。

Spring Framework 中有一个非常重要的 `MergedBeanDefinitionPostProcessor` 的实现，它就是 `AutowiredAnnotationBeanPostProcessor`，它负责给 bean 对象实现基于注解的自动注入（`@Autowired`、`@Resource`、`@Inject` 等），而注入的依据是 `postProcessMergedBeanDefinition` 方法执行后整理的一组标记，后续的注入工作会根据这组打好的标记为 bean 对象依次注入属性值/bean 对象。

## 3.6.4 BeanFactoryPostProcessor

下面我们再来了解 `BeanFactoryPostProcessor`，它是针对 `BeanDefinition` 的后置处理器，当然也可以理解为针对 `BeanFactory` 的后置处理器，只不过在 `BeanFactoryPostProcessor` 的回调阶段前后，都是在围绕着 `BeanDefinition` 做文章。

官方文档中有对 `BeanFactoryPostProcessor` 的描述，其中解释了 `BeanFactoryPostProcessor` 操作的是 Bean 的配置元信息（即 `BeanDefinition`）。而且这里还有一个关键的点：`BeanFactoryPostProcessor` 可以在 bean 对象的初始化之前修改 Bean 的定义信息。换句话说，它可以对原有的 `BeanDefinition` 进行修改。由于 Spring Framework 中设计的**所有 bean 对象在没有实例化之前都是以 `BeanDefinition` 的形式存在的**，如果提前修改了 `BeanDefinition`，那么在 Bean 的实例化时，最终创建出的 bean 对象就会受到影响。

`BeanFactoryPostProcessor` 中只定义了一个方法，就是对 `BeanFactory` 的后置处

理,它会在标准初始化之后修改 ApplicationContext 内部的 BeanFactory。在 BeanFactoryPostProcessor 触发时,所有 BeanDefinition 都已经被加载,但此时还没有实例化任何 bean 对象,在这个阶段中 BeanFactoryPostProcessor 可以给 Bean 覆盖或添加属性,甚至可以用于初始化 bean 对象(当然这种做法是不推荐的)。

```
void postProcessBeanFactory(ConfigurableListableBeanFactory beanFactory) throws BeansException;
```

> 提示:注意这里的设计,即便 ConfigurableListableBeanFactory 的最终实现类只有 DefaultListableBeanFactory,这里的入参也是接口,可见**依赖倒转**的设计在 Spring Framework 中体现得淋漓尽致。

如果用 ApplicationContext 的生命周期来体现 BeanFactoryPostProcessor 的切入时机,它出现在"解析注解配置类"与"初始化 bean 对象"之间,如图 3-13 所示。

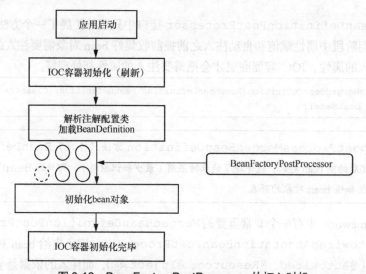

图 3-13  BeanFactoryPostProcessor 的切入时机

图 3-13 中的实线圆代表已经注册到 BeanFactory 中的 BeanDefinition,经过 BeanFactoryPostProcessor 的干预后,可能部分 BeanDefinition 会被改变(即变为虚线圆),但 BeanDefinition 的数量不会变化。

### 3.6.5 BeanDefinitionRegistryPostProcessor

如果需要在 BeanFactory 的后置处理阶段动态注册新的 BeanDefinition,除了用第 2 章我们讲到的 ImportBeanDefinitionRegistrar,另一种方案就是使用 BeanDefinitionRegistryPostProcessor。从类名上可以看出,它是针对 BeanDefinitionRegistry 的后置处理器,它的执行时机比 BeanFactoryPostProcessor 早,这就意味着 BeanDefinitionRegistryPostProcessor 允许在 BeanFactoryPostProcessor 之前**注册新的** BeanDefinition。从设计上讲,BeanFactoryPostProcessor 只用来修改、扩展 BeanDefinition 中的信息,而 BeanDefinitionRegistryPostProcessor 则可以在 BeanFactoryPostProcessor 处理 BeanDefinition 之前向 BeanFactory 注册新的

BeanDefinition，甚至注册新的 BeanFactoryPostProcessor 用于下一个阶段的回调。

BeanDefinitionRegistryPostProcessor 在 IOC 容器生命周期中应当出现在"解析注解配置类"之后、BeanFactoryPostProcessor 的集中回调之前，如图 3-14 所示。图 3-14 中 BeanDefinitionRegistryPostProcessor 执行之前，IOC 容器中只有两个 BeanDefinition，但经过 BeanDefinitionRegistryPostProcessor 的处理之后，BeanDefinition 的数量变为三个，这就充分体现了 BeanDefinitionRegistryPostProcessor 有增删 BeanDefinition 的能力。

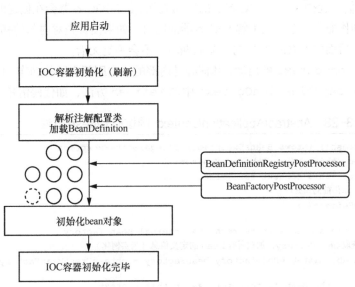

图 3-14 BeanDefinitionRegistryPostProcessor 的切入时机

> 提示：补充一点，由于实现了 BeanDefinitionRegistryPostProcessor 的类同时也实现了 BeanFactoryPostProcessor 的 postProcessBeanFactory 方法，因此在执行完所有 BeanDefinitionRegistryPostProcessor 的接口方法后，会立即执行这些类的 postProcessBeanFactory 方法，之后再执行那些普通的只实现了 BeanFactoryPostProcessor 的 postProcessBeanFactory 方法。

### 3.6.6 后置处理器对比

最后我们简单对比一下上述提到的三种核心后置处理器，总结内容如表 3-2 所示。

表 3-2　　Spring Framework 中三种后置处理器的对比

| | BeanPostProcessor | BeanFactoryPostProcessor | BeanDefinitionRegistryPostProcessor |
|---|---|---|---|
| 处理目标 | bean 对象 | BeanDefinition | BeanDefinition、.class 文件等 |
| 执行时机 | bean 对象的初始化阶段前后（已创建 bean 对象） | BeanDefinition 解析完毕并注册进 BeanFactory 之后（此时 bean 对象未实例化） | 配置文件、配置类已解析完毕并注册进 BeanFactory，但还没有被 BeanFactoryPostProcessor 处理 |

| BeanPostProcessor | BeanFactoryPostProcessor | BeanDefinitionRegistryPostProcessor |
|---|---|---|
| 可操作的空间 | 给 bean 对象的属性赋值、创建代理对象等 | 在 BeanDefinition 中增删属性、移除 BeanDefinition 等 | 向 BeanFactory 中注册新的 BeanDefinition |

## 3.7 IOC 容器的启动流程

在本节，我们一起探讨一下 IOC 容器的启动流程。由于 IOC 容器会在创建时顺便初始化好，这个初始化的动作非常复杂，在了解 IOC 容器的同时，如果读者先对内部的初始化机制有一个整体的认识，到后面学习 IOC 容器的生命周期时就不会手足无措。

ApplicationContext 的启动，其核心是内部的一个刷新容器的动作，也就是 3.1.2 节中提到的 AbstractApplicationContext 中的 refresh 方法，如代码清单 3-23 所示。

**代码清单 3-23　AbstractApplicationContext 的核心 refresh 方法**

```
public void refresh() throws BeansException, IllegalStateException {
    synchronized (this.startupShutdownMonitor) {
        // Prepare this context for refreshing.
        // 1. 初始化前的预处理
        prepareRefresh();

        // Tell the subclass to refresh the internal bean factory.
        // 2. 获取 BeanFactory，加载所有 bean 的定义信息（未实例化）
        ConfigurableListableBeanFactory beanFactory = obtainFreshBeanFactory();

        // Prepare the bean factory for use in this context.
        // 3. BeanFactory 的预处理配置
        prepareBeanFactory(beanFactory);

        try{
            // Allows post-processing of the bean factory in context subclasses.
            // 4. 准备 BeanFactory 完成后进行的后置处理
            postProcessBeanFactory(beanFactory);

            // Invoke factory processors registered as beans in the context.
            // 5. 执行 BeanFactory 创建后的后置处理器
            invokeBeanFactoryPostProcessors(beanFactory);

            // Register bean processors that intercept bean creation.
            // 6. 注册 Bean 的后置处理器
            registerBeanPostProcessors(beanFactory);

            // Initialize message source for this context.
            // 7. 初始化 MessageSource
            initMessageSource();

            // Initialize event multicaster for this context.
            // 8. 初始化事件广播器
            initApplicationEventMulticaster();

            // Initialize other special beans in specific context subclasses.
            // 9. 子类的多态 onRefresh
```

```
            onRefresh();

            // Check for listener beans and register them.
            // 10. 注册监听器
            registerListeners();

            //至此，BeanFactory 创建完成

            // Instantiate all remaining (non-lazy-init) singletons.
            // 11. 初始化所有剩下的单实例
            BeanfinishBeanFactoryInitialization(beanFactory);

            // Last step: publish corresponding event.
            // 12. 完成容器的创建工作
            finishRefresh();
        }// catch ......

        finally {
            // Reset common introspection caches in Spring's core, since we
            // might not ever need metadata for singleton beans anymore...
            // 13. 清除缓存
            resetCommonCaches();
        }
    }
}
```

整个 refresh 方法分为 13 步，本章我们只关心每一步都做了哪些事情，对于其中涉及的细节，我们统一放到本书第 2 部分详细展开。

1. prepareRefresh——初始化前的预处理
   - 这一步大多数的动作都是前置性准备，包含切换 IOC 容器启动状态、初始化属性配置、属性校验、早期事件容器准备等动作。

2. obtainFreshBeanFactory——初始化 BeanFactory
   - 该步骤在不同的 ApplicationContext 落地实现中行为不同：基于注解配置类的 ApplicationContext，在基础实现类 GenericApplicationContext 的 refreshBeanFactory 方法中有一个 CAS 判断的动作，它控制 GenericApplicationContext 不能反复刷新；而基于 XML 配置文件的 ApplicationContext 中，在该步骤会解析 XML 配置文件，封装 BeanDefinition 注册到 BeanDefinitionRegistry 中。
   - 由此可以得出一个结论：**基于 XML 配置文件的 ApplicationContext 可以反复刷新，基于注解配置类的 ApplicationContext 只能刷新一次**。

3. prepareBeanFactory——BeanFactory 的预处理动作
   - 这个方法内部处理的内容看似比较多，但总体非常有条理，主要包含三件事情：
     - 设置一些默认组件（类加载器、表达式解析器等），并注册 Environment 抽象（Environment 也是 IOC 容器中的一个 Bean）；
     - 编程式注册 ApplicationContextAwareProcessor，它负责支持 6 个 Aware 回调注入接口（包含 EnvironmentAware、ApplicationContextAware 在内的 6 个 Aware 系列接口）；
     - 绑定 BeanFactory 与 ApplicationContext 的依赖注入映射，当其他 Bean

需要注入 `BeanFactory` 时，IOC 容器会自动将当前正在处理的 `BeanFactory` 注入，当其他 Bean 需要注入 `ApplicationContext` 时，IOC 容器会自动将自身注入。
- 总的来看，这三件事情都是在为 `BeanFactory` 的初始化做一些准备，即预处理工作。

4. `postProcessBeanFactory`——`BeanFactory` 的后置处理
   - 这个方法本身是一个模板方法，在 `AbstractApplicationContext` 中没有具体实现，在基于 Web 环境的 `ApplicationContext` 实现 `GenericWebApplicationContext` 中有扩展行为：
     - 编程式注册 `ServletContextAwareProcessor`，用于支持 `ServletContext` 的回调注入；
     - 注册新的 Bean 的作用域（request、session、application），并关联绑定 `ServletRequest`、`ServletResponse` 等多个类型的依赖注入映射；
     - 注册 `ServletContext`、`ServletConfig` 对象到 IOC 容器中。
   - 在 Spring Boot 的支持嵌入式容器的 `ServletWebServer ApplicationContext` 实现基类中它覆盖了原有的实现，主要变化的是没有注册 application 作用域（因为此时嵌入式 Web 容器还没有初始化，没有 `ServletContext` 可获取）。

5. `invokeBeanFactoryPostProcessors`——执行 `BeanFactoryPostProcessor`
   - 该步骤会回调执行所有的 `BeanDefinitionRegistryPostProcessor` 与 `BeanFactoryPostProcessor`。
     - `BeanDefinitionRegistryPostProcessor` 的主要工作是，对 `BeanDefinitionRegistry` 中存放的 `BeanDefinition` 进行处理（可以注册新的 `BeanDefinition`，也可以移除现有的 `BeanDefinition`），其中一个典型的实现是注解配置类的解析与组件导入。
     - `BeanFactoryPostProcessor` 的主要工作是对 `BeanDefinitionRegistry` 中现有的 `BeanDefinition` 进行修改操作（注意只限于修改，原则上不允许再注册新的/移除现有的 `BeanDefinition`），它的执行时机较 `BeanDefinition Registry PostProcessor` 更晚。
   - 回调这两种后置处理器涉及的核心工作是组件扫描、注解配置类解析，在第 2 部分中会详细展开。

6. `registerBeanPostProcessors`——初始化 `BeanPostProcessor`
   - 该步骤会初始化所有注册的 `BeanPostProcessor`。
   - `BeanPostProcessor` 是针对 Bean 的生命周期流程中的重要扩展点，它可以干预 bean 对象的实例化、初始化等过程，一个典型的体现是给初始化好的对象织入 AOP 增强逻辑。
   - 当前步骤初始化 `BeanPostProcessor` 时，容器中还没有初始化任何业务相关的 bean 对象，所以后续初始化的所有 bean 对象都会经过 `BeanPostProcessor` 的干预。

7. `initMessageSource`——初始化国际化组件
   - 该步骤会初始化默认的国际化组件 `MessageSource`，默认的实现类 `DelegatingMessageSource` 在不作任何附加配置的情况下不会处理任何国际化的工作，只有手动向 IOC 容器注册具体的国际化组件，应用上下文才具备国际化的能力。

8. `initApplicationEventMulticaster`——初始化事件广播器
   - 该步骤默认会初始化一个 `ApplicationEventMulticaster` 的简单实现 `SimpleApplicationEventMulticaster` 并注册到 IOC 容器中。
   - `ApplicationContext` 本身具备的事件广播能力是依赖 `ApplicationEventMulticaster` 实现的（功能组合的体现）。

9. `onRefresh`——子类扩展的刷新动作
   - 该方法也是一个模板方法，默认 Spring Framework 范围的 `ApplicationContext` 都没有扩展它。
   - Spring Boot 中支持嵌入式 Web 容器的 `ApplicationContext` 在此处有扩展，它用于初始化嵌入式 Web 容器。

10. `registerListeners`——注册监听器
    - 该步骤会将 `BeanDefinitionRegistry` 中注册的所有监听器（`ApplicationListener`）的 `beanName` 取出，绑定到事件广播器 `ApplicationEventMulticaster` 中。
    - 只绑定 `beanName` 而不直接取出监听器对象的原因是，考虑到 `ApplicationListener` 作为 IOC 容器中的 Bean，应该放在一起统一创建和初始化（也就是下面的 `finishBeanFactoryInitialization` 动作，目的是希望 `BeanPostProcessor` 有机会去干预它们）。

11. `finishBeanFactoryInitialization`——初始化剩余的单实例 bean 对象
    - 该步骤主要完成两件事：
      - 初始化用于类型转换和表达式的解析器（`ConversionService` 与 `EmbeddedValueResolver`）；
      - 初始化所有非延迟加载的单实例 bean 对象。
    - 该步骤执行完毕后，`BeanFactory` 的初始化工作结束。
    - 该步骤包含一个 bean 对象的主要生命周期过程，内容很复杂，在本书第 2 部分会详细展开。

12. `finishRefresh`——刷新后的动作
    - 该步骤的工作相对比较零散，包含以下几个小动作：
      - 清除资源缓存；
      - 初始化生命周期处理器；
      - 传播生命周期动作，回调所有 `Lifecycle` 类型 Bean 的 `start` 方法；
      - 广播 `ContextRefreshedEvent` 事件。

13. `resetCommonCaches`——清除缓存
    - 最后的步骤会清除一切无用的缓存，因为 IOC 容器的刷新工作已经完成，缓存也就没有用了。

纵观整个 `refresh` 方法，每个动作的职责都很清晰，而且非常有条理。这个过程中，有对 `BeanFactory` 的处理，有对 `ApplicationContext` 的处理，有处理

BeanPostProcessor 的逻辑，有准备 ApplicationListener 的逻辑，最后它会初始化那些非延迟加载的单实例 bean 对象。refresh 方法执行完毕后，也就宣告 ApplicationContext 初始化完成。

## 3.8 小结

本章主要讲解了 Spring Boot 的 IOC 容器 BeanFactory 与 ApplicationContext 的层次化设计，以及 Spring Boot 对 IOC 容器模型的扩展。随后我们讲解了 Spring Framework 内部几个非常关键的设计——Environment、BeanDefinition、后置处理器。最后我们整体了解了 IOC 容器的启动流程。掌握整体的设计，对理解 Spring Boot 的整体流程把握非常重要，读者一定要仔细学习和探究。

# 第 4 章 Spring Boot 的核心引导：SpringApplication

本章主要内容：
◇ 理解 SpringApplication 的设计；
◇ SpringApplication 的启动阶段引导流程。

经过对第 2、3 章 Spring Framework 高级特性的回顾和温习，本章会深入 Spring Boot 的核心源码，研究具体的底层实现。Spring Boot 的主启动类中使用 SpringApplication 类来引导 Spring Boot 项目启动，这个类本身的设计就非常重要。本章内容会着重研究 SpringApplication 的作用，以及 Spring Boot 的主线应用生命周期。

## 4.1 总体设计

要想了解 SpringApplication 的设计，最好的两个切入点是 javadoc 和官方文档。本节内容将以 Spring Boot 的官方文档为主，从整体上剖析 SpringApplication 的设计。

参考 Spring Boot 2.5.x 版本的官方文档的 features 部分，第 1 节介绍的就是 SpringApplication。

> The SpringApplication class provides a convenient way to bootstrap a Spring application that is started from a main() method. In many situations, you can delegate to the static SpringApplication.run method.
>
> SpringApplication 类提供了一个非常方便的方式，利用 main 方法来引导 Spring Boot 应用程序。大多数情况下，启动 Spring Boot 应用程序只需要委托给静态的 SpringApplication.run 方法。

官方文档只是用两句话概述了 SpringApplication 的整体作用：SpringApplication 的核心是简化 Spring Boot 应用程序的启动。后面的各节中，它针对 SpringApplication 背后的非常多的特性进行比较详细的解释，本书中挑选了其中一些比较重要的特性来研究。

### 4.1.1 启动失败的错误报告

使用 Spring Boot 开发过项目的读者一定不陌生，当我们编写的 Spring Boot 应用程序启动失败时，控制台会打印关于本次启动失败的信息（诸如 Bean 不存在、端口被占用、初始化逻辑执行报错等）。下面是一个 8080 端口被占用时 Spring Boot 打印的错误报告：

```
****************************
APPLICATION FAILED TO START
****************************

Description:

Embedded servlet container failed to start. Port 8080 was already in use.

Action:

Identify and stop the process that's listening on port 8080 or configure this application to
listen on another port.
```

负责输出错误报告的是一个名为 FailureAnalyzers 的组件，这个组件实现了 SpringBootExceptionReporter，顾名思义，它就是负责异常报告输出的。注意这个类名是一个复数的概念，它的内部集成了一组 FailureAnalyzer，如代码清单 4-1 所示。

**代码清单 4-1　FailureAnalyzers 的内部集成了一组 FailureAnalyzer**

```java
final class FailureAnalyzers implements SpringBootExceptionReporter {

    private final List<FailureAnalyzer> analyzers;

    FailureAnalyzers(ConfigurableApplicationContext context, ClassLoader classLoader) {
        // ...
        this.analyzers = loadFailureAnalyzers(this.classLoader);
        prepareFailureAnalyzers(this.analyzers, context);
    }

    private List<FailureAnalyzer> loadFailureAnalyzers(ClassLoader classLoader) {
        List<String> analyzerNames = SpringFactoriesLoader
                .loadFactoryNames(FailureAnalyzer.class, classLoader);
        // ...
    }
```

由代码清单 4-1 可知，FailureAnalyzers 的集成方式是利用 Spring Framework SPI 的方式，从 spring.factories 中读取所有 FailureAnalyzer，而在 spring-boot 依赖的 spring.factories 文件中就已经有配置了，如代码清单 4-2 所示（为保证阅读体验，下面的全限定类名已作省略处理）。

**代码清单 4-2　Spring Boot 中默认已有的 FailureAnalyzer 配置**

```
# Failure Analyzers
org.springframework.boot.diagnostics.FailureAnalyzer=\
    o.s.b.c.p.NotConstructorBoundInjectionFailureAnalyzer,\
    o.s.b.d.a.BeanCurrentlyInCreationFailureAnalyzer,\
    o.s.b.d.a.BeanDefinitionOverrideFailureAnalyzer,\
    o.s.b.d.a.BeanNotOfRequiredTypeFailureAnalyzer,\
    o.s.b.d.a.BindFailureAnalyzer,\
    o.s.b.d.a.BindValidationFailureAnalyzer,\
    o.s.b.d.a.UnboundConfigurationPropertyFailureAnalyzer,\
    o.s.b.d.a.ConnectorStartFailureAnalyzer,\
    o.s.b.d.a.NoSuchMethodFailureAnalyzer,\
```

```
o.s.b.d.a.NoUniqueBeanDefinitionFailureAnalyzer,
o.s.b.d.a.PortInUseFailureAnalyzer,
o.s.b.d.a.ValidationExceptionFailureAnalyzer,
o.s.b.d.a.InvalidConfigurationPropertyNameFailureAnalyzer,
o.s.b.d.a.InvalidConfigurationPropertyValueFailureAnalyzer
```

可以发现,在默认情况下 Spring Boot 已经检查了一些常见的异常类型。如果在实际开发中发现默认检查的类型不够完善,可以自行编写 FailureAnalyzer 的实现类,自行解析异常和输出。

> 提示:即便不扩展任何 FailureAnalyzer,依然可以通过调整日志输出级别为 DEBUG,来跟踪错误。

### 4.1.2　Bean 的延迟初始化

默认情况下,如果没有特殊声明,ApplicationContext 中的 Bean 需要延迟初始化时都会在应用程序启动阶段统一预先创建好。启用 Bean 的延迟初始化,会使得 IOC 容器中绝大部分的 Bean 都变为需要时才创建(只有 SmartInitializingSingleton 类型的 Bean 依然会提早初始化),这样配置可以显著减少应用程序的启动耗时。但延迟初始化伴随的问题是,一些本应该可以在应用程序启动阶段发现的问题,由于关键组件 bean 对象的初始化延迟到应用程序运行阶段而导致真正要用它的时候才发现组件不可用。简言之,延迟初始化会导致发现问题的时机延后。

启用 Bean 的延迟初始化很简单,只需在全局配置文件中添加一行配置:

```
spring.main.lazy-initialization=true
```

> 提示:如果在全局开启 Bean 的延迟初始化的同时,又需要某些 bean 对象在应用程序启动阶段初始化就绪,可以在类上标注@Lazy(false)注解。

### 4.1.3　SpringApplication 的定制

通常在编写 Spring Boot 应用程序主启动类时,会直接使用 SpringApplication 的静态 run 方法用一行代码实现,但 SpringApplication 本身的设计很灵活,可以利用与它相关的 API 来定制化启动 Spring Boot 应用程序。Spring Boot 为我们提供了两种定制化启动方式,一种是直接创建 SpringApplication 对象,调用其 API 进行定制,另一种方式是借助 SpringApplicationBuilder 实现链式定制,如代码清单 4-3 所示。

**代码清单 4-3　定制化启动 Spring Boot 应用程序的两种方式**

```java
public static void main(String[] args) {
    // 方式1:直接操作 SpringApplication 的 API
    SpringApplication springApplication = new SpringApplication();
    springApplication.setMainApplicationClass(SpringApplicationApplication.class);
    springApplication.setBannerMode(Banner.Mode.OFF); // 关闭 Banner 打印
    springApplication.run(args);

    // 方式2:借助构建器
    new SpringApplicationBuilder(SpringApplicationApplication.class)
```

```
        .bannerMode(Banner.Mode.OFF)
        .web(WebApplicationType.NONE)  // 指定 Web 应用类型为非 Web
        .run(args);
}
```

两种操作方式基本类似，不过请留意一点，利用 `SpringApplicationBuilder` 可以构建具有层次关系的 Spring Boot 应用程序：

```
new SpringApplicationBuilder(...).parent(...).child(...).run(...);
```

这种构建完成后的效果是，应用程序的内部会形成多个 `ApplicationContext` 且彼此之间是父子关系。官方文档对此也作了说明：构建 `ApplicationContext` 层次结构时有一些限制，例如 Web 组件必须包含在子上下文中，并且 `Environment` 对于父上下文和子上下文都相同。

### 4.1.4　Web 类型推断

Spring Boot 的一大优势是整合不同场景时的简单易操作。当项目中需要整合 SpringWebMvc 时，只需引入 `spring-boot-starter-web` 依赖，整合 SpringWebFlux 时，只需引入 `spring-boot-starter-webflux` 依赖，Spring Boot 会在底层推断应用应该用哪种 IOC 容器实现来支撑整个应用的运转。Spring Boot 2.x 基于 SpringFramework 5.x，Web 场景的解决方案有 WebMvc 和 WebFlux 两种，所以在 Spring Boot 中对于 Web 应用的类型也相应分为 Servlet(WebMvc)、Reactive(WebFlux)和 None 三种。

根据官方文档的描述，可以得知 Web 类型的推断规则如下：
- 如果 SpringWebMvc 存在，则启用 **Servlet** 环境；
- 如果 SpringWebMvc 不存在并且 SpringWebFlux 存在，则启用 **Reactive** 环境；
- 如果两者都不存在，则启用最原始的没有任何 Web 概念的 **None** 环境。

注意推断规则中有一个细节：当 SpringWebMvc 与 SpringWebFlux 同时出现在项目中时，WebFlux 默认失效（除非我们手动定制 SpringApplication，指定 `WebApplicationType` 为 Reactive）。

### 4.1.5　监听与回调

原生的 Spring Framework 中已经构建了一套强大且完善的事件监听机制，开发者可以在基于 Spring Framework 的应用程序中任意编写事件，配合广播器与监听器，可以实现自定义的事件监听，而且 Spring Framework 已经预先构建了一些基于 `ApplicationContext` 的事件（如 `ContextRefreshedEvent` 等）。

Spring Boot 中将事件监听机制进一步扩展。由于 Spring Boot 的应用启动方式是通过 `SpringApplication` 这个核心 API，在这个 API 的调用期间也会有属于它自己的生命周期，Spring Boot 在该部分生命周期中又扩展了新的可切入扩展点，也就是基于 `SpringApplication` 的事件监听。

Spring Boot 中新引入的核心事件监听类是 `SpringApplicationRunListener`，它是专门用于广播 Spring Boot 事件的监听器。

**1. SpringApplicationRunListener**

官方文档中并没有给 `SpringApplicationRunListener` 留出篇幅讲解，javadoc 中也

只是简单地用一句话概括：`SpringApplicationRunListener` 是监听 Spring Application 在 run 方法内的监听器。这句话本身不太好理解，按照常理来讲一个监听器通常会对应一个或多个事件，此处监听一个方法的确有些反常。其实 `SpringApplicationRunListener` 的确是只监听 run 方法的，只不过这个 run 方法实在是太复杂了，而且整个 `SpringApplication` 的 run 方法中涉及特别多的切入点和扩展点，留有一个监听器可以在 `SpringApplication` 中预先定义好的切入点中扩展自定义逻辑。

`SpringApplicationRunListener` 本身是一个接口，它定义了 7 个事件回调的方法，如表 4-1 所示。

表 4-1　`SpringApplicationRunListener` 的接口方法详解

| 接口方法 | 可获得的组件 | 回调时机 |
| --- | --- | --- |
| starting | | 调用 `SpringApplication` 的 run 方法时立即调用 |
| environmentPrepared | `ConfigurableEnvironment` | `Environment` 构建完成，但在创建 `ApplicationContext` 之前 |
| contextPrepared | `ConfigurableApplicationContext` | 在创建和准备 `ApplicationContext` 之后，但在加载之前 |
| contextLoaded | `ConfigurableApplicationContext` | `ApplicationContext` 已加载，但尚未刷新容器时 |
| started | `ConfigurableApplicationContext` | IOC 容器已刷新，但未调用 `CommandLineRunners` 和 `ApplicationRunners` 时 |
| running | `ConfigurableApplicationContext` | 在 run 方法彻底完成之前 |
| failed | `ConfigurableApplicationContext, Throwable` | run 方法执行过程中抛出异常时 |

如果需要自定义 `SpringApplicationRunListener` 的实现类，在注册时不能直接使用 `@Component` 等常规的 Bean 注册方式，而是需要配置到 `spring.factories` 文件中，利用 SPI 机制加载。

请注意，`SpringApplicationRunListener` 的实现类要求必须显式定义一个包含两参数的构造方法（哪怕不编写构造方法的具体逻辑），如代码清单 4-4 所示。

**代码清单 4-4　`SpringApplicationRunListener` 的实现类必须定义特定的构造方法**

```
public class TestRunListener implements SpringApplicationRunListener {
    public TestRunListener(SpringApplication application, String[] args) {}
}
```

### 2. Spring Boot 新引入的事件

与 `SpringApplicationRunListener` 的各个事件方法对应，Spring Boot 给每个方法都定义了一个新的事件模型，用于支撑普通的 `ApplicationListener` 使用。方法与事件模型类对应如下。

- starting——ApplicationStartingEvent
- environmentPrepared——ApplicationEnvironmentPreparedEvent
- contextPrepared——ApplicationContextInitializedEvent

- contextLoaded——ApplicationPreparedEvent
- started——ApplicationStartedEvent
- running——ApplicationReadyEvent
- failed——ApplicationFailedEvent

请注意，部分事件在广播时，通过正常方式注册的监听器无法感知，因为这些事件广播的时候 ApplicationContext 还没有被创建出来，自然也就无法有效地初始化监听器。因此如果需要监听这些事件，需要通过调用 SpringApplication 或者 SpringApplicationBuilder 的 API 来编程式注册监听器，或者是在 spring.factories 文件中配置监听器的实现类，如代码清单 4-5 所示。

**代码清单 4-5　编程式注册监听器**

```
new SpringApplicationBuilder(SpringApplicationApplication.class)
      .listeners(new ApplicationStartingListener()).run(args);

org.springframework.context.ApplicationListener=\
    com.linkedbear.springboot.listener.ApplicationStartingListener
```

### 4.1.6　应用退出

当 Spring Boot 应用启动成功后，内部的 ApplicationContext 会连带注册一个 shutduwnhook 线程（准确的时机是在 IOC 容器刷新之前）。当 JVM 退出时，shutduwnhook 线程可以确保 IOC 容器中的 Bean 被 IOC 容器的销毁阶段生命周期回调（如被 @PreDestroy 注解标注的方法），从而合理地销毁 IOC 容器及其所有的 Bean。

从源码的角度看，它注册的关闭钩子的线程名称是固定的 SpringContextShutdownHook，并且线程中的执行逻辑就是关闭 IOC 容器本身，该方法会销毁 IOC 容器中的所有 Bean，关闭 BeanFactory，如代码清单 4-6 所示。

**代码清单 4-6　注册的关闭钩子，用于关闭 IOC 容器**

```
String SHUTDOWN_HOOK_THREAD_NAME = "SpringContextShutdownHook";

public void registerShutdownHook(){
    if (this.shutdownHook == null) {
        // 创建钩子
        this.shutdownHook = new Thread(SHUTDOWN_HOOK_THREAD_NAME) {
            @Override
            public void run() {
                synchronized (startupShutdownMonitor) {
                    doClose();
                }
            }
        };
        // 注册钩子
        Runtime.getRuntime().addShutdownHook(this.shutdownHook);
    }
}
```

## 4.2 生命周期概述

了解了 SpringApplication 的整体设计之后，下面先从宏观层面了解一个 Spring Boot 的应用从开始执行 main 方法到最终退出的整个生命周期都经历了哪些重要的环节。

本节旨在先了解大体环节，详细的原理和底层源码，我们统一放到本书第 2 部分展开讲解。

### 4.2.1 创建 SpringApplication

当执行 SpringApplication.run 方法时，底层实际上帮我们创建了一个 SpringApplication 对象并调用其 run 方法，如代码清单 4-7 所示。

**代码清单 4-7　SpringApplication#run**

```
public static ConfigurableApplicationContext run(Class<?>[] primarySources, String[] args) {
    return new SpringApplication(primarySources).run(args);
}
```

除了常规地使用 SpringApplication 的静态 run 方法，也可以自行创建 SpringApplication。无论是自行创建，还是借助 SpringApplicationBuilder 的流式 API 创建 SpringApplication 对象，最终得到的都是一个可以调用 run 方法的、完整的 SpringApplication 对象。在创建 SpringApplication 的过程中有以下关键步骤值得注意。

- Web 应用类型判断：Spring Boot 会根据当前应用的类路径确定当前应用最匹配的 Web 类型，这个 Web 类型会影响到实际创建的 ApplicationContext 落地实现类的类型；
- 初始化器&监听器的加载：在启动 SpringApplication 之前，它会先收集一些前期准备好的 ApplicationContextInitializer 以及 ApplicationListener，这些组件会在启动 SpringApplication 的环节中发挥作用（组件的具体作用放在第 6 章讲解）；
- 确定主启动类：Spring Boot 会根据当前应用的启动状态，从方法调用栈中找到主启动类并记录下来，这个主启动类会参与默认 Banner 的打印。

### 4.2.2 启动 SpringApplication

SpringApplication 创建完毕后，下一步是应用启动。整个 SpringApplication 的启动逻辑非常复杂，核心步骤大概分为以下 8 步，读者可以先大体有一个了解，详细的源码分析放到第 6 章深入探究。

1. 获取 SpringApplicationRunListener 监听器，该监听器会贯穿整个 SpringApplication 的启动过程。
2. 准备运行时环境，即 ApplicationContext 中的 Environment。
3. Banner 的加载与打印，默认情况下打印的 Banner 是以文字形式打印到控制台，可以通过定制 SpringApplication 来自定义配置。
4. 创建 ApplicationContext IOC 容器，该步骤创建的依据是创建 SpringApplication 时推断的 Web 应用类型，不同的 Web 应用类型对应不同的 ApplicationContext 落地实现。

5．初始化 IOC 容器，该步骤会应用前面步骤准备的 ApplicationContext Initializer，并获取和加载配置源（默认是主启动类）。

6．刷新 IOC 容器，该步骤会触发 ApplicationContext 的核心 refresh 方法，逻辑极其复杂，该部分内容放到第 7 章讲解。

7．启动嵌入式 Web 容器（如果有的话），当 Web 应用类型不是 None 并且以独立运行的 jar 包运行 Spring Boot 时，底层会额外创建一个嵌入式 Web 容器（默认是 Tomcat）并启动。

8．回调 Spring Boot 的运行器，包括 ApplicationRunner 和 CommandLineRunner。

### 4.2.3 应用退出

当应用被关闭时，Spring Boot 考虑到应用中可能会存在一些需要释放的资源，于是它在 SpringApplication 启动时会额外向 JVM 中注册一个"钩子线程"，这个线程会专门监听应用是否被关闭，以及关闭应用时要执行的释放资源等操作（一个典型的例子是释放数据库连接池中的连接）。源码部分在前面 4.1.6 节已经解释过，这里不再赘述。

## 4.3 小结

本章主要对 Spring Boot 的核心启动引导类 SpringApplication 有一个整体层面的介绍。SpringApplication 中的结构、逻辑本质上都是基于 Spring Framework 中 ApplicationContext 之上的扩展。ApplicationContext 中除了第 3 章讲过的 IOC 特性，另一个核心特性是 AOP，第 5 章会研究 Spring Boot 对 AOP 的支持。

# 第 5 章 Spring Boot 的 AOP 支持

**本章主要内容：**
- Spring Framework 的 AOP；
- 注解驱动 AOP 的核心组件研究。

Spring Framework 的两大核心特性中，除了 IOC，面向切面编程（AOP）也非常重要。AOP 是 OOP 的补充，OOP 的核心是对象，AOP 的核心是切面（Aspect）。AOP 可以在不修改功能代码本身的前提下，使用运行时动态代理技术对已有代码逻辑进行增强。AOP 可以实现组件化、可插拔式的功能扩展，通过简单配置即可将功能增强到指定的切入点。

## 5.1 Spring Framework 的 AOP 回顾

首先回顾 Spring Framework 阶段的 AOP 核心内容，这对于后面研究源码有很大的帮助。

### 5.1.1 AOP 术语

AOP 的基本术语包含以下 8 点。
- **Target**：目标对象，即被代理的对象。
- **Proxy**：代理对象，即经过代理后生成的对象（如 `Proxy.newProxyInstance` 返回的结果）。
- **JoinPoint**：连接点，即目标对象的所属类中定义的所有方法。
- **Pointcut**：切入点，即那些被拦截/被增强的连接点。
  - 切入点与连接点的关系应该是包含关系：切入点可以是 0 个或多个（甚至全部）连接点的组合。
  - 注意，切入点一定是连接点，连接点不一定是切入点。
- **Advice**：通知，即增强的逻辑，也就是增强的代码。
  - Proxy（代理对象） = Target（目标对象） + Advice（通知）
- **Aspect**：切面，即切入点与通知组合之后形成的产物。
  - Aspect（切面） = Pointcut（切入点） + Advice（通知）
  - 实际上切面不仅包含通知，还有一个不常见的部分是引介。
- **Weaving**：织入，这是一个动词，它是将 Advice（通知）应用到 Target（目标对象），进而生成 Proxy（代理对象）的过程。

- Proxy（代理对象） = Target（目标对象）+ Advice（通知）。这个算式中的**加号**，就是织入。
- Introduction：引介，这个概念对标的是 Advice（通知），通知是针对切入点提供增强的逻辑，而引介是针对 Class（类），它可以在不修改原有类的代码的前提下，在运行时为原始类动态添加新的属性/方法。

### 5.1.2 通知类型

Spring Framework 中支持的通知类型包含 5 种，这些通知类型是基于 AspectJ 的。下面逐一列出。

- **Before 前置通知**：目标对象的方法调用之前触发。
- **After 后置通知**：目标对象的方法调用之后触发。
- **AfterReturning 返回通知**：目标对象的方法调用成功，在返回结果值之后触发。
- **AfterThrowing 异常通知**：目标对象的方法在运行中抛出/触发异常后触发。
  - 注意，AfterReturning 与 AfterThrowing 是互斥的。如果方法调用成功无异常，则会有返回值；如果方法抛出了异常，则不会有返回值。
- **Around 环绕通知**：编程式控制目标对象的方法调用。
  - 环绕通知是所有通知类型中可操作范围最大的一种，因为它可以直接获取目标对象以及要执行的方法，所以环绕通知可以任意在目标对象的方法调用前后扩展逻辑，甚至不调用目标对象的方法。

## 5.2 Spring Boot 使用 AOP

下面简单回顾一下 Spring Boot 整合 AOP 场景。Spring Boot 整合 AOP 场景的步骤非常简单，只需在项目依赖中导入 `spring-boot-starter-aop` 依赖，并在主启动类上标注@EnableAspectJAutoProxy 注解，即可开启基于注解驱动的 AOP，如代码清单 5-1 所示。

**代码清单 5-1 导入并开启注解驱动 AOP**

```
<dependencies>
    <dependency>
        <groupId>org.springframework.boot</groupId>
        <artifactId>spring-boot-starter-web</artifactId>
    </dependency>

    <dependency>
        <groupId>org.springframework.boot</groupId>
        <artifactId>spring-boot-starter-aop</artifactId>
    </dependency>
</dependencies>

@EnableAspectJAutoProxy
@SpringBootApplication
public class SpringBootAopApplication {

    public static void main(String[] args) {
```

```
        SpringApplication.run(SpringBootAopApplication.class, args);
    }
}
```

下一步是编写测试的切面类和组件类。本章只是简单回顾基于 AspectJ 的 AOP 场景使用，所以快速编写一个示例，如代码清单 5-2 所示。

**代码清单 5-2　DemoService 与增强它的切面类 DemoAspect**

```
@Service
public class DemoService {

    public void save() {
        System.out.println("DemoService save run ......");
    }
}

@Aspect
@Component
public class ServiceAspect {

    @Before("execution(public * com.linkedbear.springboot.service.*.*(..))")
    public void beforePrint() {
        System.out.println("Service Aspect before advice run ......");
    }
}
```

要测试 AOP 是否生效，可以直接在主启动类处获取 IOC 容器，并提取出 DemoService 对象，调用其 save 方法，如代码清单 5-3 所示。

**代码清单 5-3　通过 SpringApplication 得到 IOC 容器并调用 Service**

```
public static void main(String[] args) {
    ApplicationContext ctx = SpringApplication.run(SpringBootAopApplication.class, args);
    ctx.getBean(DemoService.class).save();
}
```

运行主启动类，控制台可以正确打印两行输出，说明 Spring Boot 整合 AOP 场景顺利完成。

```
Service Aspect before advice run ......
DemoService save run ......
```

## 5.3　AOP 的开关：@EnableAspectJAutoProxy

开启注解驱动 AOP 的核心是 @EnableAspectJAutoProxy 注解，如代码清单 5-4 所示。下面深入剖析该注解的作用。有了模块装配、条件装配的基础，再来分析注解会更容易。

**代码清单 5-4　@EnableAspectJAutoProxy 的核心源码**

```
@Import(AspectJAutoProxyRegistrar.class)
public @interface EnableAspectJAutoProxy {
    boolean proxyTargetClass() default false;
    boolean exposeProxy() default false;
}
```

@EnableAspectJAutoProxy 注解中包含两个属性，分别是 proxyTargetClass 是否直接代理目标类（即强制使用 Cglib 代理），以及 exposeProxy 是否暴露当前线程的 AOP 上下文（开启后，通过 AopContext 可以获取到当前的代理对象本身）。

除了注解属性，@EnableAspectJAutoProxy 最重要的作用是使用 @Import 注解导入了一个 AspectJAutoProxyRegistrar。

## 5.3.1 AspectJAutoProxyRegistrar

从类名上可以简单理解为，它是一个基于 AspectJ 支持的自动代理注册器，那它注册了什么呢？通过 javadoc 我们可以看到端倪：

> *Registers an AnnotationAwareAspectJAutoProxyCreator against the current BeanDefinition Registry as appropriate based on a given @EnableAspectJAutoProxy annotation.*
>
> 基于给定的 @EnableAspectJAutoProxy 注解，根据当前 BeanDefinitionRegistry 在适当的位置注册 AnnotationAwareAspectJAutoProxyCreator。

从 javadoc 中可以看到，AspectJAutoProxyRegistrar 注册的组件是一个 AnnotationAwareAspectJAutoProxyCreator，这个类的作用稍后再展开讨论。首先看一下 AspectJAutoProxyRegistrar 的导入逻辑。

**1. AspectJAutoProxyRegistrar 注册核心代理组件**

AspectJAutoProxyRegistrar 注册核心代理组件的逻辑如代码清单 5-5 所示。

**代码清单 5-5　AspectJAutoProxyRegistrar 注册核心代理组件的逻辑**

```java
class AspectJAutoProxyRegistrar implements ImportBeanDefinitionRegistrar {

    @Override
    public void registerBeanDefinitions(AnnotationMetadata importingClassMetadata,
            BeanDefinitionRegistry registry) {
        // 此处会有注册新 BeanDefinition 的动作
        AopConfigUtils.registerAspectJAnnotationAutoProxyCreatorIfNecessary(registry);

        AnnotationAttributes enableAspectJAutoProxy = AnnotationConfigUtils
                .attributesFor(importingClassMetadata, EnableAspectJAutoProxy.class);
        // 给 AnnotationAwareAspectJAutoProxyCreator 设置属性值
        if (enableAspectJAutoProxy != null) {
            if (enableAspectJAutoProxy.getBoolean("proxyTargetClass")) {
                AopConfigUtils.forceAutoProxyCreatorToUseClassProxying(registry);
            }
            if (enableAspectJAutoProxy.getBoolean("exposeProxy")) {
                AopConfigUtils.forceAutoProxyCreatorToExposeProxy(registry);
            }
        }
    }
}
```

注意观察代码清单 5-5 的关键点，AspectJAutoProxyRegistrar 实现了 ImportBeanDefinitionRegistrar 接口，具备编程式注册新 BeanDefinition 的能力，核心的 registerBeanDefinitions 方法中，除后面对 @EnableAspectJAutoProxy 注解的属性

进行获取和设置以外,注册组件的核心逻辑是方法体中的第一行:`AopConfigUtils.registerAspectJAnnotationAutoProxyCreatorIfNecessary`。

### 2. 注册核心代理创建器

编程式注册核心 AutoProxyCreator 如代码清单 5-6 所示。

**代码清单 5-6　编程式注册核心 AutoProxyCreator**

```java
@Nullable
public static BeanDefinition registerAspectJAnnotationAutoProxyCreatorIfNecessary(BeanDefinition
Registry registry) {
    return registerAspectJAnnotationAutoProxyCreatorIfNecessary(registry, null);
}

@Nullable
public static BeanDefinition registerAspectJAnnotationAutoProxyCreatorIfNecessary(
        BeanDefinitionRegistry registry,@Nullable Object source) {
    // 注意在这个方法中已经把 AnnotationAwareAspectJAutoProxyCreator 的字节码类型传入方法了
    return registerOrEscalateApcAsRequired(AnnotationAwareAspectJAutoProxyCreator.class,
            registry, source);
}

public static final String AUTO_PROXY_CREATOR_BEAN_NAME =
        "org.springframework.aop.config.internalAutoProxyCreator";

private static BeanDefinition registerOrEscalateApcAsRequired(Class<?> cls,
        BeanDefinitionRegistry registry, @Nullable Object source) {
    Assert.notNull(registry, "BeanDefinitionRegistry must not be null");

    if (registry.containsBeanDefinition(AUTO_PROXY_CREATOR_BEAN_NAME)) {
        BeanDefinition apcDefinition = registry.getBeanDefinition(AUTO_PROXY_CREATOR_BEAN_NAME);
        if (!cls.getName().equals(apcDefinition.getBeanClassName())) {
            intcurrentPriority = findPriorityForClass(apcDefinition.getBeanClassName());
            int requiredPriority = findPriorityForClass(cls);
            if (currentPriority < requiredPriority) {
                apcDefinition.setBeanClassName(cls.getName());
            }
        }
        return null;
    }

    RootBeanDefinitionbeanDefinition = new RootBeanDefinition(cls);
    beanDefinition.setSource(source);
    beanDefinition.getPropertyValues().add("order", Ordered.HIGHEST_PRECEDENCE);
    beanDefinition.setRole(BeanDefinition.ROLE_INFRASTRUCTURE);
    registry.registerBeanDefinition(AUTO_PROXY_CREATOR_BEAN_NAME, beanDefinition);
    return beanDefinition;
}
```

静态的方法调用中,它已经指明了最终要注册的 AOP 代理创建器的落地实现 AnnotationAwareAspectJAutoProxyCreator,而在下面最终注册的方法 registerOrEscalateApcAsRequired 中,手动创建了一个 RootBeanDefinition,将 AOP 代理创建器的类型传入,并设置了最高级别的优先级等其他属性和配置项,随后注册到 BeanDefinitionRegistry 中。

> 提示：对于上述源码的内部组件注册逻辑，后续还会遇到类似的情况，各位读者注意整理和总结。

### 5.3.2 AnnotationAwareAspectJAutoProxyCreator

注册的核心动作本身没有太多的研究价值，重点还是 AOP 动态代理创建器本身。借助 javadoc 可以获取到以下的一段描述。

> AspectJAwareAdvisorAutoProxyCreator subclass that processes all AspectJ annotation aspects in the current application context, as well as Spring Advisors.
>
> Any AspectJ annotated classes will automatically be recognized, and their advice applied if Spring AOP's proxy-based model is capable of applying it. This covers method execution joinpoints.
>
> If the `<aop:include>` element is used, only @AspectJ beans with names matched by an include pattern will be considered as defining aspects to use for Spring auto-proxying.
>
> Processing of Spring Advisors follows the rules established in org.springframework.aop.framework.autoproxy.AbstractAdvisorAutoProxyCreator.
>
> `AnnotationAwareAspectJAutoProxyCreator` 是 `AspectJAwareAdvisorAutoProxyCreator` 的子类，用于处理当前 `ApplicationContext` 中的所有基于 `AspectJ` 注解的切面，以及 Spring 原生的 `Advisor`。
>
> 如果 Spring AOP 基于代理的模型能够应用任何被@AspectJ 注解标注的类，那么它们的增强方法将被自动识别。这涵盖了方法执行的切入点表达式。
>
> 如果使用<aop:include>元素，则只有名称与包含模式匹配的被@AspectJ 标注的 Bean 将被视为定义要用于 Spring 自动代理的切面。
>
> Spring 中内置的 `Advisor` 的处理遵循 `AbstractAdvisorAutoProxyCreator` 中建立的规则。

拆解来看，javadoc 中解释的核心内容是，`AnnotationAwareAspectJAutoProxyCreator` 可以兼顾 AspectJ 风格的切面声明，以及 Spring Framework 原生的 AOP 编程。

> 提示：目前主流的 AOP 开发都是基于 AspectJ 的切面使用，有关 Spring Framework 原生的 AOP 编程已基本不用，故此部分本书不再展开。

#### 1. 继承结构

借助 IDEA 可以很清楚地看到 `AnnotationAwareAspectJAutoProxyCreator` 的继承结构，以及其中的重要核心接口，如图 5-1 所示。

观察顶层的接口，其中有几个重要的根接口值得关注。

- `BeanPostProcessor`：用于在 `postProcessAfterInitialization` 方法中生成代理对象。
- `InstantiationAwareBeanPostProcessor`：拦截 Bean 的正常 `doCreateBean` 创建流程。
- `SmartInstantiationAwareBeanPostProcessor`：提前预测 Bean 的类型、暴露 Bean 的引用（AOP、循环依赖等在现阶段解释起来过于复杂，故此处暂时略过）。

- `AopInfrastructureBean`：实现了该接口的 Bean 永远不会被代理（防止反复被代理导致逻辑死循环）。

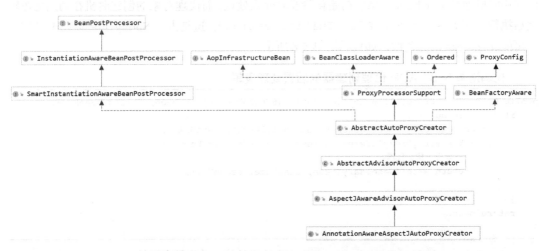

图 5-1　AnnotationAwareAspectJAutoProxyCreator 的类继承结构

除此之外，`AnnotationAwareAspectJAutoProxyCreator` 的顶层抽象实现类 **`AbstractAutoProxyCreator`** 也需要读者引起重视，在第 9 章生命周期 AOP 环节中会频繁遇到它。

### 2. 初始化时机

`AnnotationAwareAspectJAutoProxyCreator` 本身是一个后置处理器，在第 3 章中已经提到了，后置处理器的初始化时机是在 `AbstractApplicationContext` 刷新动作的第 6 步 `registerBeanPostProcessors` 方法中，如代码清单 5-7 所示。

**代码清单 5-7　AOP 代理创建器的初始化时机**

```java
public void refresh() throws BeansException, IllegalStateException {
    synchronized (this.startupShutdownMonitor) {
        // ...
        try {
            postProcessBeanFactory(beanFactory);
            invokeBeanFactoryPostProcessors(beanFactory);
            // 6. 注册、初始化 BeanPostProcessor
            registerBeanPostProcessors(beanFactory);
            initMessageSource();
            initApplicationEventMulticaster();
            // ...
        }
    }
}
```

`registerBeanPostProcessors` 方法会按照既定的排序规则初始化所有 `BeanPostProcessor`。此处有一个细节，`AnnotationAwareAspectJAutoProxyCreator` 实现了 `Ordered` 接口，并且声明了最高优先级，这就意味着它会提前于其他 `BeanPostProcessor` 创建，从而也会干预这些普通 `BeanPostProcessor` 的初始化（即也有可能被 AOP 代理增强）。

### 3. 作用时机

与其他 BeanPostProcessor 相似，AnnotationAwareAspectJAutoProxyCreator 的作用时机通常是在 bean 对象的初始化阶段时介入处理，而代理对象的创建时机在初始化逻辑之后执行（即 postProcessAfterInitialization），这是由于 Spring Framework 考虑到尽可能保证 Bean 的完整性，如代码清单 5-8 所示。

**代码清单 5-8　AOP 代理创建器的后置处理逻辑**

```
public Object postProcessAfterInitialization(@Nullable Object bean, String beanName) {
    if (bean != null) {
        Object cacheKey = getCacheKey(bean.getClass(), beanName);
        if (this.earlyProxyReferences.remove(cacheKey) != bean) {
            // 核心：构造代理
            return wrapIfNecessary(bean, beanName, cacheKey);
        }
    }
    return bean;
}
```

postProcessAfterInitialization 方法中的核心动作是中间的 wrapIfNecessary，这个动作从方法名上就很容易理解，如果有必要的话，wrapIfNecessary 方法会给当前对象包装生成代理对象。下面进入源码部分，如代码清单 5-9 所示（现阶段读者只需关注标有注释的部分）。

**代码清单 5-9　生成代理对象的主干逻辑**

```
protected Object wrapIfNecessary(Object bean, StringbeanName, Object cacheKey) {
    if (StringUtils.hasLength(beanName) && this.targetSourcedBeans.contains(beanName)) {
        return bean;
    }
    if (Boolean.FALSE.equals(this.advisedBeans.get(cacheKey))) {
        return bean;
    }
    if (isInfrastructureClass(bean.getClass()) || shouldSkip(bean.getClass(),beanName)) {
        this.advisedBeans.put(cacheKey, Boolean.FALSE);
        return bean;
    }

    // Create proxy if we have advice.
    // 如果上面的判断都没有成立，则决定是否需要进行代理对象的创建
    Object[] specificInterceptors = getAdvicesAndAdvisorsForBean(bean.getClass(), beanName, null);
    if (specificInterceptors != DO_NOT_PROXY) {
        this.advisedBeans.put(cacheKey, Boolean.TRUE);
        // 创建代理对象的动作
        Object proxy = createProxy(bean.getClass(), beanName,
                specificInterceptors, new SingletonTargetSource(bean));
        this.proxyTypes.put(cacheKey, proxy.getClass());
        return proxy;
    }
    // 记录缓存
    this.advisedBeans.put(cacheKey, Boolean.FALSE);
    return bean;
}
```

从源码逻辑中概括起来，创建代理对象的核心动作分为三个步骤。

1. 判断决定是否是一个不会被增强的 bean 对象。
2. 根据当前正在创建的 bean 对象去匹配增强器。
3. 如果有增强器，创建 bean 对象的代理对象。

> 提示：本章只是先熟悉整体流程，创建代理对象的具体动作及细节在第 9 章中会详细讲解和剖析。

## 5.4 小结

本章主要回顾了 Spring Boot 对 AOP 的支持，初步了解了注解驱动 AOP 的核心组件与生效机制。`AnnotationAwareAspectJAutoProxyCreator` 作为 AOP 的核心后置处理器，掌握其代理增强逻辑至关重要，在第 9 章中还会进一步展开分析。

# 第 2 部分
# Spring Boot 的生命周期原理分析

- ▶ 第 6 章　Spring Boot 准备容器与环境
- ▶ 第 7 章　IOC 容器的刷新
- ▶ 第 8 章　Spring Boot 容器刷新扩展：嵌入式 Web 容器
- ▶ 第 9 章　AOP 模块的生命周期

# 第 6 章 Spring Boot 准备容器与环境

**本章主要内容：**
- SpringApplication 的创建流程全解析；
- SpringApplication 的启动阶段；
- 内置 IOC 容器的创建流程；
- SpringApplication 的事件回调机制。

本书第 1 部分的内容系统地讲解了 Spring Boot 的核心容器设计，第 2 部分将针对 Spring Boot 的完整启动过程作一个全面、详细的全生命周期解析。本章先来探究 SpringApplication 这个核心引导类，它的设计在第 4 章已经详细探讨过，本章的重点是它的工作全流程。

按照 Spring Boot 推荐的主启动类编写方式，在主启动类的 main 方法中使用 SpringApplication 的静态 run 方法，以一行代码即可启动 Spring Boot 应用，如代码清单 6-1 所示。

**代码清单 6-1　经典的 Spring Boot 主启动类编写方式**

```java
@SpringBootApplication
public class SpringBootLifecycleApplication {

    public static void main(String[] args) {
        SpringApplication.run(SpringBootLifecycleApplication.class, args);
    }
}
```

本章研究的核心是 SpringApplication 的静态 run 方法，下面正式进入源码分析环节。

进入 run 方法后，可以发现最终调用的是下面重载的 run 方法，而重载的 run 方法又将启动的动作拆解为两步：创建 SpringApplication 对象；启动 SpringApplication（见代码清单 6-2）。针对这两个动作下面分别展开解析。

**代码清单 6-2　SpringApplication 的 run 方法调用**

```java
public static ConfigurableApplicationContext run(Class<?> primarySource, String... args) {
    return run(new Class<?>[] { primarySource }, args);
}

public static ConfigurableApplicationContext run(Class<?>[] primarySources, String[] args) {
    return new SpringApplication(primarySources).run(args);
}
```

# 6.1 创建 SpringApplication

静态 run 方法调用的 SpringApplication 构造方法中只传入了主启动类，但同时 SpringApplication 还支持传入特定的 ResourceLoader 以实现自定义的资源加载，通常情况下在项目开发中不会使用特殊定制的 ResourceLoader，故该部分可以忽略。由代码清单 6-3 可以看出，最终调用的是下面的两参数构造方法（关键步骤已标有注释）。

**代码清单 6-3　SpringApplication 的构造方法调用**

```java
private Set<Class<?>> primarySources;

public SpringApplication(Class<?>... primarySources) {
    this(null, primarySources);
}

public SpringApplication(ResourceLoader resourceLoader, Class<?>... primarySources) {
    this.resourceLoader = resourceLoader;
    Assert.notNull(primarySources, "PrimarySources must not be null");
    // 将传入的 SpringBootLifecycleApplication 启动类放入 primarySources 中
    // 这样 Spring Boot 应用就知道主启动类在哪里，名称是什么
    // Spring Boot 一般称呼这种主启动类叫 primarySource（主配置资源来源）
    this.primarySources = new LinkedHashSet<>(Arrays.asList(primarySources));
    // 判断应用环境
    this.webApplicationType = WebApplicationType.deduceFromClasspath();
    // 设置应用初始化器
    setInitializers((Collection) getSpringFactoriesInstances(ApplicationContextInitializer.class));
    // 设置 Spring Boot 全局监听器
    setListeners((Collection) getSpringFactoriesInstances(ApplicationListener.class));
    // 确定主启动类
    this.mainApplicationClass = deduceMainApplicationClass();
}
```

由代码清单 6-3 可以发现，SpringApplication 的构造方法中存储了传入的主配置资源（即普遍认为的主启动类），后续还有几个复杂动作，逐一拆解来看。

## 6.1.1 推断 Web 环境

SpringApplication 构造方法的第一步会从当前应用的类路径下尝试寻找一些特定的类，并以此推断当前应用适合使用哪种 Web 环境，如代码清单 6-4 所示。参考第 4 章中提到的官方文档的内容，再结合源码来阅读，这部分逻辑会更加容易理解。

**代码清单 6-4　Web 环境类型推断**

```java
// 全限定名常量有精简
private static final String[] SERVLET_INDICATOR_CLASSES = { "javax.servlet.Servlet",
        "org.springframework.web.context.ConfigurableWebApplicationContext" };
private static final String WEBMVC_INDICATOR_CLASS = "o.s.web.servlet.DispatcherServlet";
private static final String WEBFLUX_INDICATOR_CLASS = "o.s.web.reactive.DispatcherHandler";
private static final String JERSEY_INDICATOR_CLASS = "o.g.jersey.servlet.ServletContainer";
```

```
static WebApplicationType deduceFromClasspath() {
    if (ClassUtils.isPresent(WEBFLUX_INDICATOR_CLASS, null)
            && !ClassUtils.isPresent(WEBMVC_INDICATOR_CLASS, null)
            && !ClassUtils.isPresent(JERSEY_INDICATOR_CLASS, null)) {
        return WebApplicationType.REACTIVE;
    }
    for (String className : SERVLET_INDICATOR_CLASSES) {
        if (!ClassUtils.isPresent(className, null)) {
            return WebApplicationType.NONE;
        }
    }
    return WebApplicationType.SERVLET;
}
```

可以看到源码的推断规则与之前总结的顺序有一些小差别，具体如下：

- 如果 WebFlux 的核心控制器 DispatcherHandler 存在并且 WebMvc 的核心控制器 DispatcherServlet 不存在，则启用 Reactive 环境；
- 如果 Servlet 类以及 ConfigurableWebApplicationContext 中有任何一个不存在，则认为导入 WebMvc 相关的环境不全，从而启用 None 环境；
- 否则，启用 Servlet 环境（包含只导入了 WebMvc 环境，以及 WebMvc 与 WebFlux 共存的情况）。

## 6.1.2 设置初始化器

```
setInitializers((Collection) getSpringFactoriesInstances(ApplicationContextInitializer.class));
```

紧接着的第二步要利用 Spring Framework 的 SPI 机制，从 spring.factories 中加载一组 ApplicationContext 的初始化器并应用到当前项目中。首先读者要对 ApplicationContextInitializer 有一个认识。

### 1. ApplicationContextInitializer

参照文档注释可以对 ApplicationContextInitializer 有一个初步的了解：

Callback interface for initializing a Spring ConfigurableApplicationContext prior to being refreshed.

Typically used within web applications that require some programmatic initialization of the application context. For example, registering property sources or activating profiles against the context's environment.

ApplicationContextInitializer processors are encouraged to detect whether Spring's Ordered interface has been implemented or if the @Order annotation is present and to sort instances accordingly if so prior to invocation.

它是一个用于在刷新容器之前初始化 ConfigurableApplicationContext 的回调接口。

通常在需要对应用程序上下文进行某些编程初始化的 Web 应用程序中使用，例如根据上下文环境注册属性源或激活配置文件。

> 鼓励ApplicationContextInitializer处理器检测是否已实现Ordered接口或者是否标注了@Order注解，并在调用之前相应地对实例进行排序。

文档注释理解起来有一些难度，不过第一段内容就已经表明，ApplicationContextInitializer是一个在IOC容器刷新之前被回调的接口，这也意味着它可以在IOC容器创建之后、未触发刷新动作之前执行额外的逻辑处理。借助IDE可以发现ApplicationContextInitializer是一个接口，并且只有一个initialize方法，该方法需要传入一个ConfigurableApplicationContext的实现类，用于在方法体内对IOC容器进行一些前置处理（只有传入Configurable前缀的IOC容器才允许对其进行修改），如代码清单6-5所示。

**代码清单6-5　ApplicationContextInitializer**

```java
public interface ApplicationContextInitializer<C extends ConfigurableApplicationContext> {
    void initialize(C applicationContext);
}
```

下面介绍ApplicationContextInitializer的一个经典实现类，这个类来自spring-boot-starter-web的ServletContextApplicationContextInitializer，它的作用是将ApplicationContext与ServletContext互相放置于对方容器中，以便可以互相查找获取，核心源码如代码清单6-6所示。

**代码清单6-6　ServletContextApplicationContextInitializer的实现**

```java
public class ServletContextApplicationContextInitializer
        implements ApplicationContextInitializer<ConfigurableWebApplicationContext>, Ordered {

    private final ServletContext servletContext;
    public ServletContextApplicationContextInitializer(ServletContext servletContext) {
        this(servletContext, false);
    }

    //...
    @Override
    public void initialize(ConfigurableWebApplicationContext applicationContext) {
        applicationContext.setServletContext(this.servletContext);
        if (this.addApplicationContextAttribute) {
            this.servletContext.setAttribute(WebApplicationContext.ROOT_WEB_APPLICATION_CONTEXT_ATTRIBUTE, applicationContext);
        }
    }
}
```

**2. 设置初始化器的逻辑**

简单了解了ApplicationContextInitializer的设计之后，下面我们看一下Spring Boot是如何加载这些初始化器并且应用于ApplicationContext本身的。

```java
setInitializers((Collection) getSpringFactoriesInstances(ApplicationContextInitializer.class));
```

注意，setInitializers方法中传入的ApplicationContextInitializer集合是由getSpringFactoriesInstances方法获取而来的，该方法明显是通过使用Spring Framework

的 SPI 机制获取的，通过源码也可以明显看出，如代码清单 6-7 所示。

**代码清单 6-7　getSpringFactoriesInstances 的底层利用 Spring Framework 的 SPI 机制**

```
private <T> Collection<T> getSpringFactoriesInstances(Class<T> type) {
    return getSpringFactoriesInstances(type, new Class<?>[] {});
}

private <T> Collection<T> getSpringFactoriesInstances(Class<T> type,
        Class<?>[] parameterTypes, Object... args) {
    ClassLoader classLoader = getClassLoader();
    // 此处使用了 SpringFramework 的 SPI 机制
    Set<String> names = new LinkedHashSet<> (SpringFactoriesLoader.loadFactoryNames(type,
classLoader));
    List<T> instances = createSpringFactoriesInstances(type, parameterTypes,
            classLoader, args, names);
    AnnotationAwareOrderComparator.sort(instances);
    return instances;
}
```

## 6.1.3　设置监听器

ApplicationContextInitializer 设置完毕后，下一个核心动作是设置 ApplicationListener 监听器，设置的逻辑与上一步完全一致，不再赘述。下面简单介绍有关监听器和事件驱动的内容。

### 1. Spring Framework 的事件驱动模型

Spring Framework 中体现观察者模式的特性是事件驱动和监听器。在事件驱动模型中监听器充当订阅者，监听特定的事件，**事件源充当被观察的主题**，用来发布事件；IOC 容器本身可以看作事件广播器，对应的角色是**观察者**。我个人倾向于把 Spring Framework 的事件驱动核心概念分为 4 个：**事件源、事件、广播器、监听器**。

- **事件源**：发布事件的对象。
- **事件**：事件源发布的信息/作出的动作（即 ApplicationEvent）。
- **广播器**：事件真正广播给监听器的对象（即 ApplicationContext）。
    - ApplicationContext 接口实现了 ApplicationEventPublisher 接口,具备**事件广播器发布事件的能力**。
    - ApplicationEventMulticaster 组合了所有的监听器，具备事件广播器广播事件的能力。
- **监听器**：监听事件的对象（即 ApplicationListener）。

### 2. ApplicationListener

ApplicationListener 接口的设计本身继承自 JDK 原生的观察者模式接口 EventListener，如代码清单 6-8 所示（其实 Spring Framework 中的很多组件设计都基于 JDK 原生的 API）。

**代码清单 6-8　ApplicationListener**

```
public interface ApplicationListener<E extends ApplicationEvent> extends EventListener{
    void onApplicationEvent(E event);
}
```

有关 `ApplicationListener` 的描述，在 javadoc 中解释得言简意赅：

> Interface to be implemented by application event listeners. Based on the standard java.util.EventListener interface for the Observer design pattern.
>
> As of Spring 3.0, an ApplicationListener can generically declare the event type that it is interested in. When registered with a Spring ApplicationContext, events will be filtered accordingly, with the listener getting invoked for matching event objects only.
>
> 由应用程序事件监听器实现的接口直接继承自 JDK 中观察者模式的 `java.util.EventListener` 标准接口。
>
> 从 Spring 3.0 版本开始，`ApplicationListener` 可以使用泛型类型声明监听的事件类型。当监听器注册到 IOC 容器后，底层将相应地过滤事件，并且仅针对匹配的事件对象调用所有适配的监听器。

对于上述内容，读者可以先简单地对事件机制以及 `ApplicationListener` 有一个初步了解，至于事件机制在 Spring Framework 和 Spring Boot 底层如何运行，在第 7 章才能看到。

## 6.1.4　确定主启动类

创建 `SpringApplication` 的最后一步，`deduceMainApplicationClass` 方法会确定主启动类，对应的查找方式是借助方法调用栈，寻找触发方法名为 `main` 的所在类，如代码清单 6-9 所示。

**代码清单 6-9　确定主启动类的核心逻辑**

```java
private Class<?> deduceMainApplicationClass() {
    try{
        StackTraceElement[] stackTrace = new RuntimeException().getStackTrace();
        for (StackTraceElement stackTraceElement : stackTrace) {
            // 从本方法开始往上查找，哪一层调用栈上有 main 方法，方法对应的类就是主配置类
            if ("main".equals(stackTraceElement.getMethodName())) {
                return Class.forName(stackTraceElement.getClassName());
            }
        }
    } // catch ......
    return null;
}
```

借助 Debug 调试可以很清楚地发现，`stackTrace` 的内容就是正在触发的方法调用栈信息，而最终获取到的 `main` 方法所在类就是目前启动测试的主启动类 `SpringBootLifecycleApplication`，如图 6-1 所示。

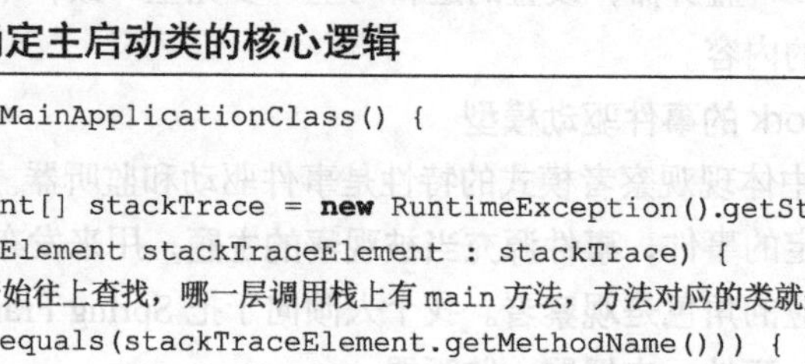

图 6-1　stackTrace 即方法调用栈本身

### 6.1.5 与 Spring Boot 1.x 的区别

补充一下与 Spring Boot 1.x 版本的对比。

> 提示：由于本书在进行源码分析时主要采用 Spring Boot 2.3.x 版本的源码，而不同的版本中底层的实现会有一些差别，为了能更好地理解版本间的一些设计上的区别，在一些特殊的位置会引用其他版本的 Spring Boot 源码进行对比。

#### 1. 主配置类概念的引入

Spring Boot 2.x 中引入了一个新的概念：**主配置类**。这个设计与 Spring Boot 1.x 有区别，在 Spring Boot 1.x 中没有这个概念，所有的配置源都被视为 `Object` 类型（读者可以先思考一下为什么会被设置为 `Object` 类型），存放在 `SpringApplication` 的成员属性中：

```
// Spring Boot 1.x
private final Set<Object> sources = new LinkedHashSet<Object>();
```

而到了 Spring Boot 2.x 版本后，Spring 团队认为项目应该全面支持注解配置类的方式，尽可能地不使用 XML 配置文件进行应用配置，于是在 `SpringApplication` 的属性成员中单独抽取了一个新的集合 `primarySources`，并将其命名为"主配置类"：

```
// Spring Boot 2.x
private Set<Class<?>> primarySources;
private Set<String> sources = new LinkedHashSet<>();
```

注意，此处的 `sources` 的集合泛型被替换为 `String` 类型，这个设计又有什么特殊之处呢？

结合上面 Spring Boot 1.x 的 `Object` 泛型，读者可以大概猜测出来，利用两个 `Set` 集合就可以分离出两种不同的配置载体，而这两种载体就是 Spring Framework 支持的 XML 配置文件与注解配置类。因为 Spring Boot 1.x 中没有主配置类的概念，所以 `sources` 里既要存放注解配置类，又要考虑存入 XML 配置文件的路径（甚至一个包扫描的路径），因此被迫将 `sources` 集合的泛型置为 `Object`；而 Spring Boot 2.x 中引入了主配置类的概念，并且 Spring Boot 本身鼓励开发者使用注解配置类，尽量避免使用 XML 配置文件作为配置源（当然如果确实仍需使用，它还可以支持），所以 Spring Boot 将重心放在 `primarySources`。

#### 2. Web 类型推断的区别

由于 Spring Boot1.x 基于 Spring Framework 4.x，而 4.x 还没有 WebFlux 模块，因此在进行 Web 类型推断时只会在 Servlet 与 None 之间选择，并且推断的方法也是直接在 `SpringApplication` 中实现的。而到了 Spring Boot 2.x 后 Web 类型推断的逻辑被单独抽取为一个静态方法，如代码清单 6-10 所示。

**代码清单 6-10　Web 类型推断的区别**

```
// Spring Boot 2.x
this.webApplicationType = WebApplicationType.deduceFromClasspath();

// Spring Boot 1.x
this.webEnvironment = deduceWebEnvironment();
```

## 6.1.6　与 Spring Boot 2.4.x 的区别

Spring Boot 版本升级到 2.4.x 之后，在创建 `SpringApplication` 的阶段引入了一个新的扩展切入机制：借助 `BootstrapRegistry` 实现重对象的预创建/共享。

### 1. BootstrapRegistry

`BootstrapRegistry` 是 Spring Boot 2.4.0 版本后引入的全新 API，从类名上理解 `BootstrapRegistry` 有注册表的概念，同时又有启动器引导的概念，借助 javadoc 可以大体了解它的用途。

> A simple object registry that is available during startup and Environment post-processing up to the point that the ApplicationContext is prepared.
>
> Can be used to register instances that may be expensive to create, or need to be shared before the ApplicationContext is available.
>
> The registry uses Class as a key, meaning that only a single instance of a given type can be stored.
>
> 它是一个简单的对象注册表，在启动和环境后置处理期间可用，直到准备好 `ApplicationContext`。
>
> 可用于注册创建成本高的实例，或需要在 `ApplicationContext` 可用之前共享的实例。
>
> 注册表使用 `Class` 作为键，这意味着只能存储给定类型的单个实例。

抽取 javadoc 中想表达的核心内容：`BootstrapRegistry` 是一个全新的对象容器，它内部存放的对象最终会传递给 `ApplicationContext`（即 IOC 容器），它可以用于提前创建一些重对象。这种设计不难理解，`BootstrapRegistry` 可以在 `ApplicationContext` 初始化之前预先初始化一些创建成本很高的对象（即重对象），等到 `ApplicationContext` 真正需要初始化时，`BootstrapRegistry` 可以直接将这些重对象共享给 `ApplicationContext`，使得 IOC 容器不再需要初始化这些重对象，从而避免 IOC 容器初始化过程过慢的问题。

`BootstrapRegistry` 的唯一实现是 `DefaultBootstrapContext`，通过观察代码清单 6-11 可以进一步体会到，`BootstrapRegistry` 本身只是一个容器。

**代码清单 6-11　DefaultBootstrapContext 只是一个对象容器**

```java
public class DefaultBootstrapContext implements ConfigurableBootstrapContext {
    private final Map<Class<?>, InstanceSupplier<?>> instanceSuppliers = new HashMap<>();
    private final Map<Class<?>, Object> instances = new HashMap<>();
```

容器中只有两个 `Map`，其中 `instances` 负责存放对象，`instanceSuppliers` 则存放对象创建器，但无论怎样设计，最终都可以根据类型获取到一个对象。

### 2. BootstrapRegistryInitializer

`BootstrapRegistry` 本身只是一个容器，要把对象注册到容器中，需要借助另一个 API：`BootstrapRegistryInitializer`（前身是 `Bootstrapper`，在 2.4.5 版本中被废弃，在 2.6.0 版本中被移除），如代码清单 6-12 所示。

**代码清单 6-12　BootstrapRegistryInitializer 接口**

```java
public interface BootstrapRegistryInitializer {
    void initialize(BootstrapRegistry registry);
}
```

第一眼看上去，BootstrapRegistryInitializer 的设计比较类似于 Application ContextInitializer，其实这两个接口的设计如出一辙，核心都是获取被初始化的目标后进行一些初始化动作（如注册对象、改变配置等）。

不过 Spring Boot 本身没有 BootstrapRegistryInitializer 的实现类，这就意味着 BootstrapRegistryInitializer 仅是 Spring Boot 为开发者预留的一个新扩展点，Spring Boot 本身并没有在此基础上扩展更多的底层逻辑实现（对比老版本的逻辑只是扩展，但没有更复杂）。

### 3. 初始化的区别

对比 2.3.x 版本 SpringApplication 的构造方法，可以发现构造方法的逻辑中额外添加了一个集合成员的赋值操作，如代码清单 6-13 所示。额外添加的方法会利用 Spring Framework 的 SPI 机制，从 spring.factories 中加载所有 BootstrapRegistryInitializer 的配置实现类，并在创建对象后保存到 SpringApplication 中。至于这些 BootstrapRegistry Initializer 何时发挥作用，下面在启动 SpringApplication 阶段马上就可以看到。

**代码清单 6-13　初始化 BootstrapRegistryInitializer**

```java
private List<BootstrapRegistryInitializer> bootstrapRegistryInitializers;

public SpringApplication(ResourceLoader resourceLoader, Class<?>... primarySources) {
    // ......
    this.bootstrapRegistryInitializers = getBootstrapRegistryInitializersFromSpringFactories();
    // ......
}

private List<BootstrapRegistryInitializer> getBootstrapRegistryInitializersFromSpringFactories
() {
    ArrayList<BootstrapRegistryInitializer> initializers = new ArrayList<>();
    // Bootstrapper 在 2.4.5 版本中被废弃，在 2.6.0 版本中被移除
    getSpringFactoriesInstances(Bootstrapper.class).stream()
            .map((bootstrapper) -> ((BootstrapRegistryInitializer) bootstrapper::initialize))
            .forEach(initializers::add);
    // 利用 SPI 加载 BootstrapRegistryInitializer
    initializers.addAll(getSpringFactoriesInstances(BootstrapRegistryInitializer.class));
    return initializers;
}
```

## 6.2　启动 SpringApplication

在 SpringApplication 的构造方法执行完毕后，SpringApplication 对象创建就绪了，下一步会执行 SpringApplication 的 run 方法，从而真正地启动 Spring Boot 应用。整个 run 方法的过程共包含 13 步，其中包含了 ApplicationContext 的最核心的刷新容器环节。为了能清楚划分整个 Spring Boot 应用启动的多个环节，接下来会在 6.2～6.4 节中解析整个 run 方法的启动过程。

本节先聚焦启动 SpringApplication 的一些准备工作，这部分涉及 run 方法的 3 个重要环节，如代码清单 6-14 所示。

## 代码清单6-14　SpringApplication#run 片段（1）

```java
public ConfigurableApplicationContext run(String... args) {
    // 前置准备
    StopWatch stopWatch = new StopWatch();
    stopWatch.start();
    ConfigurableApplicationContext context = null;
    // 配置与 awt 相关的信息
    configureHeadlessProperty();
    // 获取 SpringApplicationRunListeners，并调用 starting 方法（回调机制）
    SpringApplicationRunListeners listeners = getRunListeners(args);
    // 【回调】首次启动 run 方法时立即调用。可用于非常早期的初始化（准备运行时环境之前）
    listeners.starting();
    try {
        // 将 main 方法的 args 参数封装到一个对象中
        ApplicationArguments applicationArguments = new DefaultApplicationArguments(args);
        // 准备运行时环境
        ConfigurableEnvironment environment = prepareEnvironment(listeners, applicationArguments);
        // ......
        // StopWatch 对象在此处被调用 stop 方法
        stopWatch.stop();
        // ......
    }
```

下面逐一拆解来看。

## 6.2.1　前置准备

### 1. 计时器对象的使用

run 方法的第一个步骤会初始化一个 StopWatch 对象，这个对象看上去与计时有关，从文档注释中可以获取到的关键信息如下：

This class is normally used to verify performance during proof-of-concepts and in development, rather than as part of production applications.

常用于在概念验证和开发过程中验证性能，而不是作为生产应用的一部分。

这段注释的重点是：**仅用于验证性能**。换言之，StopWatch 是用来监控启动时间的，它本身不太重要，读者只需留意一下它的使用位置。通过观察代码清单6-15可知，在 run 方法刚启动执行其 start 方法时开始计时，当后面的 try 块即将执行完成而执行其 stop 方法时停止计时，由此可以计算出 Spring Boot 应用启动的大体耗时。

## 代码清单6-15　设置了一个特殊的系统配置变量

```java
private void configureHeadlessProperty() {
    System.setProperty(SYSTEM_PROPERTY_JAVA_AWT_HEADLESS,
        System.getProperty(SYSTEM_PROPERTY_JAVA_AWT_HEADLESS, Boolean.toString(this.headless)));
}
```

### 2. awt 的设置

第二个关键动作是 configureHeadlessProperty 方法，该方法的实现如代码清单6-15

所示。方法的实现比较诡异,它从 `System` 中提取了一个配置属性 `SYSTEM_PROPERTY_JAVA_AWT_HEADLESS`,之后又重新设置回去。

这样做的目的是什么呢?需要读者在 JDK 的 `System` 类中观察代码清单 6-16 中的几个方法。

**代码清单 6-16　JDK 中 System 源码节选**

```
public static String getProperty(String key) {
    checkKey(key);
    // JMX......
    // 从 Properties 中取值
    return props.getProperty(key);
}

public static String getProperty(String key, String def) {
    checkKey(key);
    // JMX......
    // 从 Properties 中取值,如果没有取到,返回默认值
    return props.getProperty(key, def);
}

public static String setProperty(String key, String value) {
    checkKey(key);
    // JMX ......
    return (String) props.setProperty(key, value);
}
```

通过代码清单 6-16 读者可以发现,`System` 类中有两个重载的 `getProperty` 方法,但只有一个 `setProperty` 方法。仔细观察源码可以发现,两个重载的 `getProperty` 方法有一点微妙的区别。这里要提一下 `Properties` 的机制:`setProperty` 方法中调用的是 `Properties` 的两参数,分别为 `key` 和 `value`。而两个重载的 `getProperty` 方法的唯一区别是分别调用 `Properties` 的一参数和两参数,这种区别类似于 `Map` 中的 `get` 和 `getOrDefault`。换言之,两参数的 `getProperty` 方法如果没有取到指定的 `key` 则会返回给定的默认值;一参数的 `getProperty` 方法调用时如果没有取到值则返回 `null`。

经过代码清单 6-15 的设置后,无论怎样从 `System` 中获取都能取到"`SYSTEM_PROPERTY_JAVA_AWT_HEADLESS`"对应的值。为什么 Spring Boot 要设置这个特殊的属性值?`SYSTEM_PROPERTY_JAVA_AWT_HEADLESS` 代表什么含义呢?观察源码中该常量的值:

```
// 显示器缺失
private static final String SYSTEM_PROPERTY_JAVA_AWT_HEADLESS = "java.awt.headless";
```

由此可知代码清单 6-15 的真正作用是:设置该配置属性后,在启动 Spring Boot 应用时即使没有检测到显示器也允许其继续启动(对于服务器而言,没有显示器也可以运行)。

### 3. 对比 Spring Boot 2.1.x

Spring Boot 2.0 到 2.2 版本中在前置准备阶段还预留了一个异常分析器的集合:

```
Collection<SpringBootExceptionReporter> exceptionReporters = new ArrayList<>();
```

集合泛型中的 `SpringBootExceptionReporter` 就是 Spring Boot 的异常分析器,它是一个用于支持 SpringApplication 启动错误报告的自定义异常报告回调接口,它本身也利用 Spring

Framework 的 SPI 机制加载。不过 Spring Boot 中默认只有一个实现类：FailureAnalyzers。注意观察类名，FailureAnalyzers 是一个具有复数概念的类，而且 FailureAnalyzers 的访问修饰符不是 public 类型，这就意味着 FailureAnalyzers 是一个组合的概念，它内部会集成一组 FailureAnalyzer。借助 IDE 可以发现设计确实如此，其对应的源码如代码清单 6-17 所示。

**代码清单 6-17　FailureAnalyzers 内部利用 SPI 机制集成一组 FailureAnalyzer**

```
final class FailureAnalyzers implements SpringBootExceptionReporter {
    private final List<FailureAnalyzer> analyzers;
    // ......
    FailureAnalyzers(ConfigurableApplicationContext context, ClassLoader classLoader) {
        // ......
        this.analyzers = loadFailureAnalyzers(context, this.classLoader);
    }

    private List<FailureAnalyzer> loadFailureAnalyzers(ConfigurableApplicationContext context,
            ClassLoader classLoader) {
        List<String> classNames = SpringFactoriesLoader.loadFactoryNames(FailureAnalyzer.class, classLoader);
        // ......
        return analyzers;
    }
}
```

从代码清单 6-17 中可以体会到，设计 FailureAnalyzers 的意图是希望开发者关注如何实现内部集成的 FailureAnalyzer，而不是如何访问 FailureAnalyzers 本身。Spring Boot 已经内置了不少 FailureAnalyzer 的实现类，如果有特殊需求，开发者可以自行定制 FailureAnalyzer 的实现类，并配置到项目的 spring.factories 中，Spring Boot 都可以获取并整合到 SpringApplication 中。

#### 4. 对比 Spring Boot 2.4.x

由于 Spring Boot 2.4.0 版本后引入了新的 API：BootstrapRegistry 与 BootstrapContext。在 SpringApplication 启动的前置准备中会创建 DefaultBootstrapContext，并应用所有的 BootstrapRegistryInitializer，如代码清单 6-18 所示。

**代码清单 6-18　初始化 DefaultBootstrapContext**

```
public ConfigurableApplicationContext run(String... args) {
    StopWatch stopWatch = new StopWatch();
    stopWatch.start();
    // 此处会创建 DefaultBootstrapContext
    DefaultBootstrapContext bootstrapContext = createBootstrapContext();
    ConfigurableApplicationContext context = null;
    // ......
}

private DefaultBootstrapContext createBootstrapContext() {
    DefaultBootstrapContext bootstrapContext = new DefaultBootstrapContext();
    this.bootstrapRegistryInitializers.forEach(initializer -> {
        initializer.initialize(bootstrapContext);
    });
    return bootstrapContext;
}
```

## 6.2.2 获取 SpringApplicationRunListeners

前置准备完成后，接下来 SpringApplication 会获取一组 SpringApplication RunListener，如代码清单 6-19 所示。

**代码清单 6-19　获取 SpringApplicationRunListeners 的逻辑**

```java
// 获取SpringApplicationRunListeners，并调用starting方法（回调机制）
SpringApplicationRunListeners listeners = getRunListeners(args);
listeners.starting();

private SpringApplicationRunListeners getRunListeners(String[] args) {
    Class<?>[] types = new Class<?>[] { SpringApplication.class, String[].class };
    // 依然使用SpringFramework的SPI机制加载
    return new SpringApplicationRunListeners(logger,
            getSpringFactoriesInstances(SpringApplicationRunListener.class, types, this, args));
}
```

有关 SpringApplicationRunListeners 的设计在第 4 章已经了解过，本节不再重复讲解，读者可以观察一下默认加载的监听器实现。借助 Debug 运行至此，发现在默认情况下可以加载到一个类型为 EventPublishingRunListener 的监听器，如图 6-2 所示。

```
listeners = {SpringApplicationRunListeners@1700}
    log = {LogAdapter$Slf4jLocationAwareLog@1712}
    listeners = {ArrayList@1713} size = 1
        0 = {EventPublishingRunListener@1715}
```

图 6-2　默认有一个 SpringApplicationRunListeners

### 1. EventPublishingRunListener 与 Spring Boot 扩展事件

从类名上理解 EventPublishingRunListener 具备事件发布能力，那么发布事件必定要配合事件广播器来实现，观察 EventPublishingRunListener 的类成员结构以及构造方法，如代码清单 6-20 所示。

**代码清单 6-20　EventPublishingRunListener 的结构**

```java
public class EventPublishingRunListener implements SpringApplicationRunListener, Ordered {

    private final SpringApplication application;

    private final String[] args;

    private final SimpleApplicationEventMulticaster initialMulticaster;

    public EventPublishingRunListener(SpringApplication application,String[] args) {
        this.application = application;
        this.args = args;
        // 注意此处初始化了一个事件广播器，并存储了所有监听器
        this.initialMulticaster = new SimpleApplicationEventMulticaster();
        for (ApplicationListener<?> listener : application.getListeners()) {
            this.initialMulticaster.addApplicationListener(listener);
        }
    }
```

可以发现，在 EventPublishingRunListener 构造成功之后，其内部就已经初始化了一个事件广播器，并整合了现阶段能获取到的所有 ApplicationListener。紧接着，EventPublishingRunListener 实现 SpringApplicationRunListener 的每个方法中都有一个特殊的事件被广播，如代码清单 6-21 所示。

**代码清单 6-21　EventPublishingRunListener 的事件广播实现（节选）**

```
@Override
public void starting() {
    this.initialMulticaster.multicastEvent(new ApplicationStartingEvent(this.application, this.args));
}

@Override
public void environmentPrepared(ConfigurableEnvironment environment) {
    this.initialMulticaster.multicastEvent(new ApplicationEnvironmentPreparedEvent(this.application, this.args, environment));
}

@Override
public void contextPrepared(ConfigurableApplicationContext context) {
    this.initialMulticaster.multicastEvent(new ApplicationContextInitializedEvent(this.application, this.args, context));
}

// ...
```

虽然读者之前没有见过这些事件，但通过事件的类名就可以领会到，它们分别对应 SpringApplicationRunListener 的各个事件切入点的方法。这就意味着借助 SpringApplicationRunListener 的事件扩展逻辑，除了可以直接编写 SpringApplicationRunListener 的实现类，也可以编写 ApplicationListener 的实现类监听这些 Spring Boot 扩展的事件，只不过监听 Spring Boot 事件的 ApplicationListener 同样需要将其配置到 spring.factories，或者在构造 SpringApplication 时编程式注册，仅通过组件扫描等方式注册的监听器无法监听到所有 Spring Boot 事件。

**2. 与其他版本的对比**

了解了 SpringApplicationRunListener 的实现后，按照惯例再了解一下不同版本之间的设计对比。

Spring Boot 1.x 中并没有把所有的时机都考虑到位，所以当时的 SpringApplicationRunListener 中只支持前四个事件切入，不过事件本身的设计与 Spring Boot 2.x 的相同，在第 4 章也提到过。

在 Spring Boot 2.4.x 中，由于引入了 BootstrapContext，这个组件会在 ApplicationContext 初始化之前起作用，而 SpringApplicationRunListener 的前两个事件触发时 BootstrapContext 处于有效状态，因此 SpringApplicationRunListener 中将前两个事件的触发入参中添加了 BootstrapContext，以便开发者在扩展事件监听逻辑时有更多的 API 可以获取，进而有机会做更多的事情，如代码清单 6-22 所示。

### 代码清单 6-22　2.4.x 版本之后的 SpringApplicationRunListener 源码（节选）

```java
default void starting(ConfigurableBootstrapContext bootstrapContext) {
    starting();
}

default void environmentPrepared(ConfigurableBootstrapContext bootstrapContext,
        ConfigurableEnvironment environment) {
    environmentPrepared(environment);
}
```

## 6.2.3　准备运行时环境

获取监听器之后，下一个重要环节是准备运行时环境，这个步骤对应第 3 章中讲到的 `Environment` 抽象，如代码清单 6-23 所示。`prepareEnvironment` 方法的主要步骤是先创建 `Environment`，再配置，最后绑定到 `SpringApplication` 上。下面对这 3 个步骤逐一解释。

### 代码清单 6-23　准备运行时环境 Environment

```java
ApplicationArguments applicationArguments = new DefaultApplicationArguments(args);
ConfigurableEnvironment environment = prepareEnvironment(listeners, applicationArguments);

private ConfigurableEnvironment prepareEnvironment(SpringApplicationRunListeners listeners,
        ApplicationArguments applicationArguments) {
    // 创建、配置运行时环境
    ConfigurableEnvironment environment = getOrCreateEnvironment();
    configureEnvironment(environment, applicationArguments.getSourceArgs());
    ConfigurationPropertySources.attach(environment);
    // 此处回调 SpringApplicationRunListener 的 environmentPrepared 方法
    // Environment 构建完成，但在创建 ApplicationContext 之前触发
    listeners.environmentPrepared(environment);
    // 环境与应用绑定
    bindToSpringApplication(environment);
    if (!this.isCustomEnvironment) {
        environment = new EnvironmentConverter(getClassLoader()).convertEnvironmentIfNecessary
(environment,deduceEnvironmentClass());
    }
    ConfigurationPropertySources.attach(environment);
    return environment;
}
```

#### 1．创建运行时环境

前面在了解 `Environment` 的时候，借助 IDEA 生成继承和派生关系时，图 3-6 中有 3 个 `Environment` 的落地实现类，分别对应普通环境的 `StandardEnvironment`、支持 Servlet 环境的 `StandardServletEnvironment` 和支持 Reactive 环境的 `StandardReactiveWebEnvironment`。具体到底层创建中，刚好是根据之前推断好的 Web 类型来决定使用何种 `Environment` 的实现，如代码清单 6-24 所示。

### 代码清单 6-24　根据应用类型创建运行时环境

```java
private ConfigurableEnvironment getOrCreateEnvironment() {
    if (this.environment! = null) {
```

```
            return this.environment;
    }
    switch (this.webApplicationType) {
        case SERVLET:
            return new StandardServletEnvironment();
        case REACTIVE:
            return new StandardReactiveWebEnvironment();
        default:
            return new StandardEnvironment();
    }
}
```

> 提示：注意，当前方法是 SpringApplication 内部的，所以可以直接提取 webApplicationType 的值。

#### 2. 配置运行时环境

创建好 Environment 对象后，接下来要向 Environment 中配置一些组件，包括 ConversionService、命令行参数、编程式配置的 profile，如代码清单 6-25 所示。Configure Environment 方法本身难度不大，内部调用方法的实现逻辑也比较容易理解，这里不再展开，读者可以结合源码自行阅读。

**代码清单 6-25　配置运行时环境**

```
protected void configureEnvironment(ConfigurableEnvironment environment, String[] args) {
    if (this.addConversionService) {
        // 获取全局共享的 ApplicationConversionService 实例
        ConversionService conversionService = ApplicationConversionService.getSharedInstance();
        environment.setConversionService((ConfigurableConversionService) conversionService);
    }
    // 将命令行参数封装为 PropertySource
    configurePropertySources(environment, args);
    // 支持编程式添加激活的 profile
    configureProfiles(environment, args);
}
```

#### 3. Environment 与 SpringApplication 绑定

在广播 environmentPrepared 事件后，prepareEnvironment 方法会将 Environment 中配置的一些属性值绑定到 SpringApplication 中，它的实现代码只有一行，如代码清单 6-26 所示。

**代码清单 6-26　Environment 的属性值绑定至 SpringApplication 对象**

```
protected void bindToSpringApplication(ConfigurableEnvironment environment) {
    try{
        Binder.get(environment).bind("spring.main", Bindable.ofInstance(this));
    } // catch ......
}
```

有关 Binder 的使用在此不深入研究，读者只需要知道一点：Binder 可以将 Environment 中的一些属性值映射到某个类的属性中（类似于在 Spring Boot 中学过的

@ConfigurationProperties 注解的作用）。从源码中可以看到它绑定的属性都是以 spring.main 开头的配置属性，而这部分属性刚好与 SpringApplication 中的部分属性一一对应，如代码清单 6-27 所示。

**代码清单 6-27　以 spring.main 开头的属性与 SpringApplication 的属性一一对应**

```
// application.properties
spring.main.lazy-initialization = true

public class SpringApplication {
    // ......
    private boolean lazyInitialization = false;
```

总结起来，bindToSpringApplication 方法实际要做的事情，就是把 application.properties 中的属性赋值到 SpringApplication 对象中。

## 6.3　IOC 容器的创建与初始化

经过 6.2 节中的几个关键步骤后，SpringApplication 的启动前置准备工作已经基本完成，接下来要初始化的是 SpringApplication 内部的核心 IOC 容器，也就是 ApplicationContext 的具体实现。本节涉及 run 方法的代码片段如代码清单 6-28 所示。

**代码清单 6-28　SpringApplication#run 片段（2）**

```
// ......
configureIgnoreBeanInfo(environment);
// 打印 Banner
Banner printedBanner = printBanner(environment);
// 创建空的 IOC 容器
context = createApplicationContext();
context.setApplicationStartup(this.applicationStartup);
// IOC 容器的初始化
prepareContext(bootstrapContext, context, environment, listeners, applicationArguments,
        printedBanner);
// 刷新容器
refreshContext(context);
// ......
```

下面就其中的重要环节——拆解。

### 6.3.1　打印 Banner

在深入 printBanner 方法之前，需要读者先明确 Banner 到底是什么。

> 提示：如果读者认为 Banner 只是打印到控制台/日志的那段内容，不妨继续往下阅读。当然如果读者已经足够了解 Banner 的设计，那可以跳过本节。

**1. Banner 的设计**

在 Spring Boot 的内部，Banner 被设计为一个接口，并且内置了一个枚举类型，代表 Banner 输出的模式（关闭、控制台打印、日志输出），如代码清单 6-29 所示。

### 代码清单 6-29  Banner 是一个接口

```java
public interface Banner {

    void printBanner(Environment environment, Class<?> sourceClass,PrintStream out);

    enum Mode {
        OFF,
        CONSOLE,
        LOG
    }
}
```

借助 IDE 可以发现 Banner 的几个实现类,如图 6-3 所示。其中读者第一眼关注的肯定是 SpringBootBanner,其对应的源码如代码清单 6-30 所示。

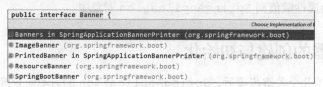

图 6-3  Banner 的实现类

### 代码清单 6-30  SpringBootBanner 的设计

```java
class SpringBootBanner implements Banner {

    private static final String[] BANNER = {"","  .   ____          _            __ _ _",
        " /\\\\ / ___'_ __ _ _(_)_ __  __ _ \\ \\ \\ \\","( ( )\\___ | '_ | '_| | '_ \\/ _` | \\ \\ \\ \\",
        " \\\\/  ___)| |_)| | | | | || (_| |  ) ) ) )","  '  |____| .__|_| |_|_| |_\\__, | / / / /",
        " =========|_|==============|___/=/_/_/_/"};

    private static final String SPRING_BOOT = " :: Spring Boot :: ";
    private static final int STRAP_LINE_SIZE = 42;

    @Override
    public void printBanner(Environment environment, Class<?> sourceClass,
            PrintStream printStream) {
        // 先打印 Banner 的内容
        for (String line : BANNER) {
            printStream.println(line);
        }
        // 打印 Spring Boot 的版本
        String version = SpringBootVersion.getVersion();
        version = (version != null) ? " (v" + version + ")" : "";
        StringBuilder padding = new StringBuilder();
        while (padding.length() < STRAP_LINE_SIZE - (version.length() + SPRING_BOOT.length())) {
            padding.append(" ");
        }

        printStream.println(AnsiOutput.toString(AnsiColor.GREEN, SPRING_BOOT, AnsiColor.DEFAULT,
                padding.toString(), AnsiStyle.FAINT, version));
        printStream.println();
    }
}
```

从代码清单 6-30 中可以发现一些奇怪但又有些似曾相识的东西,而且常量名是 BANNER,不难猜测出它就是在默认情况下打印在控制台的文字 Banner。对应实现的 printBanner 方法,就是用输出对象打印定义好的 Banner 和 Spring Boot 的版本号,逻辑上不难理解。

有关 Banner 的其他实现类,读者可以自行参考源码实现,这里不作为重点讲解。

### 2. printBanner

下面来看 printBanner 的具体实现逻辑。默认情况下 Banner 的输出模式是打印到控制台,后续它会获取一个 ResourceLoader,配合 SpringApplicationBannerPrinter 实现 Banner 的输出,如代码清单 6-31 所示。

**代码清单 6-31　打印 Banner 的底层逻辑**

```
private Banner.Mode bannerMode = Banner.Mode.CONSOLE;
private Banner printBanner(ConfigurableEnvironment environment) {
    if (this.bannerMode == Banner.Mode.OFF) {
        return null;
    }
    // Banner 文件资源加载
    ResourceLoader resourceLoader = (this.resourceLoader != null) ? this.resourceLoader
            : new DefaultResourceLoader(getClassLoader());
    // 使用 BannerPrinter 打印 Banner
    SpringApplicationBannerPrinter bannerPrinter = new SpringApplicationBannerPrinter(resourceLoader, this.banner);
    if (this.bannerMode == Mode.LOG) {
        return bannerPrinter.print(environment, this.mainApplicationClass, logger);
    }
    return bannerPrinter.print(environment, this.mainApplicationClass, System.out);
}
```

这里涉及的两个 API,读者可以简单了解。

（1）ResourceLoader

由类名不难理解,ResourceLoader 是一个资源加载器,这个组件其实在第 3 章中遇到过。ApplicationContext 本身实现了 ResourcePatternResolver 接口,而 ResourcePatternResolver 的父接口就是 ResourceLoader,所以 ApplicationContext 具备资源加载的能力。

默认情况下 SpringApplication 并没有预先设置的 ResourceLoader,所以最终它使用的是默认实现 DefaultResourceLoader,而默认实现中可以处理类路径、项目路径以及特定路径（如以 file: 为前缀）下的资源。

（2）SpringApplicationBannerPrinter

ResourceLoader 的利用要配合 SpringApplicationBannerPrinter 才能达到效果,SpringApplicationBannerPrinter 打印 Banner 的逻辑本身不复杂,它会先获取到 Banner,再调用其 printBanner 方法,最后附加一层封装代表 Banner 已经打印过。

代码清单 6-32 中展示了一个完成的 banner 打印逻辑调用,因为 Spring Boot 本身支持自定义的图片 banner 和文本 banner 两种方式,只要两者有其一,即可打印自定义的 banner。我们以自定义文本 banner 打印为例,可以看到最终加载文本 banner 的时候,会利用 `ResourceLoader` 获取 `banner.txt` 文件(为默认情况,也可以自行指定 banner 文件的路径)的内容后,封装为 `ResourceBanner` 返回,完成后续打印动作。

### 代码清单 6-32　打印 Banner 的流程逻辑

```
Banner print(Environment environment, Class<?> sourceClass, PrintStream out) {
    Banner banner = getBanner(environment);
    banner.printBanner(environment, sourceClass, out);
    return new PrintedBanner(banner, sourceClass);
}

private Banner getBanner(Environment environment) {
    Banners banners = new Banners();
    banners.addIfNotNull(getImageBanner(environment));
    banners.addIfNotNull(getTextBanner(environment));
    if (banners.hasAtLeastOneBanner()) {
        return banners;
    }
    if (this.fallbackBanner != null) {
        return this.fallbackBanner;
    }
    return DEFAULT_BANNER;
}

static final String BANNER_LOCATION_PROPERTY = "spring.banner.location";
static final String DEFAULT_BANNER_LOCATION = "banner.txt";

private Banner getTextBanner(Environment environment) {
    String location = environment.getProperty(BANNER_LOCATION_PROPERTY,
            DEFAULT_BANNER_LOCATION);
    Resource resource = this.resourceLoader.getResource(location);
    if (resource.exists()) {
        return new ResourceBanner(resource);
    }
    return null;
}
```

## 6.3.2　创建 IOC 容器

Banner 打印完毕后,下一步到了创建 `ApplicationContext` 的环节。第 3 章中读者已经了解过 `ApplicationContext` 在 Spring Boot 中的特殊扩展,在该环节也是根据 Web 类型区分创建不同的 `ApplicationContext` 落地实现类对象。创建的逻辑非常简单,如代码清单 6-33 所示。

### 代码清单 6-33　创建 IOC 容器的逻辑

```
public static final String DEFAULT_CONTEXT_CLASS = "org.springframework.context.annotation.AnnotationConfigApplicationContext";
public static final String DEFAULT_SERVLET_WEB_CONTEXT_CLASS = "org.springframework.boot.web.servlet.context.AnnotationConfigServletWebServerApplicationContext";
public static final String DEFAULT_REACTIVE_WEB_CONTEXT_CLASS = "org.springframework.boot.web.reactive.context.AnnotationConfigReactiveWebServerApplicationContext";
```

```java
protected ConfigurableApplicationContext createApplicationContext() {
    // 先检查有没有显式指定 ApplicationContext 的实现类型
    Class<?> contextClass = this.applicationContextClass;
    if (contextClass == null) {
        try {
            // 根据 Web 应用类型决定实例化哪个 IOC 容器
            switch (this.webApplicationType) {
                case SERVLET:
                    contextClass = Class.forName(DEFAULT_SERVLET_WEB_CONTEXT_CLASS);
                    break;
                case REACTIVE:
                    contextClass = Class.forName(DEFAULT_REACTIVE_WEB_CONTEXT_CLASS);
                    break;
                default:
                    contextClass = Class.forName(DEFAULT_CONTEXT_CLASS);
            }
        } // catch throw ex ......
    }
    return (ConfigurableApplicationContext) BeanUtils.instantiateClass(contextClass);
}
```

注意观察源码中的一个细节，创建的 ApplicationContext 类型均为注解驱动类型，而注解驱动类型的 ApplicationContext 在父类 GenericApplicationContext 的构造方法中，BeanFactory 已经被创建出来了。

```java
public GenericApplicationContext() {
    this.beanFactory = new DefaultListableBeanFactory();
}
```

多说一句，该部分代码在 Spring Boot 2.4.0 后被抽取出了一个全新的 API：ApplicationContextFactory，它的实现更为简洁，但本质逻辑完全一致，如代码清单 6-34 所示。

### 代码清单 6-34　Spring Boot 2.4.0 后的优化

```java
protected ConfigurableApplicationContext createApplicationContext() {
    return this.applicationContextFactory.create(this.webApplicationType);
}

ConfigurableApplicationContext create(WebApplicationType webApplicationType);

ApplicationContextFactory DEFAULT = (webApplicationType) -> {
    try {
        switch (webApplicationType) {
            case SERVLET:
                return new AnnotationConfigServletWebServerApplicationContext();
            case REACTIVE:
                return new AnnotationConfigReactiveWebServerApplicationContext();
            default:
                return new AnnotationConfigApplicationContext();
        }
    } // catch throw ex ......
};
```

### 6.3.3 初始化 IOC 容器

一个空的 IOC 容器创建完毕后，接下来是对容器进行必要的初始化处理，如代码清单 6-35 所示。prepareContext 方法的源码逻辑较多，下面有针对性地研究其中的关键逻辑部分。

**代码清单 6-35　初始化 IOC 容器的核心逻辑**

```java
private void prepareContext(ConfigurableApplicationContext context,
        ConfigurableEnvironment environment,
        SpringApplicationRunListeners listeners,
        ApplicationArguments applicationArguments, Banner printedBanner) {
    context.setEnvironment(environment);
    // IOC 容器的后置处理
    postProcessApplicationContext(context);
    // 应用 ApplicationContextInitializer
    applyInitializers(context);
    // 此处回调 SpringApplicationRunListeners 的 contextPrepared 方法
    // 在创建和准备 ApplicationContext 之后，但在加载之前触发
    listeners.contextPrepared(context);
    if (this.logStartupInfo) {
        // 打印当前应用激活的 profiles 信息
        logStartupInfo(context.getParent() == null);
        logStartupProfileInfo(context);
    }
    ConfigurableListableBeanFactory beanFactory = context.getBeanFactory();
    // 将打印过的 Banner 以及 main 方法传入的启动参数封装为组件，注册到 IOC 容器中
    beanFactory.registerSingleton("springApplicationArguments", applicationArguments);
    if (printedBanner != null) {
        beanFactory.registerSingleton("springBootBanner", printedBanner);
    }
    if (beanFactory instanceof DefaultListableBeanFactory) {
        ((DefaultListableBeanFactory) beanFactory)
                .setAllowBeanDefinitionOverriding(this.allowBeanDefinitionOverriding);
    }
    if (this.lazyInitialization) {
        context.addBeanFactoryPostProcessor(new LazyInitializationBeanFactoryPostProcessor());
    }
    // 获取配置源（主启动类）
    Set<Object> sources = getAllSources();
    Assert.notEmpty(sources, "Sources must not be empty");
    // 加载配置源
    load(context, sources.toArray(new Object[0]));
    // 此处回调 SpringApplicationRunListeners 的 contextLoaded 方法
    // ApplicationContext 已加载但在刷新之前触发
    listeners.contextLoaded(context);
}
```

整体概括一下 prepareContext 方法的逻辑：首先 prepareContext 方法会利用预先准备好的一些组件先初始化 IOC 容器本身，之后再处理内部 BeanFactory 的一些逻辑，最后加载配置源，其间也穿插着一些事件广播的动作。下面拆解其中的几个核心步骤讲解。

**1. IOC 容器的后置处理**

对 ApplicationContext 的后置处理，其实就是在底层注册和设置一些组件，包括 Bean 的名称生成器、资源加载器、类加载器和类型转换器，如代码清单 6-36 所示。默认情况下，此

处 beanNameGenerator 与 resourceLoader 都为 null,只有 addConversionService 属性为 true,对应的逻辑是设置类型转换器。除此之外的逻辑均不会触发。

**代码清单6-36 IOC容器的后置处理**

```java
public static final String CONFIGURATION_BEAN_NAME_GENERATOR =
        "org.springframework.context.annotation.internalConfigurationBeanNameGenerator";
protected void postProcessApplicationContext(ConfigurableApplicationContext context) {
    // 注册Bean的名称生成器
    if (this.beanNameGenerator != null) {
        context.getBeanFactory().registerSingleton(AnnotationConfigUtils.CONFIGURATION_BEAN_NAME_GENERATOR, this.beanNameGenerator);
    }
    // 设置资源加载器和类加载器
    if (this.resourceLoader != null) {
        if (context instanceof GenericApplicationContext) {
            ((GenericApplicationContext) context).setResourceLoader(this.resourceLoader);
        }
        if (context instanceof DefaultResourceLoader) {
            ((DefaultResourceLoader) context).setClassLoader(this.resourceLoader.getClassLoader());
        }
    }
    // 设置类型转换器
    if (this.addConversionService) {
        context.getBeanFactory().setConversionService(ApplicationConversionService
                .getSharedInstance());
    }
}
```

### 2. 应用 ApplicationContextInitializer

在 6.1.2 节中准备好的 ApplicationContextInitializer 在此处终于得到了应用,applyInitializers 方法会调用它的 initialize 方法,对 ApplicationContext 进行初始化操作,如代码清单 6-37 所示。

**代码清单6-37 应用所有的 ApplicationContextInitializer**

```java
protected void applyInitializers(ConfigurableApplicationContext context) {
    for (ApplicationContextInitializer initializer : getInitializers()) {
        Class<?> requiredType = GenericTypeResolver.resolveTypeArgument(initializer.getClass(), ApplicationContextInitializer.class);
        Assert.isInstanceOf(requiredType, context, "Unable to call initializer.");
        initializer.initialize(context);
    }
}
```

### 3. 获取配置源(主启动类)

6.1.5 节中提到,Spring Boot 2.x 引入了主配置类的概念。SpringApplication 中存储的配置源分为两部分,正常情况下使用最常用的 SpringApplication.run 方法传入的主启动类放在 primarySources 集合中,通过 XML 或者包扫描的方式传入的配置源一般不用,所以 getAllSources 方法最终返回的就是主启动类对应的 Class 类型,如代码清单 6-38 所示。

### 代码清单 6-38　获取所有的配置源

```
public Set<Object> getAllSources() {
    Set<Object> allSources = new LinkedHashSet<>();
    if (!CollectionUtils.isEmpty(this.primarySources)) {
        allSources.addAll(this.primarySources);
    }
    if (!CollectionUtils.isEmpty(this.sources)){
        allSources.addAll(this.sources);
    }
    return Collections.unmodifiableSet(allSources);
}
```

**4. 加载配置源**

获取配置源之后，下一步要先将这些配置源进行加载和解析。对应的 `load` 方法实现比较复杂，为保证读者能对整体流程有一个把握，而不在细节处停留太久，代码清单 6-39 中只展示了最值得关注的部分。

### 代码清单 6-39　加载配置源的核心方法（节选）

```
protected void load(ApplicationContext context, Object[] sources) {
    // logger ......
    BeanDefinitionLoader loader = createBeanDefinitionLoader(getBeanDefinitionRegistry
(context), sources);
    // setter ......
    loader.load();
}
```

对于抽取核心逻辑后的 `load` 方法，不难看出其加载配置源的核心动作只有两步：创建一个用于解析配置源的 `BeanDefinitionReader`，然后加载和解析配置源。其中 `createBeanDefinitionLoader` 方法的底层会调用 `BeanDefinitionLoader` 的构造方法，返回一个 `BeanDefinition` 的加载器。这个 `BeanDefinitionLoader` 的类结构比较值得关注，如代码清单 6-40 所示。

### 代码清单 6-40　BeanDefinitionLoader 的内部核心成员

```
class BeanDefinitionLoader {
    private final Object[] sources;
    private final AnnotatedBeanDefinitionReader annotatedReader;
    private final XmlBeanDefinitionReader xmlReader;
    private final ClassPathBeanDefinitionScanner scanner;
    private BeanDefinitionReader groovyReader;
    private ResourceLoader resourceLoader;
```

由 `BeanDefinitionLoader` 的成员属性可以得知，`BeanDefinitionLoader` 内部组合了几个与 `BeanDefinition` 相关的核心组件，包括 `AnnotatedBeanDefinitionReader`（注解驱动的 `BeanDefinition` 解析器）、`XmlBeanDefinitionReader`（**XML** 定义的 `BeanDefinition` 解析器）、`ClassPathBeanDefinitionScanner`（类路径下的 `BeanDefinition` 包扫描器）。这种设计明显是**外观模式**的体现，即利用一个 `BeanDefinitionLoader` 充当统一入口，针对传入的配置源选择最合适的底层组件来进行实

际处理。

具体从 `BeanDefinitionLoader` 的实现中可以看到，核心的 `load` 方法会不断测试传入的配置源类型，从而将配置源分配给合适的组件去处理。代码清单 6-41 中已经列出了不同的配置源对应的组件，感兴趣的读者可以自行借助 IDE 在源码中查看，这里不再展开解读。

**代码清单 6-41　load 方法分发不同类型的配置源给合适的组件处理**

```java
private int load(Object source) {
    Assert.notNull(source, "Source must not be null");
    // AnnotatedBeanDefinitionReader
    if (source instanceof Class<?>) {
        return load((Class<?>) source);
    }
    // XmlBeanDefinitionReader
    if (source instanceof Resource) {
        return load((Resource) source);
    }
    // ClassPathBeanDefinitionScanner
    if (source instanceof Package) {
        return load((Package) source);
    }
    // ClassPathBeanDefinitionScanner
    if (source instanceof CharSequence) {
        return load((CharSequence) source);
    }
    throw new IllegalArgumentException("Invalid source type " + source.getClass());
}
```

### 6.3.4　刷新 IOC 容器

初始化 `ApplicationContext` 后，下面到了整个启动环节中最困难、最复杂的一步：IOC 容器的刷新，如代码清单 6-42 所示。由于该部分过于复杂，本书单独将其放到第 7 章详细讲解，本章不作展开。

**代码清单 6-42　刷新 IOC 容器的最终是调用 ApplicationContext 的 refresh 方法**

```java
private void refreshContext(ConfigurableApplicationContext context){
    // 注册关闭时回调的钩子 ......
    refresh((ApplicationContext) context);
}

protected void refresh(ConfigurableApplicationContext applicationContext){
    applicationContext.refresh();
}
```

### 6.3.5　Spring Boot 2.4.x 的新特性

Spring Boot 版本升级到 2.4.x 后，底层的 Spring Framework 也升级到了 5.3.x，这里引入了一些新的特性，读者可以简单了解。

#### 1. ApplicationStartup

在创建好 IOC 容器（`createApplicationContext`）之后初始化（`prepareContext`）

之前，有这样一行代码：

```
context.setApplicationStartup(this.applicationStartup); // ApplicationStartup.DEFAULT
```

这个 ApplicationStartup 是什么？引入它可以解决什么问题呢？

（1）性能分析

在 Spring Boot 2.4（即 Spring Framework 5.3）之前，如果项目的启动速度很慢，是很难确定启动慢的根本原因的。Spring Framework 开发团队考虑到性能分析和优化的需求，于是在 Spring Framework 5.3 中引入了一个性能指标统计工具，也就是这个**启动度量** ApplicationStartup，该指标可以准确地定位到应用启动的哪个环节耗时长，对分析慢启动问题非常有帮助。

（2）ApplicationStartup 的使用

Spring Boot 考虑到 ApplicationStartup 在程序启动时的性能折损，默认情况下不会启用该性能分析器。如果需要启用则需要手动在主启动类中声明，如代码清单 6-43 所示。

### 代码清单 6-43　使用 ApplicationStartup 的方式

```java
@SpringBootApplication
public class SpringBootQuickstartApplication {

    public static void main(String[] args) {
        new SpringApplicationBuilder(SpringBootQuickstartApplication.class)
                .applicationStartup(new BufferingApplicationStartup(10000))
                .run(args);
    }
}
```

如此设置之后，还需要引入 Spring Boot 的监控依赖。

```xml
<dependency>
    <groupId>org.springframework.boot</groupId>
    <artifactId>spring-boot-starter-actuator</artifactId>
</dependency>
```

最后在 application.properties 中配置暴露的监控端点，即可开启启动度量的信息获取接口。

```
management.endpoints.web.exposure.include=startup
```

准备工作完成后，即可启动 Spring Boot 应用。使用 Postman 等 HTTP 客户端工具，以 POST 请求访问 /actuator/startup 接口，即可获得有关启动阶段的所有环节的耗时。

下面的一次实际运行生成的几段 JSON：

```
//配置类解析阶段，共耗费 0.656s，解析 104 个类
{
    "endTime": "2021-12-23T12:45:18.902Z",
    "duration": "PT0.656S",
    "startTime": "2021-12-23T12:45:18.246Z",
    "startupStep": {
        "name": "spring.context.config-classes.parse",
        "id": 9,
        "tags": [
            {
                "key": "classCount",
```

```
        }
    ],
    "parentId": 8
}
//HelloController 的创建，耗费 0.001s
{
    "endTime": "2021-12-23T12:45:19.428Z",
    "duration": "PT0.001S",
    "startTime": "2021-12-23T12:45:19.427Z",
    "startupStep": {
        "name": "spring.beans.instantiate",
        "id": 98,
        "tags": [
            {
                "key": "beanName",
                "value": "helloController"
            }
        ],
        "parentId": 4
    }
}
//整个 Bean 的后置处理阶段，共耗时 0.823s
{
    "endTime": "2021-12-23T12:45:19.014Z",
    "duration": "PT0.823S",
    "startTime": "2021-12-23T12:45:18.191Z",
    "startupStep": {
        "name": "spring.context.beans.post-process",
        "id": 5,
        "tags": [],
        "parentId": 4
    }
}
```

通过这些指标就可以精准地定位到哪些环节耗时较长，从而采取合适的针对性手段优化应用。

**2. BootstrapRegistry 的关闭**

读者已经了解到 BootstrapRegistry 与 BootstrapContext 是 Spring Boot 2.4.x 新引入的 API，之前的部分只介绍了初始化和应用逻辑的调用，那么关闭又在何时触发？

答案是在 ApplicationContext 的预初始化阶段。当 ApplicationContextInitializer 被应用并且 contextPrepared 事件触发（即 ApplicationContextInitializedEvent 事件广播）后，BootstrapRegistry 关闭，如代码清单 6-44 所示。

**代码清单 6-44　BootstrapContext 的关闭时机**

```
private void prepareContext(DefaultBootstrapContext bootstrapContext,
    ConfigurableApplicationContext context,
    ConfigurableEnvironment environment,
    SpringApplicationRunListeners listeners,
    ApplicationArguments applicationArguments,Banner printedBanner){
context.setEnvironment(environment);
postProcessApplicationContext(context);
applyInitializers(context);
```

```
        listeners.contextPrepared(context);
        // 此处关闭
        bootstrapContext.close(context);
        // ......
}
```

## 6.4 IOC 容器刷新后的回调

IOC 容器刷新完毕后仅剩余一些收尾工作,这部分的内容不多,对应 SpringApplication 的 run 方法中的源码如代码清单 6-45 所示。

**代码清单 6-45　SpringApplication#run 片段（3）**

```
        // ......
        afterRefresh(context, applicationArguments);
        stopWatch.stop();
        // 打印启动耗时的日志
        if (this.logStartupInfo) {
            new StartupInfoLogger(this.mainApplicationClass).logStarted(getApplicationLog(),
                    stopWatch);
        }
        // 此处回调 SpringApplicationRunListeners 的 started 方法
        // IOC 容器已刷新但未调用 CommandLineRunners 和 ApplicationRunners 时触发
        listeners.started(context);
        // 回调所有的运行器
        callRunners(context, applicationArguments);
} // catch 广播 failed 事件 ......

try{
        // 此处回调 SpringApplicationRunListeners 的 running 方法
        // 在 run 方法彻底完成之前触发
        listeners.running(context);
} // catch ......
return context;
}
```

可以看到 SpringApplicationRunListener 的所有方法均被调用（包括捕获异常后广播 failed 事件也在内）。除此之外,唯一值得关注的是 try-catch 块中的最后一行 callRunners 方法。

### 回调运行器

运行器是 Spring Boot 设计的用于应用完全启动后给开发者予以回调的两个扩展接口。它们本身没有对应的详细解释,甚至官方文档中也只是写明了如何使用,而没有解释设计这两个接口的缘由,好在学习 Spring Boot 时读者都知道如何使用。

观察 callRunners 方法的实现,可以发现这部分是以"妥协"的方式实现的,由于有两种不同的接口,而两种接口的设计又极其相似,因此不得不先全部收集,再分别调用,如代码清单 6-46 所示。

**代码清单 6-46　分别回调 ApplicationRunner 和 CommandLineRunner**

```
// 从容器中获取所有的 ApplicationRunner 和 CommandLineRunner
private void callRunners(ApplicationContext context, ApplicationArguments args) {
```

```java
        List<Object> runners = new ArrayList<>();
        runners.addAll(context.getBeansOfType(ApplicationRunner.class).values());
        runners.addAll(context.getBeansOfType(CommandLineRunner.class).values());
        AnnotationAwareOrderComparator.sort(runners);
        // 依次回调
        for (Object runner : new LinkedHashSet<>(runners)) {
            if (runner instanceof ApplicationRunner) {
                callRunner((ApplicationRunner) runner, args);
            }
            if (runner instanceof CommandLineRunner) {
                callRunner((CommandLineRunner) runner, args);
            }
        }
    }

    private void callRunner(ApplicationRunner runner, ApplicationArguments args) {
        try {
            (runner).run(args);
        } // catch ......
    }

    private void callRunner(CommandLineRunner runner, ApplicationArguments args) {
        try {
            (runner).run(args.getSourceArgs());
        } // catch ......
    }
```

这里多解释一句，如果翻看这两个接口组件的文档注释中的 since，会发现一个没有标注，另一个是 Spring Boot1.3.0，说明它们都来自 Spring Boot1.x。它们本来是用于在特定的时机（应用完全启动后）执行一些操作，奈何 Spring Boot2.x 后扩展了事件，可以通过监听 ApplicationStartedEvent 来实现与这两个组件一样的效果。换句话说，这两个组件已经被隐式"替代"，读者不必过多深究，平时在开发中还可以正常使用。

至此，一个完整的 Spring Boot 应用的启动流程完全结束，Spring Boot 应用启动成功。

## 6.5 小结

本章从 Spring Boot 主启动类的 main 方法开始，逐步剖析、拆解 Spring Boot 的启动全流程，并针对不同的 Spring Boot 版本对比了解不同时期的不同设计。Spring Boot 的核心还是依赖 Spring Framework 的 IOC 容器，即 ApplicationContext，而 ApplicationContext 的真正初始化是它的 refresh 刷新动作。第 7 章会针对 ApplicationContext 的核心 refresh 方法，全面剖析整个容器刷新的工作机制。

# 第 7 章 IOC 容器的刷新

**本章主要内容:**
- ApplicationContext 的容器刷新全流程机制解析;
- BeanDefinitionRegistryPostProcessor 与 BeanFactoryPostProcessor 的加载机制;
- BeanPostProcessor、ApplicationListener 等核心组件的注册;
- 一个 bean 对象的完整创建与初始化流程。

通过第 6 章的内容,想必读者已经完整地了解一个 Spring Boot 应用的启动过程的原理。整个过程中有一个关键步骤,其内部相当复杂而庞大,它就是 IOC 容器初始化环节的刷新动作,如代码清单 7-1 所示。

**代码清单 7-1  SpringApplication 中 IOC 容器刷新的环节**

```
// ......
context = createApplicationContext();
prepareContext(context, environment, listeners, applicationArguments, printedBanner);
// 刷新 ApplicationContext
refreshContext(context);
afterRefresh(context, applicationArguments);
// ......
```

这个 refreshContext 方法的核心会在底层调用 AbstractApplicationContext 的 refresh 方法, 而 refresh 方法的逻辑相当复杂, 一共包含 13 步, 如代码清单 7-2 所示。

**代码清单 7-2  AbstractApplicationContext 的 refresh 方法概览**

```
public void refresh() throws BeansException, IllegalStateException {
    synchronized (this.startupShutdownMonitor) {
        // Prepare this context for refreshing.
        // 1. 初始化前的预处理
        prepareRefresh();

        // Tell the subclass to refresh the internal bean factory.
        // 2. 获取 BeanFactory,加载所有 Bean 的定义信息(未实例化)
        ConfigurableListableBeanFactory beanFactory = obtainFreshBeanFactory();

        // Prepare the bean factory for use in this context.
        // 3. BeanFactory 的预处理配置
        prepareBeanFactory(beanFactory);
```

## 7.1 初始化前的预处理

```
    try {
        // Allows post-processing of the bean factory in context subclasses.
        // 4. 准备 BeanFactory 完成后进行的后置处理
        postProcessBeanFactory(beanFactory);

        // Invoke factory processors registered as beans in the context.
        // 5. 执行 BeanFactory 创建后的后置处理器
        invokeBeanFactoryPostProcessors(beanFactory);

        // Register bean processors that intercept bean creation.
        // 6. 注册 Bean 的后置处理器
        registerBeanPostProcessors(beanFactory);

        // Initialize message source for this context.
        // 7. 初始化 MessageSource
        initMessageSource();

        // Initialize event multicaster for this context.
        // 8. 初始化事件广播器
        initApplicationEventMulticaster();

        // Initialize other special beans in specific context subclasses.
        // 9. 供子类扩展的模板方法 onRefresh
        onRefresh();

        // Check for listener beans and register them.
        // 10. 注册监听器
        registerListeners();

        // 至此，BeanFactory 已创建并初始化完成

        // Instantiate all remaining (non-lazy-init) singletons.
        // 11. 初始化所有剩下的单实例 bean 对象
        finishBeanFactoryInitialization(beanFactory);

        // Last step: publish corresponding event.
        // 12. 完成容器的创建工作
        finishRefresh();
    }// catch ......
    finally {
        // Reset common introspection caches in Spring's core, since we
        // might not ever need metadata for singleton beans anymore...
        // 13. 清除缓存
        resetCommonCaches();
    }
  }
}
```

本章会用一整章的内容，力求把 refresh 方法彻底讲明白。读者在学习 IOC 容器的生命周期时，要注意先学习整体框架流程，再深入细节讨论，切勿本末倒置。

> 💡 提示：本章涉及的源码非常多，源码中不乏有日志打印的内容，为了在有限的篇幅内展现更多有意义的源码，本章中会尽可能地省略无关的日志打印、异常捕获的源码。

下面对方法逐个拆解、剖析。

## 7.1 初始化前的预处理

初始化前的预处理如代码清单 7-3 所示。

**代码清单 7-3　prepareRefresh——初始化前的预处理**

```java
protected void prepareRefresh() {
    // 记录刷新动作执行的时间
    this.startupDate = System.currentTimeMillis();
    // 标记当前 IOC 容器已激活
    this.closed.set(false);
    this.active.set(true);

    // 7.1.1 初始化属性配置
    initPropertySources();

    // 属性校验（通常无实际操作）
    getEnvironment().validateRequiredProperties();

    // 监听器的初始化（兼顾可以反复刷新的 IOC 容器）
    if (this.earlyApplicationListeners == null) {
        this.earlyApplicationListeners = new LinkedHashSet<>(this.applicationListeners);
    } else {
        this.applicationListeners.clear();
        this.applicationListeners.addAll(this.earlyApplicationListeners);
    }

    // 7.1.2 初始化早期事件的集合
    this.earlyApplicationEvents = new LinkedHashSet<>();
}
```

整个 `refresh` 方法的第一步是在容器刷新核心动作开始前的一些预处理，粗略地来看，`prepareRefresh` 方法中大多数的动作都是前置性的准备，有两个步骤比较关键。

### 7.1.1 初始化属性配置

初始化属性配置的方法如代码清单 7-4 所示。

**代码清单 7-4　初始化属性配置的方法**

```java
protected void initPropertySources() {
    //对于子类，默认情况下不执行任何操作
}
```

初始化 PropertySources 的动作是想预先初始化一些属性配置（properties）到 IOC 容器中（实际是存放到 Environment 中）。通过代码清单 7-4 可以明显地看出 `initPropertySources` 是一个模板方法，注释也说明了默认不会做任何事情，可以留给子类重写。借助 IDE 发现这个方法在基于 Web 环境的 ApplicationContext 子类中进行了重写，由于默认情况下引入 spring-boot-starter-web 依赖时创建的 IOC 容器实现是 AnnotationConfigServletWebServerApplicationContext，而这个实现类继承自

GenericWebApplicationContext，因此下面来看 GenericWebApplicationContext 的 initPropertySources 方法，如代码清单 7-5 所示。

**代码清单 7-5　GenericWebApplicationContext 初始化属性配置源**
```java
protected void initPropertySources() {
    ConfigurableEnvironment env = getEnvironment();
    if (env instanceof ConfigurableWebEnvironment) {
        ((ConfigurableWebEnvironment) env).initPropertySources(this.servletContext, null);
    }
}

public void initPropertySources(@Nullable ServletContext servletContext,
        @Nullable ServletConfig servletConfig) {
    WebApplicationContextUtils.initServletPropertySources(getPropertySources(),
            servletContext, servletConfig);
}
```

可以发现，initPropertySources 方法希望获取一个 ConfigurableWebEnvironment，并配置当前的 ServletContext。而 WebApplicationContextUtils 的静态 initServletPropertySources 方法中做的事情是将 ServletContext 以及 ServletConfig 当作一个属性配置源注入 Environment 中，如代码清单 7-6 所示。后续从 Environment 中获取属性时，不仅会加载 application.properties 等常规配置文件的内容，还会加载 ServletContext 以及 ServletConfig 的初始化属性（init-param）。

**代码清单 7-6　将 ServletContext 当作属性配置源注入 Environment 中**
```java
public static final String SERVLET_CONTEXT_PROPERTY_SOURCE_NAME = "servletContext Init Params";
public static final String SERVLET_CONFIG_PROPERTY_SOURCE_NAME = "servletConfigInitParams";

public static void initServletPropertySources(MutablePropertySources sources,
        @Nullable ServletContext servletContext,@Nullable ServletConfig servletConfig){
    Assert.notNull(sources, "'propertySources' must not be null");
    String name = StandardServletEnvironment.SERVLET_CONTEXT_PROPERTY_SOURCE_NAME;
    if (servletContext != null && sources.contains(name)
            && sources.get(name) instanceof StubPropertySource) {
        sources.replace(name, new ServletContextPropertySource(name, servletContext));
    }
    name = StandardServletEnvironment.SERVLET_CONFIG_PROPERTY_SOURCE_NAME;
    if (servletConfig != null && sources.contains(name)
            && sources.get(name) instanceof StubPropertySource) {
        sources.replace(name, new ServletConfigPropertySource(name, servletConfig));
    }
}

public class ServletContextPropertySource extends EnumerablePropertySource<ServletContext> {
    public ServletContextPropertySource(String name, ServletContext servletContext) {
        super(name, servletContext);
    }

    public String getProperty(String name) {
        // 获取属性的源头是 ServletContext 的初始化参数
```

```
        return this.source.getInitParameter(name);
    }
    // ……
}
```

由代码清单 7-6 可以看出，`initServletPropertySources` 方法会将 `Servlet Context` 与 `ServletConfig` 封装为一个 `PropertySource`，存入 `Environment` 内置的聚合对象 `MutablePropertySources` 中。每次从 `Environment` 中获取配置属性时，实际上是从 `MutablePropertySources` 中取值，而 `MutablePropertySources` 做的事情是遍历自身聚合的所有 `PropertySource` 并尝试获取指定的配置属性，直到找到配置属性的值时返回，或者找不到配置属性而返回 `null`（或预先指定的默认值）。当 `ServletContext` 聚合到 `MutablePropertySources` 中后，就相当于 `ServletContext` 的初始化参数也成为"属性配置源"的一部分。

### 7.1.2　初始化早期事件的集合

`prepareRefresh` 方法的最后一行代码是一个 `earlyApplicationEvent` 集合的初始化动作，如代码清单 7-7 所示。从变量名上理解，`earlyApplicationEvents` 是一个"早期事件"的集合，那么早期事件又是什么？它解决了什么问题呢？

**代码清单 7-7　earlyApplicationEvents 的初始化**

```
private Set<ApplicationEvent> earlyApplicationEvents;

this.earlyApplicationEvents = new LinkedHashSet<>();
```

其实 `earlyApplicationEvent` 的设计目的是解决 Spring Framework 3.x 的一个场景 bug。如果一个类同时实现了 `ApplicationEventPublisherAware` 与 `BeanPostProcessor`（或其他后置处理器接口），会因为 `BeanPostProcessor` 的创建时机靠前（后面会讲到）而导致提前回调 `ApplicationEventPublisherAware` 的 `setter` 方法，但是又因为此时 `ApplicationEventMulticaster` 还没有初始化，所以导致如果在此期间广播事件，则事件暂时无法广播给监听器，监听器无法接收到事件，从而产生 bug。`earlyApplicationEvent` 的设计是在所有 `ApplicationListener` 还没有创建之前把该阶段之前的事件都保存起来，等到所有 `ApplicationListener` 都创建完成后再逐个广播，如此操作就可以确保所有监听器都监听到自己本应该监听到的事件。

## 7.2　obtainFreshBeanFactory——初始化 BeanFactory

代码清单 7-8 所示为初始化 `BeanFactory` 的代码。

**代码清单 7-8　初始化 BeanFactory**

```
protected ConfigurableListableBeanFactory obtainFreshBeanFactory() {
    // 刷新 BeanFactory
    refreshBeanFactory();
    return getBeanFactory();
}
```

```
public final ConfigurableListableBeanFactory getBeanFactory() {
    return this.beanFactory;
}
```

obtainFreshBeanFactory 方法的动作非常简单：先刷新 BeanFactory 再获取它。获取的动作相当简单，该方法会返回当前 ApplicationContext 中内嵌的 BeanFactory（即 DefaultListableBeanFactory）。本节主要探究 BeanFactory 是如何刷新的。

### 7.2.1 注解驱动的 refreshBeanFactory

refreshBeanFactory 方法本身是一个抽象方法，需要子类去实现：

```
protected abstract void refreshBeanFactory() throws BeansException, IllegalStateException;
```

对于 XML 配置文件驱动的 IOC 容器和注解驱动的 IOC 容器，它们分别有一个具体的实现。借助 IDE 可以发现 GenericApplicationContext 和 AbstractRefreshableApplicationContext 分别重写了这个方法。默认引入 WebMvc 依赖时对应的 AnnotationConfigServletWebServerApplicationContext 继承自 GenericApplicationContext，所以接下来先看它的具体方法实现，如代码清单 7-9 所示。

**代码清单 7-9　GenericApplicationContext 的 refreshBeanFactory 方法实现**

```
protected final void refreshBeanFactory() throws IllegalStateException {
    if (!this.refreshed.compareAndSet(false, true)) {
        throw new IllegalStateException("GenericApplicationContext does not support multiple refresh attempts: just call 'refresh' once");
    }
    this.beanFactory.setSerializationId(getId());
}
```

通过代码清单 7-9 可以发现方法逻辑很简单，仅仅是设置了 BeanFactory 的序列化 ID 而已。

### 7.2.2 XML 驱动的 refreshBeanFactory

基于 XML 配置文件的 IOC 容器在这一步要做的事情更复杂，不过由于 Spring Boot 全系使用注解驱动的 IOC 容器，因此读者对于这部分可以参照代码清单 7-10 的内容，配合注释作简单了解，对于方法内部的具体实现本书不再展开。

**代码清单 7-10　AbstractRefreshableApplicationContext 的 refreshBeanFactory 方法实现**

```
protected final void refreshBeanFactory() throws BeansException {
    if (hasBeanFactory()) {
        // 允许重复刷新的 IOC 容器，内部的 Bean 也是可以重新加载的
        // 故此处有销毁 Bean 和关闭 BeanFactory 的动作
        destroyBeans();
        closeBeanFactory();
    }
    try {
        // 创建 BeanFactory（会组合父 BeanFactory 形成层级关系）
        DefaultListableBeanFactory beanFactory = createBeanFactory();
        beanFactory.setSerializationId(getId());
```

```
    // 自定义配置BeanFactory
    customizeBeanFactory(beanFactory);
    // 解析、加载XML中定义的BeanDefinition
    loadBeanDefinitions(beanFactory);
    synchronized (this.beanFactoryMonitor) {
        this.beanFactory = beanFactory;
    }
}// catch ......
}
```

## 7.3 prepareBeanFactory——BeanFactory 的预处理动作

BeanFactory 的预处理动作如代码清单 7-11 所示。

**代码清单 7-11  BeanFactory 的预处理动作**

```
protected void prepareBeanFactory(ConfigurableListableBeanFactory beanFactory) {
    // 设置BeanFactory的类加载器、表达式解析器等
    beanFactory.setBeanClassLoader(getClassLoader());
    beanFactory.setBeanExpressionResolver(new StandardBeanExpressionResolver(beanFactory.getBeanClassLoader()));
    beanFactory.addPropertyEditorRegistrar(new ResourceEditorRegistrar(this, getEnvironment()));

    // 7.3.1 配置一个可回调注入ApplicationContext的BeanPostProcessor
    beanFactory.addBeanPostProcessor(new ApplicationContextAwareProcessor(this));
    beanFactory.ignoreDependencyInterface(EnvironmentAware.class);
    beanFactory.ignoreDependencyInterface(EmbeddedValueResolverAware.class);
    beanFactory.ignoreDependencyInterface(ResourceLoaderAware.class);
    beanFactory.ignoreDependencyInterface(ApplicationEventPublisherAware.class);
    beanFactory.ignoreDependencyInterface(MessageSourceAware.class);
    beanFactory.ignoreDependencyInterface(ApplicationContextAware.class);

    // 7.3.2 自动注入的支持
    beanFactory.registerResolvableDependency(BeanFactory.class, beanFactory);
    beanFactory.registerResolvableDependency(ResourceLoader.class, this);
    beanFactory.registerResolvableDependency(ApplicationEventPublisher.class, this);
    beanFactory.registerResolvableDependency(ApplicationContext.class, this);

    // 7.3.3 配置一个可加载所有监听器的组件
    beanFactory.addBeanPostProcessor(new ApplicationListenerDetector(this));

    // LoadTimeWeaving的支持
    if (beanFactory.containsBean(LOAD_TIME_WEAVER_BEAN_NAME)) {
        beanFactory.addBeanPostProcessor(new LoadTimeWeaverAwareProcessor(beanFactory));
        beanFactory.setTempClassLoader(new ContextTypeMatchClassLoader(beanFactory.getBeanClassLoader()));
    }

    // 向BeanFactory中注册Environment、系统配置属性、系统环境的信息
    // Environment本身对于BeanFactory来讲也是一个Bean
    if (!beanFactory.containsLocalBean(ENVIRONMENT_BEAN_NAME)) {
        beanFactory.registerSingleton(ENVIRONMENT_BEAN_NAME, getEnvironment());
    }
```

```java
    if (!beanFactory.containsLocalBean(SYSTEM_PROPERTIES_BEAN_NAME)) {
        beanFactory.registerSingleton(SYSTEM_PROPERTIES_BEAN_NAME,
                getEnvironment().getSystemProperties());
    }
    if (!beanFactory.containsLocalBean(SYSTEM_ENVIRONMENT_BEAN_NAME)) {
        beanFactory.registerSingleton(SYSTEM_ENVIRONMENT_BEAN_NAME,
                getEnvironment().getSystemEnvironment());
    }
}
```

纵观 prepareBeanFactory 方法的逻辑,读者是否有一种感觉:prepareBeanFactory 方法做的事情看上去很多,但仔细观察又不是很多？如果你也这样认为,说明你的感觉是对的。看似很长的源码,只要做好分段和提取,prepareBeanFactory 方法就很容易理解。

### 7.3.1 ApplicationContextAwareProcessor

在设置了 BeanFactory 的类加载器等组件后,第一个关键动作是注册 ApplicationContextAwareProcessor,并且让 BeanFactory 忽略几种依赖的接口。下面我们分别拆解这两步。

#### 1. ApplicationContextAwareProcessor 的作用

ApplicationContextAwareProcessor 本身是一个 BeanPostProcessor,它的作用是给需要注入 ApplicationContext 或其他相关的 bean 对象注入对应的组件。通过分析其 postProcessBeforeInitialization 方法可以快速理解它的作用,如代码清单 7-12 所示。

**代码清单 7-12　ApplicationContextAwareProcessor 的核心源码**

```java
public Object postProcessBeforeInitialization(Object bean, String beanName)
        throws BeansException {
    // 如果被处理的 Bean 不是指定的 Aware 类型接口,则不予处理
    if (!(bean instanceof EnvironmentAware
            || bean instanceof EmbeddedValueResolverAware
            || bean instanceof ResourceLoaderAware
            || bean instanceof ApplicationEventPublisherAware
            || bean instanceof MessageSourceAware
            || bean instanceof ApplicationContextAware)) {
        return bean;
    }

    // ......
    // 执行 Aware 接口的回调注入
    invokeAwareInterfaces(bean);

    return bean;
}

private void invokeAwareInterfaces(Object bean) {
    // 判断实现的接口,进行强转,调用其 setter 方法
    if (bean instanceof EnvironmentAware) {
        ((EnvironmentAware) bean).setEnvironment(this.applicationContext.getEnvironment());
    }
    if (bean instanceof EmbeddedValueResolverAware) {
        ((EmbeddedValueResolverAware) bean).setEmbeddedValueResolver(this.embeddedValueResolver);
    }
```

```java
if (bean instanceof ResourceLoaderAware) {
    ((ResourceLoaderAware) bean).setResourceLoader(this.applicationContext);
}
if (bean instanceof ApplicationEventPublisherAware) {
    ((ApplicationEventPublisherAware) bean).setApplicationEventPublisher(this.applicationContext);
}
if (bean instanceof MessageSourceAware) {
    ((MessageSourceAware) bean).setMessageSource(this.applicationContext);
}
if (bean instanceof ApplicationContextAware) {
    ((ApplicationContextAware) bean).setApplicationContext(this.applicationContext);
}
}
```

postProcessBeforeInitialization 方法中会判断一个 bean 对象的所属类是否实现了指定的内置 Aware 系列接口。只要检测到 bean 对象所属类有一个 Aware 系列接口实现，ApplicationContextAwareProcessor 就会尝试将其强转为对应的 Aware 接口，并调用接口对应的 setter 方法完成 Aware 接口的回调注入。

> 提示：留意一点，由于 ApplicationContext 本身已经继承了 ResourceLoader、ApplicationEventPublisher 和 MessageSource 接口，因此当注入这些父接口类型时，本质上还是注入了 ApplicationContext 本身。

#### 2. ignoreDependencyInterface 的作用

ApplicationContext 的内部既有 BeanPostProcessor 的增强设计，又有 BeanFactory 负责自动依赖注入，这就意味着在 Aware 系列接口上必然会产生选择。Spring Framework 选择使用 BeanPostProcessor 作为注入的逻辑，而放弃这些 Aware 接口在 BeanFactory 中实现的自动依赖注入。换句话说，当使用 BeanFactory 的自动依赖注入，遇到 Aware 系列接口时将不予注入。

### 7.3.2 自动注入的支持

处理完 Aware 类型的接口后，下面它向 BeanFactory 中注册了几个接口与对象的对应关系，如代码清单 7-13 所示。

**代码清单 7-13　registerResolvableDependency 设置自动注入的支持**

```java
// 自动注入的支持
beanFactory.registerResolvableDependency(BeanFactory.class, beanFactory);
beanFactory.registerResolvableDependency(ResourceLoader.class, this);
beanFactory.registerResolvableDependency(ApplicationEventPublisher.class, this);
beanFactory.registerResolvableDependency(ApplicationContext.class, this);
```

从方法名和传入参数的特性可以理解，registerResolvableDependency 方法的作用是使 BeanFactory 遇到指定的类型需要注入时，直接使用指定的对象进行注入。方法的参数明显是一对映射，BeanFactory 的内部就应该有一个 Map 专门负责存储这些被指定的接口和对应实现类的对象之间的映射关系，以备后续进行依赖注入时遇到这些类型就可以直接从 Map 中提取。

进入 registerResolvableDependency 方法的内部，该方法由 BeanFactory 的最终落地 DefaultListableBeanFactory 实现，如代码清单 7-14 所示。ResolvableDependencies 集合就是负责存储类型和对应实现类的集合，该集合主要记录一些特殊的 bean 对象类型（大多是 Spring Framework 内置的核心功能型 API）。

**代码清单 7-14　注册类型与注入对象的映射关系**

```java
/**从依赖项类型映射到相应的自动装配值*/
private final Map<Class<?>, Object> resolvableDependencies = new ConcurrentHashMap<>(16);

public void registerResolvableDependency(Class<?> dependencyType,
        @Nullable Object autowiredValue) {
    // validate
    if (autowiredValue != null) {
        // validate
        this.resolvableDependencies.put(dependencyType, autowiredValue);
    }
}
```

### 7.3.3　ApplicationListenerDetector

处理完依赖类型后，prepareBeanFactory 方法最后会向 BeanFactory 中注册一个后置处理器 ApplicationListenerDetector，之前的章节中没有接触过它，但从类名上可以看出 ApplicationListenerDetector 是一个检测 ApplicationListener 接口的组件，借助 javadoc 可以获取一点信息。

> BeanPostProcessor that detects beans which implement the ApplicationListener interface. This catches beans that can't reliably be detected by getBeanNamesForType and related operations which only work against top-level beans.
>
> ApplicationListenerDetector 是一个用于检测实现 ApplicationListener 类型 Bean 的后置处理器，它可以捕获 getBeanNamesForType 以及仅对顶级 Bean 有效的相关操作无法可靠检测到的 Bean。

文档注释虽然不太容易理解，但基本信息已经清楚，ApplicationListenerDetector 负责的工作是在 bean 对象的初始化阶段检测当前 bean 对象是否为 ApplicationListener，如果是则会进行一些额外的处理。对应的处理逻辑如代码清单 7-15 所示。

**代码清单 7-15　初始化回调时检测当前 bean 对象是否为 ApplicationListener**

```java
public Object postProcessAfterInitialization(Object bean, String beanName) {
    if (bean instanceof ApplicationListener) {
        Boolean flag = this.singletonNames.get(beanName);
        if (Boolean.TRUE.equals(flag)) {
            this.applicationContext.addApplicationListener((ApplicationListener<?>) bean);
        } else if (Boolean.FALSE.equals(flag)) {
            // logger
            this.singletonNames.remove(beanName);
        }
    }
    return bean;
}
```

由代码清单 7-15 的实现逻辑可以发现，ApplicationListenerDetector 会检查被后置处理的 bean 对象是否为 ApplicationListener 类型，如果是并且当前 bean 对象同时是一个单实例 bean 对象，则会加入 ApplicationContext 的监听器集合中（ApplicationContext 本身也是事件发布器，内部组合了事件广播器）。

另外，留意一个细节，ApplicationListenerDetector 直接实现的接口是 DestructionAwareBeanPostProcessor，说明它还有对 bean 对象销毁阶段的处理。观察对应实现的 postProcessBeforeDestruction 方法可以发现，ApplicationListenerDetector 会在 bean 对象的销毁阶段将监听器类型的 bean 对象逐个从事件广播器中移除，如代码清单 7-16 所示。

**代码清单 7-16　监听器销毁时从事件广播器中移除**

```
public void postProcessBeforeDestruction(Object bean, String beanName) {
    if (bean instanceof ApplicationListener) {
        try {
            ApplicationEventMulticaster multicaster = this.applicationContext.getApplicationEventMulticaster();
            // 将监听器从事件广播器中移除
            multicaster.removeApplicationListener((ApplicationListener<?>) bean);
            multicaster.removeApplicationListenerBean(beanName);
        } // catch ex ignore ......
    }
}
```

简单总结，ApplicationListenerDetector 的核心作用是将 IOC 容器中注册的所有单实例监听器应用到事件广播机制中，在监听器创建时注册到事件广播器，监听器销毁时从事件广播器移除。

## 7.4　postProcessBeanFactory——BeanFactory 的后置处理

在 AbstractApplicationContext 中，这个方法也是一个模板方法，如代码清单 7-17 所示。

**代码清单 7-17　postProcessBeanFactory 默认是空实现**

```
protected void postProcessBeanFactory(ConfigurableListableBeanFactory beanFactory) {
}
```

在默认整合 WebMvc 时，启用的 AnnotationConfigServletWebServerApplicationContext 中重写了 postProcessBeanFactory 方法并做了三件事情：回调父类 ServletWebServerApplicationContext 的 postProcessBeanFactory 方法；组件扫描；注册解析手动传入的配置类。对应的实现如代码清单 7-18 所示。

**代码清单 7-18　AnnotationConfigServletWebServerApplicationContext 的后置处理**

```
protected void postProcessBeanFactory(ConfigurableListableBeanFactory beanFactory) {
    // 7.4.1 回调父类方法
    super.postProcessBeanFactory(beanFactory);
    // 7.4.2 组件扫描
    if (this.basePackages != null && this.basePackages.length > 0) {
```

```
            this.scanner.scan(this.basePackages);
        }
        // 7.4.2 解析配置类
        if (!this.annotatedClasses.isEmpty()) {
            this.reader.register(ClassUtils.toClassArray(this.annotatedClasses));
        }
    }
```

## 7.4.1　回调父类方法

第一步回调的父类是来自继承的 `ServletWebServerApplicationContext`，它的 post ProcessBeanFactory 方法中注册了一个后置处理器和一组基于 Web 应用的作用域，如代码清单 7-19 所示。

**代码清单 7-19　ServletWebServerApplicationContext 的后置处理**

```
protected void postProcessBeanFactory(ConfigurableListableBeanFactory beanFactory) {
    // 注册 ServletContext 回调注入器
    beanFactory.addBeanPostProcessor(new  WebApplicationContextServletContextAwareProcessor(this));
    beanFactory.ignoreDependencyInterface(ServletContextAware.class);
    // 注册 Web 相关的作用域
    registerWebApplicationScopes();
}
```

### 1. ServletContextAwareProcessor

即便读者不看 `ServletContextAwareProcessor` 的源码，仅通过类名和它下面的 `ignoreDependencyInterface` 操作也能推测出来，它的作用类似于 `ApplicationContextAwareProcessor`，通过配合 `ServletContextAware` 接口可以实现 Aware 系列接口的回调注入功能。事实上 `ServletContextAwareProcessor` 的确负责这部分工作，只不过逻辑都在其父类 `ServletContextAwareProcessor` 中，透过源码可以很清楚地看到类似于 `ApplicationContextAwareProcessor` 的实现逻辑，如代码清单 7-20 所示。

**代码清单 7-20　ServletContextAwareProcessor 的核心回调注入功能**

```
// 父类 ServletContextAwareProcessor
public Object postProcessBeforeInitialization(Object bean, String beanName)
        throws BeansException {
    if (getServletContext() != null && bean instanceof ServletContextAware) {
        ((ServletContextAware) bean).setServletContext(getServletContext());
    }
    if (getServletConfig() != null && bean instanceof ServletConfigAware) {
        ((ServletConfigAware) bean).setServletConfig(getServletConfig());
    }
    return bean;
}
```

### 2. 注册 Web 应用的作用域

注册 Web 类型的作用域实际上是让 `ApplicationContext` 认识读者熟悉的 request、session、application 等 Web 应用才会涉及的作用域，而让 `ApplicationContext` 认

识的方式是，在 registerWebApplicationScopes 方法中将这些作用域注册到 IOC 容器中。不过源码的设计可能跟读者预想的不太一样，如代码清单 7-21 所示（这部分需要详细研究一下）。

**代码清单 7-21　注册 Web 应用的作用域**

```java
private void registerWebApplicationScopes() {
    ExistingWebApplicationScopes existingScopes = new ExistingWebApplicationScopes(getBeanFactory());
    WebApplicationContextUtils.registerWebApplicationScopes(getBeanFactory());
    existingScopes.restore();
}
```

registerWebApplicationScopes 方法中涉及一个很奇怪的 API：ExistingWebApplicationScopes，由类名直译为"已存在的作用域"，这个设计可能与读者目前掌握的知识有冲突：Spring Boot 应用刚开始初始化的 IOC 容器何来"已存在的作用域"呢？带着这个问题，请读者深入 ExistingWebApplicationScopes 的源码中尝试寻求一些思路，如代码清单 7-22 所示。

**代码清单 7-22　ExistingWebApplicationScopes 的结构**

```java
public static class ExistingWebApplicationScopes {

    private static final Set<String> SCOPES;
    static {
        // 预初始化了两个内置支持的作用域 REQUEST 和 SESSION
        Set<String> scopes = new LinkedHashSet<>();
        scopes.add(WebApplicationContext.SCOPE_REQUEST);
        scopes.add(WebApplicationContext.SCOPE_SESSION);
        SCOPES = Collections.unmodifiableSet(scopes);
    }

    private final ConfigurableListableBeanFactory beanFactory;

    private final Map<String, Scope> scopes = new HashMap<>();
    public ExistingWebApplicationScopes(ConfigurableListableBeanFactory beanFactory) {
        this.beanFactory = beanFactory;
        for (String scopeName : SCOPES) {
            // 创建对象时，从 BeanFactory 中检查是否有已经定义过的默认作用域
            Scope scope = beanFactory.getRegisteredScope(scopeName);
            if (scope != null) {
                this.scopes.put(scopeName, scope);
            }
        }
    }

    public void restore() {
        // 如果前期已经定义了内置作用域，则此处重新注册
        this.scopes.forEach((key, value) -> {
            this.beanFactory.registerScope(key, value);
        });
    }
}
```

下面结合代码清单 7-22 中简要标注的注释解释 `ExistingWebApplicationScopes` 的作用。`ExistingWebApplicationScopes` 在创建完成时会检查 `BeanFactory` 中是否已经注册过 request 和 session 的作用域。由于默认情况下项目代码中不会重新定义这两种 Web 应用中的默认作用域，因此在 `ExistingWebApplicationScopes` 的构造方法中不会提取到值。但是请读者思考，如果从构造方法中真的提取到了不为空的值，说明什么？说明项目中提前注册过 request 或者 session 的作用域。换句话说，如果**提取到不为空的作用域值，说明项目中需要重写 WebMvc 中默认的 request 或者 session 作用域的策略**。`ExistingWebApplicationScopes` 的设计相当于给开发者**提供了重写默认作用域的机会**，其构造方法中从 `BeanFactory` 中提取的动作相当于暂存了项目中自定义重写的作用域策略（不然后续执行 `WebApplicationContextUtils.registerWebApplicationScopes` 时会覆盖自定义的作用域策略），等默认的 request 和 session 作用域注册完毕后，再提取出暂存的自定义作用域策略，并重新注册到 `BeanFactory` 中（即覆盖默认的作用域策略），以此过程就可以实现默认 Web 作用域的重写。

除这个特殊的设计之外，默认的 Web 应用作用域注册是在 `WebApplicationContextUtils.registerWebApplicationScopes` 静态方法中实现，对应的实现逻辑如代码清单 7-23 所示。`registerWebApplicationScopes` 方法前两行的注册 Web 类型作用域部分有一个细节需要注意，默认情况下 `ServletWebServerApplicationContext` 不会注册 application 作用域，出现这种设计的原因是考虑到 Spring Boot 应用以独立 jar 包运行时，嵌入式 Web 容器还没有初始化，`ServletContext` 也尚未创建，所以这里无法获取到，也就不会注册 application 作用域。

**代码清单 7-23　注册默认 Web 应用作用域**

```java
public static void registerWebApplicationScopes(ConfigurableListableBeanFactory beanFactory) {
    registerWebApplicationScopes(beanFactory, null);
}

public static void registerWebApplicationScopes(ConfigurableListableBeanFactory beanFactory,
        @Nullable ServletContext sc) {
    // 注册默认的作用域策略
    beanFactory.registerScope(WebApplicationContext.SCOPE_REQUEST, new RequestScope());
    beanFactory.registerScope(WebApplicationContext.SCOPE_SESSION, new SessionScope());
    // 默认情况下 application 作用域不会注册
    if (sc != null) {
        ServletContextScope appScope = new ServletContextScope(sc);
        beanFactory.registerScope(WebApplicationContext.SCOPE_APPLICATION, appScope);
        sc.setAttribute(ServletContextScope.class.getName(), appScope);
    }

    // 依赖注入的支持
    beanFactory.registerResolvableDependency(ServletRequest.class, new RequestObjectFactory());
    beanFactory.registerResolvableDependency(ServletResponse.class, new ResponseObjectFactory());
    beanFactory.registerResolvableDependency(HttpSession.class, new SessionObjectFactory());
    beanFactory.registerResolvableDependency(WebRequest.class, new WebRequestObjectFactory());
    // jsf ......
}
```

## 7.4.2 组件扫描&解析手动传入的配置类

回调父类 `ServletWebServerApplicationContext` 的方法后,接下来 `AnnotationConfigServletWebServerApplicationContext` 要处理编程式指定的组件扫描。由于 IOC 容器的内部已经集成了注解类解析器和组件包扫描器,如代码清单 7-24 所示,因此接下来的两个步骤实质上是这两个集成的组件在工作。

**代码清单 7-24　AnnotationConfigServletWebServerApplicationContext 集成了两个组件**

```
public class AnnotationConfigServletWebServerApplicationContext
        extends ServletWebServerApplicationContext
        implements AnnotationConfigRegistry {
    private final AnnotatedBeanDefinitionReader reader;
    private final ClassPathBeanDefinitionScanner scanner;
    // ......
}
```

默认情况下 `AnnotationConfigServletWebServerApplicationContext` 中 `basePackages` 属性的值为空,在项目开发中几乎不会直接获取到 IOC 容器的落地实现类进行操作(通常只能获取到 `ConfigurableApplicationContext`,而其中没有 `scan` 和 `register` 方法),所以对于这部分读者可以忽略。

## 7.5　invokeBeanFactoryPostProcessors——执行 BeanFactoryPostProcessor

`BeanFactory` 的内部编程式后置处理完成后,接下来进入 IOC 容器刷新的第一个**超级复杂的重难点**: `BeanFactoryPostProcessor` 和 `BeanDefinitionRegistryPostProcessor` 的执行。

`invokeBeanFactoryPostProcessors` 方法的实现看上去非常简单,核心动作只有一行代码,如代码清单 7-25 所示。

**代码清单 7-25　转调 invokeBeanFactoryPostProcessors 静态方法**

```
protected void invokeBeanFactoryPostProcessors(ConfigurableListableBeanFactory beanFactory) {
    // 执行 BeanFactory 的后置处理器
    PostProcessorRegistrationDelegate.invokeBeanFactoryPostProcessors(beanFactory,
            getBeanFactoryPostProcessors());
    // AOP 的支持......
}
```

注意 `invokeBeanFactoryPostProcessors` 静态方法的第二个参数中当前阶段现有的 `BeanFactoryPostProcessor` 传入(注意这组 `BeanFactoryPostProcessor` 不是通过 `@Bean` 或者 `@Component` 注解注册的,而是通过 `ConfigurableApplicationContext` 的 `addBeanFactoryPostProcessor` 方法编程式手动注册),而调用的 `PostProcessorRegistrationDelegate` 中的 `invokeBeanFactoryPostProcessors` 静态方法难度非常大,它的源码非常长,为了便于读者阅读,本节会将这部分源码尽可能地拆解开来。

## 7.5.1 现有的后置处理器分类

invokeBeanFactoryPostProcessors 方法会首先检查当前 BeanFactory 是否同时是一个 BeanDefinitionRegistry。由于目前的 BeanFactory 落地实现只有一个 DefaultListableBeanFactory，因此该逻辑判断必定为 true。判断完成后的 if 结构中需要对方法参数中传入的 beanFactoryPostProcessors 集合元素进行类型分离，并最终分离出两个集合 registryProcessors 和 regularPostProcessors，分别对应存放编程式传入的 BeanDefinitionRegistryPostProcessor 和 BeanFactoryPostProcessor，如代码清单 7-26 所示。其中 BeanDefinitionRegistryPostProcessor 的 postProcessBeanDefinitionRegistry 方法会立即执行（由此也可以了解到，通过编程式注册的 BeanDefinitionRegistryPostProcessor，其执行优先级是最高的）。

**代码清单 7-26　invokeBeanFactoryPostProcessors（1）**

```java
public static void invokeBeanFactoryPostProcessors(ConfigurableListableBeanFactory beanFactory,
        List<BeanFactoryPostProcessor> bean FactoryPostProcessors) {
    Set<String> processedBeans = new HashSet<>();
    if (beanFactory instanceof BeanDefinitionRegistry) {
        BeanDefinitionRegistry registry = (BeanDefinitionRegistry) beanFactory;
        List<BeanFactoryPostProcessor> regularPostProcessors = new ArrayList<>();
        List<BeanDefinitionRegistryPostProcessor> registryProcessors = new ArrayList<>();

        // 该部分会将方法参数中传入的 BeanFactoryPostProcessor 分离开
        for (BeanFactoryPostProcessor postProcessor : beanFactoryPostProcessors) {
            if (postProcessor instanceof BeanDefinitionRegistryPostProcessor) {
                BeanDefinitionRegistryPostProcessor registryProcessor =
                        (BeanDefinitionRegistryPostProcessor) postProcessor;
                registryProcessor.postProcessBeanDefinitionRegistry(registry);
                registryProcessors.add(registryProcessor);
            } else {
                regularPostProcessors.add(postProcessor);
            }
        }
        // ......
```

## 7.5.2 执行最高优先级的 BeanDefinitionRegistryPostProcessor

处理好编程式注入的后置处理器后，下面要处理 IOC 容器中已有的后置处理器。由于 BeanDefinitionRegistryPostProcessor 的执行优先级较 BeanFactoryPostProcessor 靠前，因此 BeanDefinitionRegistryPostProcessor 先执行，而每种后置处理器（包括 BeanPostProcessor、BeanFactoryPostProcessor、BeanDefinitionRegistryPostProcessor）都有三种不同类型的优先级，分别是 **PriorityOrdered**、**Ordered** 和普通，其优先级依次降低。

具体到某一个优先级的后置处理器回调逻辑，首先会从 BeanFactory 中提取出所有 Bean

DefinitionRegistryPostProcessor，并检查这些后置处理器是否实现了 Priority Ordered 接口，如果实现了则创建这些后置处理器，并在排序完成后按序依次调用。依照 PriorityOrdered 接口中的 getOrder 方法的返回值进行排序，返回值越小则排序越靠前，如代码清单 7-27 所示。

除了主干动作，请读者留意一个小动作：processedBeans.add(ppName);，这里会将所有执行过的后置处理器全部保存下来，这样做的原因下面马上就会看到。

**代码清单 7-27　invokeBeanFactoryPostProcessors（2）**

```
// ......
List<BeanDefinitionRegistryPostProcessor> currentRegistryProcessors = new ArrayList<>();

// 首先，执行实现了 PriorityOrdered 接口的 BeanDefinitionRegistryPostProcessors
String[] postProcessorNames = beanFactory
        .getBeanNamesForType(BeanDefinitionRegistryPostProcessor.class, true, false);
for (String ppName : postProcessorNames) {
    if (beanFactory.isTypeMatch(ppName, PriorityOrdered.class)) {
        currentRegistryProcessors.add(beanFactory.getBean(ppName,
                BeanDefinitionRegistryPostProcessor.class));
        processedBeans.add(ppName);
    }
}
sortPostProcessors(currentRegistryProcessors, beanFactory);
registryProcessors.addAll(currentRegistryProcessors);
invokeBeanDefinitionRegistryPostProcessors(currentRegistryProcessors, registry);
currentRegistryProcessors.clear();
// ......
```

> 提示：在本环节中有一个内置的极其重要的、执行时机最靠前的后置处理器 ConfigurationClassPostProcessor，在 7.5.7 节会单独研究它的作用。

### 7.5.3　执行其他 BeanDefinitionRegistryPostProcessor

执行完最高优先级的 BeanDefinitionRegistryPostProcessor 后，接下来要执行的是实现了 Ordered 接口的后置处理器，执行的逻辑与前一阶段完全一致，如代码清单 7-28 所示。注意，代码清单 7-28 的后半部分执行其余普通的 BeanDefinitionRegistryPostProcessor 时有一个循环的动作，为什么会出现这种设计呢？请读者回想一下，BeanDefinitionRegistryPostProcessor 的核心功能是向 BeanDefinitionRegistry 中注册新的 BeanDefinition，如果注册的新 BeanDefinition 对应的刚好也是 BeanDefinitionRegistryPostProcessor，则该后置处理器也应该参与到当前步骤的执行过程中。因此，此处循环的作用是穷尽当前项目中会注册的所有 BeanDefinitionRegistryPostProcessor，并执行它们的 postProcessBeanDefinitionRegistry 方法。

另外，注意观察 processedBeans 集合的使用，因为 BeanDefinitionRegistryPostProcessor 的 postProcessBeanDefinitionRegistry 方法只能执行一次，而之前执行过的后置处理器后续允许重复执行，所以 processedBeans 集合起到了防重复执行的作用。

**代码清单 7-28  invokeBeanFactoryPostProcessors（3）**

```
// ......
// 接下来执行实现了 Ordered 接口的 BeanDefinitionRegistryPostProcessors
postProcessorNames = beanFactory
        .getBeanNamesForType(BeanDefinitionRegistryPostProcessor.class, true, false);
for (String ppName : postProcessorNames) {
    if (!processedBeans.contains(ppName) && beanFactory.isTypeMatch(ppName, Ordered.class)) {
        currentRegistryProcessors.add(beanFactory
                .getBean(ppName, BeanDefinitionRegistryPostProcessor.class));
        processedBeans.add(ppName);
    }
}
sortPostProcessors(currentRegistryProcessors, beanFactory);
registryProcessors.addAll(currentRegistryProcessors);
invokeBeanDefinitionRegistryPostProcessors(currentRegistryProcessors, registry);
currentRegistryProcessors.clear();

// 最后执行所有其他 BeanDefinitionRegistryPostProcessor
boolean reiterate = true;
while (reiterate) { // 注意此处有一个循环动作
    reiterate = false;
    postProcessorNames = beanFactory
            .getBeanNamesForType(BeanDefinitionRegistryPostProcessor.class, true, false);
    for (String ppName : postProcessorNames) {
        if (!processedBeans.contains(ppName)) {
            currentRegistryProcessors.add(beanFactory
                    .getBean(ppName, BeanDefinitionRegistryPostProcessor.class));
            processedBeans.add(ppName);
            reiterate = true;
        }
    }
    sortPostProcessors(currentRegistryProcessors, beanFactory);
    registryProcessors.addAll(currentRegistryProcessors);
    invokeBeanDefinitionRegistryPostProcessors(currentRegistryProcessors, registry);
    currentRegistryProcessors.clear();
}
// ......
```

代码清单 7-28 的逻辑执行完毕后，原则上 BeanDefinitionRegistry 中的 BeanDefinition 已经全部注册完毕，不会再有新的 BeanDefinition 注册（ImportBeanDefinitionRegistrar 也在该阶段触发执行，下面会讲到）。

### 7.5.4  回调 postProcessBeanFactory 方法

所有 BeanDefinitionRegistryPostProcessor 的 postProcessBeanDefinitionRegistry 方法执行完毕后，还需要回调它们的 postProcessBeanFactory 方法（不要忘记 BeanDefinitionRegistryPostProcessor 本身继承自 BeanFactoryPostProcessor），如代码清单 7-29 所示。这部分依然按照 PriorityOrdered、Ordered 接口的优先级顺序区分

（记录到 registryProcessors 集合时已经包含了排序），而且依然是 BeanDefinition RegistryPostProcessor 的执行优先级比 BeanFactoryPostProcessor 高。

**代码清单 7-29　invokeBeanFactoryPostProcessors（4）**

```
        // ......
        // 先回调 BeanDefinitionRegistryPostProcessor 的 postProcessBeanFactory 方法
        invokeBeanFactoryPostProcessors(registryProcessors, beanFactory);
        // 再调用编程式注入的 BeanFactoryPostProcessor
        invokeBeanFactoryPostProcessors(regularPostProcessors, beanFactory);
    }
    // 如果 BeanFactory 没有实现 BeanDefinitionRegistry 接口，则进入下面的流程
    else {
        // 仅调用编程式注册的 BeanFactoryPostProcessor
        invokeBeanFactoryPostProcessors(beanFactoryPostProcessors, beanFactory);
    }
    // ......
```

### 7.5.5　BeanFactoryPostProcessor 的分类

执行完所有 BeanDefinitionRegistryPostProcessor 之后，接下来轮到所有的普通 BeanFactoryPostProcessor 了。首先它也是把所有的 BeanFactoryPostProcessor 根据是否实现了 PriorityOrdered、Ordered 接口进行优先级的区分，最终分离出三个集合，如代码清单 7-30 所示。

**代码清单 7-30　invokeBeanFactoryPostProcessors（5）**

```
    // ......
    // 下面的部分是回调 BeanFactoryPostProcessor，思路与上面的几乎一样
    String[] postProcessorNames =
            beanFactory.getBeanNamesForType(BeanFactoryPostProcessor.class, true, false);
    List<BeanFactoryPostProcessor> priorityOrderedPostProcessors = new ArrayList<>();
    List<String> orderedPostProcessorNames = new ArrayList<>();
    List<String> nonOrderedPostProcessorNames = new ArrayList<>();
    for (String ppName : postProcessorNames) {
        if (processedBeans.contains(ppName)) {
            // skip - already processed in first phase above
        } else if (beanFactory.isTypeMatch(ppName, PriorityOrdered.class)) {
            priorityOrderedPostProcessors.add(beanFactory
                    .getBean(ppName, BeanFactoryPostProcessor.class));
        } else if (beanFactory.isTypeMatch(ppName, Ordered.class)) {
            orderedPostProcessorNames.add(ppName);
        } else {
            nonOrderedPostProcessorNames.add(ppName);
        }
    }
    // ......
```

### 7.5.6　执行 BeanFactoryPostProcessor

分离好三个集合后，后续的动作依然是按照既定的顺序逐个执行，如代码清单 7-31 所示。逻辑上与 BeanDefinitionRegistryPostProcessor 的执行机制一致，都是先排序后执行。

## 7.5 invokeBeanFactoryPostProcessors——执行 BeanFactoryPost Processor

**代码清单 7-31　invokeBeanFactoryPostProcessors（6）**

```java
// ......
// 首先，执行实现 PriorityOrdered 接口的 BeanFactoryPostProcessor
sortPostProcessors(priorityOrderedPostProcessors, beanFactory);
invokeBeanFactoryPostProcessors(priorityOrderedPostProcessors, beanFactory);
// 然后，执行实现 Ordered 接口的 BeanFactoryPostProcessor
List<BeanFactoryPostProcessor> orderedPostProcessors = new ArrayList<>();
for (String postProcessorName : orderedPostProcessorNames) {
    orderedPostProcessors.add(beanFactory
            .getBean(postProcessorName, BeanFactoryPostProcessor.class));
}
sortPostProcessors(orderedPostProcessors, beanFactory);
invokeBeanFactoryPostProcessors(orderedPostProcessors, beanFactory);
// 最后，执行普通的 BeanFactoryPostProcessor
List<BeanFactoryPostProcessor> nonOrderedPostProcessors = new ArrayList<>();
for (String postProcessorName : nonOrderedPostProcessorNames) {
    nonOrderedPostProcessors.add(beanFactory
            .getBean(postProcessorName, BeanFactoryPostProcessor.class));
}
invokeBeanFactoryPostProcessors(nonOrderedPostProcessors, beanFactory);

// 清理缓存
beanFactory.clearMetadataCache();
}
```

注意一点，因为 BeanFactoryPostProcessor 原则上只有对 BeanDefinition 的修改、移除能力，不会再注册新的 BeanDefinition，所以这里不再需要循环执行。也正是因为这个机制，提醒读者最好不要在 BeanFactoryPostProcessor 的执行期间注册新的 BeanDefinition，否则会因为没有穷尽加载的逻辑，导致注册的 BeanFactoryPostProcessor 不能生效。

至此，整个后置处理器的执行流程全部结束，整体流程总结如下。

1. 回调编程式注入的 BeanDefinitionRegistryPostProcessor。
2. 回调实现 PriorityOrdered 接口的 BeanDefinitionRegistryPostProcessor。
3. 回调实现 Ordered 接口的 BeanDefinitionRegistryPostProcessor。
4. 回调普通的 BeanDefinitionRegistryPostProcessor。
5. 回调 BeanDefinitionRegistryPostProcessor 的 postProcessBeanFactory 方法。
6. 回调编程式注入的 BeanFactoryPostProcessor。
7. 回调实现 PriorityOrdered 接口的 BeanFactoryPostProcessor。
8. 回调实现 Ordered 接口的 BeanFactoryPostProcessor。
9. 回调普通的 BeanFactoryPostProcessor。

### 7.5.7　重要的后置处理器：ConfigurationClassPostProcessor

在后置处理器执行的阶段中有一个极其重要的内置 BeanDefinitionRegistryPostProcessor：**ConfigurationClassPostProcessor**。它的核心作用是解析、处理所有被

@Configuration 标注的配置类，并向 BeanDefinitionRegistry 中注册新的 Bean Definition。这个后置处理器在注解驱动的 IOC 容器中默认生效，在基于 XML 配置文件的 IOC 容器中，只要配置文件声明<context:annotation-config/>标签，该后置处理器就会注册生效。

ConfigurationClassPostProcessor 的核心功能是解析注解配置类并注册 Bean Definition，下面直接切入它的 postProcessBeanDefinitionRegistry 方法中一探究竟，如代码清单 7-32 所示。

**代码清单 7-32　ConfigurationClassPostProcessor 的核心后置处理方法（节选）**

```
public void postProcessBeanDefinitionRegistry(BeanDefinitionRegistry registry) {
    int registryId = System.identityHashCode(registry);
    // check throw ex ......
    this.registriesPostProcessed.add(registryId);

    // 【解析配置类】
    processConfigBeanDefinitions(registry);
}
```

由代码清单 7-31 可以发现，postProcessBeanDefinitionRegistry 方法的最后一行就是解析注解配置类中的 Bean 定义信息，并封装为 BeanDefinition。由于 processConfigBeanDefinitions 方法的复杂度相当高，为厘清 ConfigurationClassPostProcessor 的核心功能，本节中涉及的源码会重点展示出核心逻辑，对于其他逻辑本书不详细展开讲解。

### 1. processConfigBeanDefinitions 主体

对于 processConfigBeanDefinitions 方法的主体逻辑，本书不过多展开，读者可以对照代码清单的注释内容参考了解。代码清单 7-33 的篇幅非常长，读者可以分段理解。

**代码清单 7-33　解析注解配置类的核心方法（节选）**

```
public void processConfigBeanDefinitions(BeanDefinitionRegistry registry) {
    List<BeanDefinitionHolder> configCandidates = new ArrayList<>();
    String[] candidateNames = registry.getBeanDefinitionNames();

    // 筛选出所有的配置类
    for (String beanName : candidateNames) {
        BeanDefinition beanDef = registry.getBeanDefinition(beanName);
        if (beanDef.getAttribute(ConfigurationClassUtils.CONFIGURATION_CLASS_ATTRIBUTE) != null) {
            // 重复注册的检查，直接跳过 ......
        } else if (ConfigurationClassUtils.checkConfigurationClassCandidate(beanDef,
                this.metadataReaderFactory)) {
            // 如果检查的 bean 是一个配置类（被@Configuration 注解标注）
            // 则加入到候选配置类集合中，并添加一个特殊标记 CONFIGURATION_CLASS_ATTRIBUTE
            configCandidates.add(new BeanDefinitionHolder(beanDef, beanName));
        }
    }

    // 无配置类的返回逻辑 ......
    // 配置类排序
    configCandidates.sort((bd1, bd2) -> {
```

```java
            int i1 = ConfigurationClassUtils.getOrder(bd1.getBeanDefinition());
            int i2 = ConfigurationClassUtils.getOrder(bd2.getBeanDefinition());
            return Integer.compare(i1, i2);
    });
    // 如果有的话,应用自定义的 BeanNameGenerator(默认没有)
    SingletonBeanRegistry sbr = null;
    if (registry instanceof SingletonBeanRegistry) {
        sbr = (SingletonBeanRegistry) registry;
        if (!this.localBeanNameGeneratorSet) {
            BeanNameGenerator generator =
                    (BeanNameGenerator) sbr.getSingleton(CONFIGURATION_BEAN_NAME_GENERATOR);
            if (generator != null) {
                this.componentScanBeanNameGenerator = generator;
                this.importBeanNameGenerator = generator;
            }
        }
    }
    // ......
    // 真正解析配置类的组件:ConfigurationClassParser
    ConfigurationClassParser parser = new ConfigurationClassParser(
            this.metadataReaderFactory, this.problemReporter, this.environment,
            this.resourceLoader, this.componentScanBeanNameGenerator, registry);

    Set<BeanDefinitionHolder> candidates = new LinkedHashSet<>(configCandidates);
    Set<ConfigurationClass> alreadyParsed = new HashSet<>(configCandidates.size());
    do {
    // 【解析配置类】
        parser.parse(candidates);
        parser.validate();

        Set<ConfigurationClass> configClasses =
                new LinkedHashSet<>(parser.getConfigurationClasses());
        configClasses.removeAll(alreadyParsed);

        if (this.reader == null) {
            // 真正读取配置类的组件:ConfigurationClassBeanDefinitionReader
            this.reader = new ConfigurationClassBeanDefinitionReader(
                    registry, this.sourceExtractor, this.resourceLoader, this.environment,
                    this.importBeanNameGenerator, parser.getImportRegistry());
        }
        // 【加载配置类的内容】
        this.reader.loadBeanDefinitions(configClasses);
        alreadyParsed.addAll(configClasses);
        // 一些额外的处理......
    }
    while (!candidates.isEmpty());

    // 一些额外的处理......
}
```

通读一遍后,想必读者对 processConfigBeanDefinitions 方法的工作内容有了一个基本的了解,方法中除了一些辅助检查、控制的逻辑,核心重点由两个核心 API 完成,分别是负责解析配置类的 ConfigurationClassParser 以及读取配置类内容的 ConfigurationClassBeanDefinitionReader,以下内容就这两个组件分别来看它们负责的工作内容。

## 2. ConfigurationClassParser

ConfigurationClassParser 的作用是解析配置类，针对不同的配置源加载来源，它可以提供不同的解析逻辑（即重载的 parse 方法），其核心 parse 方法如代码清单 7-34 所示。

**代码清单 7-34　ConfigurationClassParser 的核心 parse 策略方法**

```java
private final DeferredImportSelectorHandler deferredImportSelectorHandler =
                                    new DeferredImportSelectorHandler();
public void parse(Set<BeanDefinitionHolder> configCandidates) {
    for (BeanDefinitionHolder holder : configCandidates) {
        BeanDefinition bd = holder.getBeanDefinition();
        try{
            // 注解配置类
            if (bd instanceof AnnotatedBeanDefinition) {
                parse(((AnnotatedBeanDefinition) bd).getMetadata(), holder.getBeanName());
            }
            // 编程式注入配置类
            else if (bd instanceof AbstractBeanDefinition
                        && ((AbstractBeanDefinition)bd).hasBeanClass()) {
                parse(((AbstractBeanDefinition) bd).getBeanClass(), holder.getBeanName());
            }
            // 其他情况
            else {
                parse(bd.getBeanClassName(), holder.getBeanName());
            }
        } // catch ......
    }

    // 回调特殊的 ImportSelector
    this.deferredImportSelectorHandler.process();
}
```

parse 方法的整体逻辑是，上面的 for 循环体会提取出配置类的全限定名，并根据配置类的 BeanDefinition 类型转调不同的重载 parse 方法中（注意无论执行 if-else-if 的哪个分支，最终都是执行重载的 parse 方法）；for 循环调用完成后，最后执行 deferredImportSelectorHandler 的 process 方法，对于这个组件我们之前没有接触过，这里有必要对其进行说明。

（1）ImportSelector 的扩展

在 Spring Framework 4.0 中 ImportSelector 扩展了一个子接口 DeferredImportSelector，它的执行时机比 ImportSelector 晚。普通的 ImportSelector 会在注解配置类的解析期间生效，而 DeferredImportSelector 会在注解配置类的解析工作完成后才执行（其实上面 ConfigurationClassParser 的核心 parse 方法就已经解释了这个原理）。

一般情况下，DeferredImportSelector 会跟 @Conditional 注解配合使用完成条件装配。

（2）DeferredImportSelectorHandler 的处理逻辑

DeferredImportSelectorHandler 的处理逻辑如代码清单 7-35 所示。

**代码清单 7-35　DeferredImportSelectorHandler 的 process 方法**

```java
public void process() {
    List<DeferredImportSelectorHolder> deferredImports = this.deferredImportSelectors;
    this.deferredImportSelectors = null;
    try {
        if(deferredImports != null) {
            DeferredImportSelectorGroupingHandler handler
                = new DeferredImportSelectorGroupingHandler();
            deferredImports.sort(DEFERRED_IMPORT_COMPARATOR);
            deferredImports.forEach(handler::register);
            handler.processGroupImports();
        }
    } finally {
        this.deferredImportSelectors = new ArrayList<>();
    }
}
```

由代码清单 7-35 可以发现，DeferredImportSelectorHandler 的处理逻辑相对简单，它会提取出所有解析阶段中存储好的 DeferredImportSelector 并依次执行。由于 DeferredImportSelector 的执行时机比较晚，对于 @Conditional 条件装配的处理会更有利，因此这个设计还是不错的。

**3. ConfigurationClassParser 的落地方法**

回到主线流程中，上面的 parse(Set<BeanDefinitionHolder>) 方法中，最终动作会把注解配置类传入重载的 parse 方法中，注意方法中的参数类型是一个 ConfigurationClass 包装对象，如代码清单 7-36 所示。

**代码清单 7-36　processConfigurationClass 解析配置类的动作**

```java
protected final void parse(AnnotationMetadata metadata, String beanName) throws
IOException {
    processConfigurationClass(new ConfigurationClass(metadata, beanName));
}

protected void processConfigurationClass(ConfigurationClass configClass) throws IOException {
    // 前置校验 ……
    SourceClass sourceClass = asSourceClass(configClass);
    do {
        // 真正的解析动作
        sourceClass = doProcessConfigurationClass(configClass, sourceClass);
    } while (sourceClass != null);

    this.configurationClasses.put(configClass, configClass);
}
```

可以发现解析配置类的动作是一个 "**xxx 方法转 doxxx 方法**" 的过程。Spring Framework 中这种类似的方法设计非常多，读者在阅读源码时一定要多加注意。

在进行前置校验后，do-while 循环结构中真正解析配置类的 doProcessConfigurationClass 方法实现非常复杂，好在源码的逻辑步骤都比较清晰，下面分段研究核心的逻辑。

### (1) 处理@Component 注解

doProcessConfigurationClass 方法在开始部分会判断当前被解析的类是否标注了 @Component 注解,因为所有标注了@Configuration 注解的类必定标注了@Component 注解,因而 processMemberClasses 方法必定会触发,如代码清单 7-37 所示。而 processMemberClasses 方法的核心动作是解析当前类自身以及当前类中嵌套的内部类,如果内部有定义配置类则会递归解析。经过该方法的处理,当前配置类内部定义的内容会被全部解析。

**代码清单 7-37　doProcessConfigurationClass 方法(1)**

```
protected final SourceClass doProcessConfigurationClass(ConfigurationClass configClass,
        SourceClass sourceClass) throws IOException {
    if (configClass.getMetadata().isAnnotated(Component.class.getName())) {
        processMemberClasses(configClass, sourceClass);
    }
    // ......
```

### (2) 处理@PropertySource 注解

紧接着的第二步是处理@PropertySource 注解,如代码清单 7-38 所示。借助 AnnotationConfigUtils 工具类可以很容易地提取出配置类上标注的所有注解信息,然后筛选出指定的注解属性。而 for 循环内部的 processPropertySource 方法会真正地封装 PropertySource 导入的资源文件,如代码清单 7-39 所示。该方法会在一些前置处理完毕后将资源文件的内容解析出来,并存入 Environment 中。

**代码清单 7-38　doProcessConfigurationClass 方法(2)**

```
// ......
// 处理 @PropertySource 注解
for (AnnotationAttributes propertySource : AnnotationConfigUtils.attributesForRepeatable(
        sourceClass.getMetadata(), PropertySources.class,
        org.springframework.context.annotation.PropertySource.class)) {
    if (this.environment instanceof ConfigurableEnvironment) {
        processPropertySource(propertySource);
    }
    // else logger ......
}
// ......
```

**代码清单 7-39　processPropertySource 解析 properties 文件**

```
private void processPropertySource(AnnotationAttributes propertySource) throws
IOException {
    // 前置处理 ......
    String[] locations = propertySource.getStringArray("value");
    // ......
    for (String location : locations) {
        try {
            // 处理路径,加载资源文件,并添加到 Environment 中
            String resolvedLocation = this.environment.resolveRequiredPlaceholders(location);
```

```
            Resource resource = this.resourceLoader.getResource(resolvedLocation);
            addPropertySource(factory.createPropertySource(name,
                new EncodedResource(resource, encoding)));
        }// catch ......
    }
}
```

### （3）处理@ComponentScan注解

第三个处理的注解是@ComponentScan 组件扫描，整体流程不太复杂，核心动作是检查到有@ComponentScan 注解之后进行组件扫描，并相应地封装生成 BeanDefinition，随后注册到 BeanDefinitionRegistry 中，如代码清单7-40所示。注意代码清单7-40 中的一个细节：@ComponentScan可以标注多个，并且 Spring Framework 4.3 后新引入了@ComponentScans 注解，它可以组合多个@ComponentScan 注解，所以这里会使用 for 循环解析所有的@ComponentScan 注解。

**代码清单7-40　doProcessConfigurationClass 方法（3）**

```
// ......
// 检查配置类上是否标注了@ComponentScan 或@ComponentScans
Set<AnnotationAttributes> componentScans = AnnotationConfigUtils.attributesForRepeatable
        (sourceClass.getMetadata(), ComponentScans.class, ComponentScan.class);
if (!componentScans.isEmpty() && !this.conditionEvaluator.shouldSkip(sourceClass.getMetadata(),
ConfigurationPhase.REGISTER_BEAN)) {
    // 如果有@ComponentScans，则要提取出里面所有的@ComponentScan 依次扫描
    for (AnnotationAttributes componentScan : componentScans) {
        // 【复杂】借助 ComponentScanAnnotationParser 扫描
        // private final ComponentScanAnnotationParser componentScanParser;
        Set<BeanDefinitionHolder> scannedBeanDefinitions = this.componentScanParser
                .parse(componentScan, sourceClass.getMetadata().getClassName());
        // 是否扫描到了其他的注解配置类
        for (BeanDefinitionHolder holder : scannedBeanDefinitions) {
            BeanDefinition bdCand = holder.getBeanDefinition().getOriginatingBeanDefinition();
            if (bdCand == null) {
                bdCand = holder.getBeanDefinition();
            }
            if (ConfigurationClassUtils.checkConfigurationClassCandidate(bdCand,
                    this.metadataReaderFactory)){
                // 如果扫描到了，递归解析
                parse(bdCand.getBeanClassName(), holder.getBeanName());
            }
        }
    }
}
// ......
```

扫描的核心动作会借助一个 ComponentScanAnnotationParser 来委托处理包扫描的工作，但它的内部还集成了一个 ClassPathBeanDefinitionScanner，它们分别完成不同的职责，代码清单7-41可以解释这一点。

### 代码清单 7-41　ComponentScanAnnotationParser 委托实现组件扫描的预备动作

```
public Set<BeanDefinitionHolder> parse(AnnotationAttributes componentScan,
        final String declaringClass) {
    // 构造 ClassPathBeanDefinitionScanner
    ClassPathBeanDefinitionScanner scanner = new ClassPathBeanDefinitionScanner (this.registry,
            componentScan.getBoolean("useDefaultFilters"), this.environment, this.resourceLoader);

    // 解析@ComponentScan 中的属性 ......
    Class<? extends BeanNameGenerator> generatorClass = componentScan.getClass("nameGenerator");

    // 整理要扫描的 basePackages
    Set<String> basePackages = new LinkedHashSet<>();
    String[] basePackagesArray = componentScan.getStringArray("basePackages");
    for (String pkg : basePackagesArray) {
        String[] tokenized = StringUtils.tokenizeToStringArray(this.environment.resolve
Placeholders(pkg), ConfigurableApplicationContext.CONFIG_LOCATION_DELIMITERS);
        Collections.addAll(basePackages, tokenized);
    }
    for (Class<?> clazz : componentScan.getClassArray("basePackageClasses")) {
        basePackages.add(ClassUtils.getPackageName(clazz));
    }
    // 没有声明 basePackages，则当前配置类所在的包即为根包
    if (basePackages.isEmpty()) {
        basePackages.add(ClassUtils.getPackageName(declaringClass));
    }

    // ......
    // 【扫描】执行组件扫描动作
    return scanner.doScan(StringUtils.toStringArray(basePackages));
}
```

整个方法的逻辑总结下来，可以概括为 3 个步骤。

1. 构造 `ClassPathBeanDefinitionScanner`，并封装@ComponentScan 注解中的属性。
2. 整理要进行包扫描的 `basePackages`，以及 `include` 和 `exclude` 的过滤器。
3. 调用 `ClassPathBeanDefinitionScanner` 执行实际的组件扫描的动作。

前两个步骤都相对简单，真正的组件扫描动作在 `ClassPathBeanDefinitionScanner` 的 `doScan` 方法中（方法名也体现出它的确是真正起作用的），如代码清单 7-42 所示。

### 代码清单 7-42　ClassPathBeanDefinitionScanner 的真正的组件扫描动作

```
protected Set<BeanDefinitionHolder> doScan(String... basePackages) {
    // assert ......
    Set<BeanDefinitionHolder> beanDefinitions = new LinkedHashSet<>();
    for (String basePackage : basePackages) {
        // 【真正的组件扫描动作在这里】
        Set<BeanDefinition> candidates = findCandidateComponents(basePackage);
        for (BeanDefinition candidate : candidates) {
            // 处理 scope（默认情况下是 singleton）
            ScopeMetadata scopeMetadata = this.scopeMetadataResolver
                    .resolveScopeMetadata(candidate);
```

## 7.5 invokeBeanFactoryPostProcessors——执行 BeanFactoryPost Processor

```
        candidate.setScope(scopeMetadata.getScopeName());
        // 生成 Bean 的名称
        String beanName = this.beanNameGenerator.generateBeanName(candidate,
                this.registry);
        if (candidate instanceof AbstractBeanDefinition) {
            postProcessBeanDefinition((AbstractBeanDefinition) candidate, beanName);
        }
        // 处理 Bean 中的@Lazy、@Primary 等注解
        if (candidate instanceof AnnotatedBeanDefinition) {
            AnnotationConfigUtils.processCommonDefinitionAnnotations((AnnotatedBeanDefinition) candidate);
        }
        if (checkCandidate(beanName, candidate)) {
            BeanDefinitionHolder definitionHolder =
                    new BeanDefinitionHolder(candidate,beanName);
            // 设置 AOP 相关的属性（如果支持的话）
            definitionHolder = AnnotationConfigUtils.applyScopedProxyMode(scopeMetadata,
                    definitionHolder, this.registry);
            beanDefinitions.add(definitionHolder);
            // 注册到 BeanDefinitionRegistry 中
            registerBeanDefinition(definitionHolder, this.registry);
        }
    }
}
return beanDefinitions;
}
```

代码清单 7-42 已经基本展示了组件扫描逻辑的全貌，只要符合之前设置好的匹配规则，findCandidateComponents 方法就可以成功扫描到，进而封装为 BeanDefinition（具体的实现类是 ScannedGenericBeanDefinition），随后设置 Bean 的名称、作用域、延迟加载、是否首选等信息，BeanDefinition 的上述信息设置完成后即会注册到 BeanDefinitionRegistry 中。

BeanDefinition 注册的动作全部完成后，即代表组件扫描完毕，@ComponentScan 注解的处理完成。

（4）处理@Import 注解

代码清单 7-43 所示为处理@Import 注解的代码。

**代码清单 7-43　doProcessConfigurationClass 方法（4）**

```
// ......
//处理@Import 注解
processImports(configClass, sourceClass, getImports(sourceClass), true);
// ......
```

处理@Import 注解的方法是被封装过的，需要进入 processImports 方法实现中来看。processImports 方法的实现非常规整，它会针对@Import 注解中支持的 4 种类型分别应用不同的处理逻辑（如 ImportSelector 会直接调用其 selectImports 方法加载所需加载的类的全限定名，ImportBeanDefinitionRegistrar 会在全部收集完成后依次调用），相应的源码如代码清单 7-44 所示。

> 提示：对于该部分内容，读者应该掌握的是方法本身，而不是具体的实现机制，这个方法在深入研究时通常伴随着排查问题，源码学习阶段可以仅关注方法主体架构。

### 代码清单 7-44　processImports 处理 @Import 注解

```java
private void processImports(ConfigurationClass configClass,
        SourceClass currentSourceClass, Collection<SourceClass> importCandidates,
        Predicate<String> exclusionFilter, boolean checkForCircularImports) {
    // 前置判断 ……
    // 防止循环 @Import 导入
    if (checkForCircularImports && isChainedImportOnStack(configClass)) {
        this.problemReporter.error(new CircularImportProblem(configClass, this.importStack));
    } else {
        // importStack 的控制
        for (SourceClass candidate : importCandidates) {
            // 处理 ImportSelector
            if (candidate.isAssignable(ImportSelector.class)) {
                Class<?> candidateClass = candidate.loadClass();
                ImportSelector selector = ParserStrategyUtils.instantiateClass(candidateClass,
                        ImportSelector.class, this.environment, this.resourceLoader, this.registry);
                Predicate<String> selectorFilter = selector.getExclusionFilter();
                if (selectorFilter != null) {
                    exclusionFilter = exclusionFilter.or(selectorFilter);
                }
                // DeferredImportSelector 的执行时机后延
                if (selector instanceof DeferredImportSelector) {
                    this.deferredImportSelectorHandler.handle(configClass,
                            (DeferredImportSelector) selector);
                } else {
                    // 执行 ImportSelector 的 selectImports 方法，并注册导入的类
                    String[] importClassNames = selector.selectImports
                            (currentSourceClass.getMetadata());
                    Collection<SourceClass> importSourceClasses =
                            asSourceClasses(importClassNames, exclusionFilter);
                    processImports(configClass, currentSourceClass, importSourceClasses,
                            exclusionFilter, false);
                }
            }
            // 处理 ImportBeanDefinitionRegistrar
            else if (candidate.isAssignable(ImportBeanDefinitionRegistrar.class)) {
                Class<?> candidateClass = candidate.loadClass();
                ImportBeanDefinitionRegistrar registrar =
                        ParserStrategyUtils.instantiateClass(candidateClass,
                                ImportBeanDefinitionRegistrar.class, this.environment,
                                this.resourceLoader, this.registry);
                configClass.addImportBeanDefinitionRegistrar(registrar,
                        currentSourceClass.getMetadata());
            }
            // 导入普通类/配置类
            else {
```

```
            this.importStack.registerImport(
                    currentSourceClass.getMetadata(), candidate.getMetadata().getClassName());
            processConfigurationClass(candidate.asConfigClass(configClass), exclusionFilter);
        }
    }
}
```

（5）处理@ImportResource 注解

Spring Framework 在使用注解驱动 IOC 容器的场景中，当需要导入 XML 配置文件时可以借助@ImportResource 注解进行导入，代码清单 7-45 的逻辑会将这些配置文件收集好，并在配置类解析完成后统一加载。本段源码只需要读者了解@ImportResource 注解在此处处理，由于 Spring Boot 已不推荐使用 XML 配置文件作为配置源，本书不再深入探究。

**代码清单 7-45　doProcessConfigurationClass 方法（5）**

```
// ......
//处理@ImportResource 注解
AnnotationAttributes importResource = AnnotationConfigUtils.attributesFor
        (sourceClass.getMetadata(), ImportResource.class);
if (importResource != null) {
    String[] resources = importResource.getStringArray("locations");
    Class<? extends BeanDefinitionReader> readerClass = importResource.getClass("reader");
    for (String resource : resources) {
        String resolvedResource = this.environment.resolveRequiredPlaceholders(resource);
        configClass.addImportedResource(resolvedResource, readerClass);
    }
}
// ......
```

（6）处理@Bean 注解

注意观察代码清单 7-46 中的逻辑，retrieveBeanMethodMetadata 方法会扫描所有标注了@Bean 注解的方法，但是它跟上面的大多数处理一样，没有立即封装 BeanDefinition 并注册到 BeanDefinitionRegistry 中，而是先存入 ConfigurationClass 中，这样做的原因会在后面讲解。

**代码清单 7-46　doProcessConfigurationClass 方法（6）**

```
// ......
//处理被注解@Bean 标注的方法
Set<MethodMetadata> beanMethods = retrieveBeanMethodMetadata(sourceClass);
for (MethodMetadata methodMetadata : beanMethods) {
    configClass.addBeanMethod(new BeanMethod(methodMetadata, configClass));
}
// ......
```

有关 retrieveBeanMethodMetadata 方法对标注了@Bean 注解的方法的筛选，读者可以深入源码简单探究，如代码清单 7-47 所示。

### 代码清单 7-47　retrieveBeanMethodMetadata 筛选出标注了@Bean 注解的方法

```java
private Set<MethodMetadata> retrieveBeanMethodMetadata(SourceClass sourceClass) {
    AnnotationMetadata original = sourceClass.getMetadata();
    // 获取被@Bean 注解标注的方法
    Set<MethodMetadata> beanMethods = original.getAnnotatedMethods(Bean.class.getName());
    if (beanMethods.size() >1 && original instanceof StandardAnnotationMetadata) {
        try {
            AnnotationMetadata asm = this.metadataReaderFactory
                    .getMetadataReader(original.getClassName()).getAnnotationMetadata();
            Set<MethodMetadata> asmMethods = asm.getAnnotatedMethods(Bean.class.getName());
            if (asmMethods.size() >= beanMethods.size()) {
                Set<MethodMetadata> selectedMethods = new LinkedHashSet<>(asmMethods.size());
                // 筛选每个方法
                for (MethodMetadata asmMethod : asmMethods) {
                    for (MethodMetadata beanMethod : beanMethods) {
                        if (beanMethod.getMethodName().equals(asmMethod.getMethodName())) {
                            selectedMethods.add(beanMethod);
                            break;
                        }
                    }
                }
                if (selectedMethods.size() == beanMethods.size()) {
                    beanMethods = selectedMethods;
                }
            }
        }// catch ......
    }
    return beanMethods;
}
```

retrieveBeanMethodMetadata 方法一开始会筛选出被@Bean 注解标注的方法。请注意观察源码中的变量名 asm，明明依靠反射机制就可以得到这些被@Bean 注解标注的方法，而读取方法的部分使用了 ASM（一种读取字节码的技术），Spring Framework 为什么要如此大动干戈呢？这里要多解释一下：使用 JVM 的标准反射，在不同的 JVM 或者同一个 JVM 上的不同应用中返回的方法列表顺序可能是不同的。简言之，JVM 的标准反射不保证方法列表返回的顺序一致。若想保证程序在任何 JVM 上、任何应用中加载同一个字节码文件的方法列表时都返回相同的顺序，就只能通过读取字节码来解决，而读取字节码的技术中 Spring Framework 选择了 ASM。简单总结，Spring Framework 使用 ASM 读取字节码的目的是保证加载配置类中标注了@Bean 的方法的从上到下的顺序与源文件.java 中的一致。

#### 4. ConfigurationClassBeanDefinitionReader

注解配置类全部解析完成后，可以看到有很多需要注册的 Bean 信息被记录在 ConfigurationClass 这个对象中。接下来的步骤会将这些记录的内容依次转换为 BeanDefinition，并注册到 BeanDefinitionRegistry 中。

```java
this.reader.loadBeanDefinitions(configClasses);
```

上面代码中使用的 reader 就是接下来要讲的 ConfigurationClassBeanDefini

tionReader，从类名上可以非常直观地读懂它的功能，它是一个可以读取配置类的 `BeanDefinition` 读取器。它要做的工作是将 `ConfigurationClass` 中的内容逐个读取并封装为 `BeanDefinition`，随后注册到 `BeanDefinitionRegistry` 中，对应的关键方法 `loadBeanDefinitions` 的实现如代码清单 7-48 所示。

**代码清单 7-48　ConfigurationClassBeanDefinitionReader 的核心 loadBeanDefinitions 方法**

```java
public void loadBeanDefinitions(Set<ConfigurationClass> configurationModel) {
    TrackedConditionEvaluator trackedConditionEvaluator = new TrackedConditionEvaluator();
    for (ConfigurationClass configClass : configurationModel) {
        loadBeanDefinitionsForConfigurationClass(configClass, trackedConditionEvaluator);
    }
}

private void loadBeanDefinitionsForConfigurationClass(ConfigurationClass configClass,
        TrackedConditionEvaluator trackedConditionEvaluator) {
    // 与条件装配有关的逻辑 ......

    // 如果当前配置类是被@Import 标注的,则要把配置类自身注册到 BeanDefinitionRegistry 中
    if (configClass.isImported()) {
        registerBeanDefinitionForImportedConfigurationClass(configClass);
    }
    // 注册被@Bean 注解标注的 Bean
    for (BeanMethod beanMethod : configClass.getBeanMethods()) {
        loadBeanDefinitionsForBeanMethod(beanMethod);
    }

    // 注册来自 XML 配置文件的 Bean
    loadBeanDefinitionsFromImportedResources(configClass.getImportedResources());
    // 注册来自 ImportBeanDefinitionRegistrar 的 Bean
    loadBeanDefinitionsFromRegistrars(configClass.getImportBeanDefinitionRegistrars());
}
```

注意观察代码清单 7-48 中注册 `BeanDefinition` 的部分，它共包含下面 4 个步骤，每个步骤的 `BeanDefinition` 加载源都不一样，下面分别来看。

（1）注册配置类自身

第一步是将配置类自身对应的定义信息注册到 `BeanDefinitionRegistry` 中。按照组件扫描的原则，只要一个类标注了 `@Configuration` 注解，就相当于标注了 `@Component` 注解，当该类被成功扫描时就应该注册到 `BeanDefinitionRegistry` 中。但有一种特殊情况，如果配置类通过 `@Import` 注解的方式导入，这种情况下配置类不会主动将自身注册到 `BeanDefinitionRegistry` 中，因此在 `registerBeanDefinitionForImportedConfigurationClass` 方法中需要将那些被 `@Import` 导入的配置类也全部注册到 `BeanDefinitionRegistry` 中。

从代码清单 7-49 来看，`registerBeanDefinitionForImportedConfigurationClass` 方法仅是一个普通的 `BeanDefinition` 的注册，没有任何多余的操作，读者仅需要对该机制简单了解。

## 代码清单 7-49 被 @Import 导入的配置类注册到 BeanDefinitionRegistry 中

```java
private void registerBeanDefinitionForImportedConfigurationClass(ConfigurationClass configClass) {
    AnnotationMetadata metadata = configClass.getMetadata();
    // 构造 BeanDefinition
    AnnotatedGenericBeanDefinition configBeanDef = new AnnotatedGenericBeanDefinition(metadata);
    // 作用域、Bean 名称的处理 ......
    // 包装 BeanDefinitionHolder
    BeanDefinitionHolder definitionHolder = new BeanDefinitionHolder(configBeanDef,
            configBeanName);
    definitionHolder = AnnotationConfigUtils.applyScopedProxyMode(scopeMetadata,
            definitionHolder, this.registry);
    // 注册到 BeanDefinitionRegistry 中
    this.registry.registerBeanDefinition(definitionHolder.getBeanName(),
            definitionHolder.getBeanDefinition());
    configClass.setBeanName(configBeanName);
    // logger ......
}
```

### （2）注册 @Bean 注解的 bean

紧接着的第二步是加载 ConfigurationClass 中存储的标注了 @Bean 注解的方法，对应的 loadBeanDefinitionsForBeanMethod 方法篇幅比较长，如代码清单 7-50 所示（请读者尽可能地配合源码中标注的注释来阅读）。

## 代码清单 7-50 注册 @Bean 注解的 bean

```java
private void loadBeanDefinitionsForBeanMethod(BeanMethod beanMethod) {
    ConfigurationClass configClass = beanMethod.getConfigurationClass();
    MethodMetadata metadata = beanMethod.getMetadata();
    String methodName = metadata.getMethodName();

    // 如果条件装配将其跳过，则该 @Bean 标注的方法对应的 BeanDefinition
    // 不会注册到 BeanDefinition Registry
    if (this.conditionEvaluator.shouldSkip(metadata, ConfigurationPhase.REGISTER_BEAN)) {
        configClass.skippedBeanMethods.add(methodName);
        return;
    }
    if (configClass.skippedBeanMethods.contains(methodName)) {
        return;
    }

    // 校验 ......
    // 处理 Bean 的 name、alias ......

    // 注解中配置了 @Bean，与 xml 中的 Bean 出现完全一致，会抛出异常
    if (isOverriddenByExistingDefinition(beanMethod, beanName)) {
        if (beanName.equals(beanMethod.getConfigurationClass().getBeanName())) {
            // throw ex ......
        }
        return;
    }
```

## 7.5 invokeBeanFactoryPostProcessors——执行 BeanFactoryPost Processor

```java
// 构造 BeanDefinition
ConfigurationClassBeanDefinition beanDef = new ConfigurationClassBeanDefinition(configClass,
        metadata);
beanDef.setSource(this.sourceExtractor.extractSource(metadata, configClass.getResource()));

// 【复杂】解析@Bean 所在方法的修饰符
if (metadata.isStatic()) {
    // 静态@Bean 方法
    if (configClass.getMetadata() instanceof StandardAnnotationMetadata) {
        beanDef.setBeanClass(((StandardAnnotationMetadata) configClass.getMetadata()).get
IntrospectedClass());
    } else {
        beanDef.setBeanClassName(configClass.getMetadata().getClassName());
    }
    beanDef.setUniqueFactoryMethodName(methodName);
} else {
    // 实例@Bean 方法
    beanDef.setFactoryBeanName(configClass.getBeanName());
    beanDef.setUniqueFactoryMethodName(methodName);
}

// 处理@Bean 的属性（name、initMethod 等）、额外的注解（@Lazy、@DependsOn 等）……

// 注册到 BeanDefinitionRegistry 中
this.registry.registerBeanDefinition(beanName, beanDefToRegister);
}
```

整个处理过程分为 4 步：检查→构造 BeanDefinition→封装信息→注册到 BeanDefinition Registry 中。代码清单 7-50 中靠下方的部分中有关处理@Bean 的属性、额外的注解信息解析的逻辑相对简单，读者可以自行借助 IDE 去看看，这里不展开解释。

另外，注意代码清单 7-50 的中间部分还有一个标注有【复杂】的 metadata.isStatic() 方法，它的判断逻辑中会使用 setBeanClassName/setFactoryBeanName 以及 setUnique FactoryMethodName 方法给 BeanDefinition 封装两个属性，对应的两个属性分别指定了当前@Bean 方法所在的配置类以及方法名。这个设计需要与前面通过@Component 注解配合组件扫描构建的 BeanDefinition 进行对比。这种 BeanDefinition 在构建时会指定 Bean 的全限定名、属性注入等，而且最终创建的对象一定是通过反射创建的。而在注解配置类中的@Bean 方法有实际的代码执行，属于编程式创建，无法使用（也不适合用）反射创建 bean 对象，所以为了在后面能正常创建出 bean 对象，此处就需要记录该 Bean 的定义源（包含注解配置类和方法名），以确保在创建 bean 对象时，能够使用反射调用该注解配置类的方法生成 bean 对象并返回。

（3）解析 XML 配置文件

所有用@Bean 注解标注的方法封装完成后，下一步是将之前解析@ImportResource 注解时保存的 XML 配置文件路径逐个加载并解析，对应的逻辑如代码清单 7-51 所示。对于这部分读者可以不用了解太多，只需知道负责处理 XML 配置文件的核心组件是 XmlBean DefinitionReader。至于 XML 配置文件是如何解析的，本书不再展开讲解，感兴趣的读者可以自行深入了解。

### 代码清单 7-51　借助 XmlBeanDefinitionReader 解析 XML 配置文件

```java
private void loadBeanDefinitionsFromImportedResources(
    Map<String, Class<? extends BeanDefinitionReader>> importedResources) {
    Map<Class<?>, BeanDefinitionReader> readerInstanceCache = new HashMap<>();
    importedResources.forEach((resource, readerClass) -> {
        if (BeanDefinitionReader.class == readerClass) {
            if (StringUtils.endsWithIgnoreCase(resource, ".groovy")) {
                readerClass = GroovyBeanDefinitionReader.class;
            } else {
                // 创建 XmlBeanDefinitionReader，以备下面的解析
                readerClass = XmlBeanDefinitionReader.class;
            }
        }

        BeanDefinitionReader reader = readerInstanceCache.get(readerClass);
        // reader 的缓存等

        // 调用 XmlBeanDefinitionReader 解析资源文件
        reader.loadBeanDefinitions(resource);
    });
}
```

#### （4）回调 ImportBeanDefinitionRegistrar

最后一步是回调所有的 `ImportBeanDefinitionRegistrar`，既然是回调，就只需把所有的 `ImportBeanDefinitionRegistrar` 都提取出来，逐个调用其 `registerBeanDefinitions`。源码的逻辑也非常简单，如代码清单 7-52 所示。

### 代码清单 7-52　回调所有的 ImportBeanDefinitionRegistrar

```java
private void loadBeanDefinitionsFromRegistrars(Map<ImportBeanDefinitionRegistrar, AnnotationMetadata> registrars) {
    registrars.forEach((registrar, metadata) ->
        registrar.registerBeanDefinitions(metadata, this.registry,
            this.importBeanNameGenerator));
}
```

#### 5. ConfigurationClassPostProcess 功能小结

经过上述步骤后，`ConfigurationClassPostProcess` 的工作全部完成，简单总结 `ConfigurationClassPostProcess` 中的重要功能。

- `ConfigurationClassPostProcess` 实现了 `BeanDefinitionRegistryPostProcessor`，它会向 `BeanDefinitionRegistry` 中注册新的 `BeanDefinition`。
- `ConfigurationClassPostProcess` 中组合了一个 `ConfigurationClassParser`，其具备解析配置类的能力，`ConfigurationClassParser` 会具体负责注解配置类的解析并提取关键注解的信息，封装到 `ConfigurationClass` 对象中。
- `ConfigurationClassPostProcess` 中利用 `ConfigurationClassBeanDefinitionReader`，其具备读取配置类、封装 `BeanDefinition` 的能力，经过 `ConfigurationClassParser` 处理后，`ConfigurationClassBeanDefinitionReader` 会将提取的信息转换为 `BeanDefinition`，并注册到 `BeanDefinitionRegistry` 中。

## 7.6 registerBeanPostProcessors——初始化 BeanPostProcessor

BeanFactoryPostProcessor 处理完成之后，紧接着要初始化的是所有的 BeanPostProcessor，它的初始化逻辑整体看起来与 BeanFactoryPostProcessor 极其相似，如代码清单 7-53 所示。

**代码清单 7-53　初始化所有 BeanPostProcessor**

```java
protected void registerBeanPostProcessors(ConfigurableListableBeanFactory beanFactory) {
    // 依然借助 PostProcessorRegistrationDelegate 完成
    PostProcessorRegistrationDelegate.registerBeanPostProcessors(beanFactory, this);
}

public static void registerBeanPostProcessors(ConfigurableListableBeanFactory beanFactory,
        AbstractApplicationContext applicationContext) {
    String[] postProcessorNames = beanFactory
            .getBeanNamesForType(BeanPostProcessor.class, true, false);

    // 此处会先注册一个 BeanPostProcessorChecker
    int beanProcessorTargetCount = beanFactory.getBeanPostProcessorCount()
            + 1 + postProcessorNames.length;
    beanFactory.addBeanPostProcessor(new BeanPostProcessorChecker(beanFactory,
            beanProcessorTargetCount));

    // 根据排序规则，给所有的后置处理器分类
    List<BeanPostProcessor> priorityOrderedPostProcessors = new ArrayList<>();
    List<BeanPostProcessor> internalPostProcessors = new ArrayList<>();
    List<String> orderedPostProcessorNames = new ArrayList<>();
    List<String> nonOrderedPostProcessorNames = new ArrayList<>();
    for(String ppName : postProcessorNames) {
        if(beanFactory.isTypeMatch(ppName, PriorityOrdered.class)) {
            // 注意此处，PriorityOrdered 类型的后置处理器被提前初始化了
            BeanPostProcessor pp = beanFactory.getBean(ppName, BeanPostProcessor.class);
            priorityOrderedPostProcessors.add(pp);
            // MergedBeanDefinitionPostProcessor 类型的后置处理器被单独放在一个集合中，
            // 说明该接口比较特殊
            if (pp instanceof MergedBeanDefinitionPostProcessor) {
                internalPostProcessors.add(pp);
            }
        } else if (beanFactory.isTypeMatch(ppName, Ordered.class)) {
            orderedPostProcessorNames.add(ppName);
        } else {
            nonOrderedPostProcessorNames.add(ppName);
        }
    }
    // 注册实现了 PriorityOrdered 接口的 BeanPostProcessor
    sortPostProcessors(priorityOrderedPostProcessors, beanFactory);
    registerBeanPostProcessors(beanFactory, priorityOrderedPostProcessors);

    // 注册实现了 Ordered 接口的 BeanPostProcessor
    List<BeanPostProcessor> orderedPostProcessors = new ArrayList<>();
    for(String ppName : orderedPostProcessorNames) {
        BeanPostProcessor pp = beanFactory.getBean(ppName, BeanPostProcessor.class);
```

```
            orderedPostProcessors.add(pp);
            if (pp instanceof MergedBeanDefinitionPostProcessor) {
                internalPostProcessors.add(pp);
            }
        }
    }
    sortPostProcessors(orderedPostProcessors, beanFactory);
    registerBeanPostProcessors(beanFactory, orderedPostProcessors);

    // 注册普通的 BeanPostProcessor
    List<BeanPostProcessor> nonOrderedPostProcessors = new ArrayList<>();
    for (String ppName : nonOrderedPostProcessorNames) {
        BeanPostProcessor pp = beanFactory.getBean(ppName, BeanPostProcessor.class);
        nonOrderedPostProcessors.add(pp);
        if (pp instanceof MergedBeanDefinitionPostProcessor) {
            internalPostProcessors.add(pp);
        }
    }
    registerBeanPostProcessors(beanFactory, nonOrderedPostProcessors);

    // 最后, 重新注册被单独分离出来的 MergedBeanDefinitionPostProcessor
    sortPostProcessors(internalPostProcessors, beanFactory);
    registerBeanPostProcessors(beanFactory, internalPostProcessors);

    // 手动注册 ApplicationListenerDetector (参见 7.3.3 节)
    beanFactory.addBeanPostProcessor(new ApplicationListenerDetector(applicationContext));
}
```

registerBeanPostProcessors 方法的篇幅很长，但通读下来，读者的感受应该是非常熟悉，整体步骤与 BeanFactoryPostProcessor 的执行极其相似，最大的区别是 BeanFactoryPostProcessor 需要在提取出后立即执行，而 BeanPostProcessor 是先注册到 IOC 容器中，等待 bean 对象的初始化后再执行。对于整体的流程这里不再重复讲解，读者需要重点关注方法中的几个细节。

### 7.6.1 BeanPostProcessorChecker

由类名理解 BeanPostProcessorChecker 是一个后置处理器的**检查器**。作为一个后置处理器，它还要检查后置处理器，这看起来有些奇怪。如何理解该组件的设计，可以从 javadoc 中尝试获取一点线索。

> BeanPostProcessor that logs an info message when a bean is created during BeanPostProcessor instantiation, i.e. when a bean is not eligible for getting processed by all BeanPostProcessors.
>
> 它是一个 BeanPostProcessor，当在 BeanPostProcessor 实例化期间创建 Bean，即当某个 Bean 不能被所有 BeanPostProcessor 处理时，它会记录一条信息。

通过 javadoc 可以明白，BeanPostProcessorChecker 的作用是检查 **BeanPostProcessor** 的初始化阶段中是否有 bean 对象的意外创建。注意"意外创建"这个概念，其实它并不意外，如果在 BeanPostProcessor 中注入了其他的普通 bean 对象，根据依赖注入的原则，会在 BeanPostProcessor 创建之前先把这些注入的普通 bean 对象初始化，又因为当前阶段 **BeanPostProcessor** 还没有初始化完毕，这些普通 bean 对象还没有来得及被 **BeanPostProcessor** 处理，导致出现 bean 对象的"残缺不全"。BeanPostProcessor

Checker 的作用就是用来提醒开发者对该问题引起注意。

有关"残缺不全的 bean 对象"这种情况,其实并不少见,读者在开发时可能会遇到这种日志:xxxxxx is not eligible for getting processed by all BeanPostProcessors (for example: not eligible for auto-proxying)。这就意味着你的 BeanPostProcessor 中注入了普通 bean 对象,属于不合理设计,需要对代码做出调整。

另外,在 BeanPostProcessorChecker 的 postProcessAfterInitialization 方法中会检查当前 BeanFactory 中的后置处理器数量是否少于一开始计算的预计后置处理器数量,如果是则代表有 bean 对象被提前创建。对应的逻辑如代码清单 7-54 所示。

**代码清单 7-54　postProcessAfterInitialization 中回调检查**

```java
public Object postProcessAfterInitialization(Object bean, String beanName) {
    // 此处判断是否有普通 bean 对象被提前创建
    if(!(bean instanceof BeanPostProcessor) && !isInfrastructureBean(beanName) &&
        this.beanFactory.getBeanPostProcessorCount() < this.beanPostProcessorTargetCount){
        // 打印异常警告日志 ......
    }
    return bean;
}
```

## 7.6.2　MergedBeanDefinitionPostProcessor 被重复注册

在整个 registerBeanPostProcessors 方法接近最后的部分会手动注册一个 ApplicationListenerDetector:

```java
// 手动注册 ApplicationListenerDetector (参见 7.3.3 节)
beanFactory.addBeanPostProcessor(new ApplicationListenerDetector(applicationContext));
```

仔细回忆 7.3.3 节的内容,当时 prepareBeanFactory 方法中已经注册过这个 ApplicationListenerDetector,为什么此处还要重复注册呢?重复注册是否会存在相同类型的后置处理器呢?如果读者也有这种疑问,其实是多虑了。作为普通开发者能想到的问题,Spring Framework 的作者一定也会想到。实际上 addBeanPostProcessor 方法注册相同类型的后置处理器时,在底层不是手动添加而是**重新注册**,使其位于**所有后置处理器的末尾位置**,如代码清单 7-55 所示。

**代码清单 7-55　addBeanPostProcessor 会将已经注册过的后置处理器移至末尾**

```java
private static void registerBeanPostProcessors(
        ConfigurableListableBeanFactory beanFactory, List<BeanPostProcessor> postProcessors){

    for (BeanPostProcessor postProcessor : postProcessors) {
        beanFactory.addBeanPostProcessor(postProcessor);
    }
}

public void addBeanPostProcessor(BeanPostProcessor beanPostProcessor) {
    Assert.notNull(beanPostProcessor, "BeanPostProcessor must not be null");
    // 如果后置处理器已经存在,则移除
    this.beanPostProcessors.remove(beanPostProcessor);
```

```
        if (beanPostProcessor instanceof InstantiationAwareBeanPostProcessor) {
            this.hasInstantiationAwareBeanPostProcessors = true;
        }
        if (beanPostProcessor instanceof DestructionAwareBeanPostProcessor) {
            this.hasDestructionAwareBeanPostProcessors = true;
        }
        // 添加至后置处理器列表的末尾
        this.beanPostProcessors.add(beanPostProcessor);
    }
```

由此可知，重复注册 `ApplicationListenerDetector` 以及 `internalPostProcessors` 集合中后置处理器的目的是将这些后置处理器都放到整个后置处理器列表的末尾，仅此而已。

在第 3 章中讲解 `MergedBeanDefinitionPostProcessor` 时提到了一个非常重要的实现类 `AutowiredAnnotationBeanPostProcessor`，它会根据合并后的 `BeanDefinition` 为 bean 对象进行依赖注入。这个策略是 Spring Framework 内置的既定策略，下面介绍 bean 对象依赖注入的部分会讲到。

### 7.6.3　PriorityOrdered 类型的后置处理器

7.5 节中已经提到过一个设计，`PriorityOrdered` 接口代表最高优先级，实现它的类一般都是 Spring Framework 内置的极重要的组件。本节列举两个关键的后置处理器，读者可以先简单了解。

- `AutowiredAnnotationBeanPostProcessor`——处理 @Autowired 注解。
- `CommonAnnotationBeanPostProcessor`——处理 JSR-250 规范的注解。

这两个极重要的内置后置处理器都是用于解析和处理 bean 对象上的注解，如果这两个后置处理器不先初始化好，后面的 bean 对象中使用的 @Autowired、@PostConstruct 等注解将会失效。由此可知，这些实现了 `PriorityOrdered` 接口的组件通常都是**内置的核心组件**，Spring Framework 为了确保功能的正常使用，就必须让这些核心组件都率先准备就绪。

## 7.7　initMessageSource——初始化国际化组件

在第 3 章讲解 `ApplicationContext` 的结构时提到一点，`ApplicationContext` 本身实现了 `MessageSource` 接口，具备国际化的能力。`ApplicationContext` 初始化国际化组件的逻辑如代码清单 7-56 所示。

**代码清单 7-56　初始化国际化组件**

```
public static final String MESSAGE_SOURCE_BEAN_NAME = "messageSource";

protected void initMessageSource() {
    ConfigurableListableBeanFactory beanFactory = getBeanFactory();
    // 检查是否已经存在 MessageSource 组件，如果存在，直接赋值
    if (beanFactory.containsLocalBean(MESSAGE_SOURCE_BEAN_NAME)) {
        this.messageSource = beanFactory.getBean(MESSAGE_SOURCE_BEAN_NAME, MessageSource.class);
        if (this.parent != null && this.messageSource instanceof HierarchicalMessageSource) {
```

## 7.7 initMessageSource——初始化国际化组件

```
            HierarchicalMessageSource hms = (HierarchicalMessageSource) this.messageSource;
            if (hms.getParentMessageSource() == null) {
                hms.setParentMessageSource(getInternalParentMessageSource());
            }
        }
        // logger ......
    }
    // 如果不存在,则会创建一个全新的对象并注册到BeanFactory中
    else {
        DelegatingMessageSource dms = new DelegatingMessageSource();
        dms.setParentMessageSource(getInternalParentMessageSource());
        this.messageSource = dms;
        beanFactory.registerSingleton(MESSAGE_SOURCE_BEAN_NAME, this.messageSource);
        // logger ......
    }
}
```

初始化 MessageSource 的逻辑本身不复杂,initMessageSource 方法首先会检查 BeanFactory 中是否已经存在一个 MessageSource 类型的对象,如果存在则会直接提取,并在特定情况下进行一些额外的操作;如果没有注册过 MessageSource 对象,下方的 else 结构中会初始化一个默认的 DelegatingMessageSource 实现,这个实现类本身不会进行任何国际化操作(输入即输出),这样做的原因是配合 ApplicationContext 的功能做出的兜底处理,因为 ApplicationContext 本身实现了 MessageSource,所以在底层必然需要一个可以委托完成具体工作的落地实现,即便这个落地实现不会进行任何处理。

此处读者还需要注意,如果是纯 Spring Framework 应用的话,默认没有预先注册的 MessageSource 对象,所以会创建 else 结构中的空实现 DelegatingMessageSource,但由于我们目前探究的是 Spring Boot 应用,Spring Boot 考虑到国际化相关的配置可以借助 application.properties 实现配置内容外部化,因此它会借助 Spring Boot 的全局配置文件与参数绑定特性,在特定条件下预先创建一个 MessageSource,对于具体的逻辑可以查看自动配置类 MessageSourceAutoConfiguration,如代码清单 7-57 所示。

**代码清单 7-57　MessageSource 的自动配置类会注册一个 MessageSource**

```
@ConditionalOnMissingBean(name = AbstractApplicationContext.MESSAGE_SOURCE_BEAN_NAME,
                    search = SearchStrategy.CURRENT)
@Conditional(ResourceBundleCondition.class)
public class MessageSourceAutoConfiguration {
    // ......

    @Bean
    public MessageSource messageSource(MessageSourceProperties properties) {
        ResourceBundleMessageSource messageSource = new ResourceBundleMessageSource();
        // 应用 application.properties 外部化的配置内容
        return messageSource;
    }
```

由代码清单 7-57 可以看到,MessageSourceAutoConfiguration 中定义了一个 @Bean 方法,该方法会注册一个 ResourceBundleMessageSource 对象,这就是 Spring Boot 自动

装配的 MessageSource 实现。请读者注意，若想触发该自动装配，需要满足两个条件：
- IOC 容器中不存在一个名称为"messageSource"的 Bean；
- 类路径下可以找到一个默认的名称为 messages.properties 的文件（properties 文件名称会随 spring.messages.basename 的属性值改变）。

只有满足以上两个条件，Spring Boot 才认定当前应用支持它装载 MessageSource 的实现类，从而使自动配置类生效。

## 7.8　initApplicationEventMulticaster——初始化事件广播器

处理完 ApplicationContext 支持的国际化功能后，下一个动作是初始化内置的事件广播器。ApplicationContext 接口实现了 ApplicationEventPublisher，具备事件发布的能力，注意事件发布器 ApplicationEventPublisher 与事件广播器 ApplicationEventMulticaster 不是一回事，**事件发布器用来接受事件，并交给事件广播器处理；事件广播器取得事件发布器的事件，并广播给监听器**。在观察者模式中，观察者是这两者的合体，在 Spring Framework 中将该职责拆分为两部分。

ApplicationContext 内部初始化事件广播器的逻辑如代码清单 7-58 所示。初始化 ApplicationEventMulticaster 的逻辑与初始化 MessageSource 类似，都是先判断是否已经在 BeanFactory 中有注册，如果有则直接获取并应用，否则创建默认的实现。注意，initApplicationEventMulticaster 方法默认创建的事件广播器类型为 SimpleApplicationEventMulticaster，这是 Spring Framework 中唯一具体的 ApplicationEventMulticaster 落地实现，因此在没有额外扩展的前提下，负责广播的事件广播器一定是 SimpleApplicationEventMulticaster。

**代码清单 7-58　初始化事件广播器**

```
private ApplicationEventMulticaster applicationEventMulticaster;

public static final String APPLICATION_EVENT_MULTICASTER_BEAN_NAME = "applicationEventMulticaster";

protected void initApplicationEventMulticaster() {
    ConfigurableListableBeanFactory beanFactory = getBeanFactory();
    if (beanFactory.containsLocalBean(APPLICATION_EVENT_MULTICASTER_BEAN_NAME)) {
        this.applicationEventMulticaster =
                beanFactory.getBean(APPLICATION_EVENT_MULTICASTER_BEAN_NAME,
                        ApplicationEventMulticaster.class);
        // logger ......
    } else {
        this.applicationEventMulticaster = new SimpleApplicationEventMulticaster(beanFactory);
        beanFactory.registerSingleton(APPLICATION_EVENT_MULTICASTER_BEAN_NAME,
                this.applicationEventMulticaster);
        // logger ......
    }
}
```

## 7.9 onRefresh——子类扩展的刷新动作

接下来的 `onRefresh` 方法又是一个模板方法,如代码清单 7-59 所示。在 Spring Framework 的基本 IOC 容器实现中该方法并没有值得关注的扩展,不过 Spring Boot 在此处扩展了嵌入式 Web 容器的初始化。有关嵌入式 Web 容器的内容会统一放到第 8 章中讲解。

**代码清单 7-59　onRefresh 子类扩展的刷新动作**

```java
protected void onRefresh() throws BeansException {
    // 对于子类: 默认没有任何操作
}
```

## 7.10 registerListeners——注册监听器

下面的步骤是注册事件监听器,注意此处只是**将监听器注册到事件广播器,并没有初始化这些监听器对象**,对应的逻辑如代码清单 7-60 所示。

**代码清单 7-60　注册监听器**

```java
protected void registerListeners() {
    // 把所有的 IOC 容器中以前缓存好的一组 ApplicationListener 提取出来, 添加到事件广播器中
    for (ApplicationListener<?> listener : getApplicationListeners()) {
        getApplicationEventMulticaster().addApplicationListener(listener);
    }

    // 将 BeanFactory 中定义的所有 ApplicationListener 类型的组件全部提取出, 添加到事件广播器中
    String[] listenerBeanNames = getBeanNamesForType(ApplicationListener.class, true, false);
    for (String listenerBeanName : listenerBeanNames) {
        getApplicationEventMulticaster().addApplicationListenerBean(listenerBeanName);
    }

    // 广播早期事件
    Set<ApplicationEvent> earlyEventsToProcess = this.earlyApplicationEvents;
    this.earlyApplicationEvents = null;
    if (earlyEventsToProcess != null) {
        for (ApplicationEvent earlyEvent : earlyEventsToProcess) {
            getApplicationEventMulticaster().multicastEvent(earlyEvent);
        }
    }
}

public Collection<ApplicationListener<?>> getApplicationListeners() {
    return this.applicationListeners;
}
```

所有监听器的来源包含两部分:一部分是在 ApplicationContext 初始化(刷新)之前手动注册的;另一部分是在 IOC 容器中通过组件注册的方式添加的。对于前者只需一次性提取出,并关联到事件广播器中;而对于通过组件注册的方式注册到 IOC 容器中的监听器,在关联事件广播器时关联的是 **Bean 的名称**而不是 bean 对象本身,这样做的目的是防止监听器被不合理地提早初始化(单实例 Bean 的统一初始化动作在下一步)。

另外,请读者注意 `registerListeners` 方法的最后一个环节中会广播所有的早期事件,

这在 7.1.2 节中提到过，早期事件是记录 ApplicationEventMulticaster 尚未初始化时被广播的事件，当 ApplicationEventMulticaster 被初始化完成且监听器也关联到事件广播器后，早期事件就可以被正常广播了。

## 7.11 finishBeanFactoryInitialization——初始化剩余的单实例 bean 对象

接下来我们要进入 IOC 容器刷新的第二个**超级复杂**的重难点：初始化剩余的非延迟加载的单实例 bean 对象。在该阶段中，项目中定义的所有普通的单实例 bean 对象均会被创建和初始化。本方法涉及的原理非常多且复杂，建议读者在阅读本节内容时最好配合 IDE 的 Debug 调试，借助一个实际的应用同步阅读。

首先简单介绍 finishBeanFactoryInitialization 方法的总体实现，除了最后一行方法，其余部分的所有工作都是预备性的，读者不需要在这部分投入过多精力。整个 finishBeanFactoryInitialization 方法的重中之重，当属最后一行 beanFactory.preInstantiateSingletons()，如代码清单 7-61 所示。

**代码清单 7-61　finishBeanFactoryInitialization 初始化剩余的单实例 bean 对象**

```java
protected void finishBeanFactoryInitialization(ConfigurableListableBeanFactory beanFactory) {
    // 初始化 ConversionService，这个 ConversionService 是用于类型转换的服务接口
    // 它的工作是将配置文件/properties 中的数据进行类型转换，得到真正想要的数据类型
    if (beanFactory.containsBean(CONVERSION_SERVICE_BEAN_NAME) &&
            beanFactory.isTypeMatch(CONVERSION_SERVICE_BEAN_NAME, ConversionService.class)){
        beanFactory.setConversionService(
                beanFactory.getBean(CONVERSION_SERVICE_BEAN_NAME, ConversionService.class));
    }

    // 嵌入式值解析器 EmbeddedValueResolver 的组件注册，它负责解析占位符和表达式
    if (!beanFactory.hasEmbeddedValueResolver()) {
        beanFactory.addEmbeddedValueResolver(strVal ->
                getEnvironment().resolvePlaceholders(strVal));
    }

    // 与 LoadTimeWeaverAware 有关的部分
    String[] weaverAwareNames = beanFactory
            .getBeanNamesForType(LoadTimeWeaverAware.class, false, false);
    for (String weaverAwareName : weaverAwareNames) {
        getBean(weaverAwareName);
    }
    beanFactory.setTempClassLoader(null);

    // 冻结配置，此时无论如何获取 BeanDefinition 的名称集合，
    // 获取到的都是同样的（除非增减新的 BeanDefinition）
    beanFactory.freezeConfiguration();

    // 【初始化】实例化所有非延迟加载的单实例 Bean
    beanFactory.preInstantiateSingletons();
}
```

## 7.11.1 beanFactory.preInstantiateSingletons

preInstantiateSingletons 方法的定义来自 ConfigurableListableBeanFactory，最终实现在 DefaultListableBeanFactory 中。这个方法的分支逻辑比较多，如代码清单 7-62 所示（可以只关注源码中标有注释的部分，对于没有标注注释的源码，在第一次深入底层时可以暂时忽略）。

**代码清单 7-62　preInstantiateSingletons 初始化所有单实例 Bean**

```java
public void preInstantiateSingletons() throws BeansException {
    // logger ......

    List<String> beanNames = new ArrayList<>(this.beanDefinitionNames);

    // 触发所有非延迟加载的单实例 Bean 的初始化
    // 此处循环初始化剩余的非延迟加载的单实例 Bean
    for (String beanName : beanNames) {
        // 先合并 BeanDefinition
        RootBeanDefinition bd = getMergedLocalBeanDefinition(beanName);
        // 不是抽象的、不是延迟加载的单实例 Bean 需要初始化
        if (!bd.isAbstract() && bd.isSingleton() && !bd.isLazyInit()) {
            // FactoryBean 默认不立即初始化，除非指定 isEagerInit=true
            if (isFactoryBean(beanName)) {
                Object bean = getBean(FACTORY_BEAN_PREFIX + beanName);
                if (bean instanceof FactoryBean) {
                    // FactoryBean 的处理
                }
            } else {
                // 普通的初始化就是 getBean 方法
                getBean(beanName);
            }
        }
    }

    // 初始化的最后阶段 ......
}
```

简单概括代码清单 7-62 中的逻辑，在初始化所有非延迟加载的单实例 bean 对象时会根据 bean 对象的类型分别处理。如果 bean 对象的类型是 FactoryBean，会有单独的处理逻辑；而初始化普通 bean 对象时，采用的方法是在刚开始学习 Spring Framework 时使用的 **getBean** 方法！这里读者可以体会一下：Spring Framework 的底层在进行 bean 对象的初始化时，采用的是最基本的 getBean 方法，而 getBean 方法本身来自 IOC 容器的顶层接口 BeanFactory，所以底层的初始化方法用的反而是最朴素的方式，这也告诉各位研讨源码的开发者，底层源码并不神秘，很多有趣的设计本质上还是对基础部分学过的知识的运用。

## 7.11.2 getBean

BeanFactory 中的 getBean 方法在底层会转调 doGetBean 方法，这种设计是 Spring Framework 中非常常见的方法命名风格，读者只需跟随方法调用步步跟进。

由于 doGetBean 方法的篇幅非常长，为方便读者更好地阅读和理解源码，本节将其拆分为多个片段讲解。

**1. 别名的解析处理**

在使用 Spring Framework 的注解配置类时，如果要向 IOC 容器中注册新的 Bean，可以使用 @Bean 注解标注到方法上进行定义注册。用这种方式定义时，如果显式定义 @Bean 注解中的 name 或 value 属性，可以为 Bean 指定名称，但请读者注意的是，name 和 value 属性可以传入一个数组，这就意味着一个 bean 对象可以有多个名称，默认情况下传入的第一个名称是 bean 对象的 name，其余的名称都以别名（即 alias）的方式记录在 IOC 容器中。而代码清单 7-63 中的逻辑就是针对调用 getBean 方法传入 Bean 的别名时的处理，经过 transformedBeanName 方法即可准确定位到 bean 对象的 name，进而继续向下执行获取动作。

**代码清单 7-63　doGetBean（1）**

```
protected <T> T doGetBean(Class<T> requiredType, @Nullable Object[] args,
      boolean typeCheckOnly) throws BeansException {
    String beanName = transformedBeanName(name);
    Object bean;
    // ......

protected String transformedBeanName(String name) {
    return canonicalName(BeanFactoryUtils.transformedBeanName(name));
}

public String canonicalName(String name){
    String canonicalName = name;
    String resolvedName;
    do {
        resolvedName = this.aliasMap.get(canonicalName);
        if (resolvedName != null) {
            canonicalName = resolvedName;
        }
    } while (resolvedName != null);
    return canonicalName;
}
```

**2. 循环依赖的解决处理**

紧接着的环节是尝试获取 IOC 容器中是否已经创建并缓存当前正在获取的单实例 bean 对象，如代码清单 7-64 所示。这里的 getSingleton 方法会针对循环依赖的情况进行额外处理，有关循环依赖的内容会在 7.15 节中讲解。如果可以成功获取到 bean 对象，会对 FactoryBean 类型进行额外处理。读者都清楚如果 IOC 容器中注册了一个 FactoryBean，实际获取到的对象是通过 FactoryBean 的 getObject 方法而不是 FactoryBean 本身得到的对象，此处的 getObjectForBeanInstance 方法就是调用 FactoryBean 的 getObject 方法获取真正的对象。该方法的调用链较长，读者可以借助 IDE 按照以下方法链寻找踪迹：getObjectForBeanInstance→getObjectFromFactoryBean→doGetObjectFromFactoryBean。最终可以找到代码清单 7-65 中 FactoryBean 的 getObject 方法调用。

## 7.11 finishBeanFactoryInitialization——初始化剩余的单实例 bean 对象

**代码清单 7-64　doGetBean（2）**

```
// ......
// 先尝试从之前实例化好的 Bean 中找有没有当前 Bean
// 如果能找到，说明 Bean 已经被实例化，可以直接返回
Object sharedInstance = getSingleton(beanName);
if(sharedInstance != null && args == null) {
    // logger ......
    // 如果是 FactoryBean，则会调用 getObject 获取真正创建的对象
    bean = getObjectForBeanInstance(sharedInstance, name, beanName, null);
}
// ......
```

**代码清单 7-65　FactoryBean#getObject 方法调用**

```
private Object doGetObjectFromFactoryBean(FactoryBean<?> factory,String beanName)
      throws BeanCreationException {
    Object object;
    try {
        // JMX 监控 ......
        else {
            object = factory.getObject();
        }
    }// catch throw ex ......
    // ......
    return object;
}
```

### 3. 创建前的检查

如果代码清单 7-64 中的 `getSingleton` 方法没有获取到正在创建的单实例 bean 对象，说明当前获取的 bean 对象尚未创建，则执行下面的 else 结构部分。首先 else 结构中会判断当前创建的 bean 对象是否是一个原型 Bean 并且这个 bean 对象正在创建，如果的确正在创建则说明当前原型 Bean 在一次获取中即将产生两个对象，这种现象不合理，所以会抛出异常。

> 提示：有关不合理的原因，请读者思考，如果执行一次获取 bean 对象的动作会产生两个 bean 对象，说明在创建过程中有其他机制引导 IOC 容器重新创建当前正在获取的 bean 对象，而这个机制通常是有其他原型 Bean 依赖了当前正在获取的 bean 对象，在这种情况下如果 IOC 容器不加以检查和拦截，后果是会不断创建新的 bean 对象以注入给另外的 bean 对象，最终导致内存溢出。

原型 Bean 的循环依赖检查完毕后，下一步要做的事情是检查父容器中是否包含当前 bean 对象对应的 `BeanDefinition` 信息。Spring Framework 为了确保 IOC 容器存在父子关系时各司其职，底层会让每个 `BeanFactory` 初始化自身持有的 `BeanDefinition` 对应的 bean 对象，而不是将所有的 bean 对象集中存放在顶层容器中，这可以进一步体现出 `BeanFactory` 的层次性。这种设计在源码中的体现是，在检查自身容器中不包含当前正在获取的 bean 对象时调用父容器的 `doGetBean` 方法，由父容器获取后返回，如代码清单 7-66 所示。

**代码清单 7-66　doGetBean（3）**

```
// ......
else {
```

```
// 如果原型 Bean 之间互相依赖,则一定会引发无限循环,此处会抛出循环依赖异常
if (isPrototypeCurrentlyInCreation(beanName)) {
    throw new BeanCurrentlyInCreationException(beanName);
}

// 如果本地不存在当前 Bean 的定义信息,则尝试让父容器实例化 Bean
// 此举可以确保每个 BeanFactory 持有它应该有的 Bean,而不是所有的 Bean 都集中在某一个 BeanFactory 中
BeanFactory parentBeanFactory = getParentBeanFactory();
if (parentBeanFactory != null && !containsBeanDefinition(beanName)) {
    String nameToLookup = originalBeanName(name);
    if (parentBeanFactory instanceof AbstractBeanFactory) {
        return ((AbstractBeanFactory) parentBeanFactory).doGetBean(
                nameToLookup, requiredType, args, typeCheckOnly);
    } // else if ......
}
// ......
```

### 4. 标记准备创建的 bean 对象

源码继续往下进行会有一个 bean 对象名称的标记动作,这个 `markBeanAsCreated` 方法会将当前正在获取的 bean 对象的名称放入 `alreadyCreated` 集合中,代表当前 bean 对象已经被创建。请注意代码清单 7-67 中 `markBeanAsCreated` 方法内部的设计,它使用双检锁的机制检查当前创建的 bean 对象的名称,目的是防止多线程同时进行到该步骤而引发 bean 对象多次创建的问题。

**代码清单 7-67　doGetBean(4)**

```
// ......
// 程序运行至此处,证明 Bean 确需创建
if (!typeCheckOnly) {
    markBeanAsCreated(beanName);
}
// ......

protected void markBeanAsCreated(String beanName) {
    if (!this.alreadyCreated.contains(beanName)) {
        synchronized (this.mergedBeanDefinitions) {
            if (!this.alreadyCreated.contains(beanName)) {
                clearMergedBeanDefinition(beanName);
                this.alreadyCreated.add(beanName);
            }
        }
    }
}
```

### 5. 合并 BeanDefinition

标记当前正在创建的 bean 对象名称之后,下一个环节是合并 BeanDefinition,如代码清单 7-68 所示。这个合并 BeanDefinition 的动作非常重要。通过合并 BeanDefinition 信息,IOC 容器即可知晓当前正在创建的 bean 对象需要依赖哪些 bean 对象,进而支撑后续的依赖注入动作。执行完 `getMergedLocalBeanDefinition` 方法后,返回的 RootBeanDefinition 对象中会一并收集当前创建的 bean 对象中显式标注了 @DependsOn 注解的属性。由于标注了 @DependsOn 注解的属性代表强制依赖,因此 IOC 容器在此处会优先处理这些被强

## 7.11 finishBeanFactoryInitialization——初始化剩余的单实例 bean 对象

制依赖的 bean 对象并将其初始化，而初始化的方式依然是 `getBean` 方法。

**代码清单 7-68　doGetBean（5）**

```
// ......
try {
    RootBeanDefinition mbd = getMergedLocalBeanDefinition(beanName);
    checkMergedBeanDefinition(mbd, beanName, args);

    // 处理当前 bean 对象的 bean 对象依赖（@DependsOn 注解的依赖）
    String[] dependsOn = mbd.getDependsOn();
    if (dependsOn != null) {
        for (String dep : dependsOn) {
            // 循环依赖的检查 ......
            registerDependentBean(dep, beanName);
            try {
                getBean(dep);
            } // catch throw ex ......
        }
    }
    // ......
```

### 6. bean 对象的创建

连续几步检查和前置处理后，代码清单 7-69 展示了最重要的创建对象环节。IOC 容器会根据当前正在创建的 bean 对象的作用域决定如何创建对象，默认情况下 IOC 容器只有两种作用域：单实例 singleton 和原型 prototype。而对于两者的创建机制而言，底层均通过调用 **createBean** 方法完成创建，说明实际创建对象的方法就是 createBean。请读者注意观察 createBean 的调用方法，对于原型 Bean，每次调用都会创建一个全新的对象实例，而对于单实例 Bean，无论调用多少次 getBean 方法，底层始终保持只能有一个对象实例，而 IOC 容器中控制单实例对象的方式是使用 `getSingleton` 方法配合 `ObjectFactory` 实现，读者需要先了解这个单实例对象的控制机制。

**代码清单 7-69　doGetBean（6）**

```
// ......
if(mbd.isSingleton()) {
    sharedInstance = getSingleton(beanName, () -> {
        try {
            return createBean(beanName, mbd, args);
        } // catch throw ex ......
    });
    bean = getObjectForBeanInstance(sharedInstance, name, beanName, mbd);
} else if (mbd.isPrototype()) {
    Object prototypeInstance = null;
    try {
        beforePrototypeCreation(beanName);
        prototypeInstance = createBean(beanName, mbd, args);
    } finally {
        afterPrototypeCreation(beanName);
    }
    bean = getObjectForBeanInstance(prototypeInstance, name, beanName, mbd);
} // else 其他作用域类型 bean 对象的创建 ......
// ......
```

### 7. getSingleton 控制单实例对象

IOC 容器控制单实例 bean 对象的最有效方式是使用缓存，在第一次创建好单实例 bean 对象后会将其放入 IOC 容器中的缓存区，后续再获取该 bean 对象时可直接从缓存区中取出并返回。代码清单 7-70 展示了 getSingleton 方法的大体脉络。

**代码清单 7-70　getSingleton 控制单实例对象**

```java
private final Map<String, Object> singletonObjects = new ConcurrentHashMap<>(256);

public Object getSingleton(String beanName, ObjectFactory<?> singletonFactory) {
    Assert.notNull(beanName, "Bean name must not be null");
    synchronized(this.singletonObjects) {
        // 加锁后再查一次单实例 bean 对象的缓存
        Object singletonObject = this.singletonObjects.get(beanName);
        if (singletonObject == null) {
            if(this.singletonsCurrentlyInDestruction) {
                // throw ex ......
            }
            // 控制循环依赖的关键步骤
            beforeSingletonCreation(beanName);
            // ......
            try {
                // 【createBean】如果单实例 bean 对象的缓存中没有，就创建对象
                singletonObject = singletonFactory.getObject();
                newSingleton = true;
            } // catch finally ......

            // 新创建的单实例 bean 要存入单实例 bean 的缓存中
            if (newSingleton) {
                addSingleton(beanName, singletonObject);
            }
        }
    }
    return singletonObject;
}
```

通读 getSingleton 方法的实现可以得知，底层控制单实例 bean 对象的方式是借助一个名为 singletonObjects 的 Map 充当缓存区，在获取单实例 bean 对象时，会先从缓存区中尝试获取，如果没有获取到则会调用 singletonFactory，即代码清单 7-69 中的 lambda 表达式的 getObject 方法执行实际创建对象的动作，而单实例 bean 对象的创建动作与原型 bean 对象的一致（都是 createBean 方法），所以简单总结：单实例 bean 对象在第一次创建时会调用 createBean 方法真正地创建对象，创建完毕后会存入 IOC 容器底层的 singletonObjects 缓存区中，后续再次获取时会直接从缓存区中取出 bean 对象并返回。

> 💡 提示：注意在 singletonFactory.getObject() 方法调用之前，还有一步 beforeSingletonCreation 的动作，该动作是 IOC 容器处理循环依赖的关键动作，具体处理流程会在 7.15 节中讲解。

简单了解 IOC 容器中控制单实例对象的机制后，下面的重点环节是 createBean 方法。

### 7.11.3 createBean

经过 doGetBean 方法没有实际获取到 bean 对象之后,createBean 负责真正的 bean 对象的创建工作。经过 createBean 方法之后必定会创建一个全新的对象。createBean 方法的底层源码不复杂,抽取核心逻辑后的源码如代码清单 7-71 所示。

**代码清单 7-71　createBean 的核心逻辑**

```
protected Object createBean(String beanName, RootBeanDefinition mbd, @Nullable Object[] args)
        throws BeanCreationException {
    // logger ......
    RootBeanDefinition mbdToUse = mbd;
    // 根据 BeanDefinition 获取当前正在创建的 bean 对象的类型
    Class<?> resolvedClass = resolveBeanClass(mbd, beanName);
    if (resolvedClass != null && !mbd.hasBeanClass() && mbd.getBeanClassName() != null) {
        mbdToUse = new RootBeanDefinition(mbd);
        mbdToUse.setBeanClass(resolvedClass);
    }
    // ......
    try {
        // 后置处理器拦截创建 bean 对象
        Object bean = resolveBeforeInstantiation(beanName, mbdToUse);
        if (bean != null) {
            return bean;
        }
    } // catch ......

    try {
        // 真正创建 bean 对象
        Object beanInstance = doCreateBean(beanName, mbdToUse, args);
        return beanInstance;
    } // catch ......
}
```

根据 createBean 方法的主干逻辑,可以提取出实际创建 bean 对象的两个切入点:通过 resolveBeforeInstantiation 方法创建 bean 对象;通过 doCreateBean 方法创建 bean 对象。根据以往阅读源码的经验,此处更为明显的方法是 doCreateBean,但是 resolveBeforeInstantiation 方法同样值得关注,下面先来看 resolveBeforeInstantiation 方法的实现。

#### resolveBeforeInstantiation

resolveBeforeInstantiation 方法可以理解为"实例化之前的处理",由方法名可以获取到一点额外信息:doCreateBean 的确是创建 bean 对象的核心逻辑,resolveBeforeInstantiation 只是核心逻辑之前的拦截而已。通过代码清单 7-72 可以了解到,拦截 bean 对象的创建行为通过 InstantiationAwareBeanPostProcessor 实现,如果容器中包含这类后置处理器,则会执行拦截动作尝试创建 bean 对象,反之则不会拦截。

**代码清单 7-72　resolveBeforeInstantiation**

```
protected Object resolveBeforeInstantiation(String beanName, RootBeanDefinition mbd) {
```

```java
    Object bean = null;
    if (!Boolean.FALSE.equals(mbd.beforeInstantiationResolved)) {
        if (!mbd.isSynthetic() && hasInstantiationAwareBeanPostProcessors()) {
            Class<?> targetType = determineTargetType(beanName, mbd);
            if (targetType != null) {
                // 执行所有 InstantiationAwareBeanPostProcessor
                bean = applyBeanPostProcessorsBeforeInstantiation(targetType, beanName);
                if (bean != null) {
                    // 如果成功创建了 bean 对象，则执行 BeanPostProcessor 的后置初始化
                    bean = applyBeanPostProcessorsAfterInitialization(bean, beanName);
                }
            }
        }
        mbd.beforeInstantiationResolved=(bean != null);
    }
    return bean;
}
```

着重观察拦截的动作，它共有两个回调后置处理器的方法，分别是使用 InstantiationAwareBeanPostProcessor 的 postProcessBeforeInstantiation 方法创建对象以及使用 BeanPostProcessor 的 postProcessAfterInitialization 方法增强对象。可能有读者会不理解，为什么通过 postProcessBeforeInstantiation 方法创建出对象后，还需要回调 BeanPostProcessor 的 postProcessAfterInitialization 进行后置处理呢？这个问题要结合 BeanPostProcessor 的经典应用 AOP 来解释。BeanPostProcessor 的 postProcessAfterInitialization 方法可以用于生成代理对象，如果一个 bean 对象被 InstantiationAwareBeanPostProcessor 提前创建之后，还需要被 AOP 增强，这时就必须回调 BeanPostProcessor 的 postProcessAfterInitialization 方法，触发 AutoProxyCreator 的逻辑进而生成代理对象。

实际的回调后置处理器的方法的底层逻辑基本一致，都是获取到 IOC 容器中注册的所有后置处理器并一一回调，代码清单 7-73 展示了所有回调 InstantiationAwareBeanPostProcessor 的 postProcessBeforeInstantiation 方法的实现（逻辑本身很简单，不再展开）。

**代码清单 7-73　回调 postProcessBeforeInstantiation 方法**

```java
protected Object applyBeanPostProcessorsBeforeInstantiation(Class<?> beanClass,
        String beanName) {
    for (BeanPostProcessor bp : getBeanPostProcessors()) {
        // 循环找出所有的 InstantiationAwareBeanPostProcessor
        if (bp instanceof InstantiationAwareBeanPostProcessor) {
            // 调用它们的 postProcessBeforeInstantiation 尝试实例化 bean 对象
            InstantiationAwareBeanPostProcessor ibp = (InstantiationAwareBeanPostProcessor) bp;
            Object result = ibp.postProcessBeforeInstantiation(beanClass, beanName);
            if (result != null) {
                return result;
            }
        }
    }
    return null;
}
```

### 7.11.4 doCreateBean

如果 `InstantiationAwareBeanPostProcessor` 没有创建出 bean 对象，则需要执行 `doCreateBean` 方法创建 bean 对象的实例。`doCreateBean` 方法从主干逻辑上可抽取为三大步骤。

1. 实例化 bean 对象（此时 bean 对象中所有属性均为空）。
2. 属性赋值&依赖注入。
3. bean 对象的初始化（执行完该步骤后 bean 已经完整）。

由于每一步的逻辑都非常复杂，所以本节内容会非常长，读者需要细心跟进。

**1. 实例化 bean 对象**

实例化 bean 对象是创建对象的第一步，只有创建出对象之后才能进行后续的赋值、依赖注入、初始化逻辑回调等方法。创建 bean 对象的动作在代码清单 7-74 中体现为 `createBeanInstance` 方法。由于 `createBeanInstance` 方法篇幅非常长，为便于读者更好地阅读和理解源码，本节将其拆分为多个片段讲解。

**代码清单 7-74　doCreateBean（1）**

```
protected Object doCreateBean(String beanName, RootBeanDefinition mbd,
    @Nullable Object[] args) throws BeanCreationException {
    BeanWrapper instanceWrapper = null;
    // ......
    if (instanceWrapper == null) {
        // 真正的 bean 对象创建动作
        instanceWrapper = createBeanInstance(beanName, mbd, args);
    }
    // 得到真实的 bean 对象引用
    Object bean = instanceWrapper.getWrappedInstance();
    Class<?> beanType = instanceWrapper.getWrappedClass();
    if (beanType != NullBean.class) {
        mbd.resolvedTargetType= beanType;
    }
    // ......
```

**（1）解析 bean 对象的类型**

`createBeanInstance` 方法的第一环节会先检验当前要创建的 bean 对象所属类型是否可以被正常访问，如果 bean 对象所属类型本身无法被 Spring Framework 底层正常访问，则会因为无法创建对象而抛出异常。

**代码清单 7-75　createBeanInstance（1）**

```
protected BeanWrapper createBeanInstance(String beanName, RootBeanDefinition mbd,
        @Nullable Object[] args) {
    // 解析出 bean 对象的类型
    Class<?> beanClass = resolveBeanClass(mbd, beanName);
    // 如果 bean 对象无法被访问，则抛出异常
    if (beanClass != null && !Modifier.isPublic(beanClass.getModifiers())
            && !mbd.isNonPublicAccessAllowed()) {
        // throw ex ......
```

}
// ......

### （2）工厂方法创建

代码清单 7-76 中的两段逻辑本质上是做同一件事情：如果在 BeanDefinition 中指定了工厂类型的创建逻辑，则会直接执行工厂创建逻辑。注意，这里的工厂包含两种情况，下面分别解释。

**代码清单 7-76　createBeanInstance（2）**

```
// ......
// Spring Framework 5.x 的新特性
Supplier<?> instanceSupplier = mbd.getInstanceSupplier();
if (instanceSupplier != null) {
    return obtainFromSupplier(instanceSupplier, beanName);
}

// 工厂方法创建
if (mbd.getFactoryMethodName()!= null) {
    return instantiateUsingFactoryMethod(beanName, mbd, args);
}
// ......
```

**InstanceSupplier**

InstanceSupplier 的设计来自 AbstractBeanDefinition，是 Spring Framework 5.0 之后出现的新 API。JDK 8 之后多了几个很重要很实用的函数式接口，Supplier 作为生产型接口，与工厂的思路类似，Supplier 可以完成构造 bean 对象的工作，所以在 BeanDefinition 中加入了 InstanceSupplier 的设计，作为 factory-method 的一个替代方案。不过一般情况下项目开发中不会直接操作 BeanDefinition，所以对于 InstanceSupplier 读者只需一般了解。

**factoryMethod**

基于 Spring Boot 的项目开发中更多接触到的是 factoryMethod，注解配置类中被 @Bean 注解标注的方法，其本质就是**工厂方法**。代码清单 7-84 的逻辑中会执行 instantiateUsingFactoryMethod 方法来触发 factoryMethod，底层会根据工厂名称找到对应的静态工厂/实例工厂/注解配置类对象（如果是注解配置类的话，还需要解析 @Bean 方法上的参数列表注入对应的依赖），随后反射执行工厂方法生成 bean 对象。整体方法的逻辑很复杂，感兴趣的读者可以借助 IDE 深入研究，本书不再附源码。

### （3）原型 Bean 的创建优化

当程序运行至代码清单 7-77 处，意味着当前需要创建的 bean 对象是使用普通的构造方法创建。对于单实例 Bean 而言，代码清单 7-77 中的逻辑在程序运行期间只会执行一次，这部分工作显得比较多余。而对于原型 Bean 而言，该部分工作就可以发挥其价值，由于原型 Bean 在程序运行期间每次创建的过程通常是相同的，IOC 容器考虑到执行效率的问题后权衡利弊，最终选择以空间换时间的策略，在第一次原型 bean 对象创建完成后，将创建过程中引用的构造方法参数缓存到 BeanDefinition 中，以备后续创建时可以直接取出。

## 7.11 finishBeanFactoryInitialization——初始化剩余的单实例 bean 对象

**代码清单 7-77　createBeanInstance（3）**

```
// ......
// 这一步是为原型 Bean 创建的优化
boolean resolved = false;
boolean autowireNecessary = false;
if (args == null) {
    synchronized (mbd.constructorArgumentLock) {
        if (mbd.resolvedConstructorOrFactoryMethod != null) {
            resolved = true;
            autowireNecessary = mbd.constructorArgumentsResolved;
        }
    }
}
if (resolved) {
    if (autowireNecessary) {
        return autowireConstructor(beanName, mbd, null, null);
    } else {
        return instantiateBean(beanName, mbd);
    }
}
// ......
```

（4）实例化 bean 对象的真实动作

如果程序运行至代码清单 7-78 处，代表当前是 IOC 容器第一次创建 bean 对象，需要严格遵循构造方法的创建原则。对象的基本创建流程是解析构造方法→构造方法参数注入→反射调用构造方法创建对象实例。如果当前创建的 bean 对象所属类型中有显式定义带参数的构造方法，则会依次解析构造方法参数列表，并一一进行依赖注入。

**代码清单 7-78　createBeanInstance（4）**

```
// ......
// 回调 SmartInstantiationAwareBeanPostProcessor 寻找构造方法
Constructor<?>[] ctors = determineConstructorsFromBeanPostProcessors(beanClass, beanName);
// 触发执行基于构造方法的实例化判断
if (ctors != null || mbd.getResolvedAutowireMode() == AUTOWIRE_CONSTRUCTOR
        || mbd.hasConstructorArgumentValues() || !ObjectUtils.isEmpty(args)) {
    // 此处会额外缓存创建当前 bean 对象所需的构造方法参数
    return autowireConstructor(beanName, mbd, ctors, args);
}

ctors = mbd.getPreferredConstructors();
if (ctors != null) {
    return autowireConstructor(beanName, mbd, ctors, null);
}
return instantiateBean(beanName, mbd);
}
```

请读者注意，当触发以下任意条件时，IOC 容器会选择使用显式构造方法的对象实例化方式：1）通过 SmartInstantiationAwareBeanPostProcessor 找到了构造方法；2）配置自动注入方式为 AUTOWIRE_CONSTRUCTOR；3）使用 XML 配置文件的方式定义 Bean 时指定了 constructor-arg 标签；4）调用 getBean 方法获取 bean 对象时传入了 args 参数。

如果当前正在创建的 bean 对象没有指定任何构造方法，则会使用默认的无参构造方法创建对象，具体的逻辑在最后一行代码 instantiateBean 方法中，如代码清单 7-79 所示。instantiateBean 方法的核心动作是获取 InstantiationStrategy，调用 BeanUtils.instantiateClass 方法反射实例化 bean 对象。

**代码清单 7-79　instantiateBean 使用默认构造方法创建对象**

```java
protected BeanWrapper instantiateBean(final String beanName, final RootBeanDefinition mbd) {
    try {
        Object beanInstance;
        final BeanFactory parent = this;
        if // ......
        else {
            // 借助 InstantiationStrategy
            beanInstance = getInstantiationStrategy().instantiate(mbd, beanName, parent);
        }
        BeanWrapper bw = new BeanWrapperImpl(beanInstance);
        initBeanWrapper(bw);
        return bw;
    }
    // ......
}
```

经过 createBeanInstance 方法后，即可得到一个对象内部没有任何额外注入的 bean 对象，bean 对象的实例化完毕。

**2. 属性赋值前的收集**

进入属性赋值与依赖注入的核心逻辑之前，在 doCreateBean 方法中还有一个额外的逻辑：属性赋值前的注解信息收集。这里先对该逻辑进行讲解，如代码清单 7-80 所示。之后才是核心的赋值和注入动作。

**代码清单 7-80　doCreateBean（2）**

```java
// ......
synchronized (mbd.postProcessingLock) {
    if (!mbd.postProcessed) {
        try {
            applyMergedBeanDefinitionPostProcessors(mbd, beanType, beanName);
        } // catch ......
        mbd.postProcessed = true;
    }
}
// ......
```

代码清单 7-80 中的核心方法是 try-catch 结构中的 applyMergedBeanDefinitionPostProcessors 方法，该动作会回调 IOC 容器中的所有 MergedBeanDefinitionPostProcessor，其中有几个重要的后置处理器需要了解，下面一一列举。

（1）InitDestroyAnnotationBeanPostProcessor

由类名可知，InitDestroyAnnotationBeanPostProcessor 的核心处理工作与 Bean 的初始化和销毁动作有关，它的逻辑在 postProcessMergedBeanDefinition 方法中，该

## 7.11 finishBeanFactoryInitialization——初始化剩余的单实例 bean 对象

方法会扫描和收集当前正在创建的 bean 对象中标注了 `@PostConstruct` 和 `@PreDestroy` 注解的方法。代码清单 7-81 中简要展示了 InitDestroyAnnotationBeanPostProcessor 的核心方法实现。源码中匹配注解时使用的 initAnnotationType 与 destroyAnnotationType 刚好分别对应 `@PostConstruct` 与 `@PreDestroy` 注解。

**代码清单 7-81　InitDestroyAnnotationBeanPostProcessor 源码（节选）**

```java
private LifecycleMetadata buildLifecycleMetadata(final Class<?> clazz) {
    // ......
    List<LifecycleElement> initMethods = new ArrayList<>();
    List<LifecycleElement> destroyMethods = new ArrayList<>();
    Class<?> targetClass = clazz;
    do {
        final List<LifecycleElement> currInitMethods = new ArrayList<>();
        final List<LifecycleElement> currDestroyMethods = new ArrayList<>();

        // 反射所有的 public 方法
        ReflectionUtils.doWithLocalMethods(targetClass, method -> {
            // 寻找所有被初始化注解@PostConstruct 标注的方法
            if (this.initAnnotationType != null
                    && method.isAnnotationPresent(this.initAnnotationType)) {
                LifecycleElement element = new LifecycleElement(method);
                currInitMethods.add(element);
            }
            // 寻找所有被销毁注解@PreDestroy 标注的方法
            if (this.destroyAnnotationType != null
                    && method.isAnnotationPresent(this.destroyAnnotationType)) {
                currDestroyMethods.add(new LifecycleElement(method));
            }
        });
        // ......
    } // 依次向上寻找父类
    while (targetClass != null && targetClass != Object.class);
    // return ......
}
```

**（2）CommonAnnotationBeanPostProcessor**

CommonAnnotationBeanPostProcessor 的作用是收集当前正在创建的 bean 对象中标注了 JSR-250 规范注解的元信息，这个后置处理器扩展自 InitDestroyAnnotationBeanPostProcessor，所以它也具备收集 `@PostConstruct` 和 `@PreDestroy` 注解的能力。除此之外，CommonAnnotationBeanPostProcessor 还可以额外收集 JSR-250 规范中的其他注解信息（如 `@Resource` 注解），其核心收集逻辑如代码清单 7-82 所示。

**代码清单 7-82　CommonAnnotationBeanPostProcessor 源码（节选）**

```java
do {
    final List<InjectionMetadata.InjectedElement> currElements = new ArrayList<>();

    ReflectionUtils.doWithLocalFields(targetClass, field -> {
        if (webServiceRefClass != null && field.isAnnotationPresent(webServiceRefClass)) {
            // 检查 ......
            currElements.add(new WebServiceRefElement(field, field, null));
```

```
            } else if (ejbClass != null && field.isAnnotationPresent(ejbClass)) {
                // 检查 ......
                currElements.add(new EjbRefElement(field, field, null));
            } else if (field.isAnnotationPresent(Resource.class)) {
                // 检查 ......
                if (!this.ignoredResourceTypes.contains(field.getType().getName())) {
                    currElements.add(new ResourceElement(field, field, null));
                }
            }
        });
```

### （3）AutowiredAnnotationBeanPostProcessor

由类名可知 AutowiredAnnotationBeanPostProcessor 具备的能力是收集与自动注入有关的注解信息，它的底层收集原理与上述两种后置处理器基本一致，因此这里不再展示收集相关的逻辑，而重点观察 AutowiredAnnotationBeanPostProcessor 支持的注解类型，如代码清单 7-83 所示。可以发现 AutowiredAnnotationBeanPostProcessor 在构造方法中已经指定好默认支持的两个注解，分别是 @Autowired 与 @Value 注解，此外，如果当前项目中有来自 JSR-330 规范的 @Inject 注解，则会一并支持。

**代码清单 7-83　AutowiredAnnotationBeanPostProcessor 支持的注解**

```
public AutowiredAnnotationBeanPostProcessor() {
    this.autowiredAnnotationTypes.add(Autowired.class);
    this.autowiredAnnotationTypes.add(Value.class);
    try {
        this.autowiredAnnotationTypes.add((
                Class<? extends Annotation>) ClassUtils.forName("javax.inject.Inject",
                AutowiredAnnotationBeanPostProcessor.class.getClassLoader()));
    }
```

### 3. 早期 bean 对象引用的获取与缓存

Bean 中的注解信息收集完毕后，接下来有一个获取早期 bean 对象引用的动作，如代码清单 7-84 所示。该步骤是为了解决 bean 对象之间循环依赖的问题，这里先就"早期 bean 对象的引用"这个概念简单解释，完整的循环依赖解决方案会在 7.15 节中讲解。

**代码清单 7-84　doCreateBean（3）**

```
// ......
boolean earlySingletonExposure = (mbd.isSingleton() && this.allowCircularReferences &&
        isSingletonCurrentlyInCreation(beanName));
if (earlySingletonExposure) {
    // 处理循环依赖的问题
    addSingletonFactory(beanName, () -> getEarlyBeanReference(beanName, mbd, bean));
}
// ......
```

从 doCreateBean 方法的前半段逻辑中可以发现，bean 对象被创建之后尚未进行属性赋值和依赖注入的动作，但此时的 bean 对象已经实实在在地存在。如果在此期间有另外的 bean 对象又需要依赖它时，就不应该再创建同样的一个 bean 对象，而是直接获取当前 bean 对象的引用即可，"早期 bean 对象的引用"的设计就是为了解决 bean 之间的循环依赖而产生的。

## 4. 属性赋值与依赖注入

中间的两部分逻辑执行完毕后,接下来到了创建对象的第二个核心步骤:**属性赋值与依赖注入**。从代码清单 7-85 中可以发现该步骤与下面的初始化 bean 对象步骤紧挨着。首先介绍 `populateBean` 方法的逻辑。由于 `populateBean` 方法的篇幅非常长,为便于读者更好地阅读和理解源码,这里将其拆分为多个片段讲解。

**代码清单 7-85　doCreateBean(4)**

```
// ......
Object exposedObject = bean;
try {
    populateBean(beanName, mbd, instanceWrapper);
    exposedObject = initializeBean(beanName, exposedObject, mbd);
}
// ......
```

### (1) 回调 InstantiationAwareBeanPostProcessor

第一个重点关注的源码片段是代码清单 7-86 中回调所有的 `InstantiationAwareBeanPostProcessor`,注意此处回调的方法是 `postProcessAfterInstantiation`,而且在这段源码的上方有一段注释,原文及翻译如下:

> Give any InstantiationAwareBeanPostProcessors the opportunity to modify the state of the bean before properties are set. This can be used, for example, to support styles of field injection.
>
> 在设置属性之前,让任何 `InstantiationAwareBeanPostProcessor` 都有机会修改 Bean 的状态,例如支持属性字段的注入。

从注释中可以得到的关键信息是,`postProcessAfterInstantiation` 方法是一个可以干预 Bean 的属性、状态等信息的**扩展点**。另外请读者注意一个细节,`postProcessAfterInstantiation` 方法的返回值是 `boolean` 类型,当 `postProcessAfterInstantiation` 返回 `false` 时会直接返回,**不再执行真正的属性赋值+组件依赖注入的逻辑**,这就意味着 `postProcessAfterInstantiation` 提供了流转控制的能力,它可以决定当前创建的 bean 对象是否可以执行 IOC 容器默认的属性赋值和依赖注入的逻辑。

**代码清单 7-86　populateBean(1)**

```
protected void populateBean(String beanName, RootBeanDefinition mbd, @Nullable BeanWrapper bw) {
    // 前置检查 ......

    if (!mbd.isSynthetic() && hasInstantiationAwareBeanPostProcessors()) {
        for (BeanPostProcessor bp :getBeanPostProcessors()) {
            if (bp instanceof InstantiationAwareBeanPostProcessor) {
                InstantiationAwareBeanPostProcessor ibp =
                        (InstantiationAwareBeanPostProcessor) bp;
                if (!ibp.postProcessAfterInstantiation(bw.getWrappedInstance(), beanName)){
                    return;
                }
            }
        }
    }
    // ......
```

此外可能有读者会产生疑问，该扩展点有何实际意义？请读者思考，当 bean 对象的创建逻辑执行至此，此时 bean 对象的内部所有需要赋值和注入的属性全部为空，而且尚未执行任何初始化逻辑，所以 `postProcessAfterInstantiation` 方法回调的意义是，允许开发者在 bean 对象已经实例化完毕但还没有开始属性赋值和依赖注入时切入自定义逻辑。

（2）自动注入的支持

紧接着的第二段源码是对早期 IOC 中自动注入的支持，如代码清单 7-87 所示。在 Spring Framework 5.1 之前的版本中，可以使用 @Bean 注解中的 autowire 属性指定 bean 对象的自动注入模式。自动注入模式指的是组件中的类型/属性名与需要注入的 bean 对象的类型/name 完全一致，无须标注 @Autowired 等注解就能实现依赖注入的效果。由于 Spring Framework 5.1 及之后的版本对于 @Bean 注解标注的方法提供了 @Autowired 注解的支持，自动注入的特性便被 Spring Framework 废弃，因此该特性仅供了解，感兴趣的读者可以自行借助 IDE 深入源码了解，这里不再展开。

**代码清单 7-87　populateBean（2）**

```
// ......
PropertyValues pvs = (mbd.hasPropertyValues() ? mbd.getPropertyValues() : null);

// 解析出当前 bean 支持的自动注入模式
int resolvedAutowireMode = mbd.getResolvedAutowireMode();
if (resolvedAutowireMode == AUTOWIRE_BY_NAME || resolvedAutowireMode == AUTOWIRE_BY_TYPE) {
    MutablePropertyValues newPvs = new MutablePropertyValues(pvs);
    if (resolvedAutowireMode == AUTOWIRE_BY_NAME) {
        autowireByName(beanName, mbd, bw, newPvs);
    }
    if (resolvedAutowireMode == AUTOWIRE_BY_TYPE) {
        autowireByType(beanName, mbd, bw, newPvs);
    }
    pvs = newPvs;
}
// ......
```

（3）回调 InstantiationAwareBeanPostProcessor

自动注入匹配完成后，代码清单 7-88 中的逻辑会再一次提取出所有的 InstantiationAwareBeanPostProcessor 并回调其 `postProcessProperties` 方法，该方法中包含了所有组件的依赖注入动作，而负责依赖注入的后置处理器是前面提到过的 **AutowiredAnnotationBeanPostProcessor**，对应的逻辑源码如代码清单 7-89 所示。

**代码清单 7-88　populateBean（3）**

```
// ......
for (BeanPostProcessor bp : getBeanPostProcessors()) {
    if (bp instanceof InstantiationAwareBeanPostProcessor) {
        InstantiationAwareBeanPostProcessor ibp = (InstantiationAwareBeanPostProcessor) bp;
        // 回调 postProcessProperties
        PropertyValues pvsToUse = ibp.postProcessProperties(pvs,
                bw.getWrappedInstance(), beanName);
        // ......
```

## 7.11 finishBeanFactoryInitialization——初始化剩余的单实例 bean 对象

```
    }
  }
  // ......
```

**代码清单 7-89  AutowiredAnnotationBeanPostProcessor#postProcessProperties**

```java
public PropertyValues postProcessProperties(PropertyValues pvs, Object bean, String beanName){
    InjectionMetadata metadata = findAutowiringMetadata(beanName, bean.getClass(), pvs);
    try {
        metadata.inject(bean, beanName, pvs);
    } // catch ......
    return pvs;
}

public void inject(Object target, @Nullable String beanName, @Nullable PropertyValues pvs)
        throws Throwable {
    Collection<InjectedElement> checkedElements = this.checkedElements;
    // 收集所有要注入的信息
    Collection<InjectedElement> elementsToIterate =
            (checkedElements != null ? checkedElements : this.injectedElements);
    if (!elementsToIterate.isEmpty()) {
        // 迭代，依次注入
        for (InjectedElement element : elementsToIterate){
            // logger ......
            element.inject(target, beanName, pvs);
        }
    }
}
```

观察 AutowiredAnnotationBeanPostProcessor 的 postProcessProperties 方法实现，可以发现方法内部会提取出属性赋值前收集的所有标注了 @Autowired、@Value、@Inject 注解的方法信息封装对象 InjectionMetadata，并调用其 inject 方法进行实际的依赖注入。由于 InjectionMetadata 中已经组合了所有需要被注入的元素，因此 inject 方法在执行时会对这些元素逐个进行依赖注入。具体到单个元素的注入逻辑，本质上是使用反射机制将需要注入的 bean 对象设置到对应的成员属性，或反射调用 setter 方法进行赋值。具体的底层实现比较复杂，感兴趣的读者可以借助 IDE 深入探究，这里不对单个元素的注入逻辑详细展开。

> 💡 **提示**：BeanFactoryPostProcessor 与 BeanDefinitionRegistryPostProcessor 无法被 @Autowired 等注解支持，其原因就在于，在 BeanFactoryPostProcessor 初始化阶段，容器中尚不存在 BeanPostProcessor，所以用于支持依赖注入的后置处理器 AutowiredAnnotationBeanPostProcessor 尚未初始化，自然无法提供依赖注入的支持。如果需要在 BeanFactoryPostProcessor 和 BeanDefinitionRegistryPostProcessor 中注入 BeanFactory 或 ApplicationContext 等核心组件，正确的做法是使用 **Aware** 接口的回调注入。

（4）属性赋值

经过前面几个步骤后，最终生成的是一个 PropertyValues 对象，该对象中封装了当前

正在创建的 bean 对象中需要依赖的所有属性赋值类元素，最后执行的 applyProperty
Values 方法实际上就是把前面准备好的 PropertyValues 对象封装的内容应用到当前正在
创建的 bean 对象实例中，如代码清单 7-9 所示。由于 applyPropertyValues 方法的源码篇
幅较长且可研究价值不高，感兴趣的读者可以借助 IDE 自行深入调试，这里只简单概述
applyPropertyValues 方法的作用，不再展开详细讲解。

**代码清单 7-90　populateBean（4）**

```
// ......
if (needsDepCheck){
    if (filteredPds == null) {
        filteredPds = filterPropertyDescriptorsForDependencyCheck(bw, mbd.allowCaching);
    }
    checkDependencies(beanName, mbd, filteredPds, pvs);
}

if (pvs != null) {
    // 将 PropertyValues 应用到 Bean
    applyPropertyValues(beanName, mbd, bw, pvs);
}
```

applyPropertyValues 方法在底层会依次进行预检查、缓存预处理、属性值解析器的
初始化、属性类型转换，以及最终将属性值反射注入 bean 对象的成员属性这 5 个步骤，其中属
性值解析器要依赖 Spring Framework 中的一个核心 API——TypeConverter。利用 Type
Converter 可以将一个 String 类型的数据转换为特定的所需的类型的值，而 Bean
DefinitionValueResolver 利用 TypeConverter 可以完成对 bean 对象实例中需要注入
的属性值进行解析，并适配为 bean 对象成员属性所需的类型（如 String→int、依赖 bean 对象的
名称转换为实际 bean 对象的引用等）。

经过该阶段后，bean 对象的属性赋值和依赖注入工作完成。

**5. bean 对象的初始化**

当 bean 对象的创建过程进入 initializeBean 方法时，意味着 bean 对象中的属性已基
本齐全，但生命周期相关的逻辑都还没有回调，下面展开讲解 initializeBean 方法中回调
的所有核心逻辑。

由代码清单 7-91 可以总体了解到，initializeBean 方法中包含 4 个回调逻辑，下面分
步骤讲解。

**代码清单 7-91　initializeBean**

```
protected Object initializeBean(String beanName, Object bean, @Nullable RootBeanDefinition mbd) {
    if (...) {...} else {
        // 执行 Aware 类型接口的回调
        invokeAwareMethods(beanName, bean);
    }

    Object wrappedBean = bean;
    // 执行 BeanPostProcessor 的前置回调
    if (mbd == null || !mbd.isSynthetic()) {
```

## 7.11 finishBeanFactoryInitialization——初始化剩余的单实例 bean 对象

```
        wrappedBean = applyBeanPostProcessorsBeforeInitialization(wrappedBean, beanName);
    }

    try {
        // 执行生命周期回调
        invokeInitMethods(beanName, wrappedBean, mbd);
    } // catch ......
    // 执行 BeanPostProcessor 的后置回调
    if (mbd == null || !mbd.isSynthetic()) {
        wrappedBean = applyBeanPostProcessorsAfterInitialization(wrappedBean, beanName);
    }
    return wrappedBean;
}
```

（1）invokeAwareMethods——执行 Aware 回调

第一个回调的初始化逻辑是有关 Aware 回调注入的动作。请注意，由此处可以总结一点：Bean 的所有依赖注入动作不全都是在 populateBean 方法中实现的。

从代码清单 7-92 中可以发现，invokeAwareMethods 的实现方式非常简单，只需判断当前 bean 对象是否实现了特定的 Aware 接口方法以及强转后的接口方法调用。另外请读者注意，有关 ApplicationContext 接口及相关组件的注入逻辑没有在 invokeAwareMethods 方法中体现，如果读者还记得在 7.3 节中提到的 ApplicationContextAwareProcessor 后置处理器，应当联想到它的作用，下面马上就可以看到。

**代码清单 7-92　invokeAwareMethods**

```
private void invokeAwareMethods(String beanName, Object bean) {
    if (bean instanceof Aware) {
        if (bean instanceof BeanNameAware) {
            ((BeanNameAware) bean).setBeanName(beanName);
        }
        if (bean instanceof BeanClassLoaderAware) {
            ClassLoader bcl = getBeanClassLoader();
            if (bcl != null) {
                ((BeanClassLoaderAware) bean).setBeanClassLoader(bcl);
            }
        }
        if (bean instanceof BeanFactoryAware) {
            ((BeanFactoryAware) bean).setBeanFactory(AbstractAutowireCapableBeanFactory.this);
        }
    }
}
```

（2）applyBeanPostProcessorsBeforeInitialization

紧接着执行的初始化生命周期回调逻辑是执行 BeanPostProcessor 的 postProcessBeforeInitialization 方法，如代码清单 7-93 所示。回调逻辑本身很简单，但是中间有一个分支设计需要注意。如果一个 BeanPostProcessor 处理 bean 对象后返回的结果为 null，则不再执行剩余的 BeanPostProcessor，而直接返回上一个 BeanPostProcessor 处理之后的 bean 对象并返回。由此特性可以提示开发者在设计后置处理器时，通过设计 post

ProcessBeforeInitialization 方法的返回值可以针对项目中某些特定的 bean 对象设计特殊的拦截处理。

### 代码清单 7-93　applyBeanPostProcessorsBeforeInitialization

```
public Object applyBeanPostProcessorsBeforeInitialization(Object existingBean, String beanName)
throws BeansException {
    Object result = existingBean;
    for (BeanPostProcessor processor : getBeanPostProcessors()) {
        Object current = processor.postProcessBeforeInitialization(result, beanName);
        // 注意此处做了一个分支控制
        // 如果处理完的结果返回 null，则认为停止 BeanPostProcessor 的处理，返回 bean 对象
        if (current == null) {
            return result;
        }
        result = current;
    }
    return result;
}
```

下面简单了解在此处有关键作用的两个后置处理器实现类。

#### InitDestroyAnnotationBeanPostProcessor

InitDestroyAnnotationBeanPostProcessor 不仅在属性赋值之前收集所有标注了 @PostConstruct 和 @PreDestroy 的方法，还负责在 postProcessBeforeInitialization 方法中回调 bean 对象中所有标注了 @PostConstruct 注解的方法。底层回调方法的机制是获取到 applyMergedBeanDefinitionPostProcessors 方法中缓存好的方法集合，并利用反射机制对方法逐个回调，如代码清单 7-94 所示。

### 代码清单 7-94　回调所有 @PostConstruct 注解标注的方法

```
public Object postProcessBeforeInitialization(Object bean, String beanName)
    throws BeansException {
    LifecycleMetadata metadata = findLifecycleMetadata(bean.getClass());
    try {
        metadata.invokeInitMethods(bean, beanName);
    } // catch ......
    return bean;
}

public void invokeInitMethods(Object target, String beanName) throws Throwable {
    Collection<LifecycleElement> checkedInitMethods = this.checkedInitMethods;
    Collection<LifecycleElement> initMethodsToIterate =
        (checkedInitMethods != null ? checkedInitMethods : this.initMethods);
    if (!initMethodsToIterate.isEmpty()) {
        for (LifecycleElement element : initMethodsToIterate) {
            element.invoke(target);
        }
    }
}

public void invoke(Object target) throws Throwable {
    ReflectionUtils.makeAccessible(this.method);
    this.method.invoke(target, (Object[]) null);
}
```

## 7.11 finishBeanFactoryInitialization——初始化剩余的单实例 bean 对象

这里注意两点：最终回调方法时传入的参数是空对象，这也解释了为什么在使用 `@PostConstruct` 注解标注方法时方法一定要设置为空参数方法；反射执行目标方法时会先获取其访问权，这也意味着对于方法的访问修饰符没有强限制。

**ApplicationContextAwareProcessor**

对于所有与 `ApplicationContext` 相关的 Aware 接口回调注入，在底层会由 `ApplicationContextAwareProcessor` 统一负责，它本身支持 6 个 Aware 子接口的回调注入，具体如代码清单 7-95 所示。`invokeAwareInterfaces` 方法中回调的实现与 `invokeAwareMethods` 如出一辙，这里不再展示，读者可以借助 IDE 自行翻阅源码。

**代码清单 7-95　ApplicationContextAwareProcessor 支持回调注入逻辑**

```java
public Object postProcessBeforeInitialization (Object bean, String beanName)
        throws Beans Exception {
    if (!(bean instanceof EnvironmentAware || bean instanceof EmbeddedValueResolverAware
            || bean instanceof ResourceLoaderAware || bean instanceof ApplicationEventPublisherAware
            || bean instanceof MessageSourceAware || bean instanceof ApplicationContextAware)) {
        return bean;
    }
    // JMX 相关 ......
    else {
        invokeAwareInterfaces(bean);
    }
    return bean;
}
```

（3）invokeInitMethods——执行初始化生命周期回调

`BeanPostProcessor` 的前置拦截处理完成后，下一步是读者可能最熟悉的生命周期回调，包括 `init-method` 和 `InitializingBean` 接口的初始化逻辑回调，如代码清单 7-96 所示。源码的逻辑非常简单，这里不再过多解释。

**代码清单 7-96　回调 init-method 和 InitializingBean 接口的初始化逻辑**

```java
protected void invokeInitMethods(String beanName, Object bean, @Nullable RootBeanDefinition mbd)
        throws Throwable {

    boolean isInitializingBean = (bean instanceof InitializingBean);
    if (isInitializingBean
            && (mbd == null || !mbd.isExternallyManagedInitMethod("afterPropertiesSet"))) {
        // JMX 相关 ......
        else {
            // 回调 InitializingBean 的 afterPropertiesSet 方法
            ((InitializingBean) bean).afterPropertiesSet();
        }
    }
    if (mbd != null && bean.getClass()!= NullBean.class) {
        String initMethodName = mbd.getInitMethodName();
        if (StringUtils.hasLength(initMethodName)
```

```
                && !(isInitializingBean && "afterPropertiesSet".equals(initMethodName))
                && !mbd.isExternallyManagedInitMethod(initMethodName)) {
            // 回调 init-method 方法（同样是反射调用）
            invokeCustomInitMethod(beanName, bean, mbd);
        }
    }
}
```

**（4）applyBeanPostProcessorsAfterInitialization**

最后一步是回调所有 `BeanPostProcessor` 的后置拦截处理，回调的机制与前置拦截几乎完全一致，所以 `applyBeanPostProcessorsAfterInitialization` 的源码不再展示，此处重点关注两个后置处理器的实现。

**AbstractAutoProxyCreator**

所有以 `AutoProxyCreator` 结尾的类通常都与 AOP 相关，且都是具备代理对象创建能力的后置处理器，可以在 bean 对象本身的初始化逻辑回调完成后根据需要创建代理对象。有关该部分的内容本书放到第 9 章详细讲解，此处读者只需要大概了解。

**ApplicationListenerDetector**

`ApplicationListenerDetector` 在 7.3.3 节中提到过，它的作用是关联所有监听器的引用，在 `ApplicationListener` 类型的 bean 对象创建时 `ApplicationListenerDetector` 会检测并将其添加到 `ApplicationContext` 中，关联 `ApplicationEventMulticaster` 事件广播器。

**6. 注册销毁时的回调**

`initializeBean` 方法执行后，bean 对象的创建逻辑已基本执行完毕，在 `doCreateBean` 方法的最后有一个对 `DisposableBean` 类型的 bean 对象的处理逻辑。如果一个 bean 对象的所属类型实现了 `DisposableBean` 接口，或者内部方法中标注了 `@PreDestroy` 注解，或者声明了 `destroy-method` 方法，则会在 `doCreateBean` 方法的最后阶段注册一个销毁 bean 对象的回调钩子，如代码清单 7-97 所示。

**代码清单 7-97　doCreateBean（5）**

```
// ......
try {
    registerDisposableBeanIfNecessary(beanName, bean, mbd);
} // catch ......
return exposedObject;
}

protected void registerDisposableBeanIfNecessary(String beanName, Object bean,
        RootBeanDefinition mbd) {
    AccessControlContext acc = (System.getSecurityManager() != null
            ? getAccessCont rolContext() : null);
    // 不是原型 bean，且有定义销毁类型的方法
    if (!mbd.isPrototype() && requiresDestruction(bean, mbd)) {
        if (mbd.isSingleton()) {
            registerDisposableBean(beanName, new DisposableBeanAdapter(bean,
                beanName, mbd, getBeanPostProcessors(), acc));
```

## 7.11 finishBeanFactoryInitialization——初始化剩余的单实例 bean 对象

```
    }
    // 处理特殊的 scope ……
  }
}
```

由代码清单 7-97 可知，通常情况下记录销毁 bean 对象的回调原则是**单实例 bean 对象**，**并且有定义销毁类型的方法**。该部分被记录的 bean 对象将在 IOC 容器关闭时回调其自定义销毁逻辑。

至此，`doCreateBean` 方法执行完毕，一个 bean 对象被创建完成并返回。

### 7.11.5 SmartInitializingSingleton

所有非延迟加载的单实例 bean 对象创建完成后，在 `preInstantiateSingletons` 方法的最后有一段额外的逻辑，这段逻辑是 Spring Framework 4.1 之后新添加的回调处理，用于触发所有实现了 `SmartInitializingSingleton` 接口的 bean 对象的额外初始化逻辑。从源码角度分析可知，`SmartInitializingSingleton` 的设计是为了让 `BeanFactory` 也有机会控制和回调 bean 对象的额外扩展的生命周期逻辑，而不再强依赖于 `ApplicationContext` 控制；从使用角度看，`SmartInitializingSingleton` 接口的引入实际上是在非延迟加载的单实例 bean 对象全部创建完成后提供一个统一的扩展回调时机，利用这个回调时机在 `ApplicationContext` 的初始化完成之前可以处理一些特殊的逻辑。回调 `SmartInitializingSingleton` 类型 bean 对象的逻辑非常简单，仅仅是遍历后回调其 `afterSingletonsInstantiated` 方法，如代码清单 7-98 所示。

**代码清单 7-98　回调 SmartInitializingSingleton 类型的方法**

```
        // ……
        else {
            getBean(beanName);
        }
      }
    }

    for (String beanName : beanNames) {
        Object singletonInstance = getSingleton(beanName);
        if (singletonInstance instanceof SmartInitializingSingleton) {
            SmartInitializingSingleton smartSingleton =(SmartInitializingSingleton)
            singletonInstance;
            // JMX 相关 ……
            else {
                smartSingleton.afterSingletonsInstantiated();
            }
        }
    }
}
```

经过上述一系列复杂逻辑后，`finishBeanFactoryInitialization` 方法执行完毕，所有非延迟加载的单实例 bean 对象全部完成创建并初始化。

## 7.12 finishRefresh——刷新后的动作

所有非延迟加载的单实例 bean 对象初始化完毕后，后续的动作相对比较简单。finishRefresh 方法负责的工作基本以收尾性质为主，如代码清单 7-99 所示。方法内部实现中有两个步骤值得关注，分别展开讲解。

**代码清单 7-99　刷新后的动作**

```
protected void finishRefresh() {
    // 清除资源缓存（如扫描的 ASM 元数据）
    clearResourceCaches();
    // 初始化生命周期处理器
    initLifecycleProcessor();
    // 回调 LifecycleProcessor 的刷新动作
    getLifecycleProcessor().onRefresh();
    // 发布容器刷新完成的事件，触发特定的监听器
    publishEvent(new ContextRefreshedEvent(this));
    LiveBeansView.registerApplicationContext(this);
}
```

### 7.12.1 LifecycleProcessor

对于 bean 对象的生命周期而言，除了在 initializeBean 方法中会执行的 init-method、@PostConstruct 等扩展点，Spring Framework 还通过 Lifecycle 接口提供了一个更晚的时机。这个接口为 bean 对象提供了新的生命周期回调切入时机，可以在 IOC 容器的启动、停止时自动触发 Lifecycle 接口中的 start 方法和 stop 方法。代码清单 7-100 展示了 LifecycleProcessor 的初始化动作，默认注册的生命周期处理器类型为 DefaultLifecycleProcessor。

**代码清单 7-100　初始化生命周期处理器**

```
public static final String LIFECYCLE_PROCESSOR_BEAN_NAME = "lifecycleProcessor";

protected void initLifecycleProcessor() {
    ConfigurableListableBeanFactory beanFactory = getBeanFactory();
    if (beanFactory.containsLocalBean(LIFECYCLE_PROCESSOR_BEAN_NAME)) {
        this.lifecycleProcessor =
                beanFactory.getBean(LIFECYCLE_PROCESSOR_BEAN_NAME, LifecycleProcessor.class);
        // logger ......
    } else {
        DefaultLifecycleProcessor defaultProcessor = new DefaultLifecycleProcessor();
        defaultProcessor.setBeanFactory(beanFactory);
        this.lifecycleProcessor = defaultProcessor;
        beanFactory.registerSingleton(LIFECYCLE_PROCESSOR_BEAN_NAME, this.lifecycleProcessor);
        // logger ......
    }
}
```

### 7.12.2 getLifecycleProcessor().onRefresh()

LifecycleProcessor 初始化完毕后，紧接着会执行其 onRefresh 方法，如代码清单 7-101 所示。注意这个方法的内部有一个比较难理解的设计，需要读者仔细观察和理解。

**代码清单 7-101　回调 Lifecycle 类型的 Bean**

```java
public void onRefresh() {
    startBeans(true);
    this.running = true;
}

private void startBeans(boolean autoStartupOnly) {
    Map<String, Lifecycle> lifecycleBeans = getLifecycleBeans();
    Map<Integer, LifecycleGroup> phases = new HashMap<>();
    lifecycleBeans.forEach((beanName, bean) -> {
        // 注意此处，如果 autoStartupOnly 为 true，则不会执行
        if (!autoStartupOnly
                || (bean instanceof SmartLifecycle && ((SmartLifecycle) bean).isAutoStartup())) {
            int phase = getPhase(bean);
            LifecycleGroup group = phases.get(phase);
            if (group == null) {
                group = new LifecycleGroup(phase, this.timeoutPerShutdownPhase,
                        lifecycleBeans, autoStartupOnly);
                phases.put(phase, group);
            }
            group.add(beanName, bean);
        }
    });
    if (!phases.isEmpty()) {
        List<Integer> keys = new ArrayList<>(phases.keySet());
        Collections.sort(keys);
        for (Integer key : keys) {
            // 依次调用 Lifecycle 的 start 方法
            phases.get(key).start();
        }
    }
}
```

观察代码清单 7-101 中的逻辑，由于 onRefresh 方法中调用的 startBeans(true);传入的参数是 true，lifecycleBeans.forEach 部分不会执行，因此在该阶段不会回调 Lifecycle 的 start 方法。如果要触发回调 Lifecycle 接口的 start 方法，需要显式调用 ApplicationContext 的 start 方法。由于在 Spring Boot 的启动过程中没有回调 start 方法，所以仅实现 Lifecycle 接口的 bean 对象并不会被回调。如果需要在 IOC 容器的刷新过程中自动回调，需要实现 Lifecycle 的子接口 SmartLifecycle，并确保 isAutoStartup 方法的返回值为 true（底层默认为 true），以此法编写的 bean 对象即可实现 start 方法的自动回调。

## 7.13　resetCommonCaches——清除缓存

refresh 方法的最后一步是收尾相关的动作，该方法会清除整个 IOC 容器刷新期间的缓

存，如代码清单 7-102 所示。从源码角度讲已无研究价值，故不再展开。

**代码清单 7-102　清除缓存**

```
protected void resetCommonCaches() {
    ReflectionUtils.clearCache();
    AnnotationUtils.clearCache();
    ResolvableType.clearCache();
    CachedIntrospectionResults.clearClassLoader(getClassLoader());
}
```

至此，`ApplicationContext` 的刷新动作执行完毕。

## 7.14　ApplicationContext 初始化中的扩展点

介绍了整个 `refresh` 方法后，想必读者已经清楚地理解 `ApplicationContext` 中容器刷新的逻辑。本节内容会梳理整个 `ApplicationContext` 的初始化逻辑中有哪些扩展点可供开发者切入利用。理解这些扩展点，可以在构建实际项目的底层架构和扩展时更加容易和游刃有余。

> 💡 提示：基于 Spring Boot 的应用包含 `ApplicationContextInitializer` 等组件，在第 6 章已经讲过，所以本节只讨论原生 Spring Framework 有关的部分。对于原生 Spring Framework 来讲，一般情况下不会在 `ApplicationContext` 的初始化和 `refresh` 动作之间进行太多的处理，而主要从 `refresh` 方法本身考虑，所以以下梳理的扩展点都来自 `refresh` 方法开始触发时。

### 7.14.1　invokeBeanFactoryPostProcessors

`refresh` 方法的前四个步骤中，在普通的 `ApplicationContext` 下都无法切入，所以第一个可切入的步骤是第五步的 `invokeBeanFactoryPostProcessors` 方法，这个方法中可供切入的点非常多，下面一一列举。

#### 1. ImportSelector 和 ImportBeanDefinitionRegistrar

读者看到第一个步骤可能会有一些诧异，第一个扩展点竟然不是 `BeanFactoryPostProcessor` 或者 `BeanDefinitionRegistryPostProcessor`，其原因是它们的执行时机通常都在 `ConfigurationClassPostProcessor` 之后，而 **ConfigurationClassPostProcessor** 的执行过程中会解析 **@Import** 注解，提取出其中的 **ImportBeanDefinitionRegistrar** 并执行，所以第一个扩展点是 `ImportSelector` 和 `ImportBeanDefinitionRegistrar` 接口。

注意对比这两者的区别，`ImportSelector` 在该阶段只能获取当前 `@Import` 标注的注解配置类的信息，而 `ImportBeanDefinitionRegistrar` 在该阶段除了可以获取当前 `@Import` 标注的注解配置类的信息，更重要的是能获取 `BeanDefinitionRegistry`，由此可供扩展的动作主要是给 `BeanDefinitionRegistry` 中编程式注册新的 `BeanDefinition`。

> 💡 提示：此处把 `BeanDefinitionRegistry` 看作 `DefaultListableBeanFactory` 也可以。

### 2. BeanDefinitionRegistryPostProcessor

最特殊的 `ConfigurationClassPostProcessor` 执行完成后才是普通的 `BeanDefinitionRegistryPostProcessor` 接口。使用 `BeanDefinitionRegistryPostProcessor` 可以获取 `BeanDefinitionRegistry` 对象，利用 `BeanDefinitionRegistry` 可以直接向 IOC 容器中编程式注册新的 `BeanDefinition`，以及移除容器中已有的 `BeanDefinition`。

请注意，一般情况下，**自定义的 BeanDefinitionRegistryPostProcessor** 的执行时机比内置的 **ConfigurationClassPostProcessor** 晚，这也是 Spring Framework 最开始的设计（`ConfigurationClassPostProcessor` 实现了 `PriorityOrdered` 接口，这个接口的优先级最高）。

> 提示：注意此处的措辞是"一般情况下"，说明可以出现例外情况。如果确需编写一个执行时机比 `ConfigurationClassPostProcessor` 早的 `BeanDefinitionRegistryPostProcessor`，可以让 **BeanDefinitionRegistryPostProcessor** 实现 **PriorityOrdered** 接口，声明较高执行优先级（不能是 `Ordered.LOWEST_PRECEDENCE`，否则排序规则会变成字母表顺序）。经过此设计之后，自定义的 `BeanDefinitionRegistryPostProcessor` 即可在 `ConfigurationClassPostProcessor` 之前执行。

### 3. BeanFactoryPostProcessor

`BeanFactoryPostProcessor` 的切入时机紧随 `BeanDefinitionRegistryPostProcessor` 之后。请注意，`BeanFactoryPostProcessor` 与 `BeanDefinitionRegistryPostProcessor` 的一个核心区别在于，在 `BeanFactoryPostProcessor` 的切入回调中，可以获取的参数是 `ConfigurableListableBeanFactory` 而不再是 `BeanDefinitionRegistry`，这就意味着在当前阶段原则上不应向 `BeanFactory` 中注册新的 `BeanDefinition`，只能获取和修改现有的 `BeanDefinition`。

另外还需要关注 javadoc 中提到的一点，`BeanFactoryPostProcessor` 的处理阶段中允许提早初始化 bean 对象，但是这个阶段中只有 `ApplicationContextAwareProcessor` 注册到了 `BeanFactory` 中，没有其余关键的 `BeanPostProcessor`，所以这个阶段初始化的 bean 对象有一个共同的特点：可以使用 `Aware` 回调注入，但无法使用`@Autowired` 等依赖注入的注解，且不会产生任何代理对象。

## 7.14.2 finishBeanFactoryInitialization

下面来到最复杂的环节：初始化非延迟加载的单实例 bean 对象。这里面的切入时机非常多，前面通过研究 getBean 方法的一系列调用中，可以看到 bean 对象的实例化和初始化过程中有非常多可供切入的时机，逐个来看。

> 提示：本方法中涉及的所有切入点均为针对单个 bean 对象的扩展。

### 1. InstantiationAwareBeanPostProcessor#postProcessBeforeInstantiation

在 bean 对象的实例化阶段之前，`InstantiationAwareBeanPostProcessor` 可以前

置拦截 bean 对象的实例化动作，IOC 容器中的任何 bean 对象在创建之前都会尝试使用 `InstantiationAwareBeanPostProcessor` 来代替创建，如果没有任何 `InstantiationAwareBeanPostProcessor` 可以拦截创建，则会执行真正的 bean 对象实例化流程。

请注意，在 `InstantiationAwareBeanPostProcessor` 的 `postProcessBeforeInstantiation` 方法中，只能获取 bean 对象对应的 `Class` 类型以及 bean 对象的名称（由于 `InstantiationAwareBeanPostProcessor` 实例化对象本身是凭空创建的，因此仅需 `Class` 类型就足够了）。如果 `postProcessBeforeInstantiation` 方法返回 null，则代表 `InstantiationAwareBeanPostProcessor` 不参与拦截 bean 对象创建的动作。

#### 2. SmartInstantiationAwareBeanPostProcessor#determineCandidateConstructors

如果在实例化 bean 对象之前 `InstantiationAwareBeanPostProcessor` 没有拦截创建成功，就会通过构造方法创建对象。如果一个 bean 对象的所属类型中定义了多个构造方法，那么选择合适的构造方法去创建对象就是很重要的一步。筛选构造方法的核心动作会在底层寻找所有 `SmartInstantiationAwareBeanPostProcessor`，回调 `determineCandidateConstructors` 方法获取可选择的构造方法。如果在这里打入 Debug 断点，程序运行时可以停在断点但不会有具体的返回值，该现象产生的原因是，默认情况下 `ConfigurationClassPostProcessor` 会向 IOC 容器中注册一个 `ImportAwareBeanPostProcessor`，但该后置处理器未重写 `determineCandidateConstructors` 方法，这就造成了方法会执行但无返回值的结果。

一般情况下，`SmartInstantiationAwareBeanPostProcessor` 在 Spring Framework 内部未内置使用逻辑，项目开发中只有在极特殊的场景中才可能会对该方法加以利用，所以 `SmartInstantiationAwareBeanPostProcessor` 仅供了解。

#### 3. MergedBeanDefinitionPostProcessor#postProcessMergedBeanDefinition

在 `doCreateBean` 方法中的 `createBeanInstance` 执行完毕之后，此时 bean 对象已经实例化并返回，但需要属性赋值和依赖注入的成员属性为空。之后 `doCreateBean` 方法会执行到 `applyMergedBeanDefinitionPostProcessors` 步骤，回调所有 `MergedBeanDefinitionPostProcessor` 收集 bean 对象所属 `Class` 中的注解信息。在 7.11.4 节中列出了三个关键的 `MergedBeanDefinitionPostProcessor`，它们分别是 `InitDestroyAnnotationBeanPostProcessor`（收集@PostConstruct 与@PreDestroy 注解）、`CommonAnnotationBeanPostProcessor`（收集 JSR-250 的其他注解）、`AutowiredAnnotationBeanPostProcessor`（收集自动注入相关的注解）。

在此处切入扩展，意味着可以对 bean 对象所属的 `Class` 执行一些处理或者收集的动作（也可以进行属性赋值等动作，但考虑到职责分离原则，该步骤通常不会进行赋值和注入动作）。

#### 4. InstantiationAwareBeanPostProcessor#postProcessAfterInstantiation

所有 `MergedBeanDefinitionPostProcessor` 执行完成后，此时 bean 对象所属的 `Class` 类型中的信息已收集完毕。下一个步骤需要根据 `InstantiationAwareBeanPostProcessor` 的 `postProcessAfterInstantiation` 方法的返回值决定是否继续执行后续的 `populateBean` 和 `initializeBean` 方法初始化 bean 对象。

所以在 `postProcessAfterInstantiation` 方法切入扩展逻辑只能起到流程控制的作用。

### 5. InstantiationAwareBeanPostProcessor#postProcessProperties

下一个切入点在 `InstantiationAwareBeanPostProcessor` 的 `postProcessProperties` 方法中，这个步骤会将 bean 对象对应的 `PropertyValues` 中封装赋值和注入的属性和依赖对象实际应用到 bean 对象的实例中。通常情况下，在该阶段内部起作用的后置处理器是 `AutowiredAnnotationBeanPostProcessor`，它会搜集 bean 对象所属的 `Class` 类型中标注了 `@Autowired`、`@Value`、`@Resource` 等注解的属性和方法，并反射赋值/调用。

在此处扩展逻辑相当于扩展了后置处理器的属性赋值+依赖注入逻辑。当 `postProcessProperties` 方法执行完毕后，将不再有新的属性赋值和组件注入的回调动作产生。

### 6. BeanPostProcessor

属性赋值和依赖注入完成后，下一个核心步骤是 `initializeBean` 方法，该方法中包含 `BeanPostProcessor` 的前后两个执行动作 `postProcessBeforeInitialization` 和 `postProcessAfterInitialization`。进入 `initializeBean` 方法后，bean 对象的生命周期已经到了初始化逻辑回调阶段，此时 bean 对象中注入的属性均已完备，`BeanPostProcessor` 的切入大多是给 bean 对象添加一些额外的属性的赋值、回调以及生成代理对象等动作。

在 `BeanPostProcessor` 中切入扩展逻辑相当于针对一个接近完善的 bean 对象进行扩展或包装。当后置处理器执行完 `postProcessAfterInitialization` 方法后，意味着 bean 对象的初始化动作结束。

### 7. SmartInitializingSingleton

IOC 容器中的最后一个扩展动作是 `SmartInitializingSingleton`，它会在所有非延迟加载的单实例 bean 对象全部初始化完成后回调扩展。在该阶段中会提取出所有实现了 `SmartInitializingSingleton` 接口的 bean 对象，回调其 `afterSingletonsInstantiated` 方法。请注意，`SmartInitializingSingleton` 接口的设计是为了让 BeanFactory 也能参与 bean 对象初始化完成后的扩展处理，并没有太深层面的考虑。

另外注意一个细节，`SmartInitializingSingleton` 这个扩展本身是针对单个 bean 对象的，而不是切入所有 bean 对象，所以严格意义上讲它属于 bean 对象初始化的扩展点，但是 `SmartInitializingSingleton` 的处理时机是在 BeanFactory 把单实例 bean 对象初始化完成后进行统一回调，这又属于一个整体动作，读者要仔细体会 `SmartInitializingSingleton` 的设计，理解上不要产生偏差。

## 7.15 循环依赖的解决方案

IOC 容器初始化 bean 对象的逻辑中可能会遇到 bean 对象之间循环依赖的问题。Spring Framework 不提倡在实际开发中设计 bean 对象之间的循环依赖，但是当循环依赖的场景出现时，IOC 容器内部可以恰当地予以解决。本节内容会全面讲解不同场景下的循环依赖以及 IOC 容器的处理方案。

### 7.15.1 循环依赖的产生

循环依赖，简单理解就是两个或多个 bean 对象之间互相引用（互相持有对方的引用）。以下假定一个场景，人（Person）与猫（Cat）之间相互引用，人养猫，猫依赖人，用 UML 类图可以抽象为图 7-1 所示。

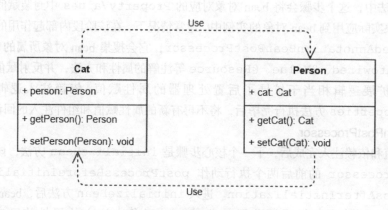

图 7-1 循环依赖的简单示意

循环依赖的产生通常与具体业务场景有关，例如在电商系统中，用户服务和订单服务之间就会存在循环依赖：用户服务需要依赖订单服务查询指定用户的订单列表，订单服务需要根据用户的详细信息对商品订单分类处理。Spring Framework 会针对不同类型的循环依赖实行不同的处理策略，下面逐个展开讲解。

### 7.15.2 循环依赖的解决模型

IOC 容器内部对于解决循环依赖主要使用了三级缓存的设计，其中的核心成员如下。

- **singletonObjects**：一级缓存，存放完全初始化好的 bean 对象容器，从这个集合中提取出来的 bean 对象可以立即返回。
- **earlySingletonObjects**：二级缓存，存放创建好但没有初始化属性的 bean 对象的容器，它用来解决循环依赖。
- **singletonFactories**：三级缓存，存放单实例 Bean 工厂的容器。
- **singletonsCurrentlyInCreation**：存放正在被创建的 bean 对象名称的容器。

上述成员均在 `DefaultListableBeanFactory` 的父类 `DefaultSingletonBeanRegistry` 中。`DefaultSingletonBeanRegistry` 本身是一个单实例 bean 对象的管理容器，`DefaultListableBeanFactory` 继承它之后可以直接获得单实例 bean 对象的管理能力而无须重复设计。

### 7.15.3 基于 setter/@Autowired 的循环依赖

Spring Framework 可以解决的循环依赖类型是基于 setter 方法或@Autowired 注解实现属性注入的循环依赖，整个 bean 对象的创建阶段和初始化阶段是分开的，这给了 IOC 容器插手处理的时机。下面通过一个具体的示例来讲解循环依赖的处理思路。

1. 编写测试代码

为了单独研究 IOC 容器处理循环依赖的场景，下面的源码不再依赖 Spring Boot，而是使用原始的 Spring Framework 注解驱动 IOC 容器测试。所有涉及的测试代码如代码清单 7-103 所示。

### 代码清单 7-103 循环依赖的测试代码

```
@Component
public class Person {

    @Autowired
    Cat cat;
}

@Component
public class Cat {

    @Autowired
    Person person;
}

public class App {

    public static void main(String[] args) {
        ApplicationContext ctx = new AnnotationConfigApplicationContext
                ("com.linkedbear.springframework.bean");
    }
}
```

由于 AnnotationConfigApplicationContext 在创建时传入了组件扫描的根包，底层会在扫描后自动刷新 IOC 容器，由此就可以触发 person 与 cat 对象的初始化动作。

#### 2. 初始化 Cat

由于在字母表中 cat 比 person 的首字母靠前，IOC 容器会先初始化 cat 对象。

（1）getSingleton 中的处理

从 7.11.2 节中的 bean 对象创建流程中可以得知，在 doGetBean 和 createBean 方法之间有一个特殊的步骤 beforeSingletonCreation，如代码清单 7-104 中的中间部分。

### 代码清单 7-104 getSingleton 方法中的关键处理步骤

```
private final Map<String, Object> singletonObjects = new ConcurrentHashMap<>(256);

public Object getSingleton(String beanName, ObjectFactory<?> singletonFactory) {
    Assert.notNull(beanName, "Bean name must not be null");
    synchronized (this.singletonObjects) {
        Object singletonObject = this.singletonObjects.get(beanName);
        if (singletonObject == null) {
            // ......
            // 【控制循环依赖的关键步骤】
            beforeSingletonCreation(beanName);
            // ......
            try {
                singletonObject = singletonFactory.getObject();
                newSingleton = true;
            }// catch finally ......
            // ......
```

beforeSingletonCreation 方法非常关键，它会检查当前正在获取的 bean 对象是否存

在于 `singletonsCurrentlyInCreation` 集合中。如果当前 bean 对象对应的名称在该集合中已经存在，说明出现了循环依赖（同一个对象在一个创建流程中被创建了两次），则抛出 `BeanCurrentlyInCreationException` 异常如代码清单 7-105 所示。

**代码清单 7-105　检查当前正在获取的 bean 对象是否已经在创建中**

```
protected void beforeSingletonCreation(String beanName) {
    if (!this.inCreationCheckExclusions.contains(beanName)
            && !this.singletonsCurrentlyInCreation.add(beanName)) {
        throw new BeanCurrentlyInCreationException(beanName);
    }
}
```

（2）对象创建完毕后的处理

第一次进入 doGetBean 方法汇总，此时 cat 对象对应的名称不在 `singletonsCurrentlyInCreation` 集合中，可以顺利进入 createBean→doCreateBean 方法中，而 doCreateBean 方法又分为 3 个步骤，其中第一个步骤 createBeanInstance 方法执行完毕后，一个空的 cat 对象已经被成功创建，如图 7-2 所示。此时这个 cat 对象被称为"早期 Bean"，而且被 `BeanWrapper` 包装。

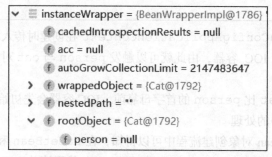

图 7-2　createBeanInstance 方法执行后产生属性均为空的 cat 对象

接下来进入 populateBean 方法之前，源码中又设计了一个逻辑，该逻辑会提前暴露当前正在创建的 bean 对象引用，如代码清单 7-106 所示。

**代码清单 7-106　earlySingletonExposure 的设计**

```
protected Object doCreateBean(finalString beanName, final RootBeanDefinition mbd,
        final @Nullable Object[] args) throws BeanCreationException {
    // ......
    boolean earlySingletonExposure = (mbd.isSingleton() && this.allowCircularReferences
            && isSingletonCurrentlyInCreation(beanName));
    if (earlySingletonExposure) {
        addSingletonFactory(beanName, () -> getEarlyBeanReference(beanName, mbd, bean));
    }
    // ......
}
```

注意源码中的 `earlySingletonExposure` 变量，它的值需要由三部分判断结果共同计算产生，包括：

- 当前创建的 bean 对象是一个单实例 bean 对象；

- IOC 容器本身允许出现循环依赖（默认为 true，在 Spring Boot 2.6.0 之后默认为 false）；
- 正在创建的单实例 bean 对象名称中存在当前 bean 对象的名称。

前两个条件的判断结果在当前测试场景中显然可知为 true，而第三个判断条件中，由于在上一个环节中看到 singletonsCurrentlyInCreation 集合中已经放入了当前正在创建的 "cat" 名称，因此第三个条件的判断结果也为 true。三个条件的判断结果全部为 true，所以会执行 if 结构中的 addSingletonFactory 方法，如代码清单 7-107 所示。

**代码清单 7-107　addSingletonFactory**

```
addSingletonFactory(beanName, () -> getEarlyBeanReference(beanName, mbd, bean));
protected void addSingletonFactory(String beanName, ObjectFactory<?> singletonFactory) {
    Assert.notNull(singletonFactory, "Singleton factory must not be null");
    synchronized (this.singletonObjects) {
        if (!this.singletonObjects.containsKey(beanName)) {
            this.singletonFactories.put(beanName, singletonFactory);
            this.earlySingletonObjects.remove(beanName);
            this.registeredSingletons.add(beanName);
        }
    }
}
```

请注意，addSingletonFactory 方法的第二个参数是一个 ObjectFactory，在方法调用时以 Lambda 表达式传入。而方法的内部实现逻辑是将当前正在创建的 bean 对象名称保存到三级缓存 singletonFactories 中，并从二级缓存 earlySingletonObjects 中移除。此处由于二级缓存中没有正在创建的 "cat" 名称，因此当前环节可以简单理解为仅将 cat 对象的名称 "cat" 放入了三级缓存 singletonFactories 中。

（3）依赖注入时的处理

下一个处理的时机是在 cat 对象的依赖注入时。由于使用 @Autowired 注解注入了 Person 对象，AutowiredAnnotationBeanPostProcessor 会在 postProcessProperties 方法回调时将 person 对象提取出并注入 cat 对象中。而注入的底层逻辑依然是使用 BeanFactory 的 getBean 方法，如代码清单 7-108 中的一系列方法调用所示。

**代码清单 7-108　被依赖对象的底层获取逻辑**

```
public PropertyValues postProcessProperties(PropertyValues pvs, Object bean, String beanName){
    InjectionMetadata metadata = findAutowiringMetadata(beanName, bean.getClass(), pvs);
    try {
        metadata.inject(bean, beanName, pvs);
    } // catch ......
    return pvs;
}
protected void inject(Object bean, @Nullable String beanName, @Nullable PropertyValues pvs)
        throws Throwable {
    Field field = (Field) this.member;
    // 从 BeanFactory 中获取被依赖对象
    // 【此处源码有调整】
    Object value = beanFactory.resolveDependency(desc, beanName,
            autowiredBeanNames, typeConverter);
    // ......
```

```java
    if (value != null) {
        // 反射注入属性
        ReflectionUtils.makeAccessible(field);
        field.set(bean, value);
    }
}

public Object resolveDependency(DependencyDescriptor descriptor,
        @Nullable String requestingBeanName, @Nullable Set<String> autowiredBeanNames,
        @Nullable TypeConverter typeConverter) throws BeansException {
    // if-else ......
    else {
        // 【此处源码有调整】
        Object result = doResolveDependency(descriptor, requestingBeanName,
                autowiredBeanNames, typeConverter);
        return result;
    }
}

public Object doResolveDependency(DependencyDescriptor descriptor, @Nullable String beanName,
        @Nullable Set<String> autowiredBeanNames, @Nullable TypeConverter typeConverter)
        throws BeansException {
    // try ......
    if (instanceCandidate instanceof Class) {
        instanceCandidate = descriptor.resolveCandidate(autowiredBeanName, type, this);
    }
    // ......
}

public Object resolveCandidate(String beanName, Class<?> requiredType,
        BeanFactory beanFactory) throws BeansException {
    return beanFactory.getBean(beanName);
}
```

当执行到最后一个方法 `resolveCandidate` 时，会触发 person 对象的初始化全流程。

### 3. 初始化 Person

创建 person 对象的过程与创建 cat 类似，都是执行 getBean→doGetBean，其中包含 getSingleton 的处理，以及对象创建完毕后将 Person 对象包装为 ObjectFactory 后放入三级缓存 singletonFactories 中，最后到了依赖注入的环节。由于 person 中使用 @Autowired 注解注入了 cat，因此 AutowiredAnnotationBeanPostProcessor 处理注入的逻辑与代码清单 7-107 中一样，从 BeanFactory 中获取 cat 对象。

（1）再次获取 cat 对象的细节

再次获取 cat 对象时执行的方法依然是 getBean→doGetBean，但是在 doGetBean 方法中有一个非常关键的环节：**getSingleton**。注意方法名与上面讲解的一致，但它是另一个重载方法，如代码清单 7-109 所示。

**代码清单 7-109　getSingleton 检查正在创建中的 bean 对象**

```java
protected Object getSingleton(String beanName, boolean allowEarlyReference) {
    Object singletonObject = this.singletonObjects.get(beanName);
    // 检查当前获取的 bean 对象是否正在创建
```

```
        if (singletonObject == null && isSingletonCurrentlyInCreation(beanName)) {
            synchronized (this.singletonObjects) {
                // 检查二级缓存中是否包含当前正在创建的 bean 对象
                singletonObject = this.earlySingletonObjects.get(beanName);
                if (singletonObject == null && allowEarlyReference) {
                    // 检查三级缓存中是否包含当前正在创建的 bean 对象
                    ObjectFactory<?> singletonFactory = this.singletonFactories.get(beanName);
                    if (singletonFactory != null) {
                        // 将 bean 对象放入二级缓存,并从三级缓存中移除
                        singletonObject = singletonFactory.getObject();
                        this.earlySingletonObjects.put(beanName, singletonObject);
                        this.singletonFactories.remove(beanName);
                    }
                }
            }
        }
        return singletonObject;
    }
```

仔细观察代码清单 7-109 中的逻辑,IOC 容器为了确保一个单实例 bean 对象不被多次创建,在此处下了非常大的检查成本,检查的范围如下:如果当前获取的 bean 对象正在创建,并且二级缓存 earlySingletonObjects 中没有正在创建的 bean 对象以及三级缓存 singletonFactories 中存在正在创建的 bean 对象,说明当前获取的 bean 对象是一个没有完成依赖注入的不完全对象。即便当前 cat 是一个不完全对象,它也真实地存在于 IOC 容器中,不会影响 cat 与 person 对象之间的依赖注入(即属性成员的引用),所以 getSingleton 方法会在判断条件成立后,将当前正在获取的 cat 对象从三级缓存 singletonFactories 中移除,并放入二级缓存 earlySingletonObjects 中,最后返回 cat 对象。

getSingleton 方法执行完成后的状态如图 7-3 所示。

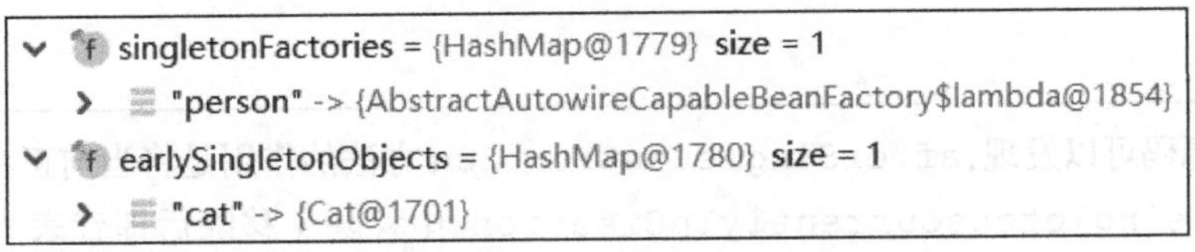

```
▼ f singletonFactories = {HashMap@1779} size = 1
  > ≡ "person" -> {AbstractAutowireCapableBeanFactory$lambda@1854}
▼ f earlySingletonObjects = {HashMap@1780} size = 1
  > ≡ "cat" -> {Cat@1701}
```

图 7-3　getSingleton 方法执行完毕后缓存的状态

> 💡 提示:注意一点,此处如果仅有一级缓存,也可以处理循环依赖,Spring Framework 在此处设计了两级缓存,是考虑到 AOP 的情况,7.15.6 节会分析 AOP 场景下第三级缓存 singletonFactories 的特殊作用。

(2) Person 初始化完成后的处理

通过 getSingleton 方法获取到 cat 对象后,doGetBean 方法的后续动作与循环依赖无关,此处不再提及。person 对象的 populateBean 方法执行完毕后,意味着 person 对象的属性赋值和依赖注入工作完成。后续的 initializeBean 方法中会对 person 对象进行初始化逻辑相关处理,该动作也与循环依赖无关,一并跳过。

当 doCreateBean→createBean 方法执行完毕后,回到 getSingleton 方法(即代码清单 7-104)中,在方法的最后有两个关键动作,如代码清单 7-110 所示。

**代码清单 7-110　getSingleton 方法的最后处理动作**

```
public Object getSingleton(String beanName, ObjectFactory<?> singletonFactory) {
    // ......
            try {
                // 此处的 singletonFactory.getObject() 即 createBean 方法
                singletonObject = singletonFactory.getObject();
                newSingleton = true;
            } // catch ......
            finally {
                // ......
                afterSingletonCreation(beanName);
            }
            if (newSingleton) {
                addSingleton(beanName, singletonObject);
            }
        }
        return singletonObject;
    }
}
protected void afterSingletonCreation(String beanName) {
    if (!this.inCreationCheckExclusions.contains(beanName)
            && !this.singletonsCurrentlyInCreation.remove(beanName)) {
        // throw ex ......
    }
}

protected void addSingleton(String beanName, Object singletonObject) {
    synchronized (this.singletonObjects) {
        this.singletonObjects.put(beanName, singletonObject);
        this.singletonFactories.remove(beanName);
        this.earlySingletonObjects.remove(beanName);
        this.registeredSingletons.add(beanName);
    }
}
```

由上述源码可以发现，`afterSingletonCreation` 方法的作用是将当前正在获取的 bean 对象名称从 `singletonsCurrentlyInCreation` 中移除（移除后即代表当前环节中该 bean 对象未正在创建），而 `addSingleton` 方法的作用则是将创建好的 bean 对象放入一级缓存 `singletonObjects` 中，且从二级缓存 `earlySingletonObjects` 和三级缓存 `singletonFactories` 中移除，并记录已经创建的单实例 bean 对象。

至此，一个 `person` 对象的创建流程完全结束。

#### 4．回到 Cat 的创建流程

`person` 对象创建完成后，回到 `cat` 对象的创建流程中，此时 `cat` 对象的依赖注入工作尚未完成，此处会将完全创建好的 `person` 对象进行依赖注入。请注意，**该动作完成后，即代表 Cat 与 Person 循环依赖的场景处理完毕。**

后续的动作与 `Person` 一致，最终会将 `cat` 对象放入一级缓存 `singletonObjects`，并从其他几个缓存集合中移除，从而完成 `cat` 对象的创建。

#### 5．小结

基于 setter 方法或 @Autowired 属性注入的循环依赖，IOC 容器的解决流程如下。

## 7.15 循环依赖的解决方案

（1）创建 bean 对象之前，将该 bean 对象的名称放入"正在创建的 bean 对象"集合 `singletonsCurrentlyInCreation` 中。

（2）`doCreateBean` 方法中的 `createBeanInstance` 方法执行完毕后，会将当前 bean 对象放入三级缓存中。注意此处放入的是经过封装后的 `ObjectFactory` 对象，在该对象中有额外的处理逻辑。

（3）对 bean 对象进行属性赋值和依赖注入时，会触发循环依赖的对象注入。

（4）被循环依赖的对象创建时，会检查三级缓存中是否包含且二级缓存中不包含正在创建的、被循环依赖的对象。如果三级缓存中存在且二级缓存不存在，则会将三级缓存的 bean 对象移入二级缓存，并进行依赖注入。

（5）被循环依赖的 bean 对象创建完毕后，会将该 bean 对象放入一级缓存，并从其他缓存中移除。

（6）所有循环依赖的 bean 对象均注入完毕后，一个循环依赖的处理流程结束。

图 7-4 中展示了上述流程，流程图中简化了源码中方法调用的级别，读者只需关注主干逻辑。

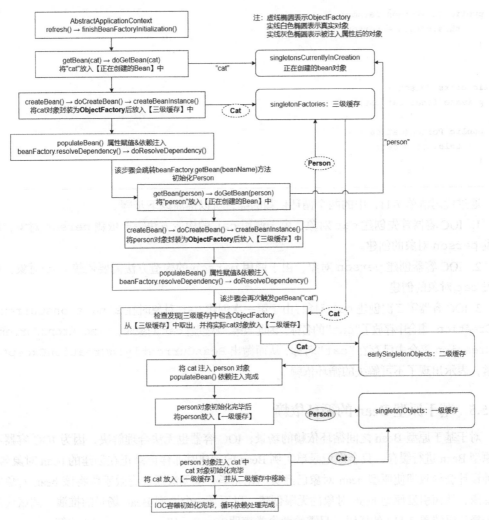

图 7-4　IOC 容器解决循环依赖的流程

### 7.15.4 基于构造方法的循环依赖

对于基于构造方法的循环依赖场景，IOC 容器无法给予合理的处理，只能抛出 `BeanCurrentlyInCreationException` 异常，原因是通过构造方法进行循环依赖，在构造方法没有执行完毕时，bean 对象尚未真正创建完毕并返回，此时若不加干预，会导致参与循环依赖的对象产生构造方法循环调用闭环，从而一直在轮流创建对象，直至内存溢出。IOC 容器为了避免对象的无限创建，采用 singletonsCurrentlyInCreation 集合记录正在创建的 bean 对象名称，当一个 bean 对象名称出现两次时，IOC 容器会认为出现了不可解决的循环依赖，从而抛出 `BeanCurrentlyInCreationException` 异常。下面通过一个例子简要分析，如代码清单 7-111 所示。

**代码清单 7-111　基于构造方法的循环依赖示例代码**

```java
public class Cat {
    private final Person person;

    public Cat(Person person) {
        this.person = person;
    }
}

public class Person {
    private final Cat cat;

    public Person(Cat cat) {
        this.cat = cat;
    }
}
```

通过代码清单 7-111 中的两个循环依赖的类，可以推演以下步骤。

1. IOC 容器首先创建 cat 对象，由于调用 cat 的构造方法需要依赖 person 对象，从而引发 person 对象的创建。

2. IOC 容器创建 person 对象，由于调用 person 的构造方法需要依赖 cat 对象，从而引发 cat 对象的创建。

3. IOC 容器第二次创建 cat 对象，由于第一次创建 cat 对象时在 singletonsCurrentlyInCreation 集合中存放了 "cat" 的名称，因此当第二次创建 cat 对象时，singletonsCurrentlyInCreation 集合中已存在 "cat" 名称，从而抛出 `BeanCurrentlyInCreationException` 异常，表示出现了不可解决的循环依赖。

### 7.15.5 基于原型 Bean 的循环依赖

对于基于原型 Bean 之间循环依赖的场景，IOC 容器也无法合理解决，因为 IOC 容器不会对原型 Bean 进行缓存，只会像记录单实例 Bean 的创建时那样记录正在创建的 bean 对象名称。这种设计会导致即使原型 bean 对象已经实例化完毕，也无法通过有效手段将该 bean 对象的引用暴露，从而引发原型 bean 对象的无限创建。以下是一个原型 Bean 场景的推演，测试代码可以选择代码清单 7-112 的代码，只需给两个类声明 `@Scope("prototype")` 注解。

1. IOC 容器首先创建 cat 对象，之后进行 person 对象的依赖注入，由于 person 被定义为原型 Bean，触发 person 对象的创建。

2. IOC 容器创建 person 对象，之后进行 cat 对象的依赖注入，由于 cat 对象也被定义为原型 Bean，触发 cat 对象的全新创建。

3. IOC 容器再次创建 cat 对象，由于第一次创建 cat 对象时 prototypesCurrentlyInCreation 原型 bean 对象名称集合（注意与 singletonsCurrentlyInCreation 集合区分）中已经存放了"cat"名称，因此当第二次创建时，prototypesCurrentlyInCreation 集合中已存在"cat"名称，从而抛出 BeanCurrentlyInCreationException 异常，表示出现了不可解决的循环依赖。

### 7.15.6 引入 AOP 的额外设计

对于在 7.15.3 节中提到的 getSingleton 方法中将三级缓存中的 bean 对象放入二级缓存的动作，其实如果读者仔细观察会发现，**三级缓存中存放的是被封装过的 ObjectFactory 对象，而二级缓存中存放的是真正的 bean 对象**，为什么会有 ObjectFactory 到 bean 对象之间的过渡呢？这就是 Spring Framework 设计的高深之处。Spring Framework 的两大核心特性中，除 IOC 之外还有一个重要特性是 AOP，在 bean 对象创建完成后，IOC 容器会指派 BeanPostProcessor 对需要进行 AOP 增强的 bean 对象进行代理对象的创建。原始的目标对象和被 AOP 增强的代理对象本质上是两个完全不同的对象，IOC 容器为了确保 bean 对象中最终注入的是 AOP 增强后的代理对象而不是原始对象，会在 **ObjectFactory** 到 bean 对象的过渡期间进行额外的检查，该环节的检查会提前创建代理对象，并替换原始对象。经过此法处理后的 bean 对象就是一个被 AOP 增强后的代理对象，即便后续执行属性赋值和依赖注入，最终也是给内层的目标对象赋值和注入，而不会有任何副作用，但是从 IOC 容器的整体角度而言，IOC 容器内部的所有 bean 对象通过依赖注入后的属性成员都是正确的 bean 对象（此处的正确是指如果一个 bean 对象的确需要被 AOP 增强，则注入的是正确的代理对象而不是错误的原始对象）。

下面从源码的角度简单了解引入 AOP 之后的额外逻辑触发。通过代码清单 7-112 可以发现，getEarlyBeanReference 方法的实现是回调 IOC 容器中所有 SmartInstantiationAwareBeanPostProcessor 的 getEarlyBeanReference 方法，该方法可以获得单实例 bean 对象的引用，也正是通过该方法 IOC 容器可以有机会将一个普通 bean 对象转化为被 AOP 增强的代理对象。

**代码清单 7-112　getEarlyBeanReference 方法的实现**

```
protected Object doCreateBean(final String beanName, final RootBeanDefinition mbd,
        final @Nullable Object[] args) throws BeanCreationException {
    // ......
    boolean earlySingletonExposure = (mbd.isSingleton() && this.allowCircularReferences
        && isSingletonCurrentlyInCreation(beanName));
    if (earlySingletonExposure) {
        addSingletonFactory(beanName, () -> getEarlyBeanReference(beanName, mbd, bean));
    }
    // ......
}
```

```java
protected Object getEarlyBeanReference(String beanName, RootBeanDefinition mbd, Object bean) {
    Object exposedObject = bean;
    if (!mbd.isSynthetic() && hasInstantiationAwareBeanPostProcessors()) {
        for (BeanPostProcessor bp :getBeanPostProcessors()) {
            if (bp instanceof SmartInstantiationAwareBeanPostProcessor) {
                SmartInstantiationAwareBeanPostProcessor ibp =
                        (SmartInstantiationAwareBeanPostProcessor) bp;
                exposedObject = ibp.getEarlyBeanReference(exposedObject, beanName);
            }
        }
    }
    return exposedObject;
}
```

目前还没有讲到具体实现 AOP 代理增强的后置处理器，这里先简单提一下，所有实现 AOP 增强的后置处理器都继承自 `AbstractAutoProxyCreator`，而它本身实现了 `SmartInstantiationAwareBeanPostProcessor` 接口，内部自然有 `getEarlyBeanReference` 方法的实现，如代码清单 7-113 所示。

**代码清单 7-113　AbstractAutoProxyCreator 中实现的 getEarlyBeanReference 方法**

```java
public Object getEarlyBeanReference(Object bean, String beanName) {
    Object cacheKey = getCacheKey(bean.getClass(), beanName);
    this.earlyProxyReferences.put(cacheKey, bean);
    // 必要时创建代理对象
    return wrapIfNecessary(bean, beanName, cacheKey);
}
```

由 `AbstractAutoProxyCreator` 实现的逻辑可以明显看出，如果当前正在创建的 bean 对象的确需要创建代理对象（即有必要），则会先行创建代理对象，并替换原始对象。由此就解释了为什么 IOC 容器解决循环依赖需要使用三级缓存而不是二级。

至此，IOC 容器解决循环依赖的方案全部讲解完毕，读者最好能自行编写一些实际的测试代码，配合 Debug 体会一遍，以加深印象。

## 7.16　小结

本章中我们全面、深入地了解了 `ApplicationContext` 的核心初始化动作 `refresh`，并重点剖析了各类后置处理器的加载、应用以及容器中 bean 对象的初始化流程。Spring Boot 的底层核心容器建立在 Spring Framework 的原生 `ApplicationContext` 之上，一切核心底层动作还是基于 Spring Framework 的原生 IOC 容器，所以掌握 `ApplicationContext` 的初始化逻辑显得格外重要。

第 8 章我们将关注 Spring Boot 的一个重要特性：嵌入式 Web 容器。从底层深入探究 Spring Boot 是如何将嵌入式 Web 容器集成到项目中，并在底层引导启动嵌入式容器的。

# 第 8 章　Spring Boot 容器刷新扩展：嵌入式 Web 容器

**本章主要内容：**
- ◇ 嵌入式 Tomcat 容器简介；
- ◇ Tomcat 的整体架构与核心工作流程；
- ◇ 嵌入式 Web 容器的模型设计；
- ◇ 嵌入式 Web 容器的初始化时机；
- ◇ 嵌入式 Tomcat 的回调启动。

对于 Web 应用的项目开发，Spring Boot 提供了 WebMvc 和 WebFlux 的支持，以帮助开发者更快速、简单地构建和开发，而 Web 应用需要部署运行时，传统的方式是将项目打包成 war 包后部署到外置的 Web 容器（如 Tomcat、Jetty、Undertow 等传统 Servlet 容器，以及支持异步非阻塞的 Netty 等容器）。Spring Boot 的一大重要特性是支持嵌入式 Web 容器，有了嵌入式 Web 容器的支持，基于 Spring Boot 的 Web 应用仅凭一个单独的 jar 包即可独立运行。本章内容会从嵌入式 Web 容器开始，逐步深入研究嵌入式 Web 容器的设计，以及在 Spring Boot 应用启动过程中嵌入式 Web 容器的构建、初始化、回调启动等重要环节。

## 8.1　嵌入式 Tomcat 简介

对于 Apache Tomcat 想必读者都不陌生，它是一个使用范围极广的 Servlet 容器。自 Tomcat 7.0 开始，普通的 Tomcat Server 与嵌入式 Tomcat（Embedded Tomcat）同步发行。从 Tomcat 的官方网站下载页面中可以看到，普通的 Tomcat Server 与 Embedded Tomcat 都可以下载，如图 8-1 所示。

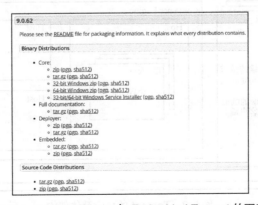

图 8-1　Tomcat Server 与 Embedded Tomcat 的下载

### 8.1.1 嵌入式 Tomcat 与普通 Tomcat

嵌入式 Tomcat 是一种可以嵌入 Java Web 应用中，而无须单独部署的 Tomcat 容器。整合嵌入式 Tomcat 后的应用可以在编写少量代码后，不借助外部容器和资源即可独立启动 Web 应用。也正因如此，嵌入式 Tomcat 成为 Spring Boot 予以支持和首选的嵌入式 Web 容器实现。

普通的外置 Tomcat 与嵌入式 Tomcat 从核心、本质上看没有任何区别，它们都可以承载 Web 应用的运行。但是 Spring Boot 整合嵌入式 Tomcat 容器时考虑到一些特殊的因素，在底层设定了一些额外的限制，对此读者也需要简单了解。

- 部署应用的限制：由于嵌入式 Tomcat 不是独立的 Web 容器，而是嵌入特定应用中的，此特性使得嵌入式 Tomcat 一次只能部署一个 Web 应用。基于该限制也可以推理延伸一些其他限制，例如数据库连接池无法在多个 Web 应用中共享等。
- web.xml 的限制：Spring Boot 整合嵌入式 Tomcat 后不再对 web.xml 文件予以支持，转而替代的方案是使用 @Bean 注解配合 ServletRegistrationBean 实现 Servlet、Filter、Listener 的编程式注册。
- Servlet 原生三大组件的限制：原生的基于 Servlet 3.0 及以上规范的 Web 项目，其类路径下的 Servlet、Filter、Listener 可以被自动扫描并注册，而 Spring Boot 整合嵌入式 Tomcat 后该特性会消失，如果需要开启此特性，需要配合 @ServletComponentScan 注解使用。
- JSP 的限制：Spring Boot 整合嵌入式 Tomcat 后，如果以独立 jar 包的方式启动项目，则项目中编写的 JSP 页面会失效；如果以 war 包的方式部署到外置 Tomcat 容器，则 JSP 可以正常运行（出现该现象的原因是嵌入式 Tomcat 没有引入 JSP 引擎依赖 tomcat-embed-jasper）。

### 8.1.2 Tomcat 整体架构

Tomcat 的核心整体架构如图 8-2 所示。

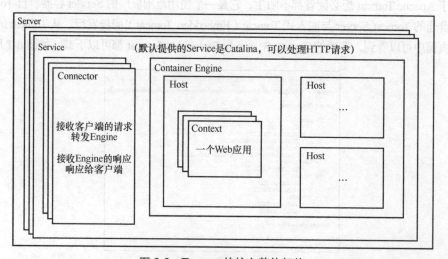

图 8-2 Tomcat 的核心整体架构

从图 8-2 中可以提取出以下关键信息：

- 一个 Tomcat 服务器是一个 `Server`；
- 一个 `Server` 中包含多个服务 `Service`，其中默认提供 HTTP 请求响应服务的是 **Catalina**；
- 一个 `Service` 中包含一组 `Connector`，用于与客户端交互，实现 HTTP 接收请求和响应结果；
- 一个 `Service` 中还包含一个 `Container Engine`，用于真正处理客户端的请求，并响应结果；
- 一个 `Container Engine` 中包含一组 `Host`，每个 `Host` 可以装载一组 Web 应用；
- 一个 Web 应用对应一个 `Context`，一个 `Context` 中包含多个 `Servlet`。

## 8.1.3 Tomcat 的核心工作流程

Tomcat 作为一个 Web 服务器，它的核心工作是接收客户端发起的 HTTP 请求，转发给服务器端的 Web 应用处理，处理完成后将结果响应给客户端。大致的工作流程分步骤解析如下。

1. 当请求进入 Tomcat 容器后，Tomcat 内部根据请求的 URL，判断该请求应该由哪个应用处理（一个 Tomcat 中可以部署多个 Web 应用），并将请求封装为 `ServletRequest` 对象，转发至对应 Web 应用中的 `Context`。

> 提示：从 Tomcat 的 `Server` 接收到请求到转发给 `Context` 的过程中还包含几个步骤，由于这几个步骤对于理解和把控 Tomcat 整体工作流程不会有太大影响，因此本节没有提及，感兴趣的读者可以自行查阅相关资料。

2. `Context` 接收到 `ServletRequest` 后，根据请求的 URI 定位可以接收当前请求的 `Servlet`，并将请求转发给具体的 `Servlet` 进行处理。该阶段定位 `Servlet` 的依据是请求的 URI 以及 `Servlet` 的 URI 映射关系。

3. `Servlet` 容器在转发 `Servlet` 之前会检查 `Servlet` 是否已经加载，如果没有加载，则会利用反射机制创建 `Servlet` 的对象，并调用其 `init` 方法完成初始化，之后再调用其 `service` 方法进行逻辑处理。

4. `Servlet` 处理完成后，将响应结果以 `ServletResponse` 对象响应给 `Service` 中的 `Connector`，由 `Connector` 响应给客户端，至此完成一次请求处理。

整体的工作流程图如图 8-3 所示。

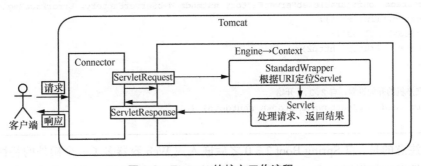

图 8-3　Tomcat 的核心工作流程

## 8.2 Spring Boot 中嵌入式容器的模型

Spring Boot 对嵌入式 Web 容器的默认支持包含 Tomcat、Jetty、Undertow、Netty 等。为确保对不同嵌入式 Web 容器的统一支持，Spring Boot 制定了有关嵌入式 Web 容器的一些接口和实现类，本节简单介绍 Spring Boot 提供的支持嵌入式 Web 容器的模型。

### 8.2.1 WebServer

WebServer 是 Spring Boot 针对所有嵌入式 Web 容器制定的顶级接口，它本身仅定义了嵌入式 Web 容器的启动和停止动作。实现 WebServer 接口的实现类通常会在内部组合一个真正的嵌入式容器（例如 `TomcatWebServer` 中包含一个 Tomcat 对象），并且会在 start 方法中实现嵌入式 Web 容器的启动逻辑，以及在 stop 方法中实现停止和销毁逻辑，如代码清单 8-1 所示。

**代码清单 8-1　WebServer 接口与实现类 TomcatWebServer 的核心成员**

```java
public interface WebServer {
    void start() throws WebServerException;
    void stop() throws WebServerException;
    int getPort();
    //......
}

public class TomcatWebServer implements WebServer {
    private final Tomcat tomcat;
    //......
}
```

### 8.2.2 WebServerFactory

WebServerFactory 是所有具备创建 WebServer 能力的工厂根接口，该接口没有定义任何方法，仅为标记性接口。ConfigurableWebServerFactory 为 WebServerFactory 的扩展接口，它具备对 WebServerFactory 的配置能力（包括配置端口、SSL 等），如代码清单 8-2 所示。

**代码清单 8-2　ConfigurableWebServerFactory 定义的部分方法**

```java
public interface ConfigurableWebServerFactory extends WebServerFactory, ErrorPageRegistry {
    void setPort(int port);
    void setSsl(Ssl ssl);
    void setHttp2(Http2 http2);
    // ......

    // 优雅停机的开关设置，自 2.3.0 开始
    default void setShutdown(Shutdown shutdown) { }
}
```

值得注意的是，自 Spring Boot 2.3.0 之后嵌入式 Web 容器多了一个额外的特性：优雅停机。该特性可以使得嵌入式 Web 容器在关闭时不直接终止进程，而是预留一些时间使容器

## 8.2.3 ServletWebServerFactory 和 ReactiveWebServerFactory

Spring Boot 2.0 之后，构建 Web 应用的基础框架可选择 WebMvc 和 WebFlux，对应的 Web 容器也就分别变成了传统 Servlet 容器和异步非阻塞 Web 容器。由于不同类型的 Web 容器在创建时传入的初始化组件不同，因此 Spring Boot 定义了两个平级的接口，分别处理 Servlet 和 Reactive 场景下的嵌入式容器创建工厂，如代码清单 8-3 所示。

**代码清单 8-3　ServletWebServerFactory 和 ReactiveWebServerFactory**

```java
@FunctionalInterface
public interface ServletWebServerFactory {
    WebServer getWebServer(ServletContextInitializer... initializers);
}

@FunctionalInterface
public interface ReactiveWebServerFactory {
    WebServer getWebServer(HttpHandler httpHandler);
}
```

由上述源码可以发现，两个接口的方法定义仅是入参不同，返回的都是 `WebServer` 对象。

## 8.2.4 ConfigurableServletWebServerFactory

当 `ConfigurableWebServerFactory` 接口与 `ServletWebServerFactory` 组合后产生的子接口会具备更多能力。从代码清单 8-4 中可以发现，`ConfigurableServletWebServerFactory` 中可以设置访问 Web 应用所需的 `context-path`，以及设置 `ServletContextInitializer`、初始化参数 `init-paran` 等。

**代码清单 8-4　ConfigurableServletWebServerFactory 定义的部分方法**

```java
public interface ConfigurableServletWebServerFactory extends ConfigurableWebServerFactory,
                                                              ServletWebServerFactory {
    void setContextPath(String contextPath);
    void setInitializers(List<? extends ServletContextInitializer> initializers);
    void addInitializers(ServletContextInitializer... initializers);
    void setInitParameters(Map<String, String> initParameters);
    //......
}
```

再往下扩展就到了具体某个嵌入式 Web 容器的实现类，这部分内容会在下面分析嵌入式 Tomcat 的初始化、回调启动等流程时展开讲解。

# 8.3　嵌入式 Web 容器的初始化时机

在 7.9 节中提到了 `ApplicationContext` 的 `refresh` 方法的第 9 步。该方法本身是一个模板方法，`AbstractApplicationContext` 对此没有逻辑实现（见代码清单 8-5），而在

Spring Boot 中支持嵌入式 Web 容器的高级 IOC 容器实现中，该步骤是触发嵌入式 Web 容器初始化的核心步骤。

**代码清单 8-5　AbstractApplicationContext#onRefresh**

```
protected void onRefresh() throws BeansException {
    //对于子类：默认没有任何操作
}
```

而 3.2.2 节中重点提到了 AnnotationConfigServletWebServerApplicationContext，它就是基于 WebMvc 的支持嵌入式 Web 容器的 IOC 容器最终实现类，在它的父类 ServletWebServerApplicationContext 中有 onRefresh 方法的重写逻辑，其内部的实现就是嵌入式 Web 容器的创建，如代码清单 8-6 所示（请注意源码中标注的注释）。

**代码清单 8-6　onRefresh 触发 WebServer 的创建**

```
protected void onRefresh() {
    super.onRefresh();
    try {
        createWebServer();
    } // catch throw ex ......
}

private void createWebServer() {
    WebServer webServer = this.webServer;
    ServletContext servletContext = getServletContext();
    // 如果 WebServer 和 ServletContext 均为 null，则需要创建嵌入式 Web 容器
    if (webServer == null && servletContext == null) {
        // 获取 WebServerFactory
        ServletWebServerFactory factory = getWebServerFactory();
        // 使用 WebServerFactory 创建 WebServer
        this.webServer = factory.getWebServer(getSelfInitializer());
        // 回调优雅停机的钩子
        getBeanFactory().registerSingleton("webServerGracefulShutdown",
                new WebServerGracefulShutdownLifecycle(this.webServer));
        // 回调嵌入式 Web 容器启停的生命周期钩子
        getBeanFactory().registerSingleton("webServerStartStop",
                new WebServerStartStopLifecycle(this, this.webServer));
    } // else if ......
    initPropertySources();
}
```

createWebServer 方法中包含 WebServer 的创建，还有两个与生命周期回调相关的钩子，下面逐一展开讲解。

### 8.3.1　创建 WebServer

创建嵌入式 Tomcat 容器的入口是 ServletWebServerFactory 的实现类 TomcatServletWebServerFactory，在 getWebServerFactory 方法中可以获取到，如代码清单 8-7 所示。

**代码清单 8-7　获取 ServletWebServerFactory**

```
protected ServletWebServerFactory getWebServerFactory() {
```

## 8.3 嵌入式 Web 容器的初始化时机

```
    String[] beanNames = getBeanFactory().getBeanNamesForType(ServletWebServerFactory.class);
    // 检查……
    return getBeanFactory().getBean(beanNames[0], ServletWebServerFactory.class);
}
```

从源码中观察，getWebServerFactory 方法仅是一个 getBean 的动作，但这其中还暗藏玄机。在第 2 章介绍 TomcatServletWebServerFactory 的注册时提到了 WebServerFactoryCustomizer 的定制器，其通过向 IOC 容器中添加 WebServerFactoryCustomizer，可以编程式地对 WebServerFactory 的配置进行修改，而 WebServerFactoryCustomizer 的执行位置在后置处理器 WebServerFactoryCustomizerBeanPostProcessor 中，对应的处理逻辑如代码清单 8-8 所示（源码逻辑很简单，不再展开）。

**代码清单 8-8　应用所有 WebServerFactoryCustomizer**

```
public Object postProcessBeforeInitialization(Object bean, String beanName)
        throws BeansException {
    if (bean instanceof WebServerFactory) {
        postProcessBeforeInitialization((WebServerFactory) bean);
    }
    return bean;
}

private void postProcessBeforeInitialization(WebServerFactory webServerFactory) {
    LambdaSafe.callbacks(WebServerFactoryCustomizer.class,getCustomizers(), webServerFactory)
            .withLogger(WebServerFactoryCustomizerBeanPostProcessor.class)
            // 注意看这一行：customizer.customize
            .invoke((customizer) -> customizer.customize(webServerFactory));
}
```

获取到 WebServerFactory 后，下一步会执行其 getWebServer 方法创建嵌入式 Tomcat，方法实现如代码清单 8-9 所示。纵观整个方法实现，核心步骤分为三步：创建 Tomcat 对象，并初始化基础的 Connector 和 Engine（该步骤包含整个 getWebServer 方法的大部分内容）；prepareContext 方法初始化 Context，用于构建当前 Spring Boot 应用的上下文；getTomcatWebServer 方法创建最终的 TomcatWebServer 对象。

**代码清单 8-9　TomcatServletWebServerFactory#getWebServer**

```
public static final String DEFAULT_PROTOCOL = "org.apache.coyote.http11.Http11NioProtocol";
private String protocol = DEFAULT_PROTOCOL;

public WebServer getWebServer(ServletContextInitializer... initializers) {
    // JMX 相关……
    Tomcat tomcat = new Tomcat();
    // 给嵌入式 Tomcat 创建一个临时文件夹，用于存放 Tomcat 运行中需要的文件
    File baseDir = (this.baseDirectory != null) ? this.baseDirectory : createTempDir("tomcat");
    tomcat.setBaseDir(baseDir.getAbsolutePath());
    // Connector 中默认放入的 protocol 为 NIO 模式
    Connector connector = new Connector(this.protocol);
    connector.setThrowOnFailure(true);
    // 向 Service 中添加 Connector，并执行定制规则（修改端口号等）
    tomcat.getService().addConnector(connector);
```

239

```
    customizeConnector(connector);
    tomcat.setConnector(connector);
    // 关闭热部署（嵌入式 Tomcat 不存在修改 web.xml、war 包等情况）
    tomcat.getHost().setAutoDeploy(false);
    configureEngine(tomcat.getEngine());
    for (Connector additionalConnector : this.additionalTomcatConnectors) {
        tomcat.getService().addConnector(additionalConnector);
    }
    // 该动作会生成 TomcatEmbeddedContext
    prepareContext(tomcat.getHost(), initializers);
    // 创建 TomcatWebServer
    return getTomcatWebServer(tomcat);
}
```

其中第一步比较容易理解，第三步会真正创建嵌入式 Tomcat 并初始化，而中间的第二步 `prepareContext` 方法会根据当前的 Spring Boot 应用加载其中的 Servlet 三大核心组件、配置生命周期监听器、应用 `ServletContextInitializer` 等，注意该环节中的 `ServletContextInitializer`，这个组件相当关键，Spring Boot 中注册的所有 Servlet 三大核心组件均会通过该组件与嵌入式 Tomcat 对接。

#### 1. ServletContextInitializer 的设计

`ServletContextInitializer` 本身不是 Servlet 相关规范中定义的 API，它是 Spring Boot 1.4.0 以后定义的 API 接口（这也说明 `ServletContextInitializer` 与某个具体的 Web 容器没有任何关系）。借助 IDE 搜索 `ServletContextInitializer` 的实现类，可以发现有一个抽象类 `RegistrationBean`，而它又有对应 Servlet 三大核心组件的注册实现类 `ServletRegistrationBean`、`FilterRegistrationBean` 以及 `ServletListenerRegistrationBean`。从接口设计上看，`ServletContextInitializer` 发挥作用的时机是 `ServletContext` 的初始化阶段，即 Servlet 容器加载 `ServletContainerInitializer` 的阶段（有关 `ServletContainerInitializer` 的内容会在第 14 章讲解），如代码清单 8-10 所示。

**代码清单 8-10　ServletContextInitializer**

```
@FunctionalInterface
public interface ServletContextInitializer {
    void onStartup(ServletContext servletContext) throws ServletException;
}
```

#### 2. ServletContextInitializer 应用于嵌入式 Tomcat

`ServletContextInitializer` 作为注册 Servlet 原生三大核心组件的切入点，其应用时机在初始化 Tomcat 内部 Context 的阶段，即 `prepareContext` 方法中。代码清单 8-11 展示了调用流程，`prepareContext` 方法的最后有对 `ServletContextInitializer` 的配置应用，其调用的 `configureContext` 方法会在内部实例化一个 `TomcatStarter` 对象，而 `TomcatStarter` 对象本身是一个 `ServletContainerInitializer`，它就是上面提到的可以被 Servlet 容器加载的组件，加载后调用的逻辑刚好是下面的 `onStartUp` 方法，该方法会循环调用所有 `ServletContextInitializer` 的 `onStartUp` 方法，由此完成 Servlet 原生三大核心组件的注册（也包含 `DispatcherServlet`）。

### 代码清单 8-11　ServletContextInitializer 被触发调用的逻辑

```java
protected void prepareContext(Host host, ServletContextInitializer[] initializers) {
    // ......
    ServletContextInitializer[] initializersToUse = mergeInitializers(initializers);
    host.addChild(context);
    configureContext(context, initializersToUse);
    postProcessContext(context);
}

protected void configureContext(Context context, ServletContextInitializer[] initializers) {
    TomcatStarter starter = new TomcatStarter(initializers);
    // ......
}

class TomcatStarter implements ServletContainerInitializer {
    private final ServletContextInitializer[] initializers;

    @Override
    public void onStartup(Set<Class<?>> classes, ServletContext servletContext)
            throws ServletException {
        try {
            for (ServletContextInitializer initializer : this.initializers) {
                initializer.onStartup(servletContext);
            }
        } // catch ......
    }
}
```

**3. Context 的初始化细节**

注意 prepareContext 方法中的细节，在方法的开始会创建一个 TomcatEmbeddedContext，它就是 Host 中的 Context 落地实现，而且从方法的最后也可以看到，Tomcat EmbeddedContext 对象被添加到 Host 中（这个添加的 Context 在 8.4 节中会被提取出），如代码清单 8-12 所示。

### 代码清单 8-12　prepareContext 中创建 Context 并放入 Host

```java
protected void prepareContext(Host host, ServletContextInitializer[] initializers) {
    File documentRoot = getValidDocumentRoot();
    // 创建 Context
    TomcatEmbeddedContext context = new TomcatEmbeddedContext();
    // ......
    // 添加到 Host 中
    host.addChild(context);
    // ......
}
```

#### 8.3.2　Web 容器关闭相关的回调

回到代码清单 8-6 的 createWebServer 方法中。除了创建 TomcatWebServer，createWebServer 方法还会向 BeanFactory 中注册两个与生命周期相关的回调钩子，这部分也需要简单了解一下。

### 1. WebServerGracefulShutdownLifecycle

WebServerGracefulShutdownLifecycle 用于触发嵌入式 Web 容器优雅停机的核心生命周期回调，它实现了 SmartLifecycle 接口，可以在 IOC 容器销毁阶段回调其 stop 方法以触发销毁逻辑。WebServerGracefulShutdownLifecycle 中实现的 stop 方法会回调 WebServer 的 shutDownGracefully 方法，实现优雅停机，如代码清单 8-13 所示。

**代码清单 8-13　WebServerGracefulShutdownLifecycle 可使嵌入式 Web 容器优雅停机**

```java
class WebServerGracefulShutdownLifecycle implements SmartLifecycle {

    private final WebServer webServer;
    WebServerGracefulShutdownLifecycle(WebServer webServer) {
        this.webServer = webServer;
    }

    @Override
    public void stop(Runnable callback) {
        this.running = false;
        this.webServer.shutDownGracefully((result) -> callback.run());
    }
}
```

### 2. WebServerStartStopLifecycle

WebServerStartStopLifecycle 的作用是启动和关闭嵌入式 Web 容器，它会在 IOC 容器刷新即将完成/销毁时被回调，从而回调 WebServer 的 start/stop 方法，真正启动/关闭嵌入式 Web 容器，如代码清单 8-14 所示。

**代码清单 8-14　WebServerStartStopLifecycle 关闭嵌入式 Web 容器**

```java
class WebServerStartStopLifecycle implements SmartLifecycle {

    private final WebServer webServer;

    @Override
    public void start() {
        this.webServer.start();
        this.running = true;
        this.applicationContext.publishEvent(
                new ServletWebServerInitializedEvent(this.webServer, this.applicationContext));
    }

    @Override
    public void stop() {
        this.webServer.stop();
    }
}
```

## 8.4　嵌入式 Tomcat 的初始化

当 createWebServer 方法的前置逻辑执行完毕后，最后一行的 return 会调用 get

TomcatWebServer 方法，以创建 TomcatWebServer 对象，而这个创建逻辑内部蕴藏着嵌入式 Tomcat 的内部初始化逻辑，如代码清单 8-15 中最后一行的 initialize 方法所示。这个 initialize 方法至关重要，且内部源码篇幅较长，为了便于读者更好地阅读和理解源码，本节将其拆分为多个片段讲解。

**代码清单 8-15　创建 TomcatWebServer**

```java
protected TomcatWebServer getTomcatWebServer(Tomcat tomcat) {
    return new TomcatWebServer(tomcat, getPort() >= 0, getShutdown());
}

public TomcatWebServer(Tomcat tomcat, boolean autoStart, Shutdown shutdown) {
    Assert.notNull(tomcat, "Tomcat Server must not be null");
    this.tomcat = tomcat;
    this.autoStart = autoStart;
    this.gracefulShutdown =(shutdown == Shutdown.GRACEFUL)
            ? new GracefulShutdown(tomcat) : null;
    initialize();
}
```

## 8.4.1　获取 Context

initialize 方法的第一个核心动作是获取 Tomcat 中第一个可用的 Context，如代码清单 8-16 所示。获取逻辑本身很简单，此处仅是从 Host 中获取第一个类型为 Context 的子元素，如代码清单 8-17 所示。

**代码清单 8-16　initialize（1）**

```java
private void initialize() throws WebServerException {
    logger.info("Tomcat initialized with port(s): " + getPortsDescription(false));
    synchronized (this.monitor) {
        try {
            addInstanceIdToEngineName();
            // 获取第一个可用的 Context
            Context context = findContext();
            //......
```

**代码清单 8-17　获取第一个可用的 Context**

```java
private Context findContext() {
    for (Container child : this.tomcat.getHost().findChildren()) {
        if (child instanceof Context) {
            return (Context) child;
        }
    }
    throw new IllegalStateException("The host does not contain a Context");
}
```

请注意，在该环节获取的 Context 是在 getWebServer 环节中 prepareContext 方法内创建的 TomcatEmbeddedContext，在此处会被提取并返回，如图 8-4 所示。

# 第 8 章 Spring Boot 容器刷新扩展：嵌入式 Web 容器

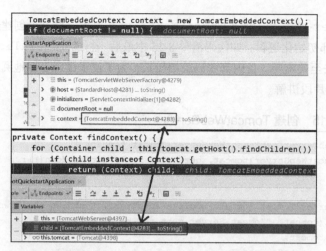

图 8-4 提取的 Context 是 prepareContext 创建的 Context

## 8.4.2 阻止 Connector 初始化

获取到 Context 后，接下来执行的这个步骤是添加一个 LifecycleListener。这个监听器的功能是移除 Service 中的 Connector，如代码清单 8-18 所示。读者读到此处可能会产生疑惑：在 getWebServer 方法中创建 TomcatWebServer 之前已经创建了 Connector，并且将 Connector 放入了 Service 中，为何此处还要把这些 Connector 移除呢？

**代码清单 8-18　initialize（2）**

```
//......
context.addLifecycleListener((event) -> {
    if (context.equals(event.getSource()) && Lifecycle.START_EVENT.equals(event.getType())) {
        removeServiceConnectors();
    }
});
//......
```

要消除这个疑惑，需要读者回想一下创建嵌入式 Web 容器的时机，此时进行到 ApplicationContext 的 refresh 的第几步？第 9 步 onRefresh 方法。由于**执行 onRefresh 方法早于 finishBeanFactoryInitialization 方法**，此时 IOC 容器中的绝大多数单实例 bean 对象尚未初始化，而 Connector 的功能是与客户端交互，**一旦 Connector 初始化完成**，就意味着 Tomcat 可以对外提供服务，即客户端可以成功访问到 Tomcat 的服务。但是显然初始化嵌入式 Tomcat 的时机下当前应用还不具备提供服务的能力，所以需要先将 Connector 移除，以防止客户端成功访问到 Tomcat 服务。

回到代码清单 8-18 中，Spring Boot 对该情况的处理是在嵌入式 Tomcat 广播事件时检查事件类型是否为 "START_EVENT"，如果是则会将所有 Connector 移除。

## 8.4.3 启动 Tomcat

Connector 的准备工作完成后，下一个核心动作是启动 Tomcat 服务，如代码清单 8-19 所示。该步骤会从嵌入式 Tomcat 中的 Server 层级开始，逐个组件进行初始化和启动这两个步骤的执行。源码的内部比较复杂，读者可以先参照图 8-5，通过时序图对 Tomcat 内部的组件初始化和启动有一个整体的认识。

**代码清单 8-19　initialize（3）**

```
//......
// 启动服务器以触发初始化监听器
this.tomcat.start();
//......
```

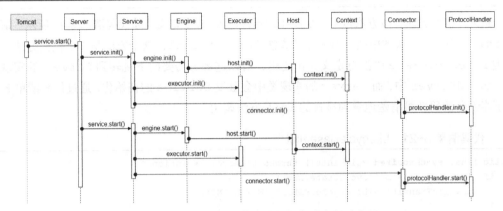

图 8-5　Tomcat 的组件初始化与启动流程时序图

熟悉 Tomcat 内部启动流程的读者可能看到图 8-5 的起点时会感到困惑：外置 Tomcat 通过 `startup.bat` 或 `startup.sh` 启动时内部会初始化一个 `Catalina` 对象，由它引导 `Server` 的初始化和启动，为什么在图 8-5 中是直接由 Tomcat 来引导启动呢？其实这就是外置 Tomcat 与嵌入式 Tomcat 的区别。Spring Boot 整合嵌入式 Tomcat 时，考虑到 Tomcat 只会为当前项目提供服务，加上嵌入式 Tomcat 本身的机制，于是在内部创建 `TomcatWebServer` 对象时，直接使用 Tomcat 对象的 `start` 方法来引导内部组件的初始化和启动流程。

下面以 `tomcat.start()` 方法为例，分析 Tomcat 引导 `Server` 初始化和启动的源码执行逻辑，如代码清单 8-20 所示。Tomcat 类的内部会调用 `Server` 的 `start` 方法，而 `start` 方法定义在所有 Tomcat 核心组件的共同父类 `LifecycleBase` 上。从方法实现上看，`start` 方法更像是制定所有组件生命周期流程的规范型方法，其内部的很多方法都是抽象的，这跟 WebMvc 的设计非常类似：父类制定流程的抽象定义，子类负责具体的步骤实现。

**代码清单 8-20　tomcat.start()方法引发 Server 的初始化和启动**

```java
public void start() throws LifecycleException {
    getServer();
    server.start();
}

// LifecycleBase
public final synchronized void start() throws LifecycleException {
    // 前置判断 ......

    if (state.equals(LifecycleState.NEW)) {
        // 初始化
```

```
        init();
    } // else if ......

    try {
        setStateInternal(LifecycleState.STARTING_PREP, null, false);
        // 启动自身
        startInternal();
        //......
    } //catch ......
}
```

进入 init 方法,该方法依然由父类 LifecycleBase 定义,如代码清单 8-21 所示。请注意 init 方法中 try 块的中间一行调用的方法 initInternal,这个方法是一个模板方法,在父类 LifecycleBase 中没有定义关键逻辑,所以这部分需要向下跳转到 Server 的实现类 StandardServer 中,而 Server 的实现类中会触发 Service 的初始化,通过代码清单 8-21 的下半部分源码即可清楚地看到显式的 init 方法调用。

**代码清单 8-21　LifecycleBase#init**

```
public final synchronized void init() throws LifecycleException {
    if (!state.equals(LifecycleState.NEW)) {
        invalidTransition(Lifecycle.BEFORE_INIT_EVENT);
    }

    try {
        setStateInternal(LifecycleState.INITIALIZING, null, false);
        initInternal();
        setStateInternal(LifecycleState.INITIALIZED, null, false);
    } // catch ......
}

// StandardServer
protected void initInternal() throws LifecycleException {
    super.initInternal();

    //......
    for (int i = 0; i < services.length; i++) {
        services[i].init();
    }
}
```

对于后续的所有核心组件的初始化和启动,源码中的逻辑几乎完全相同,读者可以借助 IDE 自行翻阅源码,本节不再过多展开。

### 8.4.4　阻止 Tomcat 结束

TomcatWebServer 的 initialize 方法中最后一个关键步骤是 startDaemonAwaitThread,如代码清单 8-22 所示。该步骤会启动一个新的 awaitThread 线程,以阻止 Tomcat 的进程结束。注意观察代码清单 8-23 中创建 awaitThread 线程的细节,它内部实现的 run 方法是回调 Server 的 await 方法,并且还设置了 Daemon 属性为 **false**。这其中涉及两个关键点,下面一一解读。

### 代码清单 8-22　initialize（4）

```
    // ......
    startDaemonAwaitThread();
    // catch throw ex ......
    }
}
```

### 代码清单 8-23　startDaemonAwaitThread

```
private void startDaemonAwaitThread() {
    Thread awaitThread = new Thread("container-" + (containerCounter.get())) {
        @Override
        public void run() {
            TomcatWebServer.this.tomcat.getServer().await();
        }
    };
    awaitThread.setContextClassLoader(getClass().getClassLoader());
    awaitThread.setDaemon(false);
    awaitThread.start();
}
```

其一是 Daemon 线程，Tomcat 中的所有进程都是 Daemon 线程。在一个 Java 应用中，只要有一个非 Daemon 线程在运行，Daemon 线程就不会停止，整个应用也不会终止。如果 Tomcat 需要一直运行以接收客户端请求，就必须让 Tomcat 内部的 Daemon 线程都存活，至少需要一个能阻止 Tomcat 进程停止的非 Daemon 线程，代码清单 8-23 中新创建的 `awaitThread` 线程就是负责阻止 Tomcat 进程停止的线程。

其二是 `await` 方法，这里调用的是 `Server` 中的 `await` 方法，核心源码如代码清单 8-24 所示。重点关注退出 Tomcat 的端口的判断逻辑为-1 的分支，退出端口-1 代表当前启动的是一个嵌入式 Tomcat，阻塞 Tomcat 进程结束的方式是每隔 10s 检查一次 `stopAwait` 的值，只要该值一直为 false，Tomcat 就不会退出，否则反之。

### 代码清单 8-24　await 阻塞 Tomcat 进程结束的实现

```
public void await() {
    //-2 时的处理 ......

    // 如果关闭 Tomcat 的端口是-1，代表是嵌入式 Tomcat
    if (getPortWithOffset() == -1) {
        try {
            awaitThread = Thread.currentThread();
            while(!stopAwait) {
                try {
                    Thread.sleep(10000);
                } // catch ignore ......
            }
        } finally {
            awaitThread = null;
        }
        return;
    }

    // 退出端口正常的处理（属于外置 Tomcat 逻辑）......
}
```

可能熟悉 Tomcat 的读者会好奇，默认情况下 Tomcat 的退出端口是 8005，为什么在此处是判断-1 呢？其实在 Tomcat 调用 getServer 方法中给 Server 设置的端口号就是-1，如代码清单 8-25 所示。

**代码清单 8-25　嵌入式 Tomcat 的端口号为-1**

```java
public Server getServer() {
    if (server != null) {
        return server;
    }
    //......
    server = new StandardServer();
    //......
    // 设置端口号为 -1，代表嵌入式
    server.setPort(-1);
    //......
    return server;
}
```

通过上述一系列核心组件的初始化和启动，嵌入式 Tomcat 容器初始化完成。请注意，onRefresh 方法执行完毕后，Tomcat 还不能提供服务（因为 Connector 在该阶段被移除，无法与客户端建立有效连接）。

## 8.5　嵌入式 Tomcat 的启动

当 onRefresh 方法执行完毕，并且后续 finishBeanFactoryInitialization 方法执行完毕，IOC 容器中所有非延迟加载的单实例 bean 对象均初始化完毕，此时会执行 refresh 方法的第 12 步 finishRefresh 方法，该方法中会回调所有 SmartLifecycle，其中包含 8.3.2 节中提到的回调钩子 WebServerStartStopLifecycle，它会在该阶段回调嵌入式 Web 容器的 start 方法，从而让容器可以正常接收客户端的请求，如代码清单 8-26 所示。

**代码清单 8-26　WebServerStartStopLifecycle#start**

```java
public void start() {
    this.webServer.start();
    this.running = true;
    this.applicationContext.publishEvent(
        new ServletWebServerInitializedEvent(this.webServer, this.applicationContext));
}
```

进入实现类 TomcatWebServer 的 start 方法，由于在之前初始化时移除了 Connector，此处会使用 addPreviouslyRemovedConnectors 方法将之前移除的 Connector 还原并启动，如代码清单 8-27 所示。请注意，在 addPreviouslyRemovedConnectors 方法中，向 Service 中添加 Connector 的 addConnector 方法会顺带启动 Connector 使其生效并工作。

**代码清单 8-27　TomcatWebServer#start**

```java
public void start() throws WebServerException {
    synchronized (this.monitor) {
        // started 的检查 ......
```

```java
        try {
            // 还原、启动 Connector
            addPreviouslyRemovedConnectors();
            Connector connector = this.tomcat.getConnector();
            // 老框架兼容、启动后检查等 ......
            this.started = true;
            // logger ......
        }// catch finally ......
    }
}

private void addPreviouslyRemovedConnectors() {
    Service[] services = this.tomcat.getServer().findServices();
    for (Service service : services){
        // 之前移除的 Connector 在 serviceConnectors 中
        Connector[] connectors = this.serviceConnectors.get(service);
        if (connectors != null) {
            for (Connector connector : connectors){
                // 此处添加并启动
                service.addConnector(connector);
                if (!this.autoStart){
                    stopProtocolHandler(connector);
                }
            }
            this.serviceConnectors.remove(service);
        }
    }
}
```

至此，嵌入式 Tomcat 完整启动。

## 8.6 小结

本章主要围绕 Spring Boot 的可独立运行特性的底层支撑——嵌入式 Web 容器，展开讲解了嵌入式 Tomcat 的设计、模型、架构、工作流程，以及 Spring Boot 如何引导启动一个嵌入式 Tomcat。Tomcat 作为经典的 Servlet 容器，它的架构非常值得深入探究。Spring Boot 在面对不同类型的嵌入式 Web 容器时，通过构建全新的嵌入式容器模型以适配不同的容器实现，并配合 `ApplicationContext` 的生命周期，对嵌入式 Web 容器予以合理的初始化和启动处理。

第 9 章会围绕 Spring Framework 的另一个核心特性 AOP 展开讲解，探讨 Spring Boot 中的 AOP 相关生命周期原理。

# 第 9 章 AOP 模块的生命周期

**本章主要内容：**
◇ AOP 的核心后置处理器 AnnotationAwareAspectJAutoProxyCreator；
◇ AOP 底层收集切面类的机制；
◇ Bean 被 AOP 代理的过程原理；
◇ 代理对象执行的全流程分析。

前面三章的内容从 Spring Boot 的主线引导入手，逐步剖析 `SpringApplication` 的工作原理、`ApplicationContext` IOC 容器的初始化流程，以及在独立运行时启用的嵌入式 Web 容器的底层机制。本章重点研究 Spring Framework 中 AOP 模块的全生命周期。Spring Framework 的两大核心特性是 IOC 和 AOP，前面三章主要是围绕 IOC 容器展开的，本章就 AOP 的启用、生效、运行机制进行全面讲解。

## 9.1 @EnableAspectJAutoProxy

对于启用 AOP 特性的核心注解读者都非常清楚，在 Spring Boot 主启动类上标注 `@EnableAspectJAutoProxy` 即可开启基于 AspectJ 的 AOP 动态代理。借助 IDE 查看 `@EnableAspectJAutoProxy` 注解的源码，可以发现它使用 `@Import` 注解导入了一个注册器，如代码清单 9-1 所示。

**代码清单 9-1 @EnableAspectJAutoProxy**

```
@Import(AspectJAutoProxyRegistrar.class)
public @interface EnableAspectJAutoProxy {
    boolean proxyTargetClass() default false;
    boolean exposeProxy() default false;
}
```

这个 `AspectJAutoProxyRegistrar` 的文档注释中已经解释了它的核心工作内容：
> Registers an AnnotationAwareAspectJAutoProxyCreator against the current BeanDefinition Registry as appropriate based on a given @EnableAspectJAutoProxy annotation.

对于给定的 `@EnableAspectJAutoProxy` 注解，根据当前 `BeanDefinitionRegistry` 在适当的位置注册 `AnnotationAwareAspectJAutoProxyCreator`。

由此可知实现 AOP 特性的核心后置处理器是 **AnnotationAwareAspectJAutoProxy**

## 9.1 @EnableAspectJAutoProxy

Creator。在深入该后置处理器之前,先请读者快速了解 AspectJAutoProxyRegistrar 注册 AnnotationAwareAspectJAutoProxyCreator 的时机。由于 AspectJAutoProxyRegistrar 实现了 ImportBeanDefinitionRegistrar 接口,可以直接定位到 registerBeanDefinitions 方法去寻找注册的逻辑,如代码清单 9-2 所示。

**代码清单 9-2　AspectJAutoProxyRegistrar 的核心方法**

```
public void registerBeanDefinitions(
        AnnotationMetadata importingClassMetadata, BeanDefinitionRegistry registry) {
    // 核心注册后置处理器的动作
    AopConfigUtils.registerAspectJAnnotationAutoProxyCreatorIfNecessary(registry);

    // 解析@EnableAspectJAutoProxy 的属性并配置
    AnnotationAttributes enableAspectJAutoProxy = AnnotationConfigUtils
            .attributesFor(importingClassMetadata, EnableAspectJAutoProxy.class);
    if (enableAspectJAutoProxy != null) {
        if (enableAspectJAutoProxy.getBoolean("proxyTargetClass")) {
            AopConfigUtils.forceAutoProxyCreatorToUseClassProxying(registry);
        }
        if (enableAspectJAutoProxy.getBoolean("exposeProxy")) {
            AopConfigUtils.forceAutoProxyCreatorToExposeProxy(registry);
        }
    }
}
```

注意 registerBeanDefinitions 方法的第一行逻辑就已经表明,它要注册一个 **AutoProxyCreator** 后置处理器,也就是上面提到的代理对象创建器,从代码清单 9-3 中可以看到,这个方法也明确指定了注册的 bean 对象类型就是 AnnotationAwareAspectJAutoProxyCreator。后续的方法调用就是根据这个类型构建 BeanDefinition 并注册到 BeanDefinitionRegistry 中的动作,这部分逻辑相对简单,读者可以简单浏览一遍,主要的核心是注册的代理对象创建器的类型 AnnotationAwareAspectJAutoProxyCreator。

**代码清单 9-3　注册 AnnotationAwareAspectJAutoProxyCreator 的动作**

```
public static BeanDefinition registerAspectJAnnotationAutoProxyCreatorIfNecessary(
        BeanDefinitionRegistry registry) {
    return registerAspectJAnnotationAutoProxyCreatorIfNecessary(registry, null);
}

public static BeanDefinition registerAspectJAnnotationAutoProxyCreatorIfNecessary(
        BeanDefinitionRegistry registry, @NullableObject source) {
    // 此处已指定类型
    return registerOrEscalateApcAsRequired(AnnotationAwareAspectJAutoProxyCreator.class,
            registry, source);
}

private static BeanDefinition registerOrEscalateApcAsRequired(
        Class<?> cls, BeanDefinitionRegistry registry, @Nullable Object source){
    Assert.notNull(registry, "BeanDefinitionRegistry must not be null");
    // 后置处理器的等级升级机制(关联 10.3.3 节)
    if (registry.containsBeanDefinition(AUTO_PROXY_CREATOR_BEAN_NAME)) {
```

```
        BeanDefinition apcDefinition =
                registry.getBeanDefinition(AUTO_PROXY_CREATOR_BEAN_NAME);
        if (!cls.getName().equals(apcDefinition.getBeanClassName())) {
            int currentPriority = findPriorityForClass(apcDefinition.getBeanClassName());
            int requiredPriority = findPriorityForClass(cls);
            if (currentPriority < requiredPriority) {
                apcDefinition.setBeanClassName(cls.getName());
            }
        }
        return null;
    }
    // 构建 BeanDefinition,注册到 BeanDefinitionRegistry 中
    RootBeanDefinition beanDefinition = new RootBeanDefinition(cls);
    beanDefinition.setSource(source);
    // 注意代理对象创建器的优先级是 BeanPostProcessor 中最高的
    beanDefinition.getPropertyValues().add("order", Ordered.HIGHEST_PRECEDENCE);
    beanDefinition.setRole(BeanDefinition.ROLE_INFRASTRUCTURE);
    registry.registerBeanDefinition(AUTO_PROXY_CREATOR_BEAN_NAME, beanDefinition);
    return beanDefinition;
}
```

## 9.2　AnnotationAwareAspectJAutoProxyCreator

由类名可知,`AnnotationAwareAspectJAutoProxyCreator` 是基于注解驱动的 AspectJ 的代理对象创建器。类名涵盖的要素比较多,下面借助 javadoc 对其有一个初步的了解。

> AspectJAwareAdvisorAutoProxyCreator subclass that processes all AspectJ annotation aspects in the current application context, as well as Spring Advisors.
>
> Any AspectJ annotated classes will automatically be recognized, and their advice applied if Spring AOP's proxy-based model is capable of applying it. This covers method execution joinpoints.
>
> If the `<aop:include>` element is used, only @AspectJ beans with names matched by an include pattern will be considered as defining aspects to use for Spring auto-proxying.
>
> Processing of Spring Advisors follows the rules established in org.springframework.aop.framework.autoproxy.AbstractAdvisorAutoProxyCreator.
>
> 它是 `AspectJAwareAdvisorAutoProxyCreator` 的子类,用于处理当前 ApplicationContext 中的所有基于@AspectJ 注解的切面,以及 Spring 原生的 `Advisor`。
>
> 如果 Spring AOP 基于代理的模型能够应用任何被@AspectJ 注解标注的类,那么它们的增强方法将被自动识别。这涵盖了方法执行的切入点表达式。
>
> 如果使用`<aop:include>`元素,则只有名称与包含模式匹配的被@AspectJ 标注的 Bean 将被视为定义要用于 Spring Framework 自动代理的切面。
>
> Spring Framework 中内置 `Advisor` 的处理遵循 `AbstractAdvisorAutoProxyCreator` 中建立的规则。

提取 javadoc 中解释的核心内容:`AnnotationAwareAspectJAutoProxyCreator` 兼顾 AspectJ 风格的切面声明和 Spring Framework 原生的 AOP。

## 9.2.1 类继承结构

借助 IDEA 可以很清楚地看到 `AnnotationAwareAspectJAutoProxyCreator` 的继承结构,以及其中的重要核心,如图 9-1 所示。

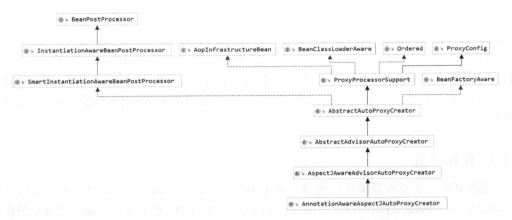

图 9-1 AnnotationAwareAspectJAutoProxyCreator 的类继承结构

注意观察顶级接口 `AnnotationAwareAspectJAutoProxyCreator`,它实现了几个重要的接口。

- `BeanPostProcessor`:用于在 `postProcessAfterInitialization` 方法中生成代理对象。
- `InstantiationAwareBeanPostProcessor`:拦截 Bean 的正常 `doCreateBean` 创建流程。
- `SmartInstantiationAwareBeanPostProcessor`:提前预测 Bean 的类型、暴露 Bean 的引用(AOP、循环依赖等),其过于复杂,此处不作解释。
- `AopInfrastructureBean`:实现了该接口的 Bean 永远不会被代理(防止无限套用)。

除此之外还有一点要注意的是,`AnnotationAwareAspectJAutoProxyCreator` 继承的顶级的抽象实现类是 **`AbstractAutoProxyCreator`**,它本身的重要程度非常高,后面的源码分析中会频繁遇到它。

## 9.2.2 初始化时机

既然在 `AspectJAutoProxyRegistrar` 中已经把 `AnnotationAwareAspectJAutoProxyCreator` 的 `BeanDefinition` 信息注册到 `BeanDefinitionRegistry` 中,那么在接下来的后置处理器初始化部分中它一定会被创建,而这个时机在第 7 章讲过了,读者可以快速回忆一下,如代码清单 9-4 所示。

**代码清单 9-4　refresh 的部分源码**

```
public void refresh() throws BeansException, IllegalStateException {
    synchronized (this.startupShutdownMonitor) {
        // ......
        try {
```

```
        postProcessBeanFactory(beanFactory);
        invokeBeanFactoryPostProcessors(beanFactory);
        // 6. 注册、初始化 BeanPostProcessor
        registerBeanPostProcessors(beanFactory);
        initMessageSource();
        initApplicationEventMulticaster();
        // ......
    }
}
```

注意设计上的一个细节，AnnotationAwareAspectJAutoProxyCreator 实现了 Ordered 接口（并且设置了最高优先级），它会提前于普通的 BeanPostProcessor 创建，那么普通的 BeanPostProcessor 是否也会被 AOP 代理呢？答案是肯定的。读者如果感兴趣，可以自行测试一下。

### 9.2.3 作用时机

既然 IOC 容器的刷新动作中的第六步 registerBeanPostProcessors 方法已经把 AnnotationAwareAspectJAutoProxyCreator 创建就绪，接下来的 bean 对象在初始化阶段中它就会干预。我们以第 5 章中编写的示例代码为例，观察 DemoService 对象的创建流程中 AnnotationAwareAspectJAutoProxyCreator 是如何切入逻辑进行干预的。

#### 1. getBean → doCreateBean

bean 对象的创建流程在第 7 章中已经讲过，从 getBean 开始依次是 doGetBean、createBean 和 doCreateBean，在 doCreateBean 方法中会真正地创建对象、属性赋值、依赖注入，以及初始化流程的执行。在 bean 对象本身的初始化流程全部执行完毕后，下一步要执行的是 BeanPostProcessor 的 postProcessAfterIntialization 方法。在这之前有一个小插曲，即 createBean 到 doCreateBean 的环节，该环节有一个 InstantiationAwareBeanPostProcessor 的拦截初始化动作，读者需要先注意一下这个动作。

#### 2. postProcessBeforeInstantiation

PostProcessBeforeInstantiation 如代码清单 9-5 所示。

**代码清单 9-5　AnnotationAwareAspectJAutoProxyCreator#postProcessBeforeInstantiation**

```java
public Object postProcessBeforeInstantiation(Class<?> beanClass, String beanName) {
    Object cacheKey = getCacheKey(beanClass, beanName);

    // 决定是否要提前增强当前 bean 对象
    if (!StringUtils.hasLength(beanName) || !this.targetSourcedBeans.contains(beanName)) {
        // 被增强过的 bean 对象不会再次被增强
        if (this.advisedBeans.containsKey(cacheKey)) {
            return null;
        }
        // 基础类的 bean 对象不会被提前增强，被跳过的 bean 不会被提前增强
        if (isInfrastructureClass(beanClass) || shouldSkip(beanClass, beanName)) {
            this.advisedBeans.put(cacheKey, Boolean.FALSE);
            return null;
        }
    }
```

```
// 原型 bean 对象的额外处理：TargetSource
// 此处的设计与自定义 TargetSource 相关，单实例 bean 对象必定返回 null
TargetSource targetSource = getCustomTargetSource(beanClass, beanName);
if (targetSource != null) {
    if (StringUtils.hasLength(beanName)) {
        this.targetSourcedBeans.add(beanName);
    }
    Object[] specificInterceptors = getAdvicesAndAdvisorsForBean(beanClass,
            beanName, targetSource);
    Object proxy = createProxy(beanClass, beanName, specificInterceptors, targetSource);
    this.proxyTypes.put(cacheKey, proxy.getClass());
    return proxy;
}

return null;
}
```

通读整个方法后，读者可能会有一种感觉：整体方法的逻辑大体清晰，但具体到某一个细节会比较迷茫。出现该感觉是正常的，因为这里面有几个小概念在之前的学习中没有接触过，了解这些概念和设计之后，读者再理解这部分逻辑就容易多了。

（1）InfrastructureClass

InfrastructureClass 直译为"基础类型"，它指代的是 IOC 容器中注册的基础类，包括切面类、切入点、增强器等 bean 对象。通过代码清单 9-6 就可以了解到，如果一个 bean 对象本身是切面类/切入点/增强器等，那么它本身是参与 AOP 底层的成员，不应该参与具体的被增强对象中。

**代码清单 9-6　isInfrastructureClass 过滤基础类的 bean 对象**

```
protected boolean isInfrastructureClass(Class<?> beanClass) {
    return (super.isInfrastructureClass(beanClass) ||
            (this.aspectJAdvisorFactory != null && this.aspectJAdvisorFactory.isAspect (beanClass)));
}

protected boolean isInfrastructureClass(Class<?> beanClass) {
    boolean retVal = Advice.class.isAssignableFrom(beanClass)
            || Pointcut.class.isAssignableFrom(beanClass)
            || Advisor.class.isAssignableFrom(beanClass)
            || AopInfrastructureBean.class.isAssignableFrom(beanClass);
    // logger ......
    return retVal;
}
```

（2）被跳过的 bean 对象

检查完基础类的 bean 对象，紧接着要检查准备增强的 bean 对象是否需要被跳过，如何理解"被跳过的 bean 对象"？读者可以先结合代码清单 9-7 简单梳理一遍逻辑。

**代码清单 9-7　shouldSkip 检查一个 bean 对象是否需要被跳过**

```
// 父类 AspectJAwareAdvisorAutoProxyCreator
protected boolean shouldSkip(Class<?> beanClass, String beanName) {
    // 加载增强器
    List<Advisor> candidateAdvisors = findCandidateAdvisors();
```

```
    for (Advisor advisor : candidateAdvisors) {
        // 逐个匹配，如果发现当前 bean 对象的名称与增强器的名称一致，则认为不能被增强
        if (advisor instanceof AspectJPointcutAdvisor &&
                ((AspectJPointcutAdvisor) advisor).getAspectName().equals(beanName)){
            return true;
        }
    }
    return super.shouldSkip(beanClass, beanName);
}

// AbstractAutoProxyCreator
protected boolean shouldSkip(Class<?> beanClass, String beanName) {
    // 检查 beanName 代表的是不是原始对象（以.ORIGINAL 结尾）
    return AutoProxyUtils.isOriginalInstance(beanName, beanClass);
}
```

代码清单 9-7 中已经辅助有代码注释，想必读者可以大体把控这段逻辑的用意，它要检查的点是当前对象的名称是否与增强器的名称一致，换句话说，shouldSkip 方法要检查当前准备增强的 bean 对象是不是一个还没有经过任何代理的原始对象，而检查的规则是观察 bean 对象的名称是否带有.ORIGINAL 的后缀，一般情况下项目中创建的 bean 不可能带有.ORIGINAL 后缀，所以 shouldSkip 方法相当于判断当前创建的 bean 对象名称是否与增强器名称一致。

可能有部分读者不了解增强器的概念，这里简单解释一下。一个 Advisor 可以视为一个**切入点+一个通知方法的结合体**，对于 Aspect 切面类中定义的通知方法，**方法体+方法上的通知注解**就可以看作一个 Advisor 增强器，关于更详细的解释会在 9.3 节中展开。

注意，实际在此处 Debug 时可以发现，findCandidateAdvisors 加载增强器的方法执行完成后能获取到一个增强器，也就是 ServiceAspect 中定义的 beforePrint 方法，如图 9-2 所示，请读者对这个细节先有一个印象。

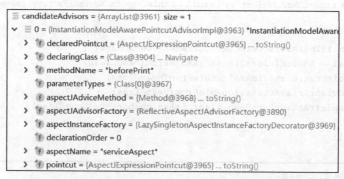

图 9-2 candidateAdvisors 中存放了 Aspect 切面类的方法

（3）TargetSource

有关 TargetSource 的设计，这里只简单解释一句：AOP 的代理其实不是代理目标对象本身，而是目标对象经过包装后的 **TargetSource** 对象。关于 Spring Framework 这样做的理由以及这样做的好处，在 9.4 节中我们会展开讲解。

### 3. postProcessAfterInitialization

前面的 postProcessBeforeInstantiation 方法拦截判断结束后，Annotation

AwareAspectJAutoProxyCreator 再发挥作用就要等到最后一步的 `postProcessAfterInitialization` 方法，该方法是真正地生成代理对象，如代码清单 9-8 所示。

**代码清单 9-9 核心后置处理动作：创建代理对象**

```
public Object postProcessAfterInitialization(@Nullable Object bean, String beanName) {
    if (bean != null) {
        Object cacheKey = getCacheKey(bean.getClass(), beanName);
        If (this.earlyProxyReferences.remove(cacheKey) != bean) {
            // 核心：构造代理
            return wrapIfNecessary(bean, beanName, cacheKey);
        }
    }
    return bean;
}
```

`postProcessAfterInitialization` 方法内部的代码量看上去不多，但这个方法的难度非常高，核心的动作在中间的 `wrapIfNecessary` 中，这部分在 9.5 节中会详细展开，这里先简单概括一下创建代理对象核心动作的三个步骤。

（1）判断决定是否是不会被增强的 bean 对象。
（2）根据当前正在创建的 bean 对象去匹配增强器。
（3）如果有增强器，创建 bean 对象的代理对象。

整体了解了 AnnotationAwareAspectJAutoProxyCreator 的设计后，下面分阶段讲解整个 AOP 生命周期中的核心步骤。

## 9.3 Advisor 与切面类的收集

上面在分析 shouldSkip 方法时提到了一个非常重要的名为 **Advisor** 增强器的概念，如何理解"增强器"的概念以及 Spring Framework 中设计的增强器的类型是我们探讨的重点。

### 9.3.1 收集增强器的逻辑

回到 AnnotationAwareAspectJAutoProxyCreator 的 postProcessBeforeInstantiation 方法中，前面已经提到了 findCandidateAdvisors 方法是收集候选增强器的动作，而这个方法本身又分为两个部分，分别是收集 Spring Framework 原生的增强器以及 BeanFactory 中所有 AspectJ 形式的切面并封装为增强器，如代码清单 9-9 所示。下面结合源码分别来看。

**代码清单 9-9 findCandidateAdvisors 收集候选的增强器**

```
// AspectJAwareAdvisorAutoProxyCreator
protected boolean shouldSkip(Class<?> beanClass, String beanName) {
    List<Advisor> candidateAdvisors = findCandidateAdvisors();
    // ......
}

protected List<Advisor> findCandidateAdvisors() {
    // 根据父类的规则添加所有找到的 Spring 原生的增强器
```

```java
    List<Advisor> advisors = super.findCandidateAdvisors();
    // 解析 BeanFactory 中所有的 AspectJ 切面,并构建增强器
    if (this.aspectJAdvisorsBuilder != null) {
        advisors.addAll(this.aspectJAdvisorsBuilder.buildAspectJAdvisors());
    }
    return advisors;
}
```

## 9.3.2 收集原生增强器

通过观察父类 `AbstractAdvisorAutoProxyCreator`,可以发现它委托了一个 `advisorRetrievalHelper` 来处理 Spring Framework 原生的 AOP 增强器,这个 `findAdvisorBeans` 方法的篇幅比较长,下面拆解出核心的主干逻辑研究,如代码清单 9-10 所示。

**代码清单 9-10　收集原生增强器借助了 BeanFactoryAdvisorRetrievalHelper**

```java
private BeanFactoryAdvisorRetrievalHelper advisorRetrievalHelper;

protected List<Advisor>findCandidateAdvisors() {
    Assert.state(this.advisorRetrievalHelper != null,
            "No BeanFactoryAdvisorRetrievalHelper available");
    return this.advisorRetrievalHelper.findAdvisorBeans();
}
```

### 1. 检查现有的增强器 bean 对象

`findAdvisorBeans` 方法的第一部分源码主要是将 IOC 容器中所有类型为 `Advisor` 的实现类对象都获取到,并检查 IOC 容器内部是否注册有增强器,如果没有注册增强器则不会执行后续逻辑,如代码清单 9-11 所示。注意这个 `BeanFactoryUtils` 的 `beanNamesForTypeIncludingAncestors` 方法,底层会使用 `getBeanNamesForType` 方法来寻找 bean 对象的名称(单纯地寻找 bean 对象的名称不会创建具体的 bean 对象,Spring Framework 在此设计得很谨慎),感兴趣的读者可以借助 IDE 自行研究,这里不展开探讨。

**代码清单 9-11　findAdvisorBeans(1)**

```java
public List<Advisor> findAdvisorBeans() {
    // 确定增强器 bean 对象名称的列表(如果尚未缓存)
    String[] advisorNames = this.cachedAdvisorBeanNames;
    if (advisorNames == null) {
        // 不要在这里初始化 FactoryBeans:
        // 我们需要保留所有未初始化的常规 bean 对象,以使自动代理创建者对其应用
        advisorNames = BeanFactoryUtils.beanNamesForTypeIncludingAncestors(
                this.beanFactory, Advisor.class, true, false);
        this.cachedAdvisorBeanNames = advisorNames;
    }
    // 如果当前 IOC 容器中没有任何增强器类型的 bean 对象,直接返回
    if (advisorNames.length == 0) {
        return new ArrayList<>();
    }
    // ......
```

## 2. 初始化原生增强器

如果上面检查到 IOC 容器中注册有原生的 `Advisor` 增强器，则下面会使用 `BeanFactory` 的 `getBean` 方法初始化这些增强器，如代码清单 9-12 所示。Spring Framework 原生的增强器模型因其设计和编写比较复杂，目前已经被淘汰，主流的通知编写还是以 AspectJ 形式为主，所以这里读者仅需一般了解。

**代码清单 9-12　findAdvisorBeans（2）**

```
// ......
List<Advisor> advisors = new ArrayList<>();
for (String name : advisorNames) {
    if (isEligibleBean(name)) {
        if (this.beanFactory.isCurrentlyInCreation(name)) {
            // logger ......
        } else {
            try {
                advisors.add(this.beanFactory.getBean(name, Advisor.class));
            } // catch ......
        }
    }
}
return advisors;
```

实际通过 Debug 也可以发现并没有原生的增强器被创建，如图 9-3 所示。

```
protected List<Advisor> findCandidateAdvisors() {
    // Add all the Spring advisors found according to superclass rules.
    List<Advisor> advisors = super.findCandidateAdvisors();  advisors:  size = 0
    // Build Advisors for all AspectJ aspects in the bean factory.
    if (this.aspectJAdvisorsBuilder != null) {  aspectJAdvisorsBuilder: Annotati
        advisors.addAll(this.aspectJAdvisorsBuilder.buildAspectJAdvisors());
    }
}
```

**图 9-3　默认情况下没有原生的增强器**

以上就是在父类 `AbstractAdvisorAutoProxyCreator` 中的 `findCandidateAdvisors` 方法的逻辑，下面来看另一部分的委托 `aspectJAdvisorsBuilder`。

### 9.3.3　解析 AspectJ 切面封装增强器

从方法名上理解，`aspectJAdvisorsBuilder.buildAspectJAdvisors` 方法可以将 **Aspect** 切面类转换为一个个增强器。既然是转换切面类，就必然有通知方法的解析、增强器的构造等步骤，这部分逻辑更加复杂，下面逐步探究。

#### 1. 逐个解析 IOC 容器中所有的 Bean 类型

> 💡 提示：`buildAspectJAdvisors` 方法的源码缩进层数比较多，为保证源码的阅读体验，代码清单 9-13 的缩进会以两个空格为主。

`buildAspectJAdvisors` 方法的第一部分核心逻辑是将 IOC 容器以及父容器中所有 bean 对象的名称全部提取出（直接声明父类为 `Object`，显然是全部提取），之后会逐个解析这些 bean 对象对应的 `Class`，如代码清单 9-13 所示。通过 Debug 可以发现，获取的 bean 对象名称中包括 IOC 容器内部的一些组件（主启动类、自动配置类等）在内的所有 bean 对象名称，如图 9-4 所示。

### 代码清单 9-13　buildAspectJAdvisors（1）

```java
public List<Advisor> buildAspectJAdvisors() {
  List<String> aspectNames = this.aspectBeanNames;

  if (aspectNames == null) {
    synchronized (this) {
      aspectNames = this.aspectBeanNames;
      if (aspectNames == null) {
        List<Advisor> advisors = new ArrayList<>();
        aspectNames = new ArrayList<>();
        // 获取 IOC 容器中的所有 bean 对象
        String[] beanNames = BeanFactoryUtils.beanNamesForTypeIncludingAncestors(
            this.beanFactory, Object.class, true, false);
        for (String beanName : beanNames) {
          if (!isEligibleBean(beanName)) {
            continue;
          }
          // 原文注释：我们必须小心，不要急于实例化 bean，因为在这种情况下，IOC 容器会缓存它们，但不会被
          // 织入增强器
          Class<?> beanType =this.beanFactory.getType(beanName);
          if (beanType == null) {
            continue;
          }
          // ......
```

```
beanNames = {String[140]@3889}
> ≡ 0 = "org.springframework.context.annotation.internalConfigurationAnnotationProcessor"
> ≡ 1 = "org.springframework.context.annotation.internalAutowiredAnnotationProcessor"
> ≡ 2 = "org.springframework.context.annotation.internalCommonAnnotationProcessor"
> ≡ 3 = "org.springframework.context.event.internalEventListenerProcessor"
> ≡ 4 = "org.springframework.context.event.internalEventListenerFactory"
> ≡ 5 = "springBootAopApplication"
> ≡ 6 = "org.springframework.boot.autoconfigure.internalCachingMetadataReaderFactory"
> ≡ 7 = "serviceAspect"
> ≡ 8 = "demoService"
> ≡ 9 = "org.springframework.aop.config.internalAutoProxyCreator"
> ≡ 10 = "org.springframework.boot.autoconfigure.AutoConfigurationPackages"
> ≡ 11 = "org.springframework.boot.autoconfigure.context.PropertyPlaceholderAutoConfiguration"
> ≡ 12 = "propertySourcesPlaceholderConfigurer"
```

图 9-4　获取 IOC 容器中的所有 bean 对象名称

请读者注意一个细节，Spring Framework 在这里控制得很好，它借助 BeanFactory 提取 bean 对象的类型，而不是先 getBean 后再提取类型，这样可以确保 bean 对象不会被提前创建。而要在没有初始化 bean 对象的前提下获取 bean 对象的 Class，只能依靠 BeanDefinition 中封装的信息，所以在 AbstractBeanFactory 的 getType 方法中可以看到合并 RootBean Definition 的动作，随后调用 RootBeanDefinition 的 getBeanClass 方法获取 bean 对象的 Class（关于具体源码读者可以自行借助 IDE 翻阅，这里不再展开）。

**2. 解析 Aspect 切面类，构造增强器**

> 💡 提示：以下源码的缩进过大，为方便读者阅读，代码清单 9-14 中的缩进有调整。

紧接着第二部分逻辑要完成的工作是判断当前解析的 bean 对象的所属类型是否为切面类，如果是则会进入 if 结构内部，将这个类中的通知方法都转换为 Advisor 增强器，如代码清单 9-14 所示。这里面的一些细节值得继续深入探究。

**代码清单 9-14　buildAspectJAdvisors（2）**

```java
// ......
if (this.advisorFactory.isAspect(beanType)) {
    // 当前解析 bean 对象的所属类型是一个切面类
    aspectNames.add(beanName);
    AspectMetadata amd = new AspectMetadata(beanType, beanName);
    // 下面是单实例切面 bean 对象会执行的流程
    if (amd.getAjType().getPerClause().getKind() == PerClauseKind.SINGLETON) {
        MetadataAwareAspectInstanceFactory factory =
                new BeanFactoryAspectInstanceFactory(this.beanFactory, beanName);
        // 解析生成增强器
        List<Advisor> classAdvisors = this.advisorFactory.getAdvisors(factory);
        if (this.beanFactory.isSingleton(beanName)) {
            this.advisorsCache.put(beanName, classAdvisors);
        } else {
            this.aspectFactoryCache.put(beanName, factory);
        }
        advisors.addAll(classAdvisors);
    }
// ......
```

### （1）判断 Class 是否是通知类

代码清单 9-14 中最上面的 if 结构中 advisorFactory.isAspect(beanType) 用来判断当前 Class 是不是通知类（切面类），除了检查类上是否标注了 @Aspect 注解，源码中还多判断了一步，如代码清单 9-15 所示。

**代码清单 9-15　isAspect 判断一个 Class 是否为通知类**

```java
public boolean isAspect(Class<?> clazz) {
    // @Aspect 注解并且不是被 ajc 编译器编译的
    return (hasAspectAnnotation(clazz) && !compiledByAjc(clazz));
}
```

请注意，isAspect 方法额外判断了是不是被 ajc 编译器编译，为什么要额外多一步判断？从文档注释中可以获取到一些信息：

> We consider something to be an AspectJ aspect suitable for use by the Spring AOP system if it has the @Aspect annotation, and was not compiled by ajc. The reason for this latter test is that aspects written in the code-style (AspectJ language) also have the annotation present when compiled by ajc with the -1.5 flag, yet they cannot be consumed by Spring AOP.
>
> 只有当它有@Aspect 注解并且不是由 ajc 编译的，我们才认为这个 Class 是适合 SpringAOP 系统使用的 AspectJ 切面。不用 ajc 编译的原因是，以代码风格（AspectJ 语言）编写的切面在由带有-1.5 标志的 ajc 编译时也存在注解，但它们不能被 Spring AOP 使用。

简单地说，Spring Framework 的 AOP 有整合 AspectJ 的部分，而原生的 AspectJ 也可以编写 Aspect 切面类，而这种切面在特殊的编译条件下生成的字节码中在类上也会标注 @Aspect

注解，但是 Spring Framework 并不能利用它，所以 `isAspect` 方法做了一个额外的判断处理，避免了这种 `Class` 被误加载。

（2）构造增强器

构造增强器的调用动作是 if 结构中的 `advisorFactory.getAdvisors`，这部分需要跳转到 `ReflectiveAspectJAdvisorFactory` 中来看，如代码清单 9-16 所示。`getAdvisors` 方法的源码比较长，读者在阅读时要结合注释来理解。

**代码清单 9-16　ReflectiveAspectJAdvisorFactory#getAdvisors**

```java
public List<Advisor> getAdvisors(MetadataAwareAspectInstanceFactory aspectInstanceFactory) {
    // Aspect 切面类的 Class
    Class<?> aspectClass = aspectInstanceFactory.getAspectMetadata().getAspectClass();
    String aspectName = aspectInstanceFactory.getAspectMetadata().getAspectName();
    // 校验 ......

    // 此处利用 Decorator 装饰者模式，目的是确保 Advisor 增强器不会被多次实例化
    MetadataAwareAspectInstanceFactory lazySingletonAspectInstanceFactory =
            new LazySingletonAspectInstanceFactoryDecorator(aspectInstanceFactory);

    List<Advisor> advisors = new ArrayList<>();
    // 逐个解析通知方法，并封装为增强器
    for (Method method : getAdvisorMethods(aspectClass)) {
        Advisor advisor = getAdvisor(method, lazySingletonAspectInstanceFactory,0, aspectName);
        if (advisor != null) {
            advisors.add(advisor);
        }
    }

    // 通过在装饰者内部的开始加入 SyntheticInstantiationAdvisor 增强器，
    // 达到延迟初始化切面 bean 对象的目的
    if (!advisors.isEmpty()
            && lazySingletonAspectInstanceFactory.getAspectMetadata().isLazilyInstantiated()) {
        Advisor instantiationAdvisor =
                new SyntheticInstantiationAdvisor(lazySingletonAspectInstanceFactory);
        advisors.add(0, instantiationAdvisor);
    }

    // 对@DeclareParent 注解功能的支持（AspectJ 的引介）
    for (Field field : aspectClass.getDeclaredFields()) {
        Advisor advisor = getDeclareParentsAdvisor(field);
        if (advisor != null) {
            advisors.add(advisor);
        }
    }
    return advisors;
}
```

整段源码阅读下来，方法执行的核心逻辑是**解析 Aspect 切面类中的通知方法**，只不过在上下文的逻辑中补充了一些额外的校验、处理等逻辑。这里面重点关注两个小动作：通知方法是如何收集的；增强器的创建需要哪些关键信息。

### (3) 收集切面类中的通知方法

对于通知方法的收集动作，需要进入 `getAdvisorMethods` 方法中分析，如代码清单 9-17 所示。`getAdvisorMethods` 方法本身不难，就是把开发者定义的切面类中除通用的切入点表达式以外的所有方法都提取出来，并且在取出之后进行排序（排序的原则是按照 Unicode 编码），最后将通知方法返回。

**代码清单 9-17　获取一个切面类的所有通知方法**

```java
private List<Method> getAdvisorMethods(Class<?> aspectClass) {
    final List<Method> methods = new ArrayList<>();
    ReflectionUtils.doWithMethods(aspectClass, method -> {
        // 除 pointcut 之外
        if (AnnotationUtils.getAnnotation(method, Pointcut.class) == null) {
            methods.add(method);
        }
    }, ReflectionUtils.USER_DECLARED_METHODS);
    if (methods.size() > 1) {
        methods.sort(METHOD_COMPARATOR);
    }
    return methods;
}
```

### (4) 创建增强器

`getAdvisor` 方法对应的逻辑是创建 Advisor 增强器。读者可能在此处会产生疑惑，上面的 `getAdvisorMethods` 方法仅提取了当前类中定义的非 @Pointcut 方法，对于没有声明切入点表达式的方法是否会一并返回？对于这个问题其实读者不必多虑，源码在此处又做了一次过滤，如代码清单 9-18 所示。

**代码清单 9-18　创建增强器时进行一次额外的过滤**

```java
public Advisor getAdvisor(Method candidateAdviceMethod,
        MetadataAwareAspectInstanceFactory aspectInstanceFactory,
        int declarationOrderInAspect, String aspectName) {
    validate(aspectInstanceFactory.getAspectMetadata().getAspectClass());

    // 解析切入点表达式
    AspectJExpressionPointcut expressionPointcut = getPointcut(candidateAdviceMethod,
            aspectInstanceFactory.getAspectMetadata().getAspectClass());
    // 没有声明通知注解的方法也会被过滤
    if (expressionPointcut == null) {
        return null;
    }
    return new InstantiationModelAwarePointcutAdvisorImpl(expressionPointcut,
            candidateAdviceMethod, this, aspectInstanceFactory,
            declarationOrderInAspect, aspectName);
}
```

Spring Framework 本身设计得很严谨，注意看最后一行的构造方法中传入的关键参数，分别如下。

- `expressionPointcut`：AspectJ 切入点表达式的封装。

- candidateAdviceMethod：通知方法本体。
- this：当前的 ReflectiveAspectJAdvisorFactory。
- aspectInstanceFactory：上面的装饰者 MetadataAwareAspectInstanceFactory。

去掉工厂本身后，其实增强器的结构就是**一个切入点表达式 + 一个通知方法**，与之前的推测完全一致。

（5）解析通知注解上的切入点表达式

至此其实增强器本身已经没有什么问题，除此之外读者可以再关注一下创建增强器时 getPointcut 方法对应的切入点表达式解析，进入 getPointcut 方法中，如代码清单 9-19 所示。

**代码清单 9-19　获取、解析切入点表达式**

```java
private AspectJExpressionPointcut getPointcut(Method candidateAdviceMethod,
        Class<?> candidateAspectClass){
    // 检索通知方法上的注解
    AspectJAnnotation<?> aspectJAnnotation =
            AbstractAspectJAdvisorFactory.findAspectJAnnotationOnMethod(candidateAdviceMethod);
    if (aspectJAnnotation == null) {
        return null;
    }

    // 根据注解的类型，构造切入点表达式模型
    AspectJExpressionPointcut ajexp =
            new AspectJExpressionPointcut(candidateAspectClass, new String[0], new Class<?>[0]);
    ajexp.setExpression(aspectJAnnotation.getPointcutExpression());
    if (this.beanFactory != null) {
        ajexp.setBeanFactory(this.beanFactory);
    }
    return ajexp;
}

private static final Class<?>[] ASPECTJ_ANNOTATION_CLASSES = new Class<?>[] {
        Pointcut.class, Around.class, Before.class, After.class,
        AfterReturning.class, AfterThrowing.class};

protected static AspectJAnnotation<?> findAspectJAnnotationOnMethod(Method method){
    for (Class<?> clazz : ASPECTJ_ANNOTATION_CLASSES) {
        AspectJAnnotation<?> foundAnnotation =
                findAnnotation(method, (Class<Annotation>) clazz);
        if (foundAnnotation != null) {
            return foundAnnotation;
        }
    }
    return null;
}
```

getPointcut 方法的步骤很简单，它首先会搜寻通知方法上标注的注解，然后提取切入点表达式的信息并返回，而寻找 AspectJ 注解的 findAspectJAnnotationOnMethod 方法逻辑，则是直接在 AbstractAspectJAdvisorFactory 中提前定义好了所有可以声明切入点表达

式的注解（@Around、@Before、@After 等），并在此处一一寻找，如果可以成功找到则返回。

在调用 getPointcut 方法的位置打入 Debug 断点，待程序停在断点时可以发现，getPointcut 方法只是把切入点表达式的内容以及参数等信息封装到 AspectJExpressionPointcut 中而已，如图 9-5 所示。

```
> p candidateAdviceMethod = {Method@3907} ... toString()
> p aspectInstanceFactory = {LazySingletonAspectInstanceFactoryDecorator@3932} ... toString()
  p declarationOrderInAspect = 0
> p aspectName = "serviceAspect"
v ≡ expressionPointcut = {AspectJExpressionPointcut@3926} ... toString()
    > f pointcutDeclarationScope = {Class@3883} ... Navigate
      f pointcutParameterNames = {String[0]@3927}
      f pointcutParameterTypes = {Class[0]@3928}
    > f beanFactory = {DefaultListableBeanFactory@3910} ... toString()
      f pointcutClassLoader = null
      f pointcutExpression = null
      f shadowMatchCache = {ConcurrentHashMap@3929} size = 0
      f location = null
    > f expression = "execution(public * com.linkedbear.springboot.service.*.*(..))"
```

图 9-5　封装好的切入点表达式模型

**3. 原型切面 Bean 的处理**

上面的逻辑只是对单实例切面 Bean 的处理和解析，下面的 else 部分是原型切面 Bean 的处理逻辑，如代码清单 9-20 所示。对于原型切面 Bean 的解析，核心解析动作依然是 advisorFactory.getAdvisors 方法，只是原型切面 Bean 的解析不会再使用 advisorsCache 这个缓存区，这也说明原型切面 Bean 的解析是多次执行的。

**代码清单 9-20　原型切面 Bean 的处理**

```
// ......
    else {
        // 检查单实例 Bean 并抛出异常 ......
        MetadataAwareAspectInstanceFactory factory =
                new PrototypeAspectInstanceFactory(this.beanFactory, beanName);
        this.aspectFactoryCache.put(beanName, factory);
        // 解析 Aspect 切面类，构造增强器
        advisors.addAll(this.advisorFactory.getAdvisors(factory));
    }
}
}
this.aspectBeanNames = aspectNames;
return advisors;
}
// ......
```

**4. 增强器汇总**

最后一部分是整理的环节，前面已经把所有的切面类都解析完毕，最后只需把这些构造好

的增强器都集中到一个 `List` 中返回，如代码清单 9-21 所示。经过以上一系列的步骤后，切入点表达式解析完毕，通知方法收集完毕，`Advisor` 增强器也顺利地创建完毕。创建好的增强器将会在普通 bean 对象的初始化阶段中待命，等待织入的动作。

**代码清单 9-21　收集好的增强器汇总**

```java
// ......
if (aspectNames.isEmpty()) {
    return Collections.emptyList();
}
List<Advisor> advisors = new ArrayList<>();
for (String aspectName : aspectNames) {
    List<Advisor> cachedAdvisors = this.advisorsCache.get(aspectName);
    if (cachedAdvisors != null) {
        advisors.addAll(cachedAdvisors);
    } else {
        MetadataAwareAspectInstanceFactory factory = this.aspectFactoryCache.get(aspectName);
        advisors.addAll(this.advisorFactory.getAdvisors(factory));
    }
}
return advisors;
```

在整个方法最后打入 Debug 断点，待程序停在断点时可以发现 `Logger` 中的 5 个增强器都封装好了，如图 9-6 所示。

```
> ≡ aspectNames = {ArrayList@3944} size = 1
∨ ≡ advisors = {ArrayList@3945} size = 1
    ∨ ≡ 0 = {InstantiationModelAwarePointcutAdvisorImpl@3955} ... toString()
        > ƒ declaredPointcut = {AspectJExpressionPointcut@3956} ... toString()
        > ƒ declaringClass = {Class@3883} ... Navigate
          ƒ methodName = "beforePrint"
        > ƒ parameterTypes = {Class[0]@3958}
        > ƒ aspectJAdviceMethod = {Method@3959} ... toString()
        > ƒ aspectJAdvisorFactory = {ReflectiveAspectJAdvisorFactory@3946}
        > ƒ aspectInstanceFactory = {LazySingletonAspectInstanceFactoryDecorator@3960}
          ƒ declarationOrder = 0
        > ƒ aspectName = "serviceAspect"
        > ƒ pointcut = {AspectJExpressionPointcut@3956} ... toString()
          ƒ lazy = false
        > ƒ instantiatedAdvice = {AspectJMethodBeforeAdvice@3962} ... toString()
          ƒ isBeforeAdvice = null
          ƒ isAfterAdvice = null
```

图 9-6　封装好的增强器就是项目中定义好的

## 9.4　TargetSource 的设计

在 9.2.3 节中提到过一点，**AOP 的代理其实代理的不是目标对象本身，而是目标对象经过包装后的 `TargetSource` 对象**。Spring Framework 为什么要这样做？这样做有什么好处？本节来深入研究。

### 9.4.1 TargetSource 的设计

在 Java 原生的动态代理中，代理对象中直接组合了原始对象，可以通过图 9-7 直观地理解。

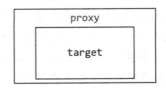

图 9-7 动态代理中 target 与 proxy 的关系

但是在 Spring Framework 的 AOP 中，代理对象并没有直接代理 target，而是给 target 加了一个"壳"，而这个"壳"就是 **TargetSource**，如图 9-8 所示。

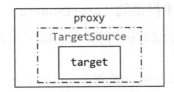

图 9-8 AOP 中 target 与 proxy 的关系

将图 9-7 与图 9-8 对比来看，想必读者可以更直观地理解两者的差异。TargetSource 可以看作目标对象 target 的一个包装、容器，原本代理对象在执行 `method.invoke(target,args)` 这样的逻辑时，要获取的是目标对象，但被 TargetSource 包装之后，就只能改用 `method.invoke(targetSource.getTarget(),args)` 进行方法的反射执行。

### 9.4.2 TargetSource 的好处

既然每次调用代理对象的方法时最终会调用 TargetSource 的 getTarget 方法，而 getTarget 方法决定了 TargetSource 如何返回目标对象，这就出现了可供扩展的切入点。举个最简单的例子：每次 getTarget 的值可以不一样，或者每次执行 getTarget 的时候可以从一个对象池中获取（读者是否突然想到了数据库连接池），其实 TargetSource 也有基于池的实现。

简单总结起来，让 AOP 代理 TargetSource 的好处是**可以控制每次方法调用时作用的具体对象实例，从而让方法的调用更加灵活**。

### 9.4.3 TargetSource 的结构

通过查看 TargetSource 的源码可以发现它是一个接口，如代码清单 9-22 所示。

**代码清单 9-22　TargetSource 本身是一个接口**

```
public interface TargetSource extends TargetClassAware {
    Class<?> getTargetClass();
    boolean isStatic();
    Object getTarget() throws Exception;
```

```
        void releaseTarget(Object target) throws Exception;
}
```

除了 getTarget 方法，TargetSource 接口还有一个 releaseTarget 方法，仅从方法名上就能很快地理解该方法的作用是交回/释放目标对象，也用于那些基于对象池的 TargetSource，在目标对象调用方法完成后紧接着调用 releaseTarget 方法释放目标对象。

另外 TargetSource 接口还有一个 isStatic 方法，对此读者可能会疑惑：如何理解 bean 对象的动态和静态之分？对于 Java 来讲，通过 Class 与对象的作用域可以区分出静态和非静态（Class 级的成员是静态的，对象级的成员是非静态的）。对于 Spring Framework 来讲，通过单实例 bean 对象与原型 bean 对象的作用域也可以区分出静态和非静态：单实例 bean 对象是一个 ApplicationContext 中只有一个实例，原型 bean 对象是每次获取都会得到一个全新的实例，所以单实例 bean 对象就为"静态 bean 对象"，原型 bean 对象则为非静态 bean 对象。

### 9.4.4　Spring Framework 中提供的 TargetSource

Spring Framework 中针对不同场景和不同需求，预设了几个 TargetSource 的实现，读者可以简单了解一下。

- SingletonTargetSource：每次调用 getTarget 方法都会返回同一个目标 bean 对象（与直接代理 target 无任何区别）。
- PrototypeTargetSource：每次调用 getTarget 方法都会从 BeanFactory 中创建一个全新的 bean 对象（被它包装的 bean 对象必须为原型 bean 对象）。
- CommonsPool2TargetSource：内部维护了一个对象池，每次调用 getTarget 方法时从对象池中提取出对象（底层使用 Apache 的 ObjectPool）。
- ThreadLocalTargetSource：每次调用 getTarget 方法都会从它所处的线程中提取目标对象（由于每个线程都有一个 TargetSource，因此被它包装的 bean 对象也必须是原型 bean 对象）。
- HotSwappableTargetSource：内部维护了一个可以热替换的目标对象引用，每次调用 getTarget 方法时都返回它（它提供了一个线程安全的 swap 方法，以热替换 TargetSource 中被代理的目标对象）。

以上设计在底层都不算复杂，感兴趣的读者可以自行借助 IDE 阅读源码，这里不再展开。

## 9.5　代理对象生成的核心：wrapIfNecessary

在 9.2.3 节中读者已经了解到，最终 bean 对象的初始化会被 BeanPostProcessor 的 postProcessAfterInitialization 方法处理，进入 AnnotationAwareAspectJAutoProxyCreator 的 postProcessAfterInitialization 方法中，它又要调用 wrapIfNecessary 方法来尝试创建代理，本节内容就深入探究 wrapIfNecessary 方法的脉络，如代码清单 9-23 所示。

**代码清单 9-23　wrapIfNecessary 决定是否生成代理对象**

```
protected Object wrapIfNecessary(Object bean, String beanName, Object cacheKey) {
    // 判断决定是否是不会被增强的 bean 对象
```

## 9.5 代理对象生成的核心：wrapIfNecessary

```java
    if (StringUtils.hasLength(beanName) && this.targetSourcedBeans.contains(beanName)) {
        return bean;
    }
    if (Boolean.FALSE.equals(this.advisedBeans.get(cacheKey))) {
        return bean;
    }
    if (isInfrastructureClass(bean.getClass()) || shouldSkip(bean.getClass(), beanName)) {
        this.advisedBeans.put(cacheKey, Boolean.FALSE);
        return bean;
    }

    // 如果上面的判断条件都不成立，则决定是否需要进行代理对象的创建
    Object[] specificInterceptors = getAdvicesAndAdvisorsForBean(bean.getClass(), beanName, null);
    if(specificInterceptors != DO_NOT_PROXY){
        this.advisedBeans.put(cacheKey,Boolean.TRUE);
        // 创建代理对象的动作
        // 注意此处它创建了一个 SingletonTargetSource，将 bean 对象包装起来了
        Object proxy = createProxy(bean.getClass(),
                beanName, specificInterceptors, new SingletonTargetSource(bean));
        this.proxyTypes.put(cacheKey, proxy.getClass());
        return proxy;
    }

    this.advisedBeans.put(cacheKey, Boolean.FALSE);
    return bean;
}
```

由代码清单 9-23 可以看出，整个 `wrapIfNecessary` 方法的主干逻辑是非常清晰的，读者可以简单梳理一遍，重点关注中间的两个重要步骤：1）获取增强器；2）创建代理对象。

### 9.5.1 getAdvicesAndAdvisorsForBean

仅从方法名上就可以理解 `getAdvicesAndAdvisorsForBean` 方法要完成的工作，该方法会根据当前正在初始化的 bean 对象，匹配可供织入通知的增强器，如代码清单 9-24 所示。这个方法又会调用下面的 `findEligibleAdvisors` 方法，而 `findEligibleAdvisors` 方法又分为 3 个步骤：获取增强器；筛选对当前 bean 对象有效的增强器；附加一些额外的增强器。虽然 `findEligibleAdvisors` 方法的篇幅不长，但其中调用的每个方法都是被封装过的，下面逐个展开。

**代码清单 9-24　getAdvicesAndAdvisorsForBean 获取可以切入逻辑的增强器**

```java
protected Object[] getAdvicesAndAdvisorsForBean(
        Class<?> beanClass, String beanName, @Nullable TargetSource targetSource) {

    List<Advisor> advisors = findEligibleAdvisors(beanClass, beanName);
    if (advisors.isEmpty()) {
        return DO_NOT_PROXY;
    }
    return advisors.toArray();
}

protected List<Advisor> findEligibleAdvisors(Class<?> beanClass, String beanName) {
```

```java
// 获取所有增强器（9.3.1 节）
List<Advisor> candidateAdvisors = findCandidateAdvisors();
// 筛选出可以切入当前 bean 对象的增强器
List<Advisor> eligibleAdvisors = findAdvisorsThatCanApply(candidateAdvisors,
        beanClass, beanName);
// 添加额外的增强器
extendAdvisors(eligibleAdvisors);
if (!eligibleAdvisors.isEmpty()) {
    // 增强器排序
    eligibleAdvisors = sortAdvisors(eligibleAdvisors);
}
return eligibleAdvisors;
```

#### 1. findCandidateAdvisors

findCandidateAdvisors 方法就是 9.3.1 节中刚研究过的收集增强器的逻辑，这个方法返回的结果即所谓的"候选增强器"。

#### 2. findAdvisorsThatCanApply

获取到所有候选增强器之后，下面需要匹配可以切入当前 bean 对象的增强器。进入 findAdvisorsThatCanApply 方法中可以发现该方法又将参数移交到工具类 AopUtils 的 findAdvisorsThatCanApply 方法中，如代码清单 9-25 所示。

**代码清单 9-25　findAdvisorsThatCanApply 匹配可以织入通知的增强器**

```java
protected List<Advisor> findAdvisorsThatCanApply(
        List<Advisor> candidateAdvisors, Class<?> beanClass, String beanName) {
    ProxyCreationContext.setCurrentProxiedBeanName(beanName);
    try {
        return AopUtils.findAdvisorsThatCanApply(candidateAdvisors, beanClass);
    } // finally ......
}

// AopUtils
public static List<Advisor> findAdvisorsThatCanApply(List<Advisor> candidateAdvisors,
        Class<?> clazz) {
    if (candidateAdvisors.isEmpty()) {
        return candidateAdvisors;
    }
    List<Advisor> eligibleAdvisors = new ArrayList<>();
    // 先匹配引介增强器
    for (Advisor candidate : candidateAdvisors){
        if (candidate instanceof IntroductionAdvisor && canApply(candidate, clazz)) {
            eligibleAdvisors.add(candidate);
        }
    }
    boolean hasIntroductions = !eligibleAdvisors.isEmpty();
    // 再匹配普通方法增强器
    for (Advisor candidate : candidateAdvisors) {
        if (candidate instanceof IntroductionAdvisor) {
            // already processed
            continue;
        }
        if (canApply(candidate, clazz, hasIntroductions)) {
```

```
                eligibleAdvisors.add(candidate);
            }
        }
        return eligibleAdvisors;
    }
```

整个 findAdvisorsThatCanApply 方法的逻辑分为两个部分且非常明确，前一部分是针对引介通知的增强器进行筛选，后一部分则是过滤普通的方法通知封装的增强器。有关引介增强器的部分本书不作过多探究，在实际的项目开发时几乎不会用到引介通知，读者可以把重点放在普通方法通知封装的增强器，观察源码如何进行匹配，如代码清单 9-26 所示。

**代码清单 9-26　canApply 匹配方法增强器的逻辑**

```java
public static boolean canApply(Advisor advisor, Class<?> targetClass, boolean hasIntroductions) {
    // 对于引介增强器，它会直接强转，使用类级的过滤器去匹配
    if (advisor instanceof IntroductionAdvisor) {
        return ((IntroductionAdvisor) advisor).getClassFilter().matches(targetClass);
    } else if (advisor instanceof PointcutAdvisor) {
        // 方法切入点的增强器匹配逻辑
        PointcutAdvisor pca = (PointcutAdvisor) advisor;
        return canApply(pca.getPointcut(), targetClass, hasIntroductions);
    } else {
        // 未知类型，默认可以使用
        return true;
    }
}

public static boolean canApply(Pointcut pc, Class<?> targetClass, boolean hasIntroductions) {
    Assert.notNull(pc, "Pointcut must not be null");
    // 如果切入点无法应用于当前类，则直接返回 false
    if (!pc.getClassFilter().matches(targetClass)) {
        return false;
    }

    MethodMatcher methodMatcher = pc.getMethodMatcher();
    // ......
    // 收集继承的父类和实现的接口（有可能切入点表达式切的是父类/接口）
    Set<Class<?>> classes = new LinkedHashSet<>();
    if (!Proxy.isProxyClass(targetClass)) {
        classes.add(ClassUtils.getUserClass(targetClass));
    }
    classes.addAll(ClassUtils.getAllInterfacesForClassAsSet(targetClass));

    for (Class<?> clazz : classes) {
        // 逐个判断每个方法能否被当前切入点表达式切入，能则立即返回 true
        Method[] methods = ReflectionUtils.getAllDeclaredMethods(clazz);
        for (Method method : methods) {
            if (introductionAwareMethodMatcher != null
                    ? introductionAwareMethodMatcher.matches(method, targetClass, hasIntroductions)
                    : methodMatcher.matches(method, targetClass)) {
                return true;
```

```
            }
        }
    }
    return false;
}
```

匹配的核心方法是 if 结构中的 canApply 方法，从方法的实现中可以看出，底层会针对不同类型的增强器分别采用不同的判断逻辑。对于方法增强器的判断会转调下面的重载 canApply 方法，而重载 canApply 方法中的整体思路比较清晰，各位要理解的核心部分是最下面的双重循环，利用 MethodMatcher 方法匹配器就能根据切入点表达式，判断出一个增强器是否能对当前 bean 对象进行增强。

实际测试时，在整个 findAdvisorsThatCanApply 方法的最后一行打入 Debug 断点，可以发现示例代码中编写的 AspectJ 通知已经被成功筛选出来，如图 9-9 所示。

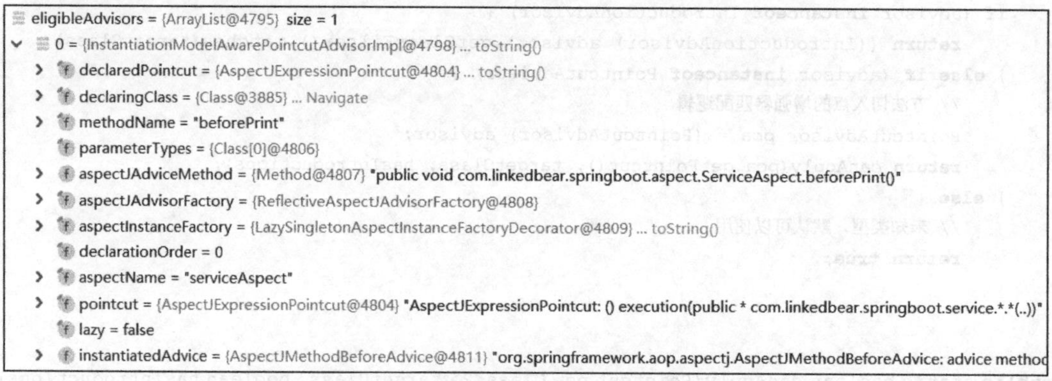

图 9-9　筛选后的可以应用于 DemoService 的增强器

### 3. extendAdvisors

接着上面的 Debug 继续进行，当执行完 extendAdvisors 方法后可以发现 eligibleAdvisors 集合中多了一个增强器，如图 9-10 所示。

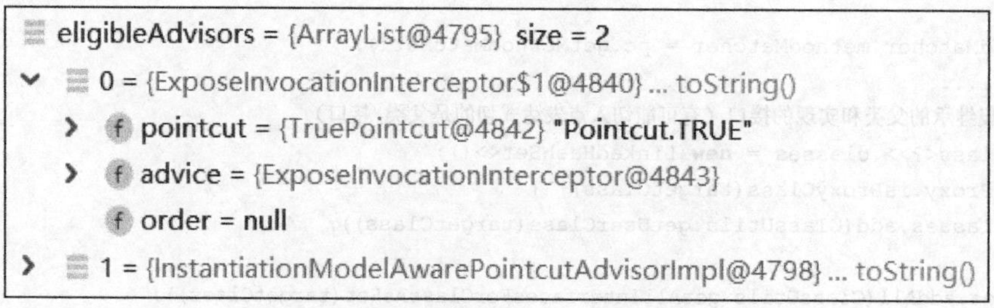

图 9-10　额外添加了一个增强器

为什么会添加额外的增强器呢？这需要深入 extendAdvisors 方法中一探究竟，如代码清单 9-27 所示。

**代码清单 9-27　extendAdvisors 添加额外的增强器**

```
protected void extendAdvisors(List<Advisor> candidateAdvisors) {
    AspectJProxyUtils.makeAdvisorChainAspectJCapableIfNecessary(candidateAdvisors);
```

```java
}

// AspectJProxyUtils
public static boolean makeAdvisorChainAspectJCapableIfNecessary(List<Advisor> advisors) {
    if (!advisors.isEmpty()) {
        boolean foundAspectJAdvice = false;
        for (Advisor advisor : advisors) {
            if (isAspectJAdvice(advisor)) {
                foundAspectJAdvice = true;
                break;
            }
        }
        // 如果发现有 AspectJ 封装的增强器，则添加一个 ExposeInvocationInterceptor
        if (foundAspectJAdvice && !advisors.contains(ExposeInvocationInterceptor.ADVISOR)){
            advisors.add(0, ExposeInvocationInterceptor.ADVISOR);
            return true;
        }
    }
    return false;
}
```

extendAdvisors 方法的逻辑不算复杂，关键是判断当前可用的增强器中是否存在 AspectJ 类型的增强器，如果有则会在整个增强器的列表最前端添加一个 ExposeInvocationInterceptor.ADVISOR。这个 ADVISOR 是什么？ExposeInvocationInterceptor 又是什么呢？读者也需要简单了解一下，如代码清单 9-28 所示。

**代码清单 9-28　ExposeInvocationInterceptor.ADVISOR**

```java
public static final ExposeInvocationInterceptor INSTANCE = new ExposeInvocationInterceptor();

public static final Advisor ADVISOR = new DefaultPointcutAdvisor(INSTANCE) {
    @Override
    public String toString() {
        return ExposeInvocationInterceptor.class.getName() +".ADVISOR";
    }
};
```

观察代码清单 9-28 中 ADVISOR 的初始化，可以发现它是一个对所有 bean 对象都生效的增强器，传入的 INSTANCE 很明显是单实例设计，核心的通知逻辑由 ExposeInvocationInterceptor 本身实现，继续往下阅读源码，如代码清单 9-29 所示。

**代码清单 9-29　ExposeInvocationInterceptor 的核心 invoke 方法**

```java
private static final ThreadLocal<MethodInvocation> invocation =
        new NamedThreadLocal<>("Current AOP method invocation");

public Object invoke(MethodInvocation mi) throws Throwable {
    MethodInvocation oldInvocation = invocation.get();
    invocation.set(mi);
    try {
        return mi.proceed();
    }
    finally {
        invocation.set(oldInvocation);
```

        }
    }

在 ExposeInvocationInterceptor 的核心 invoke 方法中，可以发现 ExposeInvocationInterceptor 的核心功能是向当前线程的 ThreadLocal 中放入当前正在执行的代理对象的方法执行包装，它每次都是增强器链中的第一个被执行的，并且放入 ThreadLocal 变量中，由此可见 ExposeInvocationInterceptor 的意图非常明显：可以让后面的增强器都获取当前正在执行的 MethodInvocation 对象。该设计通常是为 Spring Framework 内部使用，读者不必过多关注，仅作了解即可。

注意此处读者可能会产生疑惑：MethodInvocation oldInvocation = invocation.get();这行代码的设计意图是什么？正常情况下成员属性中 invocation 本身没有任何 MethodInvocation，为什么还要在方法的第一行代码中获取一次？其实这是由于 Spring Framework 考虑到多个代理对象之间互相调用的场景，如果是一个 AOP 代理对象调用了另一个 AOP 代理对象，整个过程在一个线程中执行，当第二个代理对象的方法被调用时，底层也会经过 ExposeInvocationInterceptor 的逻辑处理，此时从 ThreadLocal 中就可以提取出 MethodInvocation，以暂存上一个执行的 MethodInvocation。

### 9.5.2 createProxy

整个 getAdvicesAndAdvisorsForBean 方法执行完毕后，所有可以应用的增强器也都准备就绪，接下来是创建代理对象的部分。创建代理对象的整个方法逻辑比较长，如代码清单 9-30 所示（读者只需要关注源码中标有注释的部分）。

**代码清单 9-30　createProxy 创建代理对象**

```java
protected Object createProxy(Class<?> beanClass, @NullableString beanName,
        @Nullable Object[] specificInterceptors, TargetSource targetSource) {
    // ......
    // 代理工厂的初始化
    ProxyFactory proxyFactory = new ProxyFactory();
    proxyFactory.copyFrom(this);
    // 根据 AOP 的设计，决定是否强制使用 Cglib
    if (!proxyFactory.isProxyTargetClass()) {
        if (shouldProxyTargetClass(beanClass, beanName)) {
            // Cglib 动态代理直接记录被代理 bean 对象的所属类即可
            proxyFactory.setProxyTargetClass(true);
        } else {
            // 解析被代理 bean 对象所属类的所有实现的接口
            evaluateProxyInterfaces(beanClass, proxyFactory);
        }
    }

    // 构造整合所有增强器
    Advisor[] advisors = buildAdvisors(beanName, specificInterceptors);
    proxyFactory.addAdvisors(advisors);
    proxyFactory.setTargetSource(targetSource);
    customizeProxyFactory(proxyFactory);
    // ......
```

```
        // 创建代理对象
        return proxyFactory.getProxy(getProxyClassLoader());
}
```

从代理对象的构造流程来看,当前步骤其实是整个 wrapIfNecessary 方法的最后一步,该方法的核心动作有两步:**收集整理要织入目标对象的通知增强器;创建代理对象**。前面的细节处理相对比较简单,感兴趣的读者可以自行研究,下面重点讲解这两个核心动作。

### 1. buildAdvisors

buildAdvisors 方法中一开始整合 Spring Framework 原生 AOP 的 `MethodInterceptor` 部分,对于这部分,因为年代过于久远且主流开发中已不再使用,故本书不再展开讲解,读者需要重点关注的是下面 AspectJ 的部分,如代码清单 9-31 所示。有一个细节要注意,buildAdvisors 方法的参数中传进来的 `specificInterceptors` 就是收集好的增强器,中间部分会将这部分放入 `allInterceptors` 中。最下面有一个 `advisorAdapterRegistry.wrap` 的方法调用,这个方法内部会将"可以支持转换/包装为 Advisor 类型的对象"适配成 Advisor,所谓"可以支持的类型"有两种:`MethodInterceptor` 和 `AdvisorAdapter`。读者只需要了解,不必过度深入研究。

**代码清单 9-31 buildAdvisors 构造增强器**

```java
protected Advisor[] buildAdvisors(@Nullable String beanName,
        @Nullable Object[] specificInterceptors){
    // Handle prototypes correctly...
    // 此处是为适配 Spring 原生 AOP 的 MethodInterceptor,感兴趣的读者可自行研究
    Advisor[] commonInterceptors = resolveInterceptorNames();

    List<Object> allInterceptors = new ArrayList<>();
    if (specificInterceptors != null) {
        allInterceptors.addAll(Arrays.asList(specificInterceptors));
        // 组合原生的方法拦截器,共同作为 AOP 的通知织入
        if (commonInterceptors.length > 0) {
            if (this.applyCommonInterceptorsFirst) {
                allInterceptors.addAll(0, Arrays.asList(commonInterceptors));
            } else {
                allInterceptors.addAll(Arrays.asList(commonInterceptors));
            }
        }
    }
    // logger ......

    Advisor[] advisors = new Advisor[allInterceptors.size()];
    for (int i = 0; i < allInterceptors.size(); i++) {
        // 此处有一个原生 AOP 的适配动作
        advisors[i] = this.advisorAdapterRegistry.wrap(allInterceptors.get(i));
    }
    return advisors;
}
```

### 2. proxyFactory.getProxy

经历了增强器的构造后,下面是真正创建代理对象的逻辑,如代码清单 9-32 所示。这个方法又分为两个步骤,继续分解来看。

## 代码清单 9-32　getProxy 创建代理对象

```
public Object getProxy(@Nullable ClassLoader classLoader) {
    return createAopProxy().getProxy(classLoader);
}
```

**（1）createAopProxy**

创建 AOP 代理的方法如代码清单 9-33 所示。

## 代码清单 9-33　createAopProxy 创建 AOP 代理

```
protected final synchronized AopProxy createAopProxy() {
    // 监听器的通知动作
    // ......
    return getAopProxyFactory().createAopProxy(this);
}
```

createAopProxy 方法的上面有一个监听器的通知动作，由于这个动作涉及的监听器 AdvisedSupportListener 只在 ProxyCreatorSupport 这个类中使用，在日常开发中几乎不会触及，故对于该部分直接跳过。下面 getAopProxyFactory().createAopProxy(this); 方法的执行是关键，其中 getAopProxyFactory 方法返回的是当前 ProxyCreatorSupport 的成员 aopProxyFactory，借助 Debug 可以发现它的类型是 DefaultAopProxyFactory，如图 9-11 所示。

```
this = {ProxyFactory@2058} ... toString()
  aopProxyFactory = {DefaultAopProxyFactory@2062}
```

图 9-11　默认的 AopProxyFactory 类型

进入 DefaultAopProxyFactory 的 createAopProxy 方法中，可以发现之前在学习动态代理中读者熟悉的 JDK、Cglib 等，如代码清单 9-34 所示。至此可以了解到，AOP 底层使用 jdk 动态代理或者 Cglib 动态代理的选择动作在此处进行。

## 代码清单 9-34　根据配置与原始对象的情况选择动态代理方式

```
public AopProxy createAopProxy(AdvisedSupport config) throws AopConfigException {
    if (config.isOptimize() || config.isProxyTargetClass()
            || hasNoUserSuppliedProxyInterfaces(config)) {
        Class<?> targetClass = config.getTargetClass();
        if (targetClass == null) {
            // throw ex ......
        }
        // 如果要代理的本身就是接口，或者已经是被JDK动态代理的代理对象，则使用JDK动态代理
        if (targetClass.isInterface() || Proxy.isProxyClass(targetClass)) {
            return new JdkDynamicAopProxy(config);
        }
        // 否则使用Cglib动态代理
        return new ObjenesisCglibAopProxy(config);
    } else {
        return new JdkDynamicAopProxy(config);
    }
}
```

## （2）AopProxy#getProxy

创建完 `AopProxy` 对象后，下面是创建代理对象的动作。分别来看使用 JDK 动态代理和 Cglib 动态代理的底层创建动作。首先是基于 JDK 动态代理的 `JdkDynamicAopProxy`，如代码清单 9-35 所示。注意观察方法的最后一行，它调用的 API 就是读者熟悉的 `Proxy.newProxyInstance` 方法。所以读者应该体会到，代理对象的底层创建还是依赖 JDK 动态代理或者 Cglib 动态代理的核心 API。

**代码清单 9-35　getProxy 执行最终的代理对象创建方法**

```java
public Object getProxy(@Nullable ClassLoader classLoader) {
    // logger ......
    Class<?>[] proxiedInterfaces =
            AopProxyUtils.completeProxiedInterfaces(this.advised, true);
    findDefinedEqualsAndHashCodeMethods(proxiedInterfaces);
    // jdk动态代理的API
    return Proxy.newProxyInstance(classLoader, proxiedInterfaces, this);
}
```

至于 Cglib 动态代理的创建逻辑，这里不深入探究，底层源码中添加的额外处理比较多，不过核心的动作还是对 Cglib 中的 `Enhancer` 进行操作，如代码清单 9-36 所示。源码核心的思路不变，只是在框架上比普通开发者考虑得更多、更周全。

**代码清单 9-36　Cglib 创建代理对象的核心逻辑**

```java
public Object getProxy(@Nullable ClassLoader classLoader) {
    // ......
    // Configure CGLIB Enhancer
    Enhancer enhancer = createEnhancer();
    // 设置类加载器、父类、接口等信息 ......

    // 获取到的Callback即增加器
    Callback[] callbacks = getCallbacks(rootClass);
    // 对callback进行处理 ......

    // 调用enhancer.create创建代理
    return createProxyClassAndInstance(enhancer, callbacks);
}
```

至此，代理对象被成功创建，整个 AOP 通知织入的流程结束。随着 IOC 容器刷新完成，所有代理对象也全部创建完毕。

## 9.6　代理对象的底层执行逻辑

IOC 容器刷新完毕后，下一部分的内容会探讨程序运行期间，被 AOP 代理的对象执行被增强的方法时内部的调用机制。本节内容依然使用第 5 章中的示例代码，研究当 `DemoService` 的 `save` 方法执行时，内部执行了哪些重要环节的操作。

### 9.6.1　DemoService#save

将断点打在 `SpringBootAopApplication` 的 `save` 方法调用上，Debug 启动后可以发

现程序会先运行到 `CglibAopProxy` 的内部类 `DynamicAdvisedInterceptor` 中。由于执行的核心 `intercept` 方法很长，为体现方法内部的核心逻辑，代码清单 9-37 中已把不太重要的注释删掉，只留下重要的几个环节，读者在阅读时重点关注即可。

**代码清单 9-37　DynamicAdvisedInterceptor#intercept**

```java
public Object intercept(Object proxy, Method method, Object[] args, MethodProxy methodProxy)
        throws Throwable {
    Object oldProxy = null;
    boolean setProxyContext = false;
    Object target = null;
    TargetSource targetSource = this.advised.getTargetSource();
    try {
        // 如果在@EnableAspectJAutoProxy注解上配置了exposeProxy属性为true,
        // 则会把当前代理对象放入AOP上下文中
        if (this.advised.exposeProxy) {
            oldProxy = AopContext.setCurrentProxy(proxy);
            setProxyContext = true;
        }
        // 从 TargetSource 中提取出目标对象
        target = targetSource.getTarget();
        Class<?> targetClass =(target != null ? target.getClass() : null);
        // 根据当前执行的方法，获取要执行的增强器，并以列表返回（链的思想）
        List<Object> chain = this.advised.getInterceptorsAndDynamicInterceptionAdvice(
                method, targetClass);
        Object retVal;

        // 如果没有要执行的增强器，则直接执行目标方法
        if (chain.isEmpty() && Modifier.isPublic(method.getModifiers())) {
            Object[] argsToUse = AopProxyUtils.adaptArgumentsIfNecessary(method, args);
            retVal = methodProxy.invoke(target, argsToUse);
        } else {
            // 否则，构造增强器链，执行增强器的逻辑
            retVal = new CglibMethodInvocation(proxy, target, method, args,
                    targetClass, chain, methodProxy).proceed();
        }
        retVal = processReturnType(proxy, target, method, retVal);
        return retVal;
    }// finally ......
}
```

通读整段源码，提取两个核心步骤：获取增强器链、执行增强器（同时这里也发现了前面讲过的一些细节：`exposeProxy` 的实现、`TargetSource` 的使用等）。下面就这两个重要的核心方法分别讲解。

### 9.6.2　获取增强器链

获取增强器链的方法如代码清单 9-38 所示。

**代码清单 9-38　getInterceptorsAndDynamicInterceptionAdvice 获取增强器链**

```java
public List<Object> getInterceptorsAndDynamicInterceptionAdvice(Method method,
        @Nullable Class<?> targetClass) {
```

```java
MethodCacheKey cacheKey = new MethodCacheKey(method);
List<Object> cached = this.methodCache.get(cacheKey);
if (cached == null) {
    // 核心逻辑
    cached = this.advisorChainFactory.getInterceptorsAndDynamicInterceptionAdvice(
            this, method, targetClass);
    this.methodCache.put(cacheKey, cached);
}
return cached;
}
```

getInterceptorsAndDynamicInterceptionAdvice 方法中使用了非常经典的**缓存设计**，当方法第一次调用完成后，当前方法涉及的增强器链就会被缓存，后续再执行被增强的方法时无须再次获取和解析。而中间的核心逻辑会将获取增强器链的工作交给 advisorChainFactory 完成，继续向下深入。

**1. 前置准备**

在循环处理增强器之前，getInterceptorsAndDynamicInterceptionAdvice 方法的开始先进行一些基本的前置处理，首先初始化一个 AdvisorAdapterRegistry，意为**增强器适配器的注册器**，它的主要作用是将 AspectJ 类型的增强器转换为 MethodInterceptor（AOP 联盟的 MethodInterceptor）并返回，如代码清单 9-39 所示。这一处理的目的会在接下来的 CglibMethodInvocation 中得以体现，读者在此处先有印象即可。

**代码清单 9-39　advisorChainFactory 获取增强器链（1）**

```java
public List<Object> getInterceptorsAndDynamicInterceptionAdvice(
        Advised config, Method method, @Nullable Class<?> targetClass) {
    // 增强器适配器的注册器，它会根据增强器来解析，返回拦截器数组
    AdvisorAdapterRegistry registry = GlobalAdvisorAdapterRegistry.getInstance();
    Advisor[] advisors = config.getAdvisors();
    List<Object> interceptorList = new ArrayList<>(advisors.length);
    Class<?> actualClass =(targetClass != null ? targetClass : method.getDeclaringClass());
    Boolean hasIntroductions = null;
    // ......
```

**2. 匹配增强器**

> 💡 提示：接下来的源码部分缩进非常大，为确保阅读效果，下面几部分的源码缩进会被适当调整。

前置准备完成后开始循环匹配增强器，匹配逻辑本身不难，会把上面取出的增强器依次与当前正在调用的目标对象进行匹配，匹配的方式与之前一样，也是借助 MethodMatcher 进行，如代码清单 9-40 所示。

**代码清单 9-40　advisorChainFactory 获取增强器链（2）**

```java
// ......
for (Advisor advisor : advisors) {
    if (advisor instanceof PointcutAdvisor) {
        // 此处获取的就是 AspectJ 形式的通知方法封装
        PointcutAdvisor pointcutAdvisor = (PointcutAdvisor) advisor;
        if (config.isPreFiltered() ||
                pointcutAdvisor.getPointcut().getClassFilter().matches(actualClass)) {
```

```
// 根据通知方法上的切入点表达式，判断是否可以匹配当前要执行的目标对象所属类
MethodMatcher mm = pointcutAdvisor.getPointcut().getMethodMatcher();
boolean match;
// 引介匹配
if (mm instanceof IntroductionAwareMethodMatcher) {
  if (hasIntroductions == null) {
    hasIntroductions = hasMatchingIntroductions(advisors, actualClass);
  }
  match = ((IntroductionAwareMethodMatcher) mm)
      .matches(method, actualClass, hasIntroductions);
} else {
  // 方法匹配
  match = mm.matches(method, actualClass);
}
// ......
```

实际在 Debug 时可以发现，此处把 `ServiceAspect` 中声明的前置通知方法以及 Spring Framework 内置的 `ADVISOR` 增强器一起收集为一个 `Advisor` 数组，如图 9-12 所示。

```
▼ ≡ advisors = {Advisor[2]}@5447
  ▶ ≡ 0 = {ExposeInvocationInterceptor$1@5449} "org.springframework.aop.interceptor.ExposeInvocationInterceptor.ADVISOR"
  ▶ ≡ 1 = {InstantiationModelAwarePointcutAdvisorImpl@5450} "InstantiationModelAwarePointcutAdvisor: expression [execution(public * com.linkedbear.springboot.service.*.*(..))]"
```

图 9-12 DemoService 获取的增强器有两个

### 3. 匹配后的处理

匹配到增强器之后，接下来是决定如何封装为 `MethodInterceptor`，如代码清单 9-41 所示。这部分本身并不难，不过这段源码中有一个 runtime 的概念，这里解释一下。通常情况下，`MethodMatcher` 都是静态匹配器，但 Spring Framework 在此处做了一个设计，如果 `MethodMatcher` 被设置为动态匹配器，则每次调用匹配方法时，可以提前获取方法调用的参数值列表。这样解释可能有些难以理解，我们可以先看两个方法的签名，如代码清单 9-42 所示。

**代码清单 9-41　advisorChainFactory 获取增强器链（3）**

```
// ......
if (match) {
    MethodInterceptor[] interceptors = registry.getInterceptors(advisor);
    // runtime 的概念
    if (mm.isRuntime()) {
      for (MethodInterceptor interceptor : interceptors) {
        interceptorList.add(new InterceptorAndDynamicMethodMatcher(interceptor, mm));
      }
    } else {
      interceptorList.addAll(Arrays.asList(interceptors));
    }
  }
}
// ......
```

**代码清单 9-42　两个重载的 matches 方法签名**

```
boolean matches(Method method, Class<?> targetClass);
boolean matches(Method method, Class<?> targetClass, Object... args);
```

由上述两个方法的参数列表，想必读者应该可以理解上面的解释：静态匹配器只会做基本的方法匹配，而动态匹配器可以提前获取方法调用的参数值列表并进行深度匹配（注意是参数值）。具体的使用会在 9.6.3 节中得以体现。

**4．其他增强器的处理**

获取增强器链的最后部分还要处理引介增强器以及其他类型的增强器，如代码清单 9-43 所示。由于 AspectJ 风格的声明均执行上面的 PointcutAdvisor 判断逻辑，引介以及其他特殊类型的增强器在项目开发中一般不会触及，因此对于这部分本书不再展开，感兴趣的读者可以自行了解。

**代码清单 9-43　advisorChainFactory 获取增强器链（4）**

```java
// ......
else if (advisor instanceof IntroductionAdvisor) {
    // 处理引介增强器
    IntroductionAdvisor ia = (IntroductionAdvisor) advisor;
    if (config.isPreFiltered() || ia.getClassFilter().matches(actualClass)) {
        Interceptor[] interceptors = registry.getInterceptors(advisor);
        interceptorList.addAll(Arrays.asList(interceptors));
    }
} else {
    // 处理其他类型的增强器
    Interceptor[] interceptors = registry.getInterceptors(advisor);
    interceptorList.addAll(Arrays.asList(interceptors));
}
}

return interceptorList;
}
```

经过 getInterceptorsAndDynamicInterceptionAdvice 方法的处理后，当前目标对象要执行的方法被全部筛选出来，接下来的环节是构建方法执行器。

## 9.6.3　执行增强器

回到 DynamicAdvisedInterceptor 的 intercept 方法，下面的步骤是调用构造好的 CglibMethodInvocation 的 proceed 方法。不过在进入 proceed 方法之前，先请读者了解一下 CglibMethodInvocation 的设计，对应的类继承层次如代码清单 9-44 所示。

**代码清单 9-44　CglibMethodInvocation 的继承结构**

```java
private static class CglibMethodInvocation extends ReflectiveMethodInvocation
public class ReflectiveMethodInvocation implements ProxyMethodInvocation, Cloneable
public interface ProxyMethodInvocation extends MethodInvocation
```

从类继承层次上看，CglibMethodInvocation 最终是一个 AOP 联盟定义的 MethodInvocation，由于前面的步骤中已经完成了不同类型到 MethodInvocation 的适配，因此这里可以依次执行增强器链中的每个 MethodInvocation 对象。

简单了解之后，下面 Debug 往下执行 proceed 方法，在 CglibMethodInvocation 内部只是单纯地调用父类的 proceed 方法，而向上调用父类的 proceed 方法则会进入 ReflectiveMethodInvocation 类中，如代码清单 9-45 所示。

### 代码清单 9-45　CglibMethodInvocation 的 proceed 直接调用的父类

```
public Object proceed() throws Throwable {
    try {
        return super.proceed();
    }// catch ......
}
```

接下来的调用过程可能会很复杂，本节尝试用过程的方式记录每一步的执行动作和逻辑，以方便读者体会这里面的高深设计。

先简单通读 proceed 方法的设计，如代码清单 9-46 所示（重要的逻辑环节在代码中均已标有注释）。

### 代码清单 9-46　ReflectiveMethodInvocation 的 proceed 方法执行拦截器链

```
protected final List<?> interceptorsAndDynamicMethodMatchers;
private int currentInterceptorIndex = -1;

public Object proceed() throws Throwable {
    if (this.currentInterceptorIndex == this.interceptorsAndDynamicMethodMatchers.size() - 1) {
        // 增强器全部执行完毕后，会执行目标方法
        return invokeJoinpoint();
    }

    // 依次取出增强器封装的拦截器并执行
    Object interceptorOrInterceptionAdvice =
        this.interceptorsAndDynamicMethodMatchers.get(++this.currentInterceptorIndex);
    if (interceptorOrInterceptionAdvice instanceof InterceptorAndDynamicMethodMatcher){
        // 此处是动态匹配器构造的特殊逻辑
        InterceptorAndDynamicMethodMatcher dm =
            (InterceptorAndDynamicMethodMatcher) interceptorOrInterceptionAdvice;
        Class<?> targetClass = (this.targetClass != null
            ? this.targetClass : this.method.getDeclaringClass());
        // 动态匹配器必须将参数一并纳入匹配规则中
        if (dm.methodMatcher.matches(this.method, targetClass,this.arguments)){
            return dm.interceptor.invoke(this);
        } else {
            return proceed();
        }
    } else {
        // 此处会调用增强器的逻辑
        return ((MethodInterceptor) interceptorOrInterceptionAdvice).invoke(this);
    }
}
```

这里额外解释一下 9.6.2 节中提到的动态匹配器。观察中间的 if 结构部分，如果 MethodMatcher 被封装为动态匹配器，则此处会调用 dm.methodMatcher.matches 方法继续匹配，如果动态匹配器在匹配方法调用的参数值列表时发现匹配不上，则这个增强器不会执行。所以上面提到的动态匹配器会深度匹配方法的参数，作用时机就在这里。

其余的逻辑部分，结合 Debug 的效果一一来看。

### 1. 执行 proceed 方法

第一次进入 proceed 方法，首先执行第一个 if 判断，这里它需要比对当前已经执行过的增强器在增强器链的下标位置，如代码清单 9-47 所示。

**代码清单 9-47　判断 currentInterceptorIndex 与当前执行增强器链下标**

```
public Object proceed() throws Throwable {
    if (this.currentInterceptorIndex == this.interceptorsAndDynamicMethodMatchers.size() - 1) {
        return invokeJoinpoint();
    }
```

此时 Debug 可以发现 currentInterceptorIndex 值为-1，如图 9-13 所示。

```
oo this.currentInterceptorIndex = -1
oo this.interceptorsAndDynamicMethodMatchers = {ArrayList@5321} size = 2
  > ≡ 0 = {ExposeInvocationInterceptor@5327}
  ∨ ≡ 1 = {MethodBeforeAdviceInterceptor@5328}
      ∨ (f) advice = {AspectJMethodBeforeAdvice@5329} ... toString()
          > (f) declaringClass = {Class@3883} ... Navigate
            (f) methodName = "beforePrint"
            (f) parameterTypes = {Class[0]@5331}
          > (f) aspectJAdviceMethod = {Method@5332} ... toString()
          > (f) pointcut = {AspectJExpressionPointcut@5333} ... toString()
          > (f) aspectInstanceFactory = {LazySingletonAspectInstanceFactoryDec
            (f) aspectName = "serviceAspect"
```

图 9-13　第一次 Debug 至 proceed 方法时索引下标为-1

判断-1≠2-1，因此不进入 invokeJoinpoint 方法，继续往下执行。

### 2. 下标值++

接下来执行下面的 this.interceptorsAndDynamicMethodMatchers.get 动作，如代码清单 9-48 所示。注意此时 this.currentInterceptorIndex 变量执行了一次自增操作，自增后的 currentInterceptorIndex 值为 0。

**代码清单 9-48　currentInterceptorIndex 自增后取出增强器**

```
    if (this.currentInterceptorIndex == this.interceptorsAndDynamicMethodMatchers.size() - 1) {
        return invokeJoinpoint();
    }
    Object interceptorOrInterceptionAdvice =
            this.interceptorsAndDynamicMethodMatchers.get(++this.currentInterceptorIndex);
    // ......
```

### 3. 执行第一个增强器

从 9.5.1 节的内容以及图 9-13 中可以了解到，第一次获取到的增强器的类型是 ExposeInvocationInterceptor.ADVISOR，此处会执行它对应的通知逻辑，而 ExposeInvocationInterceptor 内部的 invoke 方法可以妥善处理好多个代理对象之间互相调用的问题，此处不讨论有关话题，直接进入 MethodInvocation 的 proceed 方法，如代码清单 9-49 所示。

### 代码清单9-49　ExposeInvocationInterceptor 的执行源码

```
public Object invoke(MethodInvocation mi) throws Throwable {
    MethodInvocation oldInvocation = invocation.get();
    invocation.set(mi);
    try {
        return mi.proceed();
    } finally {
        invocation.set(oldInvocation);
    }
}
```

#### 4. 继续执行 proceed 方法

保存好 `MethodInvocation` 后继续向下执行 proceed 方法，Debug 执行后发现程序回到了上面的第一步，不同的是此时 `currentInterceptorIndex` 值不再是–1，而是 0，如图 9-14 所示。

```
∞ this.currentInterceptorIndex = 0
∞ this.interceptorsAndDynamicMethodMatchers = {ArrayList@5321} size = 2
  > ≡ 0 = {ExposeInvocationInterceptor@5327}
  ∨ ≡ 1 = {MethodBeforeAdviceInterceptor@5328}
      ∨ 🛈 advice = {AspectJMethodBeforeAdvice@5329} ... toString()
          > 🛈 declaringClass = {Class@3883} ... Navigate
          > 🛈 methodName = "beforePrint"
```

图 9-14　第二次 Debug 至 proceed 方法时索引下标为 0

判断 0≠2-1，因此不进入 `invokeJoinpoint` 方法，继续向下执行增强器的逻辑。

#### 5. 进入 MethodBeforeAdviceInterceptor 中

下一个要执行的通知方法是 `ServiceAspect` 的 `beforePrint` 方法，它本身是一个前置通知，具体到 `MethodInterceptor` 的实现类在 `MethodBeforeAdviceInterceptor` 中。具体的底层实现非常简单，`MethodBeforeAdviceInterceptor` 会直接调用 advice 的 before 方法，而追踪调用的底层方法，则会一直调用到 `invokeAdviceMethodWithGivenArgs` 方法中，最终借助反射调用 Aspect 切面类的通知方法，如代码清单 9-50 所示。通知方法执行完成后，会继续执行 `MethodInvocation` 的 proceed 方法，推动增强器链的执行流程。

### 代码清单9-50　MethodBeforeAdviceInterceptor 的执行源码

```
public Object invoke(MethodInvocation mi) throws Throwable {
    this.advice.before(mi.getMethod(), mi.getArguments(), mi.getThis());
    return mi.proceed();
}

public void before(Method method, Object[] args, @NullableObject target) throws Throwable {
    invokeAdviceMethod(getJoinPointMatch(), null, null);
}

protected Object invokeAdviceMethod(@Nullable JoinPointMatch jpMatch,
        @Nullable Object returnValue, @Nullable Throwable ex) throws Throwable {
    return invokeAdviceMethodWithGivenArgs(argBinding(getJoinPoint(),
        jpMatch, returnValue, ex));
}
```

```
}
protected Object invokeAdviceMethodWithGivenArgs(Object[] args) throws Throwable {
    Object[] actualArgs = args;
    if (this.aspectJAdviceMethod.getParameterCount() == 0) {
        actualArgs = null;
    }
    try {
        // 反射执行通知方法
        ReflectionUtils.makeAccessible(this.aspectJAdviceMethod);
        return this.aspectJAdviceMethod.invoke(
            this.aspectInstanceFactory.getAspectInstance(), actualArgs);
    } // catch throw ex ......
}
```

#### 6. 执行目标对象方法

`MethodBeforeAdviceInterceptor` 执行完毕后，又回到 proceed 方法，此时 currentInterceptorIndex 值为 1，如图 9-15 所示。

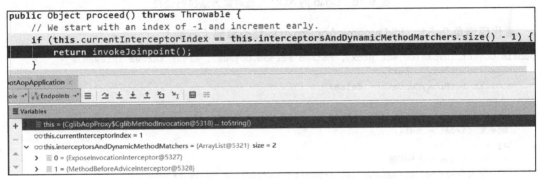

图 9-15 第三次 Debug 至 proceed 方法时索引下标为 1，进入目标方法

判断 1=2-1，因此进入内部的 `invokeJoinpoint` 方法，而 `invokeJoinpoint` 方法会借助反射执行目标对象的目标方法。

目标方法执行完毕后，一个代理对象的方法调用流程结束。

#### 7. 流程小结

经过一轮代理对象的方法调用全流程后，可以总结出代理对象的底层执行算法逻辑：**利用一个全局索引值决定每次执行的拦截器，当所有拦截器都执行完时，索引值刚好等于 `size()-1`，此时就可以执行真正的目标方法。**

最后使用一个执行流程图来更好地理解这段逻辑，如图 9-16 所示。

### 9.6.4 JDK 动态代理的执行底层

下面继续研究 JDK 动态代理的执行。要想在 Spring Boot 2.x 中激活 JDK 的动态代理，必须在 application.properties 中**显式配置 spring.aop.proxy-target-class=false**，才会禁用默认全 Cglib 的配置，即激活 JDK 动态代理的方式。执行 JDK 动态代理的底层 InvocationHandler 是 `JdkDynamicAopProxy`，它的核心 invoke 方法很长，如代码清单 9-51 所示，读者只需关注标注了注释的核心源码。

# 第 9 章　AOP 模块的生命周期

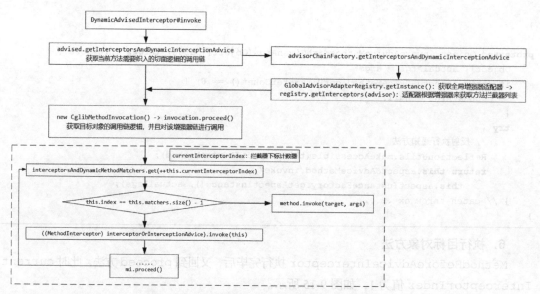

图 9-16　Cglib 的 AOP 代理对象方法执行流程

## 代码清单 9-51　JdkDynamicAopProxy 的核心源码（节选）

```
public Object invoke(Object proxy, Method method,Object[] args) throws Throwable {
    Object oldProxy = null;
    boolean setProxyContext = false;

    TargetSource targetSource = this.advised.targetSource;
    Object target = null;
    try {
        // equals 方法不代理
        if (!this.equalsDefined && AopUtils.isEqualsMethod(method)) {
            return equals(args[0]);
        } else if (!this.hashCodeDefined && AopUtils.isHashCodeMethod(method)) {
            // hashCode 方法不代理
            return hashCode();
        } else if (method.getDeclaringClass() == DecoratingProxy.class) {
            // 方法来自 DecoratingProxy 接口的也不代理
            return AopProxyUtils.ultimateTargetClass(this.advised);
        } else if (!this.advised.opaque && method.getDeclaringClass().isInterface()
                && method.getDeclaringClass().isAssignableFrom(Advised.class)) {
            // 目标对象本身实现了 Advised 接口的也不代理
            return AopUtils.invokeJoinpointUsingReflection(this.advised, method, args);
        }

        Object retVal;
        if (this.advised.exposeProxy) {
            oldProxy = AopContext.setCurrentProxy(proxy);
            setProxyContext = true;
        }

        target = targetSource.getTarget();
        Class<?> targetClass =(target != null? target.getClass() : null);

        // 根据当前执行的方法获取要执行的增强器，并以列表返回（链的思想）
```

```
        List<Object> chain = this.advised.getInterceptorsAndDynamicInterceptionAdvice(
                method, targetClass);
        if (chain.isEmpty()) {
            Object[] argsToUse = AopProxyUtils.adaptArgumentsIfNecessary(method, args);
            retVal = AopUtils.invokeJoinpointUsingReflection(target, method, argsToUse);
        } else {
            MethodInvocation invocation = new ReflectiveMethodInvocation(proxy, target,
                    method, args, targetClass, chain);
            // 构造增强器链，执行增强器的逻辑
            retVal = invocation.proceed();
        }
        // 返回值的处理 ......
        return retVal;
    } // finally ......
}
```

先看下面的部分，读者可以发现 JdkDynamicAopProxy 跟 CglibAopProxy 的实现逻辑完全一致，两者在底层实现中也都一样，包括创建的方法执行器对象直接是 ReflectiveMethodInvocation，与上面的父类是同一个。由此可以推断，JDK 动态代理的核心底层执行逻辑也大同小异，感兴趣的读者可以自行编写接口+实现类的方式，配合 Debug 加以体会，本节只提供一个类似的执行流程图供读者参考，如图 9-17 所示。

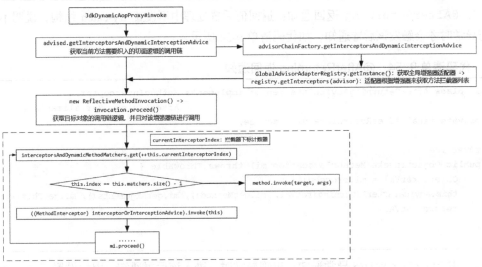

图 9-17　JDK 的 AOP 代理对象方法执行流程

### 9.6.5　AspectJ 中通知的底层实现

最后还需要读者了解 AspectJ 中提到的 5 种声明式的通知方法的底层执行逻辑。

1. @Before 前置通知：先执行前置通知，再执行目标方法，如代码清单 9-52 所示。

**代码清单 9-52　@Before 前置通知**

```
public class MethodBeforeAdviceInterceptor implements MethodInterceptor, Serializable {
    private MethodBeforeAdvice advice;

    @Override
```

```java
public Object invoke(MethodInvocation mi) throws Throwable {
    this.advice.before(mi.getMethod(), mi.getArguments(), mi.getThis());
    return mi.proceed();
}
```

2. @After 后置通知：执行目标方法后，在 finally 中执行后置方法（由此也说明了它的通知时机更靠后），如代码清单 9-53 所示。

### 代码清单 9-53　@After 后置通知

```java
public class AspectJAfterAdvice extends AbstractAspectJAdvice
        implements MethodInterceptor, AfterAdvice, Serializable {

    @Override
    public Object invoke(MethodInvocation mi) throws Throwable {
        try {
            return mi.proceed();
        } finally {
            invokeAdviceMethod(getJoinPointMatch(), null, null);
        }
    }
}
```

3. @AfterReturning 返回通知：返回值后置处理中不设置 try-catch 结构，说明不出现任何异常时才会触发该后置通知，如代码清单 9-54 所示。

### 代码清单 9-54　@AfterReturning 返回通知

```java
public class AfterReturningAdviceInterceptor implements MethodInterceptor,
                                                AfterAdvice, Serializable {
    private final AfterReturningAdvice advice;

    @Override
    public Object invoke(MethodInvocation mi) throws Throwable {
        Object retVal = mi.proceed();
        this.advice.afterReturning(retVal, mi.getMethod(), mi.getArguments(), mi.getThis());
        return retVal;
    }
}
```

4. @AfterThrowing 异常通知：出现异常时，进入该后置通知，因为设置了 try-catch 结构，所以这里 catch 中根据是否标注了异常通知进行相应的后置处理，如代码清单 9-55 所示。

### 代码清单 9-55　@AfterThrowing 异常通知

```java
public class AspectJAfterThrowingAdvice extends AbstractAspectJAdvice
        implements MethodInterceptor, AfterAdvice, Serializable {

    @Override
    public Object invoke(MethodInvocation mi) throws Throwable {
        try {
            return mi.proceed();
        } catch (Throwable ex) {
            if (shouldInvokeOnThrowing(ex)) {
```

```
            invokeAdviceMethod(getJoinPointMatch(), null, ex);
        }
        throw ex;
    }
  }
}
```

5. `@Around` 环绕通知：通过构建 `ProceedingJoinPoint` 对象，直接以参数形式传入通知方法的方法参数中（如果需要的话），反射执行通知方法，如代码清单 9-56 所示。

**代码清单 9-56　@Around 环绕通知**

```
public class AspectJAroundAdvice extends AbstractAspectJAdvice
        implements MethodInterceptor, Serializable {
    // ......
    public Object invoke(MethodInvocation mi) throws Throwable {
        if (!(mi instanceof ProxyMethodInvocation)) {
            throw new IllegalStateException("MethodInvocation is not a Spring ProxyMethod
Invocation: "+ mi);
        }
        ProxyMethodInvocation pmi = (ProxyMethodInvocation) mi;
        ProceedingJoinPoint pjp = lazyGetProceedingJoinPoint(pmi);
        JoinPointMatch jpm = getJoinPointMatch(pmi);
        return invokeAdviceMethod(pjp, jpm, null, null);
    }
}
```

## 9.7　AOP 通知的执行顺序对比

本章的最后一节我们再扩展一个知识点，即不同的 Spring Framework 版本中 AOP 通知的执行顺序对比。由于 Spring Boot 2.x 底层基于 Spring Framework 5.x，而 Spring Boot 1.x 底层基于 Spring Framework 4.x，不同的大版本之间 AOP 通知的顺序是有差异的，本节通过一个示例来探究该差异。

### 9.7.1　测试代码编写

为了能测试出尽可能多的通知执行顺序差别，在编写 Aspect 切面类时，需要覆盖所有类型的通知，如代码清单 9-57 所示。

**代码清单 9-57　覆盖全部类型通知的切面类 ServiceAspect**

```
@Aspect
@Component
public class ServiceAspect {

    @Before("execution(public * com.linkedbear.springboot.service.*.*(..))")
    public void beforePrint() {
        System.out.println("Service Aspect before advice run ......");
    }
```

```java
@After("execution(public * com.linkedbear.springboot.service.*.*(..))")
public void afterPrint(){
    System.out.println("Service Aspect after advice run ......");
}

@AfterReturning("execution(public * com.linkedbear.springboot.service.*.*(..))")
public void afterReturningPrint() {
    System.out.println("Service Aspect afterReturning advice run ......");
}

@AfterThrowing("execution(public * com.linkedbear.springboot.service.*.*(..))")
public void afterThrowingPrint() {
    System.out.println("Service Aspect afterThrowing advice run ......");
}

@Around("execution(public * com.linkedbear.springboot.service.*.*(..))")
public Object aroundPrint(ProceedingJoinPoint joinPoint) throws Throwable {
    System.out.println("Service Aspect around before run ......");
    Object ret = joinPoint.proceed();
    System.out.println("Service Aspect around after run ......");
    return ret;
}
```

Spring Boot 主启动类中,需要在 `SpringApplication.run` 方法后接收 IOC 容器本身,获取 `DemoService` 对象并调用其 `save` 方法,以触发所有可以织入 `DemoService` 的 AOP 切面通知逻辑,如代码清单 9-58 所示。

### 代码清单 9-58　主启动类

```java
@SpringBootApplication
@EnableAspectJAutoProxy
public class SpringBootAopApplication {

    public static void main(String[] args) {
        ApplicationContext ctx = SpringApplication.run(SpringBootAopApplication.class, args);
        ctx.getBean(DemoService.class).save();
    }
}
```

准备工作完成后,下面开始测试。

### 9.7.2　Spring Framework 5.x 的顺序

运行 `SpringBootAopApplication` 的 `main` 方法,观察控制台的输出如下所示。

```
Service Aspect around before run ......
Service Aspect before advice run ......
DemoService save run ......
Service Aspect afterReturning advice run ......
Service Aspect after advice run ......
Service Aspect around after run ......
```

由输出结果可以总结出 Spring Framework 5.x 中通知的执行顺序：
1. 环绕通知中 `joinPoint.proceed();` 之前的逻辑；
2. 前置通知；
3. 目标对象的目标方法；
4. 返回通知；
5. 后置通知；
6. 环绕通知中 `joinPoint.proceed();` 之后的逻辑。

### 9.7.3　Spring Framework 4.x 的顺序

要测试基于 Spring Framework 4.x 的 AOP 顺序，需要改变当前项目依赖的 Spring Boot 版本。Spring Boot 1.x 的底层依赖 Spring Framework 4.x，所以只需将 pom.xml 中 `spring-boot-starter-parent` 的版本调整为 `1.5.22.RELEASE`，如代码清单 9-59 所示。

**代码清单 9-59　调整 pom.xml**

```xml
<parent>
    <groupId>org.springframework.boot</groupId>
    <artifactId>spring-boot-starter-parent</artifactId>
    <version>1.5.22.RELEASE</version>
</parent>
<!-- ... -->
<dependencies>
    <dependency>
        <groupId>org.springframework.boot</groupId>
        <artifactId>spring-boot-starter-aop</artifactId>
        <version>1.5.22.RELEASE</version>
    </dependency>
</dependencies>
```

修改 Spring Boot 版本后，不需要修改其他任何代码，等待 IDE 重新导入依赖后，再次运行 `SpringBootAopApplication` 的 main 方法即可，控制台输出的切面执行顺序如下所示。

```
Service Aspect around before run ......
Service Aspect before advice run ......
DemoService save run ......
Service Aspect around after run ......
Service Aspect after advice run ......
Service Aspect afterReturning advice run ......
```

由输出结果可以总结出 Spring Framework 4.x 中通知的执行顺序：
1. 环绕通知中 `joinPoint.proceed();` 之前的逻辑；
2. 前置通知；
3. 目标对象的目标方法；
4. **环绕通知中 `joinPoint.proceed();` 之后的逻辑；**
5. 后置通知；
6. 返回通知。

重点总结：Spring Framework 5.x 与 4.x 中 AOP 通知顺序的不同之处在于**环绕通知可以覆盖**

的范围。Spring Framework 5.x 中环绕通知会包含其他所有通知，而 Spring Framework 4.x 中环绕通知只会包含前置通知，后置通知和返回通知会在环绕通知执行完成后执行。

## 9.8 小结

本章从 AOP 的开启注解 @EnableAspectJAutoProxy 出发，详细探讨注解驱动 AOP 底层的支撑组件，并针对核心组件 AnnotationAwareAspectJAutoProxyCreator 的初始化时机、作用机制进行深入剖析。AOP 代理对象的工作离不开增强器、代理对象执行链等核心组件的支撑，掌握核心组件的工作时机和作用机制可以更容易地理解 AOP 的工作原理。

# 第 3 部分
# Spring Boot 整合常用开发场景

▶ 第 10 章　Spring Boot 整合 JDBC

▶ 第 11 章　Spring Boot 整合 MyBatis

▶ 第 12 章　Spring Boot 整合 WebMvc

▶ 第 13 章　Spring Boot 整合 WebFlux

# 第 10 章 Spring Boot 整合 JDBC

**本章主要内容：**
- ◇ Spring Boot 整合 JDBC 的核心自动装配内容；
- ◇ 声明式事务的生效原理；
- ◇ 声明式事务的控制原理；
- ◇ 声明式事务的事务传播行为原理。

在日常项目开发中，用到 Spring Boot 的场景通常都离不开与数据层的交互，而交互的基础组件是来自 Spring Framework 的 `spring-jdbc` 包。本章重点研究的内容是 Spring Boot 整合 JDBC 后，生效的自动装配中做了哪些核心工作，以及引入的事务模块对于注解声明式事务是如何生效、控制的。

## 10.1 Spring Boot 整合 JDBC 项目搭建

首先搭建一个最基础的 Spring Boot 整合 JDBC 的项目。Spring Boot 整合原生 JDBC 的步骤很简单，导入 `spring-boot-starter-jdbc` 以及连接数据库的驱动即可完成最基本的整合。

### 10.1.1 初始化数据库

本书选择使用 MySQL 作为演示用数据库。首先创建一个新的数据库和一张用于演示的表 `tbl_user`，对应的 SQL 脚本如代码清单 10-1 所示。

**代码清单 10-1　初始化数据库的 SQL 脚本**

```sql
CREATE DATABASE springboot-dao CHARACTER SET 'utf8mb4';

CREATE TABLE tbl_user(
    id int(11) NOT NULL AUTO_INCREMENT,
    name varchar(20) NOT NULL,
    tel varchar(20) NULL,
     PRIMARY KEY (id)
);
```

> 提示：为了便于演示和调试代码，对于数据库部分建议读者从简设计，切勿本末倒置。

## 10.1.2 整合项目

整合 JDBC 模块以及编写对应的配置都非常简单,如代码清单 10-2 ~ 代码清单 10-4 所示,不过多讲解。

**代码清单 10-2　导入整合 JDBC 的依赖**

```xml
<dependencies>
    <dependency>
        <groupId>org.springframework.boot</groupId>
        <artifactId>spring-boot-starter-jdbc</artifactId>
    </dependency>
    <dependency>
        <groupId>mysql</groupId>
        <artifactId>mysql-connector-java</artifactId>
        <version>5.1.47</version>
    </dependency>
</dependencies>
```

**代码清单 10-3　配置数据源**

```
spring.datasource.driver-class-name=com.mysql.jdbc.Driver
spring.datasource.url=jdbc:mysql://localhost:3306/spring-dao?characterEncoding=utf8
spring.datasource.username=root
spring.datasource.password=123456
```

**代码清单 10-4　开启注解式声明式事务**

```java
@SpringBootApplication
@EnableTransactionManagement
public class SpringBootJdbcApplication {

    public static void main(String[] args) {
        SpringApplication.run(SpringBootJdbcApplication.class, args);
    }
}
```

## 10.1.3 编写测试代码

整合项目完毕后,下一步可以编写相应的实体类、Dao 层和 Service 层代码,本节仅快速实现几个简单方法,如代码清单 10-5 ~ 代码清单 10-7 所示。

**代码清单 10-5　与 tbl_user 对应的实体类 User**

```java
public class User {
    private Integer id;
    private String name;
    private String tel;
    // getter setter toString ......
}
```

**代码清单 10-6　数据访问层 UserDao**

```
@Repository
```

```java
public class UserDao {

    @Autowired
    private JdbcTemplate jdbcTemplate;

    public void save(User user) {
        jdbcTemplate.update("insert into tbl_user (name, tel) values (?, ?)",
                user.getName(), user.getTel());
    }
    public List<User> findAll() {
        return jdbcTemplate.query("select * from tbl_user",
                BeanPropertyRowMapper.newInstance(User.class));
    }
}
```

**代码清单 10-7　业务层 UserService**

```java
@Service
public class UserService {

    @Autowired
    private UserDao userDao;

    @Transactional(rollbackFor = Exception.class)
    public void test() {
        User user = new User();
        user.setName("test dao");
        user.setTel("1234567");
        userDao.save(user);
        List<User> userList = userDao.findAll();
        userList.forEach(System.out::println);
    }
}
```

要想测试整合是否成功，可以在主启动类中获取 IOC 容器，并提取出 UserService 类调用其 test 方法，测试是否可以成功保存并打印，如代码清单 10-8 所示。

**代码清单 10-8　通过 SpringApplication 获取到 IOC 容器后测试整合**

```java
public static void main(String[] args) {
    ApplicationContext ctx = SpringApplication.run(SpringBootJdbcApplication.class, args);
    UserService userService = ctx.getBean(UserService.class);
    userService.test();
}
```

运行主启动类，控制台可以正确打印出一条用户信息，证明 Spring Boot 整合 JDBC 场景顺利完成。

```
User{id=1, name='test dao', tel='1234567'}
```

## 10.2　整合 JDBC 后的自动装配

对于原生 JDBC 的整合后，主启动类中并没有声明与之相关的注解，所以有关 JDBC 的组件装配都是以自动配置类的方式实现。借助 IDE 通过 `spring-boot-autoconfigure` 依赖的 `spring.factories` 文件可以找到有关 JDBC 的自动配置类，如代码清单 10-9 所示（为确保阅读体验，全限定名已经过简化处理）。

**代码清单 10-9　有关 JDBC 的自动配置类**

```
o.s.b.autoconfigure.EnableAutoConfiguration=\
    o.s.b.a.jdbc.DataSourceAutoConfiguration,\
    o.s.b.a.jdbc.JdbcTemplateAutoConfiguration,\
    o.s.b.a.jdbc.JndiDataSourceAutoConfiguration,\
    o.s.b.a.jdbc.XADataSourceAutoConfiguration,\
    o.s.b.a.jdbc.DataSourceTransactionManagerAutoConfiguration,
......
```

可以发现 Spring Boot 默认支持的自动配置包含数据源、JdbcTemplate、事务管理器，以及对 JNDI 和 XA 协议的支持。本书只对日常开发中最常用的场景进行讲解，对于其余部分感兴趣的读者可以自行探究。

### 10.2.1　配置数据源

Spring Boot 除了拥有最原始的 `DriverManagerDataSource`，还默认支持 Hikari、DBCP、EmbeddedTomcat 等数据库连接池，Spring Boot 2.x 在整合 JDBC 场景时，默认会导入 HikariCP 数据库连接池，所以在 `DataSourceAutoConfiguration` 的静态内部类中 `PooledDataSourceConfiguration` 会生效，并导入 `DataSourceConfiguration` 下的一些子配置类，这些配置类同时只会有一个生效。代码清单 10-10 中列举了默认场景下激活 HikariCP 的关键源码。

**代码清单 10-10　数据库连接池的配置激活**

```
@Configuration(proxyBeanMethods = false)
@Conditional(PooledDataSourceCondition.class)
@ConditionalOnMissingBean({ DataSource.class, XADataSource.class })
@Import({DataSourceConfiguration.Hikari.class, DataSourceConfiguration.Tomcat.class,
        DataSourceConfiguration.Dbcp2.class, DataSourceConfiguration.Generic.class,
        DataSourceJmxConfiguration.class})
protected static class PooledDataSourceConfiguration {
}

@Configuration(proxyBeanMethods = false)
@ConditionalOnClass(HikariDataSource.class)
@ConditionalOnMissingBean(DataSource.class)
@ConditionalOnProperty(name = "spring.datasource.type",
        havingValue = "com.zaxxer.hikari.HikariDataSource", matchIfMissing = true)
static class Hikari {

    @Bean
    @ConfigurationProperties(prefix = "spring.datasource.hikari")
    HikariDataSource dataSource(DataSourceProperties properties){
        HikariDataSource dataSource = createDataSource(properties, HikariDataSource.class);
        if (StringUtils.hasText(properties.getName())) {
            dataSource.setPoolName(properties.getName());
```

```
        }
        return dataSource;
    }
}
```

此外 DataSourceAutoConfiguration 中还使用 @Import 注解导入了一个 DataSourceInitializationConfiguration 配置类,而这个配置类又导入了一个 DataSourceInitializerInvoker 和一个 Registrar 注册器,注册器中又向 BeanDefinitionRegistry 中注册了一个 DataSourceInitializerPostProcessor,如代码清单 10-11 所示。

**代码清单 10-11  注册 DataSourceInitializerInvoker**

```
@Configuration(proxyBeanMethods = false)
@Import({ DataSourceInitializerInvoker.class,
        DataSourceInitializationConfiguration.Registrar.class})
class DataSourceInitializationConfiguration {
    static class Registrar implements ImportBeanDefinitionRegistrar {
        private static final String BEAN_NAME = "dataSourceInitializerPostProcessor";

        @Override
        public void registerBeanDefinitions(AnnotationMetadata importingClassMetadata,
                BeanDefinitionRegistry registry) {
            if (!registry.containsBeanDefinition(BEAN_NAME)) {
                GenericBeanDefinition beanDefinition = new GenericBeanDefinition();
                beanDefinition.setBeanClass(DataSourceInitializerPostProcessor.class);
                beanDefinition.setRole(BeanDefinition.ROLE_INFRASTRUCTURE);
                beanDefinition.setSynthetic(true);
                registry.registerBeanDefinition(BEAN_NAME, beanDefinition);
            }
        }
    }
}
```

下面简单探究 DataSourceInitializerInvoker 与 DataSourceInitializerPostProcessor 的作用。

#### 1. DataSourceInitializerInvoker

由类名理解 DataSourceInitializerInvoker 是一个执行器,而这个执行器刚好就是类名前面的 DataSourceInitializer。借助 IDE 查看 DataSourceInitializerInvoker 的源码,可以发现它是一个监听器,监听的事件是 DataSourceSchemaCreatedEvent。此外,DataSourceInitializerInvoker 还实现了 InitializingBean 接口,它会在对象创建后回调 afterPropertiesSet 方法执行初始化逻辑,如代码清单 10-12 所示。

**代码清单 10-12  DataSourceInitializerInvoker 的重要结构**

```
class DataSourceInitializerInvoker implements ApplicationListener<DataSourceSchemaCreatedEvent>,
                                              InitializingBean {
    // ......
    // 内部组合了一个 DataSourceInitializer
    private DataSourceInitializer dataSourceInitializer;

    @Override
```

```java
public void afterPropertiesSet() {
    DataSourceInitializer initializer = getDataSourceInitializer();
    if (initializer != null) {
        boolean schemaCreated = this.dataSourceInitializer.createSchema();
        if (schemaCreated) {
            initialize(initializer);
        }
    }
}
```

重点关注 afterPropertiesSet 方法的实现,当初始化回调逻辑执行时,afterPropertiesSet 方法会先调用 getDataSourceInitializer 方法获取 DataSourceInitializer 实例,随后执行 DataSourceInitializer 的 createSchema 方法,如果 createSchema 方法执行成功,则继续执行后续的 initialize 方法以初始化数据。整个方法中有两个重要的初始化动作,分别是 createSchema 与 initialize 方法,逐一来看。

(1) createSchema

createSchema 方法名直译为"创建约束",它要完成的创建工作主要是对于数据库中的表结构,即执行 DDL 语句。createSchema 方法的执行机制是,通过读取 Spring Boot 全局配置文件 application.properties 中的 spring.datasource.schema 属性或者在没有任何配置的情况下执行 getScripts 方法的下半部分逻辑,加载名为 schema.sql 或 schema-all.sql 的文件,如代码清单 10-13 所示。

**代码清单 10-13　createSchema 执行 DDL 语句**

```java
boolean createSchema() {
    List<Resource> scripts = getScripts("spring.datasource.schema", this.properties.getSchema(),
            "schema");
    if (!scripts.isEmpty()) {
        if (!isEnabled()) {
            return false;
        }
        String username = this.properties.getSchemaUsername();
        String password = this.properties.getSchemaPassword();
        runScripts(scripts, username, password);
    }
    return !scripts.isEmpty();
}

private List<Resource> getScripts(String propertyName, List<String> resources, String fallback) {
    if (resources != null) {
        return getResources(propertyName, resources, true);
    }
    // 默认返回字符串"all"
    String platform = this.properties.getPlatform();
    List<String> fallbackResources = new ArrayList<>();
    fallbackResources.add("classpath*:" + fallback + "-" + platform + ".sql");
    fallbackResources.add("classpath*:" + fallback + ".sql");
    return getResources(propertyName, fallbackResources, false);
}
```

### （2）initialize

initialize 方法的执行时机在 createSchema 后，它要做的事情是执行 DML 语句，初始化数据库中实际的数据。触发 initialize 方法的方式是通过广播 DataSourceSchemaCreatedEvent 事件，回调下方 onApplicationEvent 方法执行，而最终执行的逻辑是 DataSourceInitializer 中的 initSchema 方法。从方法实现上看，initSchema 方法的逻辑与 createSchema 大同小异，但是 initSchema 方法寻找 SQL 文件的依据是全局配置文件中的 spring.datasource.data 配置项，或者在没有任何配置的情况下加载名为 data.sql 或 data-all.sql 的文件，如代码清单 10-14 所示。

**代码清单 10-14　initialize 执行 DML 语句**

```
private void initialize(DataSourceInitializer initializer) {
    try {
        this.applicationContext.publishEvent(
            new DataSourceSchemaCreatedEvent(initializer.getDataSource()));
        // ......
    } // catch throw ex ......
}

public void onApplicationEvent(DataSourceSchemaCreatedEvent event) {
    DataSourceInitializer initializer = getDataSourceInitializer();
    if (!this.initialized && initializer != null) {
        initializer.initSchema();
        this.initialized = true;
    }
}

void initSchema() {
    List<Resource> scripts = getScripts("spring.datasource.data", this.properties.getData(),
            "data");
    if (!scripts.isEmpty()) {
        if (!isEnabled()) {
            return;
        }
        String username = this.properties.getDataUsername();
        String password = this.properties.getDataPassword();
        runScripts(scripts, username, password);
    }
}
```

简单总结，有了 DataSourceInitializerInvoker 的设计，使得在项目开发中可以通过自定义 DDL 语句和 DML 语句并封装为 SQL 文件放置于项目的 resources 目录下，达到项目启动时自动初始化数据库表结构和数据的效果。

### 2. DataSourceInitializerPostProcessor

由类名理解，DataSourceInitializerPostProcessor 是专门为 DataSourceInitializer 定制的后置处理器，但是从源码中可以得知，DataSourceInitializerPostProcessor 的作用是在 DataSource 的初始化阶段即时创建 DataSourceInitializerInvoker 对象，如代码清单 10-15 所示。

**代码清单 10-15　DataSourceInitializerPostProcessor 的核心后置处理方法**

```java
class DataSourceInitializerPostProcessor implements BeanPostProcessor, Ordered {
    @Autowired
    private BeanFactory beanFactory;

    @Override
    public Object postProcessAfterInitialization(Object bean, String beanName)
            throws BeansException {
        if (bean instanceof DataSource) {
            this.beanFactory.getBean(DataSourceInitializerInvoker.class);
        }
        return bean;
    }
}
```

注意观察 postProcessAfterInitialization 方法的逻辑，当 DataSourceInitializerPostProcessor 检测到当前正在初始化的对象类型是 DataSource 时，会主动调用 BeanFactory 的 getBean 方法立即加载 DataSourceInitializerInvoker，此举的目的是使预先定义好的 SQL 脚本立即执行，以确保 DataSource 与数据库表结构、数据的同步初始化。

### 10.2.2　创建 JdbcTemplate

默认情况下 Spring Boot 会在项目中创建一个 JdbcTemplate，用于与数据库简单交互。之后还会向容器中注册一个 NamedParameterJdbcTemplate，用于支持参数命名化的 JdbcTemplate 增强，如代码清单 10-16 所示（源码逻辑很简单，不再展开）。

**代码清单 10-16　默认的 JdbcTemplate 创建（部分注解已省略）**

```java
@Configuration(proxyBeanMethods = false)
@Import({ JdbcTemplateConfiguration.class, NamedParameterJdbcTemplateConfiguration.Class })
public class JdbcTemplateAutoConfiguration {
}

class JdbcTemplateConfiguration {
    @Bean
    @Primary
    JdbcTemplate jdbcTemplate(DataSource dataSource, JdbcProperties properties) {
        JdbcTemplate jdbcTemplate = new JdbcTemplate(dataSource);
        // ......
        return jdbcTemplate;
    }
}

class NamedParameterJdbcTemplateConfiguration {
    @Bean
    @Primary
    NamedParameterJdbcTemplate namedParameterJdbcTemplate(JdbcTemplate jdbcTemplate) {
        return new NamedParameterJdbcTemplate(jdbcTemplate);
    }
}
```

## 10.2.3 配置事务管理器

引入 JDBC 数据访问场景后,必不可少地要用到事务。DataSource 创建完成后 Spring Boot 还会默认创建一个 `DataSourceTransactionManager`,用于支持基于数据源的事务控制,如代码清单 10-17 所示。

**代码清单 10-17　配置事务管理器(部分注解已省略)**

```java
@Configuration(proxyBeanMethods = false)
@ConditionalOnClass({JdbcTemplate.class, PlatformTransactionManager.class})
public class DataSourceTransactionManagerAutoConfiguration {

    static class DataSourceTransactionManagerConfiguration {
        @Bean
        @ConditionalOnMissingBean(PlatformTransactionManager.class)
        DataSourceTransactionManager transactionManager(DataSource dataSource,
                ObjectProvider<TransactionManagerCustomizers> transactionManagerCustomizers) {
            DataSourceTransactionManager transactionManager =
                    new DataSourceTransactionManager(dataSource);
            transactionManagerCustomizers.ifAvailable((customizers)->
                    customizers.customize(transactionManager));
            return transactionManager;
        }
    }
}
```

## 10.3　声明式事务的生效原理

Spring Boot 整合 JDBC 的场景中,除了引入 `spring-jdbc`,还会引入 `spring-tx` 实现事务控制。默认情况下 Spring Boot 会自动配置并开启注解声明式事务,本节内容将详细探究 Spring Boot 配置的注解声明式事务是如何生效的。

回顾 10.1 节的示例项目中,在 `SpringBootJdbcApplication` 主启动类中显式标注了 `@EnableTransactionManagement` 注解用于开启注解声明式事务,但即便不进行标注,底层仍然会使用自动配置类的方式开启,具体的开启位置在自动配置类 `TransactionAutoConfiguration` 中。

### 10.3.1　TransactionAutoConfiguration

`TransactionAutoConfiguration` 如代码清单 10-18 所示。

**代码清单 10-18　TransactionAutoConfiguration**

```java
@Configuration(proxyBeanMethods = false)
// ......
public class TransactionAutoConfiguration {
    // ......

    @Configuration(proxyBeanMethods = false)
    @ConditionalOnBean(TransactionManager.class)
    @ConditionalOnMissingBean(AbstractTransactionManagementConfiguration.class)
```

```java
    public static class EnableTransactionManagementConfiguration {

        @Configuration(proxyBeanMethods = false)
        @EnableTransactionManagement(proxyTargetClass = false)
        @ConditionalOnProperty(prefix = "spring.aop", name = "proxy-target-class",
                            havingValue = "false", matchIfMissing = false)
        public static class JdkDynamicAutoProxyConfiguration {
        }

        @Configuration(proxyBeanMethods = false)
        @EnableTransactionManagement(proxyTargetClass = true)
        @ConditionalOnProperty(prefix = "spring.aop", name = "proxy-target-class",
                            havingValue = "true", matchIfMissing =true)
        public static class CglibAutoProxyConfiguration {
        }
    }
}
```

通过代码清单 10-18 可以发现，即使项目中没有显式地标注@EnableTransactionManagement 注解，底层的配置类中也会协助开启，唯一的可变项是根据项目中配置的 AOP 是否强制代理目标类对象（proxyTargetClass）来决定如何创建事务代理对象。

既然注解声明式事务的最终开关是@EnableTransactionManagement 注解，根据前面几章的内容就可以断定，@EnableTransactionManagement 注解的内部一定是使用@Import 注解导入了一些特殊的组件，以此实现模块装配。借助 IDE 打开@EnableTransactionManagement 的源码，可以发现它导入了一个 TransactionManagementConfigurationSelector，而且这个注解中还包含三个属性，如代码清单 10-19 所示。

**代码清单 10-19　@EnableTransactionManagement 注解**

```java
@Import(TransactionManagementConfigurationSelector.class)
public @interface EnableTransactionManagement {
    boolean proxyTargetClass() default false;
    AdviceMode mode() default AdviceMode.PROXY;
    int order() default Ordered.LOWEST_PRECEDENCE;
}
```

注解属性中 proxyTargetClass 与 order 都是我们比较熟悉的属性，本节不再解释。有关 mode 属性的设计，读者可以简单了解一下。

mode 属性的取值有两个，分别是 PROXY 和 ASPECTJ。注意这两个概念不同于 AOP 部分中的原生 AOP 和基于 AspectJ 的 AOP。PROXY 的含义是事务通知会在程序运行期使用动态代理的方式向目标对象织入通知，而 ASPECTJ 代表的是在类加载期织入事务通知（类似于 load-time-weaving）。一般情况下，使用注解声明式事务时都是使用运行期的动态代理来实现 AOP 或者事务控制，实际上也可以手动调整通知织入的时机为类加载期。如果要把 mode 改为 ASPECTJ 的话，需要额外导入一个依赖，并开启 load-time-weaving，如代码清单 10-20 所示。

**代码清单 10-20　支持类加载期的事务通知需要额外导入 spring-aspects 依赖**

```xml
<dependency>
    <groupId>org.springframework</groupId>
    <artifactId>spring-aspects</artifactId>
</dependency>
```

> 提示：由于目前的主流项目开发中不会使用@EnableLoadTimeWeaving 进行类加载期的通知织入，因此对于这个知识点读者只需简单了解。

### 10.3.2　TransactionManagementConfigurationSelector

简单了解了@EnableTransactionManagement 注解的属性后，下面重点关注导入的 TransactionManagementConfigurationSelector 组件。从类名上不难得知该组件是一个 ImportSelector，从代码清单 10-21 中的 selectImports 方法中可以得知，它会根据@EnableTransactionManagement 注解的 mode 属性的值决定导入哪些组件，对于 PROXY，导入的两个组件是 AutoProxyRegistrar 和 ProxyTransactionManagementConfiguration，下面我们重点研究这两个组件的作用。

**代码清单 10-21　TransactionManagementConfigurationSelector**

```java
public class TransactionManagementConfigurationSelector
        extends AdviceModeImportSelector<EnableTransactionManagement> {

    @Override
    protected String[] selectImports(AdviceMode adviceMode) {
        switch (adviceMode) {
            case PROXY:
                return new String[] {AutoProxyRegistrar.class.getName(),
                        ProxyTransactionManagementConfiguration.class.getName() };
            case ASPECTJ:
                return new String[] {determineTransactionAspectClass() };
            default:
                return null;
        }
    }
    // ......
}
```

### 10.3.3　AutoProxyRegistrar

由源码可知 AutoProxyRegistrar 本身是一个 ImportBeanDefinitionRegistrar，它的作用是向 BeanDefinitionRegistry 中编程式注册新的 BeanDefinition。从核心实现方法 registerBeanDefinitions 中可以看出，AutoProxyRegistrar 会根据@EnableTransactionManagement 注解中的属性决定是否注册额外的组件。通常情况下项目中使用的都是运行期 AOP 的声明式事务，因此提取的 mode 值一定是 PROXY，后续会调用

`AopConfigUtils.registerAutoProxyCreatorIfNecessary` 方法注册有关 AOP 的组件,继续往下深入可以发现,实际注册的组件类型是 `InfrastructureAdvisorAutoProxyCreator`,如代码清单 10-22 所示。

**代码清单 10-22　AutoProxyRegistrar 中注册新 BeanDefinition 的核心逻辑**

```java
public void registerBeanDefinitions(AnnotationMetadata importingClassMetadata,
        BeanDefinitionRegistry registry) {
    boolean candidateFound = false;
    Set<String> annTypes = importingClassMetadata.getAnnotationTypes();
    for (String annType : annTypes) {
        // 搜寻类上所有标注的注解
        // 目的是找到@EnableTransactionManagement
        AnnotationAttributes candidate = AnnotationConfigUtils.attributesFor(
                importingClassMetadata, annType);
        if (candidate == null) {
            continue;
        }
        // 获取注解上的 mode 和 proxyTargetClass 属性值
        Object mode = candidate.get("mode");
        Object proxyTargetClass = candidate.get("proxyTargetClass");
        if (mode != null && proxyTargetClass != null && AdviceMode.class == mode.getClass()
                && Boolean.class == proxyTargetClass.getClass()) {
            candidateFound = true;
            // 当 mode 为 PROXY 时,会注册额外的 BeanDefinition
            if (mode == AdviceMode.PROXY) {
                AopConfigUtils.registerAutoProxyCreatorIfNecessary(registry);
                if ((Boolean) proxyTargetClass) {
                    AopConfigUtils.forceAutoProxyCreatorToUseClassProxying(registry);
                    return;
                }
            }
        }
    }
    // logger ......
}

public static BeanDefinition registerAutoProxyCreatorIfNecessary(
        BeanDefinitionRegistry registry) {
    return registerAutoProxyCreatorIfNecessary(registry, null);
}

public static BeanDefinition registerAutoProxyCreatorIfNecessary(
    BeanDefinitionRegistry registry, @Nullable Object source) {
    return registerOrEscalateApcAsRequired(InfrastructureAdvisorAutoProxyCreator.class,
            registry, source);
}
```

这个 `InfrastructureAdvisorAutoProxyCreator` 与第 9 章中学习的 AOP 核心代理对象创建器 `AnnotationAwareAspectJAutoProxyCreator` 有些相似,它们的父类都是 `AbstractAdvisorAutoProxyCreator`,所以它们都可以创建代理对象,不过两者最大的区别在于,`InfrastructureAdvisorAutoProxyCreator` 只会组合"基础类型"的增强器,而不会整合项目中自定义的增强器。

要理解"基础类型"这个概念,需要读者先了解 `BeanDefinition` 中给 Bean 定义的 3 种角色,如代码清单 10-23 所示。"基础类型"实际指的是 `BeanDefinition` 中的角色为 `ROLE_INFRASTRUCTURE`。通常情况下只有 Spring Framework 内部定义的 Bean 才可能被标注为 `ROLE_INFRASTRUCTURE` 角色,而且这些 Bean 在应用程序中起到基础支撑的作用。

**代码清单 10-23　BeanDefinition 中定义的 3 种角色**

```
int ROLE_APPLICATION = 0;
int ROLE_SUPPORT = 1;
int ROLE_INFRASTRUCTURE = 2;
```

由上述理论,读者可以大胆猜想:事务控制的核心是 AOP 中的一个 `MethodInterceptor`,它的角色刚好是 `ROLE_INFRASTRUCTURE`。`InfrastructureAdvisorAutoProxyCreator` 可以在 bean 对象的初始化期间搜寻到这个 `MethodInterceptor` 并包装为 `Advisor`,给需要进行注解事务控制的 bean 对象构造代理对象。

> 💡 提示:可能有读者在此处会产生疑惑:如果 AOP 与事务在项目中同时存在,是否会导入两个代理对象创建器?答案是否定的。Spring Framework 在底层为两种代理对象创建器定义了优先级,当 `AnnotationAwareAspectJAutoProxyCreator` 先注册到 IOC 容器后再注册 `InfrastructureAdvisorAutoProxyCreator` 时,会比对两者的优先级,由于 AOP 的核心代理对象创建器可以处理所有角色的通知,因此它的优先级更高,在这种场景下 `InfrastructureAdvisorAutoProxyCreator` 就不会再注册(相关源码可参照代码清单 9-3)。

### 10.3.4　ProxyTransactionManagementConfiguration

`TransactionManagementConfigurationSelector` 导入的另一个组件是 `ProxyTransactionManagementConfiguration`,它本身是一个配置类,其内部注册了 3 个与事务控制相关的核心组件,逐一来看。

**1. transactionAttributeSource:事务配置源**

第一个注册的 Bean 类型为 `TransactionAttributeSource`,我们有必要先了解一下这个接口。`TransactionAttributeSource` 接口在 Spring Framework 5.2 版本之前只有一个 `getTransactionAttribute` 方法。随着 Spring Framework 5.2 版本之后引入响应式事务控制,`TransactionAttributeSource` 接口才多了另一个 `isCandidateClass` 方法。不过这些细节不是需要重点关注的,`getTransactionAttribute` 方法才是重中之重,它可以将一个类+方法解析转换为 `TransactionAttribute`,而 `TransactionAttribute` 本身是 `TransactionDefinition`,所以可以这样理解:`TransactionAttributeSource` 可以根据一个具体的类中的方法解析并转换为一个 `TransactionDefinition`(`TransactionAttribute`)。

借助 IDE 可以找到 `TransactionAttributeSource` 的几个实现类,如图 10-1 所示。实现类中包含了在配置类中创建的 `AnnotationTransactionAttributeSource`,如代码清单 10-24 所示。除此之外,`TransactionAttributeSource` 还有一些其他的实现,感兴趣的读者可以自行了解,本节只讲解 `AnnotationTransactionAttributeSource`。

### 代码清单 10-24  注册 AnnotationTransactionAttributeSource

```
@Bean
@Role(BeanDefinition.ROLE_INFRASTRUCTURE)
public TransactionAttributeSource transactionAttributeSource() {
    return new AnnotationTransactionAttributeSource();
}

public interface TransactionAttributeSource {
    // isCandidateClass ......

    @Nullable
    TransactionAttribute getTransactionAttribute(Method method, @Nullable Class<?> targetClass);
}
```

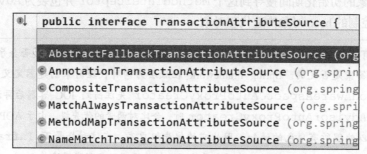

图 10-1  TransactionAttributeSource 的实现类

从 `AnnotationTransactionAttributeSource` 的 javadoc 中可以获取到关键信息：*This class reads Spring's JDK 1.5+ Transactional annotation*。这说明它解析事务信息的依据是读取 `@Transactional` 注解，这就是注解声明式事务的标注读取器。至于其中如何读取、解析，将在 10.4 节中讲解。

**2. transactionInterceptor：事务拦截器**

第二个注册的组件是事务切面的核心通知：`TransactionInterceptor` 事务拦截器，如代码清单 10-25 所示。它本身是一个 `MethodInterceptor`，通过第 9 章的学习，我们知道 `MethodInterceptor` 接口是 AOP 增强的核心拦截器接口，利用 AOP 生成的代理对象中都会包含一组 `MethodInterceptor` 接口的实现类对象。除此之外，`TransactionInterceptor` 还有一个父类 `TransactionAspectSupport`，这个父类中有一些与 Spring Framework 基础事务 API 的集成（如执行事务的核心方法 `invokeWithinTransaction`、创建事务、提交事务、回滚事务等）。有关这些事务控制动作的触发位置，同样也放到 10.4 节中讲解。

### 代码清单 10-25  注册 TransactionInterceptor

```
// 参数名有精简
@Bean
@Role(BeanDefinition.ROLE_INFRASTRUCTURE)
public TransactionInterceptor transactionInterceptor(TransactionAttributeSource attr) {
    TransactionInterceptor interceptor = new TransactionInterceptor();
    interceptor.setTransactionAttributeSource(attr);
    if (this.txManager != null) {
        interceptor.setTransactionManager(this.txManager);
```

```
        }
        return interceptor;
}
```

### 3. transactionAdvisor：事务增强器

最后一个注册的组件是事务增强器。AOP 中除了有通知，还有一个核心要素就是增强器，而注册的 `BeanFactoryTransactionAttributeSourceAdvisor` 从类名上看就是一个增强器，注意其内部组合了 `TransactionInterceptor` 事务拦截器和 `TransactionAttributeSource` 事务配置源，如代码清单 10-26 所示。

**代码清单 10-26　注册 TransactionAttributeSourceAdvisor**

```
@Bean(name = TransactionManagementConfigUtils.TRANSACTION_ADVISOR_BEAN_NAME)
@Role(BeanDefinition.ROLE_INFRASTRUCTURE)
public BeanFactoryTransactionAttributeSourceAdvisor transactionAdvisor(
        TransactionAttributeSource transactionAttributeSource,
        TransactionInterceptor transactionInterceptor) {
    BeanFactoryTransactionAttributeSourceAdvisor advisor =
            new BeanFactoryTransactionAttributeSourceAdvisor();
    advisor.setTransactionAttributeSource(transactionAttributeSource);
    advisor.setAdvice(transactionInterceptor);
    //提取@EnableTransactionManagement 的 order 属性
    if (this.enableTx != null) {
        advisor.setOrder(this.enableTx.<Integer>getNumber("order"));
    }
    return advisor;
}
```

既然是 AOP 中的增强器，那么切入点也必不可少。`BeanFactoryTransactionAttributeSourceAdvisor` 判断一个类是否可以被增强的依据是利用 `TransactionAttributeSource`，检查类和方法中是否标注有 `@Transactional` 注解，这个逻辑与读者所熟知的事务控制一致，即如果 Service 类上或者方法上标注了 `@Transactional` 注解，则事务切面会介入控制。

简单总结，`ProxyTransactionManagementConfiguration` 中注册了注解声明式事务中必备的组件，有了这些组件，就可以支撑注解事务的控制。

## 10.4　声明式事务的控制全流程

了解了声明式事务的生效原理，本节结合 10.1 节的整合项目，以 Debug 的方式研究事务控制的底层全流程。

借助 IDE 测试，将断点打在 `userService.test();` 方法上，以 Debug 的形式运行 `SpringBootJdbcApplication`，等断点停在此处时 Debug 进入。

### 10.4.1　CglibAopProxy#intercept

由于默认情况下 Spring Boot 会使用代理目标类的方式创建代理对象，因此这里首先会进入 `CglibAopProxy` 的 `intercept` 方法，通过第 9 章的内容可以快速定位到代理对象中的增强器，即上面提到的 `BeanFactoryTransactionAttributeSourceAdvisor`，而这个增强器中组合的通知,刚好是上面提到的事务拦截器 `TransactionInterceptor`,如图 10-2 所示。

```
this = {CglibAopProxy$DynamicAdvisedInterceptor@3910}
  advised = {ProxyFactory@3914} ... toString()
    aopProxyFactory = {DefaultAopProxyFactory@3915}
    listeners = {LinkedList@3916} size = 0
    active = true
    targetSource = {SingletonTargetSource@3917} ... toString()
    preFiltered = true
    advisorChainFactory = {DefaultAdvisorChainFactory@3918}
    methodCache = {ConcurrentHashMap@3919} size = 0
    interfaces = {ArrayList@3920} size = 0
    advisors = {ArrayList@3921} size = 1
      0 = {BeanFactoryTransactionAttributeSourceAdvisor@3925} ... toString()
        transactionAttributeSource = {AnnotationTransactionAttributeSource@3926}
        pointcut = {BeanFactoryTransactionAttributeSourceAdvisor$1@3927} ... toString()
        adviceBeanName = null
        beanFactory = {DefaultListableBeanFactory@3184} ... toString()
        advice = {TransactionInterceptor@3928}
        adviceMonitor = {ConcurrentHashMap@3929} size = 86
        order = {Integer@3930} 2147483647
    advisorArray = {Advisor[1]@3922}
    proxyTargetClass = true
    optimize = false
```

图 10-2　UserService 代理对象中有一个增强器

明确了通知方法的位置,下面直接将断点打在 `TransactionInterceptor` 的 `invoke` 方法上,并放行当前的断点调试,使程序停在 `invoke` 方法。

### 10.4.2　TransactionInterceptor

进入 `TransactionInterceptor` 的 `invoke` 方法,可以发现方法中直接调用了 `invokeWithinTransaction` 方法,而这个 `invokeWithinTransaction` 方法定义在 `TransactionInterceptor` 的父类 `TransactionAspectSupport` 中,如代码清单 10-27 所示。

**代码清单 10-27　TransactionInterceptor 的核心通知方法**

```java
@Override
public Object invoke(MethodInvocation invocation) throws Throwable {
    Class<?> targetClass = (invocation.getThis() != null
            ? AopUtils.getTargetClass(invocation.getThis()) : null);
    return invokeWithinTransaction(invocation.getMethod(), targetClass, invocation::proceed);
}
```

由于 `invokeWithinTransaction` 方法的篇幅很长,为便于读者更好地阅读和理解源码,本节将其拆分为多个片段讲解。

## 1. 获取 TransactionAttribute

invokeWithinTransaction 方法的第一步工作是通过 TransactionAttributeSource 获取到 TransactionAttribute 事务定义信息，如代码清单 10-28 所示。getTransactionAttribute 方法的内部有一个缓存机制，方法的核心工作是**根据方法和方法所在的类获取并缓存对应的事务定义信息**，如果没有获取到事务定义信息则会缓存 NULL_TRANSACTION_ATTRIBUTE 空定义并返回，如代码清单 10-29 所示。

**代码清单 10-28　invokeWithinTransaction（1）**

```
@Nullable
protected Object invokeWithinTransaction(Method method, @Nullable Class<?> targetClass,
        final InvocationCallback invocation) throws Throwable {
    //如果事务属性为空，则方法非事务性的
    TransactionAttributeSource tas = getTransactionAttributeSource();
    final TransactionAttribute txAttr = (tas != null
            ? tas.getTransactionAttribute(method, targetClass) : null);
    // ......
```

**代码清单 10-29　getTransactionAttribute 获取事务定义信息**

```
private final Map<Object, TransactionAttribute> attributeCache = new ConcurrentHashMap<>(1024);

public TransactionAttribute getTransactionAttribute(Method method,
        @Nullable Class<?> targetClass) {
    if (method.getDeclaringClass() == Object.class) {
        return null;
    }

    // 根据 method 和 targetClass 构造一个缓存 key
    Object cacheKey = getCacheKey(method, targetClass);
    // 如果能从缓存中获取到事务定义信息，则直接返回
    TransactionAttribute cached = this.attributeCache.get(cacheKey);
    if (cached != null) {
        // （此处简化了 if 结构）
        return cached == NULL_TRANSACTION_ATTRIBUTE ? null : cached;
    } else {
        // 如果没有获取到事务定义信息，则需要根据方法和方法所在类的信息构造事务定义信息
        TransactionAttribute txAttr = computeTransactionAttribute(method, targetClass);
        // 无论是否构造成功，最终都会将该方法缓存至 attributeCache 中
        if (txAttr == null) {
            this.attributeCache.put(cacheKey, NULL_TRANSACTION_ATTRIBUTE);
        } else {
            String methodIdentification = ClassUtils.getQualifiedMethodName(method, targetClass);
            if (txAttr instanceof DefaultTransactionAttribute) {
                ((DefaultTransactionAttribute) txAttr).setDescriptor(methodIdentification);
            }
            // logger ......
            this.attributeCache.put(cacheKey, txAttr);
        }
        return txAttr;
    }
}
```

注意这里有一个问题，当 Debug 至此处，可以发现缓存中已经成功获取到 Transaction

Attribute，如图 10-3 所示。出现这种现象的原因是，事务通知在织入之前需要对每个正在创建的 bean 对象进行匹配，而匹配时需要使用 `TransactionAttributeSource` 检查方法或方法所在类上是否标注有@`Transactional` 注解，以此来判断是否需要对当前正在创建的 bean 对象织入事务通知。而不管最终事务通知是否需要织入目标对象，`TransactionAttributeSource` 中都会留下被检查方法的判断痕迹。

```
// First, see if we have a cached value.
Object cacheKey = getCacheKey(method, targetClass);   cacheKey: MethodClas
TransactionAttribute cached = this.attributeCache.get(cacheKey);   cached:
if (cached != null) {   cached: "PROPAGATION_REQUIRED,ISOLATION_DEFAULT"
    // Value will either be canonical value indicating there is no transa
    // or an actual transaction attribute.
    if (cached == NULL_TRANSACTION_ATTRIBUTE) {
```

图 10-3　第一次获取事务定义信息即可成功获取

既然所有的事务定义信息已经在 bean 对象的初始化阶段被后置处理器扫描并封装过，如果读者想探究封装的过程，需要重新 Debug 启动程序，并使用条件断点打在 `AbstractFallbackTransactionAttributeSource` 的第 112 行，如图 10-4 所示。

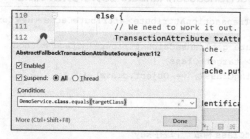

图 10-4　使用条件断点调试 DemoService 的事务定义信息封装

标注好断点后，重新 Debug 运行 `SpringBootJdbcApplication`，待程序停在条件断点处时就可以获得此时正在创建的 `UserService` 对象，如图 10-5 所示。

```
▼ ● method = {Method@4477} "public void com.linkedbear.springboot.service.UserService.test()"
   ▶ ● clazz = {Class@3538} ... Navigate
     ● slot = 1
   ▶ ● name = "test"
   ▶ ● returnType = {Class@4494} ... Navigate
```

图 10-5　使用条件断点的效果

有了业务层对象，下一步就可以进入 `computeTransactionAttribute` 方法实际地解析和封装事务定义信息。方法逻辑本身不难，它会寻找方法上是否标注了@`Transactional` 注解，以及寻找类上是否标注了@`Transactional` 注解，从源码的搜寻顺序上明显可知是先找方法后找类，由此可见方法级的优先级更高，如代码清单 10-30 所示。

**代码清单 10-30　computeTransactionAttribute 解析和封装事务定义信息**

```
protected TransactionAttribute computeTransactionAttribute(Method method,
      @Nullable Class<?> targetClass) {
   // 非 public 方法不处理
   if (allowPublicMethodsOnly() && !Modifier.isPublic(method.getModifiers())) {
      return null;
```

```java
}
Method specificMethod = AopUtils.getMostSpecificMethod(method, targetClass);
// 首先寻找方法上是否标注了@Transactional注解
TransactionAttribute txAttr = findTransactionAttribute(specificMethod);
if (txAttr != null) {
    return txAttr;
}

// 如果方法上没有，则接下来寻找类上是否标注了@Transactional注解
txAttr = findTransactionAttribute(specificMethod.getDeclaringClass());
if (txAttr != null && ClassUtils.isUserLevelMethod(method)) {
    return txAttr;
}
// ......
return null;
}
```

下面 `findTransactionAttribute` 方法的实现会将@Transactional 注解中的属性逐一获取，封装为 `RuleBasedTransactionAttribute` 对象并返回，以完成整个事务定义信息的解析和封装。

简单总结，当应用启动时，由于自动配置类中默认标注有@EnableTransactionManagement 注解，该注解会向 IOC 容器中注册 InfrastructureAdvisorAutoProxyCreator 事务通知增强器，这个增强器会参与 bean 对象初始化的 AOP 后置处理逻辑中，检查被创建的 bean 对象中是否可以织入事务通知（标注@Transactional 注解），检查的动作会同时保存到 AbstractFallbackTransactionAttributeSource 的本地缓存中，因此在真正触发事务拦截器的逻辑而需要取出事务定义信息时，可以直接从缓存中取出而不需要重新解析。

#### 2. 获取 TransactionManager

获取到事务定义信息之后，接下来要做的是获取事务管理器。执行 `determineTransactionManager` 方法，而方法的最终实现是在源码最后的最深层的 `if` 结构中从 BeanFactory 中获取 TransactionManager，如代码清单 10-31 所示。

**代码清单 10-31　invokeWithinTransaction（2）**

```java
// ......
TransactionAttributeSource tas = getTransactionAttributeSource();
final TransactionAttribute txAttr = (tas != null
        ? tas.getTransactionAttribute(method, targetClass) : null);
final TransactionManager tm = determineTransactionManager(txAttr);
// ......

protected TransactionManager determineTransactionManager(@Nullable TransactionAttribute txAttr) {
    // 不触发的部分已省略 ......
    else {
        TransactionManager defaultTransactionManager = getTransactionManager();
        if (defaultTransactionManager == null) {
            defaultTransactionManager = this.transactionManagerCache
                    .get(DEFAULT_TRANSACTION_MANAGER_KEY);
            if (defaultTransactionManager == null) {
                defaultTransactionManager = this.beanFactory.getBean(TransactionManager.class);
                this.transactionManagerCache.putIfAbsent(
```

```
                    DEFAULT_TRANSACTION_MANAGER_KEY, defaultTransactionManager);
        }
    }
    return defaultTransactionManager;
}
```

Debug 至此处,可以从 BeanFactory 中成功获取到基于数据源的 `DataSource TransactionManager`,如图 10-6 所示。

```
▼  ≡ defaultTransactionManager = {DataSourceTransactionManager@3927}
   ▶  ⓕ dataSource = {HikariDataSource@3928} ... toString()
      ⓕ enforceReadOnly = false
   ▶  ⓕ logger = {LogAdapter$Slf4jLocationAwareLog@3929}
```

图 10-6 从 BeanFactory 中获取到的事务管理器

### 3. 响应式事务管理器的处理

下一部分源码针对响应式事务,如代码清单 10-32 所示。Spring Framework 5.2 版本之后引入了响应式事务的概念,所以源码中有响应式事务管理器的判断和处理,如代码清单 10-32 所示(本书不对响应式事务展开讨论,故跳过)。

**代码清单 10-32 invokeWithinTransaction(3)**

```java
//......
if (this.reactiveAdapterRegistry != null && tm instanceof ReactiveTransactionManager) {
    ReactiveTransactionSupport txSupport = transactionSupportCache.computeIfAbsent (method, key ->{
        // kotlin ......
        ReactiveAdapter adapter = this.reactiveAdapterRegistry.getAdapter(method.getReturnType());
        if (adapter == null) {
            // throw ex ......
        }
        return new ReactiveTransactionSupport(adapter);
    });
    return txSupport.invokeWithinTransaction(
            method, targetClass, invocation, txAttr,(ReactiveTransactionManager) tm);
}
//......
```

### 4. 事务控制核心

下面的部分是事务控制的核心源码。通过代码清单 10-33 可以看出,注解声明式事务的核心是一个环绕通知。这部分核心源码的动作有 4 步:**开启事务、执行 Service 方法、遇到异常就回滚事务和没有异常就提交事务**。这个步骤对于原生 JDBC 事务同样适用,Spring Framework 也完全遵循该流程实现。

**代码清单 10-33 invokeWithinTransaction(4)**

```java
//......
PlatformTransactionManager ptm = asPlatformTransactionManager(tm);
final String joinpointIdentification = methodIdentification(method, targetClass, txAttr);

if (txAttr == null || !(ptm instanceof CallbackPreferringPlatformTransactionManager)) {
    // 1. 开启事务
    TransactionInfo txInfo = createTransactionIfNecessary(ptm, txAttr, joinpointIdentification);
```

```
    Object retVal;
    try {
        // 2. 环绕通知执行 Service 方法
        retVal = invocation.proceedWithInvocation();
    } catch (Throwable ex) {
        // 3. 捕获到异常，回滚事务
        completeTransactionAfterThrowing(txInfo, ex);
        throw ex;
    } finally {
        cleanupTransactionInfo(txInfo);
    }

    // ......

    // 4. Service 方法执行成功，提交事务
    commitTransactionAfterReturning(txInfo);
    return retVal;
}
// 下面的逻辑基本一致，不再展开 ......
```

下面通过整合项目中的正常执行代码以及抛出异常的代码分别测试事务的成功提交与失败回滚。

（1）成功的事务提交

以当前的代码继续 Debug，当执行完 `createTransactionIfNecessary` 方法后，观察 `txInfo` 的属性，如图 10-7 所示。此时事务的定义信息、事务的状态已经封装完毕，`completed` 属性为 `false`。

事务开启后，由于 `UserService` 的代码可以正常执行，会触发下面的 `commitTransactionAfterReturning` 方法提交事务。提交事务的逻辑是获取到事务管理器后执行提交逻辑。注意，事务管理器的 `commit` 方法并不会直接提交，而是会先进行一些异常情况的检查，确保无误后再执行 `processCommit` 方法提交事务，如代码清单 10-34 所示。

```
txInfo = {TransactionAspectSupport$TransactionInfo@4140} ... toString()
    transactionManager = {DataSourceTransactionManager@3917}
    transactionAttribute = {TransactionAspectSupport$1@4141} ... toString()
    joinpointIdentification = "com.linkedbear.springboot.service.UserService.test"
    transactionStatus = {DefaultTransactionStatus@4142}
        transaction = {DataSourceTransactionManager$DataSourceTransactionObject@4143}
        newTransaction = true
        newSynchronization = true
        readOnly = false
        debug = false
        suspendedResources = null
        rollbackOnly = false
        completed = false
        savepoint = null
    oldTransactionInfo = null
```

图 10-7 开启事务后的 TransactionInfo

### 代码清单 10-34 调用事务管理器提交事务

```java
protected void commitTransactionAfterReturning(@Nullable TransactionInfo txInfo) {
    if (txInfo != null && txInfo.getTransactionStatus() != null) {
        txInfo.getTransactionManager().commit(txInfo.getTransactionStatus());
    }
}

// AbstractPlatformTransactionManager
public final void commit(TransactionStatus status) throws TransactionException {
    // 如果事务已经完成，则无法提交，抛出异常
    if (status.isCompleted()) {
        // throw ex ......
    }
    DefaultTransactionStatus defStatus = (DefaultTransactionStatus) status;
    // 如果事务已被标记为需要回滚，回滚事务
    if (defStatus.isLocalRollbackOnly()) {
        processRollback(defStatus, false);
        return;
    }
    // 全局事务的回滚标识判断(JTA) ......

    // 正常的待提交事务可以执行提交
    processCommit(defStatus);
}
```

提交事务的 processCommit 方法内部实现比较复杂，代码清单 10-35 中只列举了最关键的几行源码，整个方法的核心提交动作是一个以 do 开头的 doCommit 方法，这个方法的实现在落地实现类 DataSourceTransactionManager 中，它会获取到原生 JDBC 的 Connection，执行 commit 方法完成事务提交。至此就可以看到 Spring Framework 封装的事务抽象在底层操控原生 JDBC 的核心实现。

### 代码清单 10-35 提交事务的底层核心源码

```java
private void processCommit(DefaultTransactionStatus status) throws TransactionException {
    try {
        // ......
        if (status.hasSavepoint()) {
            // 如果当前事务存在保存点，则处理保存点的逻辑
        }
        // 对于新事务，直接提交事务即可
        else if (status.isNewTransaction()) {
            unexpectedRollback = status.isGlobalRollbackOnly();
            doCommit(status);
        }
        // ......
    }
}

@Override
protected void doCommit(DefaultTransactionStatus status) {
    DataSourceTransactionObject txObject =
            (DataSourceTransactionObject) status.getTransaction();
    Connection con = txObject.getConnectionHolder().getConnection();
    try {
```

```
        con.commit();
    } // catch
}
```

**（2）异常的事务回滚**

如果需要测试异常的事务回滚，需要在 `UserService` 的 `test` 方法中人为构造一个异常（如使用 `int i = 1 / 0;` 构造除零异常），并在 `invokeWithTransaction` 方法的 try-catch 块中调用 `completeTransactionAfterThrowing` 方法处打断点，之后重新 Debug 执行 `SpringBootJdbcApplication`。等程序运行至 `completeTransactionAfterThrowing` 方法即将调用时，除零异常已经抛出，如代码清单 10-36 所示。

**▍代码清单 10-36　打断点的位置是下面的 catch 部分**

```
try {
    // 2. 环绕通知执行 Service 方法
    retVal = invocation.proceedWithInvocation();
} catch (Throwable ex) {
    // 3. 捕获到异常，回滚事务
    completeTransactionAfterThrowing(txInfo, ex);
    throw ex;
}
```

此时 Debug 的状态中，`TransactionStatus` 中已经封装了异常信息，如图 10-8 所示。

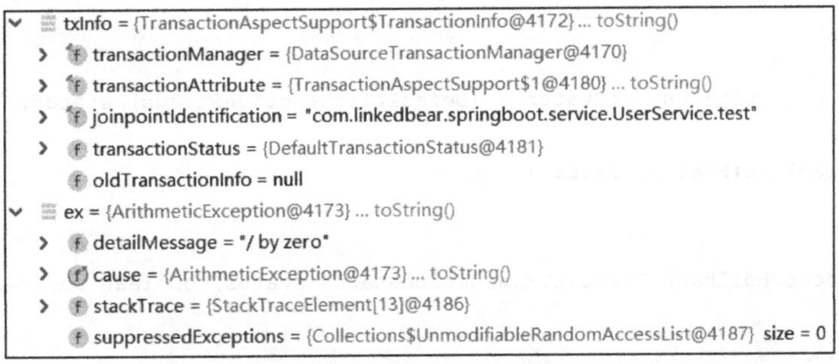

图 10-8　触发异常时异常信息已被捕获并记录

获取到异常后，底层会根据异常类型决定是否回滚异常。默认情况下 `@Transactional` 注解控制回滚的异常类型包括 `Error` 和 `RuntimeException`，对于普通的 `Exception`，默认策略不会回滚，而是选择继续提交事务，如代码清单 10-37 所示。这也提示读者在日常开发中，**给方法标注 `@Transactional` 注解时一定要显式地声明事务回滚的异常类型**。

**▍代码清单 10-37　根据抛出异常类型和事务定义信息决定是否回滚事务**

```
protected void completeTransactionAfterThrowing(@Nullable TransactionInfo txInfo, Throwable ex) {
    if (txInfo != null && txInfo.getTransactionStatus() != null) {
        // logger ......
        // 如果当前异常在回滚范围之内，则会调用事务管理器，回滚事务
        if (txInfo.transactionAttribute != null && txInfo.transactionAttribute.rollbackOn(ex)) {
            try {
                txInfo.getTransactionManager().rollback(txInfo.getTransactionStatus());
            } // catch ......
        } else {
```

```
            // 如果不在回滚范围内，则依然会提交事务
            try {
                txInfo.getTransactionManager().commit(txInfo.getTransactionStatus());
            } // catch ......
        }
    }
}

// 默认的事务回滚捕获异常类型
public boolean rollbackOn(Throwable ex) {
    return (ex instanceof RuntimeException|| ex instanceof Error);
}
```

由于测试代码 `UserService` 中标注的@`Transactional` 注解已经显式声明了 `rollbackFor=Exception.class`，故此处可以正常回滚事务，而回滚事务时底层也有一些简单的判断，在判断无异常时最终依然调用 `DataSourceTransactionManager` 的方法，不过这里调用的是 `doRollback` 方法，它的实现方式与 `doCommit` 方法如出一辙，都是获取到原生 JDBC 的 `Connection` 后执行提交/回滚动作，如代码清单 10-38 所示。

**代码清单 10-38　回滚的底层源码逻辑**

```
public final void rollback(TransactionStatus status) throws TransactionException {
    // 如果事务已经完成，则无法继续回滚
    if (status.isCompleted()) {
        // throw ex ......
    }

    DefaultTransactionStatus defStatus = (DefaultTransactionStatus) status;
    // 回滚事务
    processRollback(defStatus, false);
}

private void processRollback(DefaultTransactionStatus status, boolean unexpected) {
    try {
        // ......
            // 如果存在保存点，则直接回滚到保存点位置
            if (status.hasSavepoint()) {
                status.rollbackToHeldSavepoint();
            }
            // 对于新事务，直接回滚
            else if (status.isNewTransaction()) {
                doRollback(status);
            }
        // ......
}

@Override
protected void doRollback(DefaultTransactionStatus status) {
    DataSourceTransactionObject txObject =
            (DataSourceTransactionObject) status.getTransaction();
    Connection con = txObject.getConnectionHolder().getConnection();
    try {
        con.rollback();
```

```
        } // catch ......
}
```

至此，异常的事务回滚也测试完成。

**5. 事务执行的后处理**

无论是事务的提交还是回滚，在整个方法的最后都要执行一个 cleanupAfter
Completion 方法，这个方法用来清除整个事务执行过程中所需的 ConnectionHolder 等组件以及解除线程中同步的事务信息，如代码清单 10-39 所示。cleanupAfterCompletion 方法中的前两个 if 结构都是与组件资源清除相关的工作，逻辑比较简单，感兴趣的读者可以自行深入研究；最后一个 if 结构中有一个 resume 的动作，它的作用是释放挂起的事务，这个逻辑在 10.5 节会讲到。

**代码清单 10-39　invokeWithinTransaction（5）**

```
// ......
finally {
    cleanupAfterCompletion(status);
}
}

private void cleanupAfterCompletion(DefaultTransactionStatus status) {
    status.setCompleted();
    if (status.isNewSynchronization()) {
        TransactionSynchronizationManager.clear();
    }
    if (status.isNewTransaction()) {
        doCleanupAfterCompletion(status.getTransaction());
    }
    if (status.getSuspendedResources() != null) {
        // logger ......
        Object transaction = (status.hasTransaction() ? status.getTransaction() : null);
        resume(transaction,(SuspendedResourcesHolder) status.getSuspendedResources());
    }
}
```

至此，整个事务控制全流程执行完毕，底层的控制原理也完整介绍了一遍。建议读者实际动手结合代码 Debug 跟进几遍逻辑，以加深印象。

## 10.5　声明式事务的传播行为控制

事务传播行为指的是外层事务传播到内层事务后内层事务做出的行为（持有的态度）。实际的项目开发中难免会出现 Service 之间嵌套的场景，Spring Framework 通过事务传播行为可以控制一个 Service 中的事务传播到另一个 Service 方法中的方式。Spring Framework 中定义了 7 种传播行为，如表 10-1 所示。

**表 10-1　Spring Framework 中定义的事务传播行为**

| 传播行为 | 含义 |
| --- | --- |
| PROPAGATION_REQUIRED：必需的 | 如果当前没有事务运行，则会开启一个新的事务；如果当前已经有事务运行，则方法会运行在当前事务中 |

续表

| 传播行为 | 含义 |
|---|---|
| PROPAGATION_REQUIRES_NEW：新事务 | 如果当前没有事务运行，则会开启一个新的事务；如果当前已经有事务运行，则会将原事务挂起（暂停），重新开启一个新的事务。当新的事务运行完毕后，再将原来的事务释放 |
| PROPAGATION_SUPPORTS：支持 | 如果当前有事务运行，则方法会运行在当前事务中；如果当前没有事务运行，则不会创建新的事务（即不运行在事务中） |
| PROPAGATION_NOT_SUPPORTED：不支持 | 如果当前有事务运行，则会将该事务挂起（暂停）；如果当前没有事务运行，则方法也不会运行在事务中 |
| PROPAGATION_MANDATORY：强制 | 当前方法必须运行在事务中，如果没有事务，则抛出异常 |
| PROPAGATION_NEVER：不允许 | 当前方法不允许运行在事务中，如果当前已经有事务运行，则抛出异常 |
| PROPAGATION_NESTED：嵌套 | 如果当前没有事务运行，则开启一个新的事务；如果当前已经有事务运行，则会记录一个保存点，并继续运行在当前事务中。如果子事务运行中出现异常，则不会全部回滚，而是回滚到上一个保存点 |

> 提示：本书不会讲解全部事务传播行为，而是选择最常见的 PROPAGATION_REQUIRED、比较有研究价值的 PROPAGATION_REQUIRES_NEW 讲解，对于比较简单的传播行为，在剖析 PROPAGATION_REQUIRED 传播行为时会一并讲解。

### 10.5.1 修改测试代码

为了演示事务传播行为在不同 Service 之间嵌套的效果，下面在示例项目中创建一个新的 DeptService，并使其依赖 UserService，如代码清单 10-40 所示。

**代码清单 10-40　DeptService 用于测试事务传播行为**

```
@Service
public class DeptService {

    @Autowired
    private UserService userService;

    @Transactional(rollbackFor = Exception.class)
    public void save() {
        System.out.println("DeptService save run ......");
        userService.test();
    }
}
```

随后修改 SpringBootJdbcApplication 的 main 方法，从 IOC 容器中获取 DeptService 并调用其 save 方法，如代码清单 10-41 所示。

**代码清单 10-41　获取 DeptService 触发事务控制**

```
public static void main(String[] args) {
    ApplicationContext ctx = SpringApplication.run(SpringBootJdbcApplication.class, args);
    DeptService deptService = ctx.getBean(DeptService.class);
    deptService.save();
}
```

## 10.5.2 PROPAGATION_REQUIRED

默认情况下，标注 @Transactional 注解对应的事务传播行为是 PROPAGATION_REQUIRED，这种传播行为的特征是执行 Service 方法时必定确保事务的开启，而创建事务的位置是 10.4 节中提到的 createTransactionIfNecessary 方法，如代码清单 10-42 所示。

**代码清单 10-42　创建/获取事务的核心动作 createTransactionIfNecessary**

```
// ......
PlatformTransactionManager ptm = asPlatformTransactionManager(tm);
final String joinpointIdentification = methodIdentification(method, targetClass, txAttr);

if (txAttr == null || !(ptm instanceof CallbackPreferringPlatformTransactionManager)) {
    // 此处会创建/获取事务
    TransactionInfo txInfo = createTransactionIfNecessary(ptm, txAttr,
            joinpointIdentification);

    Object retVal;
    try {
        retVal = invocation.proceedWithInvocation();
    } // ......
```

本节的研究重点是 createTransactionIfNecessary 方法中底层逻辑面对事务的不同传播行为时会如何进行处理。进入 createTransactionIfNecessary 方法，可以发现整个方法有两个核心动作：根据事务定义信息获取事务状态，以及获取事务状态后的事务信息构建，如代码清单 10-43 所示。下面分别深入研究。

**代码清单 10-43　createTransactionIfNecessary 的方法实现**

```
protected TransactionInfo createTransactionIfNecessary(
        @Nullable PlatformTransactionManager tm,
        @Nullable TransactionAttribute txAttr, final String joinpointIdentification) {
    // 如果事务定义中没有 name，则将方法名作为事务定义标识名
    if (txAttr != null && txAttr.getName() == null) {
        txAttr = new DelegatingTransactionAttribute(txAttr) {
            @Override
            public String getName() {
                return joinpointIdentification;
            }
        };
    }

    TransactionStatus status = null;
    if (txAttr != null) {
        if (tm != null) {
            // 获取事务定义信息对应的事务状态
            status = tm.getTransaction(txAttr);
        } // else logger ......
    }
    // 构建事务信息
    return prepareTransactionInfo(tm, txAttr, joinpointIdentification, status);
}
```

### 1. tm.getTransaction

事务管理器的 getTransaction 方法定义在 DataSourceTransactionManager 的父类 AbstractPlatformTransactionManager 中，这个方法篇幅略长，为便于读者更好地阅读和理解源码，本节将其拆分为多个片段讲解。

（1）doGetTransaction

getTransaction 方法依然是一个需要转调 do 开头的 doGetTransaction 方法，而 doGetTransaction 方法本身是一个模板方法，它的实现又要回到 DataSourceTransactionManager 中。从代码清单 10-44 的下半部分来看，即使大部分 API 都是陌生的，整体思路也比较清晰：doGetTransaction 方法会创建并返回一个 **DataSourceTransactionObject**，其中包含一个 **Connection** 对象。ConnectionHolder 的设计比较类似于 BeanDefinition 的持有者 BeanDefinitionHolder，它们的本质都是内部组合了一个对象。

**代码清单 10-44　getTransaction（1）**

```
public final TransactionStatus getTransaction(@Nullable TransactionDefinition definition)
        throws TransactionException {
    TransactionDefinition def = (definition != null
            ? definition : TransactionDefinition.withDefaults());
    Object transaction = doGetTransaction();
    // ......
}

protected Object doGetTransaction() {
    DataSourceTransactionObject txObject = new DataSourceTransactionObject();
    txObject.setSavepointAllowed(isNestedTransactionAllowed());
    ConnectionHolder conHolder = (ConnectionHolder) TransactionSynchronizationManager
            .getResource(obtainDataSource());
    txObject.setConnectionHolder(conHolder, false);
    return txObject;
}
```

doGetTransaction 方法执行完毕后，可以简单地理解为返回了一个 Connection 对象。

（2）传播行为处理（1）

紧接着的 if 结构看似简单，实际上这个 handleExistingTransaction 方法大有作为，它里面的处理逻辑很复杂，都是当外部事务已经存在时的处理逻辑，如代码清单 10-45 所示。

**代码清单 10-45　getTransaction（2）**

```
// ......
if (isExistingTransaction(transaction)) {
    return handleExistingTransaction(def, transaction, debugEnabled);
}
// ......
```

具体的传播行为如何处理，在下面再展开讲解。

（3）超时检测

接着的 if 结构实际上是校验 @Transactional 注解中的 timeout 属性，如代码清单 10-46 所示。Spring Framework 支持的事务模型可以设置事务方法最长执行时间，如果在 Service 方法中标注的 @Transactional 注解显式设置了 timeout 的最长耗时，则当方法执行时间过长时会由底层抛出异常。默认情况下 timeout 的值是-1，表示不限制事务方法执行时间，但如果显式设置的 timeout 值比 -1 小，说明设置的值本身不合理，底层也会抛出异常。

**代码清单 10-46　getTransaction（3）**

```
// ......
if (def.getTimeout() < TransactionDefinition.TIMEOUT_DEFAULT) {
    throw ex ......
}
// ......
```

（4）传播行为处理（2）

getTransaction 方法的最后一个片段中，可以看到部分事务传播行为的处理逻辑，如代码清单 10-47 所示。请注意，如果程序进入该部分源码，说明线程中没有开启事务，此时如果方法的事务定义信息配置的传播行为是 MANDATORY，则直接抛出异常；如果传播行为配置了 REQUIRED、REQUIRES_NEW 或 NESTED，则会直接开启一个新的事务；如果是最后的几种情况，即使没有显式列举其他传播行为，也可以推断出来，除去上面列举的 4 种传播行为，还剩下 3 种：SUPPORTS、NOT_SUPPORTED 和 NEVER，它们都可以在没有事务的情况下运行。如果代码运行至此处，说明目前的确没有事务，无须做任何处理，因此 else 部分中不会有任何多余逻辑执行。

**代码清单 10-47　getTransaction（4）**

```
// ......
// 当前没有事务 -> 检查传播行为以决定如何向下执行
if (def.getPropagationBehavior()== TransactionDefinition.PROPAGATION_MANDATORY) {
    throw ex ......
} else if (def.getPropagationBehavior() == TransactionDefinition.PROPAGATION_REQUIRED
    || def.getPropagationBehavior() == TransactionDefinition.PROPAGATION_REQUIRES_NEW
    || def.getPropagationBehavior() == TransactionDefinition.PROPAGATION_NESTED) {
    SuspendedResourcesHolder suspendedResources = suspend(null);
    try {
        return startTransaction(def, transaction, debugEnabled, suspendedResources);
    } // catch ......
} else {
    boolean newSynchronization = (getTransactionSynchronization() == SYNCHRONIZATION_ALWAYS);
    return prepareTransactionStatus(def, null, true, newSynchronization,
        debugEnabled, null);
}
}
```

实际 Debug 至此处，程序会进入 else-if 分支，如图 10-9 所示。不难理解，此时线程中没有事务，而 DeptService 的 save 方法中事务传播行为是默认的 **REQUIRED**，所以此处需要开启一个新的事务。

```
else if (def.getPropagationBehavior() == TransactionDefinition.PROPAGATION_REQUIRED ||
        def.getPropagationBehavior() == TransactionDefinition.PROPAGATION_REQUIRES_NEW ||
        def.getPropagationBehavior() == TransactionDefinition.PROPAGATION_NESTED) {
    SuspendedResourcesHolder suspendedResources = suspend(null);  suspendedResources: null
    if (debugEnabled) {...}
    try {
        return startTransaction(def, transaction, debugEnabled, suspendedResources);  def: TransactionAspectSupport$
    }
    catch (RuntimeException | Error ex) {
```

图 10-9　DeptService 的 save 方法执行时进入 else-if 块

下面继续深入开启事务的 startTransaction 方法。

### 2. startTransaction：开启事务

从代码清单 10-48 中可以发现一个特别显眼的方法 **doBegin**。根据探究前面源码的经验，想必读者可以马上反应过来，这个方法就是开启事务的核心逻辑，如代码清单 10-48 所示。

**代码清单 10-48　startTransaction 开启一个新的事务**

```java
private TransactionStatus startTransaction(TransactionDefinition definition,
        Object transaction, boolean debugEnabled,
        @Nullable SuspendedResourcesHolder suspendedResources) {
    boolean newSynchronization = (getTransactionSynchronization() != SYNCHRONIZATION_NEVER);
    DefaultTransactionStatus status = newTransactionStatus(definition,
            transaction, true, newSynchronization, debugEnabled, suspendedResources);
    // 开启事务
    doBegin(transaction, definition);
    prepareSynchronization(status, definition);
    return status;
}
```

进入 doBegin 方法中可以发现源码篇幅很长，代码清单 10-49 中只截取最吸引各位的一段，可以发现 doBegin 方法中的主要工作是从 DataSourceTransactionObject 内部组合的 ConnectionHolder 中提取出真正的 Connection 对象，随后执行 **setAutoCommit(false)** 方法关闭自动提交，即开启事务，由此可以了解到原生 JDBC 事务开启的位置。

**代码清单 10-49　doBegin 开启事务的核心**

```java
@Override
protected void doBegin(Object transaction, TransactionDefinition definition) {
    DataSourceTransactionObject txObject = (DataSourceTransactionObject) transaction;
    Connection con = null;
    try {
        // ......
        txObject.getConnectionHolder().setSynchronizedWithTransaction(true);
        // 获取到真正的 Connection 对象
        con = txObject.getConnectionHolder().getConnection();
        // ......

        if (con.getAutoCommit()) {
            txObject.setMustRestoreAutoCommit(true);
            // 开启事务
            con.setAutoCommit(false);
        }
        // ......
}
```

## 3. prepareTransactionInfo

从事务管理器中获取到 `TransactionStatus` 后,下一步要执行的是 `prepareTransactionInfo` 方法,而这个方法仅创建一个 `TransactionInfo`,把事务状态信息放入其中,并将其绑定到当前线程中,如代码清单 10-50 所示。当创建 `TransactionInfo` 并返回之后,整个 `createTransactionIfNecessary` 方法执行完毕,事务也就开启了。

**代码清单 10-50　prepareTransactionInfo 封装事务状态信息到当前线程中**

```
    // ......
    if (tm != null) {
        status = tm.getTransaction(txAttr);
    } // else logger ......
    }
    return prepareTransactionInfo(tm, txAttr, joinpointIdentification, status);
}

protected TransactionInfo prepareTransactionInfo(@Nullable PlatformTransactionManager tm,
        @Nullable TransactionAttribute txAttr, String joinpointIdentification,
        @Nullable TransactionStatus status) {
    TransactionInfo txInfo = new TransactionInfo(tm, txAttr, joinpointIdentification);
    if (txAttr != null) {
        txInfo.newTransactionStatus(status);
    } // else logger ......

    txInfo.bindToThread();
    return txInfo;
}
```

## 4. UserService.test

`DeptService#save` 方法的事务开启之后,它的内部会调用 `UserService` 的 `test` 方法,由于 `userService` 也是被 AOP 代理过的代理对象,在调用 `test` 方法时会再次进入 `TransactionInterceptor`,而调用逻辑还是之前的那些环节,为了快速定位到事务传播行为的部分,这里直接进入 `tm.getTransaction->doGetTransacation` 的步骤。由于此时线程中已存在事务,因此 `doGetTransacation` 方法中的这个 if 分支会触发进入,如代码清单 10-51 所示。

**代码清单 10-51　UserService 的 test 方法执行时线程中已有事务**

```
if (isExistingTransaction(transaction)) {
    // 当前没有事务 -> 检查传播行为以决定如何向下执行
    return handleExistingTransaction(def, transaction, debugEnabled);
}
```

## 5. handleExistingTransaction

`handleExistingTransaction` 方法中的所有分支均为事务传播行为的判断与行为动作,逐一来看。

(1) NEVER 的处理

对于 **NEVER** 的传播行为,要求线程中不能有事务,所以如果线程中检测到事务,则抛出异常,如代码清单 10-52 所示。

**代码清单 10-52　NEVER 的处理**

```java
private TransactionStatus handleExistingTransaction(TransactionDefinition definition,
        Object transaction, boolean debugEnabled) throws TransactionException {

    if (definition.getPropagationBehavior() == TransactionDefinition.PROPAGATION_NEVER) {
        throw ex ......
    }
    // ......
```

（2）NOT_SUPPORTED 的处理

对于 **NOT_SUPPORTED** 的传播行为，要求方法执行不在事务中，所以此处会把当前事务挂起，并执行目标方法，如代码清单 10-53 所示。

**代码清单 10-53　NOT_SUPPORTED 的处理**

```java
// ......
if (definition.getPropagationBehavior() == TransactionDefinition.PROPAGATION_NOT_SUPPORTED) {
    // logger ......
    // 挂起当前事务
    Object suspendedResources = suspend(transaction);
    boolean newSynchronization = (getTransactionSynchronization() == SYNCHRONIZATION_ALWAYS);
    return prepareTransactionStatus(
            definition, null, false, newSynchronization, debugEnabled, suspendedResources);
}
// ......
```

（3）REQUIRES_NEW 的处理

对于 **REQUIRES_NEW** 的传播行为，会开启一个新的事务，不过在此之前还要挂起已有的外层事务。除此之外，从源码中可以发现，如果新事务抛出异常，会恢复之前的外层事务。

**代码清单 10-54　REQUIRES_NEW 的处理**

```java
// ......
if (definition.getPropagationBehavior() == TransactionDefinition.PROPAGATION_REQUIRES_NEW) {
    // logger ......
    // 挂起当前外层事务
    SuspendedResourcesHolder suspendedResources = suspend(transaction);
    try {
        // 开启一个全新的事务
        return startTransaction(definition, transaction, debugEnabled, suspendedResources);
    } catch (RuntimeException | Error beginEx) {
        // 还原外层事务
        resumeAfterBeginException(transaction, suspendedResources, beginEx);
        throw beginEx;
    }
}
// ......
```

（4）NESTED 的处理

对于 **NESTED** 嵌套事务的处理，除了需要允许嵌套事务在程序中启用，还需要连接的数据库支持基于保存点的嵌套事务。如果这两个条件都成立，则会创建保存点，否则会降级为

REQUIRES_NEW 的传播行为，开启一个全新的事务，如代码清单 10-55 所示。

**代码清单 10-55　NESTED 的处理**

```
// ......
if (definition.getPropagationBehavior() == TransactionDefinition.PROPAGATION_NESTED) {
    // 检查是否允许嵌套事务
    if (!isNestedTransactionAllowed()) {
        throw ex ......
    }
    // logger ......
    // 判断是否支持保存点
    if (useSavepointForNestedTransaction()) {
        DefaultTransactionStatus status = prepareTransactionStatus(definition,
                transaction, false, false, debugEnabled, null);
        status.createAndHoldSavepoint();
        return status;
    } else {
        return startTransaction(definition, transaction, debugEnabled, null);
    }
}
// ......
```

### （5）SUPPORTS&REQUIRED

最后的部分是处理 REQUIRED 和 SUPPORTS 的逻辑。由于 `isValidateExistingTransaction()` 方法在默认情况下返回 `false`，因此不会进入中间的 if 结构，而最后的两行源码仅是把现有的事务信息封装并返回，没有任何额外的逻辑，如代码清单 10-56 所示。

**代码清单 10-56　SUPPORTS&REQUIRED 的处理**

```
// Assumably PROPAGATION_SUPPORTS or PROPAGATION_REQUIRED.
if (isValidateExistingTransaction()) {
    // ......
}
boolean newSynchronization =(getTransactionSynchronization() != SYNCHRONIZATION_NEVER);
return prepareTransactionStatus(definition, transaction, false,
        newSynchronization, debugEnabled, null);
}
```

由上述两部分传播行为的源码分析，想必读者应该对大部分传播行为的底层逻辑有了一个大概的认识。下面再着重研究 PROPAGATION_REQUIRES_NEW 的传播行为。

## 10.5.3　PROPAGATION_REQUIRES_NEW

将 UserService 的 `test` 方法上的 `@Transactional` 注解事务传播行为改为 REQUIRES_NEW，并重新 Debug 测试。由上面的分析可知，REQUIRES_NEW 会在 handleExistingTransaction 方法中暂停并检查，如代码清单 10-57 所示。

**代码清单 10-57　检查 PROPAGATION_REQUIRES_NEW**

```
// ......
if (definition.getPropagationBehavior() == TransactionDefinition.PROPAGATION_REQUIRES_NEW) {
    // logger ......
```

```
        // 挂起当前外层事务
        SuspendedResourcesHolder suspendedResources = suspend(transaction);
        try {
            // 开启一个全新的事务
            return startTransaction(definition, transaction, debugEnabled, suspendedResources);
        }// catch 还原外层事务 throw ex ......
    }
    // ......
```

看似只是简单地开启和处理新事务,实际上里面有几个需要注意的细节,下面逐一讲解。

**1. 新事务的创建细节**

新事务在创建之前会将外层的原事务挂起,挂起的逻辑中有一个细节值得关注,如代码清单 10-58 所示。仔细观察源码中的两行注释,标注的都是 doSuspend 方法,这就意味着真正挂起的动作在 doSuspend 方法中。

**代码清单 10-58　suspend 方法转调 doSuspend 方法**

```
protected final SuspendedResourcesHolder suspend(@Nullable Object transaction)
        throws TransactionException {
    if (TransactionSynchronizationManager.isSynchronizationActive()){
        List<TransactionSynchronization> suspendedSynchronizations =
                doSuspendSynchronization();
        try {
            Object suspendedResources = null;
            if (transaction != null) {
                // 注意此处: doSuspend
                suspendedResources = doSuspend(transaction);
            }
            // ......
        } catch (RuntimeException|Error ex) {
            // doSuspend failed - original transaction is still active...
            doResumeSynchronization(suspendedSynchronizations);
            throw ex;
        }
    } else if (transaction != null) {
        // Transaction active but no synchronization active.
        // 注意此处: doSuspend
        Object suspendedResources =doSuspend(transaction);
        return new SuspendedResourcesHolder(suspendedResources);
    } // else return null
}
```

由于 doxxx 方法一般都是模板方法,具体的实现需要最终的落地实现类,从 DataSourceTransactionManager 中可以找到 doSuspend 方法的实现,如代码清单 10-59 所示。方法的具体逻辑是移除 DataSourceTransactionObject 中的 Connection 对象,并且解除 TransactionSynchronizationManager 中的数据源绑定。

**代码清单 10-59　DataSourceTransactionManager 中 doSuspend 方法的实现**

```
@Override
protected Object doSuspend(Object transaction) {
    DataSourceTransactionObject txObject = (DataSourceTransactionObject) transaction;
    txObject.setConnectionHolder(null);
```

```
    return TransactionSynchronizationManager.unbindResource(obtainDataSource());
}
```

实际 Debug 时可以发现,解绑之后返回的是组合了 Connection 的 ConnectionHolder,如图 10-10 所示。

```
Object suspendedResources = null;    suspendedResources: ConnectionHolder@2950
if (transaction != null) {
    suspendedResources = doSuspend(transaction);    suspendedResources: ConnectionHolder@2950
}
```

图 10-10　解除绑定后返回的是 ConnectionHolder

随后回到上面的 suspend 方法中,注意观察最后方法的返回值,它将 ConnectionHolder 包装为一个 SuspendedResourcesHolder 之后返回,这个 SuspendedResourcesHolder 对象会在下面的两个环节发挥作用。

**2. 开启事务的细节**

注意观察代码清单 10-60 中执行的 startTransaction 方法中参数列表的最后一个参数,上一步返回的 SuspendedResourcesHolder 对象会在此处被一并传入,并且存入 TransactionStatus 中。这样做的目的是,当内层新事务执行完成后清理相关的线程同步等信息时可以获得事务状态信息,从中提取出被挂起的事务,然后继续恢复外层事务。

**代码清单 10-60　startTransaction 方法的参数中包含 SuspendedResourcesHolder**

```
private TransactionStatus startTransaction(TransactionDefinition definition,Object transaction,
        boolean debugEnabled, @Nullable SuspendedResourcesHolder suspendedResources) {

    boolean newSynchronization = (getTransactionSynchronization() != SYNCHRONIZATION_NEVER);
    DefaultTransactionStatus status = newTransactionStatus(definition,
            transaction, true, newSynchronization, debugEnabled, suspendedResources);
    doBegin(transaction, definition);
    prepareSynchronization(status, definition);
    return status;
}
```

如果读者没有立即理解也不要紧张,下面马上讲解。

**3. 内层新事务执行完成后的细节**

当事务执行完成之后,会执行 cleanupAfterCompletion 方法,以清除其中的线程同步信息等,如代码清单 10-61 所示。

**代码清单 10-61 cleanupAfterCompletion 方法清除信息**

```
private void cleanupAfterCompletion(DefaultTransactionStatus status) {
    status.setCompleted();
    if (status.isNewSynchronization()) {
        TransactionSynchronizationManager.clear();
    }
    if (status.isNewTransaction()) {
        doCleanupAfterCompletion(status.getTransaction());
    }
    if (status.getSuspendedResources() != null) {
```

```
        // logger ......
        Object transaction = (status.hasTransaction() ? status.getTransaction() : null);
        resume(transaction, (SuspendedResourcesHolder) status.getSuspendedResources());
    }
}
```

注意观察最下面的 if 部分，这个动作是恢复挂起事务的核心动作，10.4 节中未展开，此处展开解读一下源码的核心内容，如代码清单 10-62 所示。resume 方法的逻辑非常像前面 suspend 方法的逆动作，上面的 doResume 方法对应 suspend 中的 doSuspend 方法，下面的 set 动作对应 suspend 中的 set(null)。另外，doResume 方法的逻辑更为简单，且刚好跟代码清单 10-59 一样，一个是 bindResource，另一个是 unbindResource。

**代码清单 10-62  resume 释放事务**

```
protected final void resume(@Nullable Object transaction,
        @Nullable SuspendedResourcesHolder resourcesHolder) throws TransactionException {

    if (resourcesHolder != null) {
        Object suspendedResources = resourcesHolder.suspendedResources;
        if (suspendedResources != null) {
            doResume(transaction, suspendedResources);
        }
        List<TransactionSynchronization> suspendedSynchronizations =
                resourcesHolder.suspendedSynchronizations;
        if (suspendedSynchronizations != null) {
            TransactionSynchronizationManager.setActualTransactionActive(
                    resourcesHolder.wasActive);
            TransactionSynchronizationManager.setCurrentTransactionIsolationLevel(
                    resourcesHolder.isolationLevel);
            TransactionSynchronizationManager.setCurrentTransactionReadOnly(
                    resourcesHolder.readOnly);
            TransactionSynchronizationManager.setCurrentTransactionName(resourcesHolder.name);
            doResumeSynchronization(suspendedSynchronizations);
        }
    }
}

protected void doResume(@Nullable Object transaction,Object suspendedResources) {
    TransactionSynchronizationManager.bindResource(obtainDataSource(), suspendedResources);
}
```

等 resume 方法执行完毕后，原来的外层事务又重新被绑定到线程上，相当于恢复了被挂起之前的状态，这就体现了 REQUIRES_NEW 的处理逻辑。

## 10.6  小结

本章全方位研究了 Spring Boot 整合 JDBC 场景下的组件装配，以及注解声明式事务的生效原理、控制流程、事务传播行为等场景。Spring Boot 整合的事务场景底层依然是 Spring Framework 已有的功能，Spring Boot 做的事情仅是对默认场景下的组件自动进行装配。事务管理是 AOP 技术的经典应用，掌握事务管理底层的模型离不开 AOP 部分后置处理器、增强器、通知等重要组件的支撑。

大多数的项目开发中，更多的选择是使用持久层框架 MyBatis 或 Spring Data（Hibernate），而不是原生的 spring-jdbc。第 11 章会重点研究 Spring Boot 整合 MyBatis 持久层框架后底层完成的装配工作。

# 第 11 章 Spring Boot 整合 MyBatis

本章主要内容：
◇ MyBatis 框架概述与工程整合；
◇ Spring Boot 整合 MyBatis 的核心自动装配。

第 10 章中系统地讲解了 Spring Boot 整合 JDBC 场景下的自动装配，以及注解声明式事务的底层原理。尽管 Spring Framework 和 Spring Boot 在简化与数据库的交互方面已经做了很大的努力，但是在实际的项目开发中更多的场景是整合成熟的持久层框架来完成与数据库的交互。目前市面上比较流行的持久层框架包括 Spring Data JPA（底层默认依赖 Hibernate）和 MyBatis。Spring Data 本身属于 Spring 生态的一部分，与 Spring Boot 的整合非常简单，本书不对此展开讲解。MyBatis 作为第三方框架，它与 Spring Boot 的整合方式是通过第三方框架编写 starter 场景启动器并配置相应的组件自动装配。本章内容会以 MyBatis 整合 Spring Boot 的核心场景启动器为切入点，研究 MyBatis 如何完成与 Spring Boot 的场景整合。

## 11.1 MyBatis 框架概述

MyBatis 是一款优秀的持久层框架，它支持自定义 SQL、存储过程以及高级映射。MyBatis 免除了几乎所有的 JDBC 代码以及设置参数和获取结果集的工作。MyBatis 可以通过简单的 XML 或注解来配置和映射原始类型、接口和普通老式 Java 对象（Plain Old Java Object，POJO）为数据库中的记录。

从整体上讲，MyBatis 的架构可以分为三层，如图 11-1 所示。

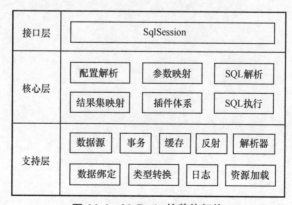

图 11-1 MyBatis 的整体架构

- 接口层：`SqlSession` 是平时与 MyBatis 完成交互的核心接口（包括整合 Spring Framework 和 Spring Boot 后用到的 `SqlSessionTemplate`）。
- 核心层：`SqlSession` 执行的方法，包括底层需要经过配置文件的解析、SQL 解析，以及执行 SQL 时的参数映射、SQL 执行、结果集映射，另外还有穿插其中的扩展插件。
- 支持层：核心层的功能实现，它基于底层的各个模块，以共同协调完成。

总体来讲，使用 MyBatis 可以灵活、完整、相对轻量化地与数据库进行交互。

## 11.2 Spring Boot 整合 MyBatis 项目搭建

下面来搭建一个 Spring Boot 整合 MyBatis 的项目。由于 MyBatis 整合 Spring Boot 的场景启动器中已经对 Spring Boot 原生的 starter 做了整合，因此在导入依赖时，只需要引入 mybatis-spring-boot-starter 和具体的数据库连接驱动，如代码清单 11-1 所示。

**代码清单 11-1　引入 MyBatis 整合 Spring Boot 的必要依赖**

```xml
<dependencies>
    <dependency>
        <groupId>org.mybatis.spring.boot</groupId>
        <artifactId>mybatis-spring-boot-starter</artifactId>
        <version>2.1.3</version>
    </dependency>
    <dependency>
        <groupId>mysql</groupId>
        <artifactId>mysql-connector-java</artifactId>
        <version>5.1.47</version>
    </dependency>
</dependencies>
```

> 提示：mybatis-spring-boot-starter 2.1.3 版本底层依赖 Spring Boot 2.3.0，与本书研究的 Spring Boot 主要版本同属一个中版本，整合的可靠性高。

有关数据库搭建的部分，可以直接使用第 10 章的 `springboot-dao` 作为连接库，本章不再重新创建；连接数据源的内容，同样参照第 10 章配置，具体可见代码清单 10-3，本章不再赘述。

有关测试代码，部分内容与第 10 章类似，区别是 Dao 层的实现由依赖 `JdbcTemplate` 的 `UserDao` 替换为 Mapper 动态代理接口 `UserMapper` 以及对应的 XML 映射文件 `UserMapper.XML`，具体的代码如代码清单 11-2 和代码清单 11-3 所示。

**代码清单 11-2　UserMapper 接口**

```java
@Mapper
public interface UserMapper {

    void save(User user);

    List<User> findAll();
}
```

### 代码清单 11-3　UserMapper.XML 映射文件

```xml
<?xml version="1.0" encoding="UTF-8"?>
<!DOCTYPE mapper PUBLIC "-//mybatis.org//DTD Mapper 3.0//EN"
        "http://mybatis.org/dtd/mybatis-3-mapper.dtd">
<mapper namespace="com.linkedbear.springboot.mapper.UserMapper">
    <insert id="save" parameterType="User">
        insert into tbl_user (name, tel) values (#{name}, #{tel})
    </insert>
    
    <select id="findAll" resultType="User">
        select * from tbl_user
    </select>
</mapper>
```

另外，为了能使 MyBatis 完成实体类别名的预处理以及 mapper.xml 文件的扫描，需要在 Spring Boot 全局配置文件中编写一些额外的配置，如代码清单 11-4 所示。

### 代码清单 11-4　有关 MyBatis 的配置

```
mybatis.type-aliases-package=com.linkedbear.springboot.entity
mybatis.mapper-locations=classpath:mapper/*.xml
mybatis.configuration.map-underscore-to-camel-case=true
```

最后编写 Spring Boot 主启动类，它的编写与第 10 章的内容几乎完全一致，仅是主启动类的类名不同而已，如代码清单 11-5 所示。

### 代码清单 11-5　主启动类中获取 UserService 并调用

```java
@SpringBootApplication
@EnableTransactionManagement
public class MyBatisSpringBootApplication {
    
    public static void main(String[] args) {
        ApplicationContext ctx = SpringApplication.run(MyBatisSpringBootApplication.class, args);
        UserService userService = ctx.getBean(UserService.class);
        userService.test();
    }
}
```

经过上述代码编写，就完成了 Spring Boot 与 MyBatis 的整合。

## 11.3　自动装配核心

在上面的项目搭建中，仅编写了少量的几行配置，便完成了与 MyBatis 框架的整合，而底层整合的核心部分，一定是自动配置类完成的大多数组件装配和默认配置的应用。接下来的内容会分析 MyBatis 的场景启动器，研究 MyBatis 在自动装配的部分做了哪些工作。

### 11.3.1　场景启动器的秘密

借助 IDE，观察 `mybatis-spring-boot-starter` 依赖，可以发现它仅是一个空的 jar

包，没有具体的内容，而这个依赖本身又依赖 `mybatis-spring-boot-autoconfigure`。通过前面章节的学习，可以得知这个依赖中通常包含一个 `spring.factories` 文件。打开 jar 包依赖，可以发现它的确包含一个 `spring.factories` 文件，这个文件中仅定义了两个自动配置类，如代码清单 11-6 所示。

**代码清单 11-6　MyBatis 整合 SpringBoot 的自动配置类**

```
org.springframework.boot.autoconfigure.EnableAutoConfiguration=
    org.mybatis.spring.boot.autoconfigure.MybatisLanguageDriverAutoConfiguration,
    org.mybatis.spring.boot.autoconfigure.MybatisAutoConfiguration
```

由此可知，MyBatis 整合 SpringBoot 的核心装配就在这两个自动配置类上，下面逐一展开研究。

### 11.3.2　MybatisLanguageDriverAutoConfiguration

从 `MybatisLanguageDriverAutoConfiguration` 的类名中可以得知，它的配置与"语言驱动"有关。要理解"语言驱动"这个概念，需要读者先了解 MyBatis 中的设计。MyBatis 默认使用 XML 映射文件作为载体，且 XML 映射文件中的内容是固定的几个标签，从 MyBatis 3.2 版本开始支持使用第三方模板引擎框架作为编写映射文件的实现，而使用 XML 文件编写仅是 MyBatis 默认提供的映射文件实现。

从 `MybatisLanguageDriverAutoConfiguration` 的源码中可以发现，MyBatis 支持的第三方模板引擎框架包含 FreeMarker、Velocity、Thymeleaf 等，但是默认场景下 MyBatis 不会引入过多的依赖，这也使得开发者仅熟悉原生的 XML 映射文件编写即可。这个配置类本身不重要，读者不必在此耗费过多的精力。

### 11.3.3　MybatisAutoConfiguration

`spring.factories` 中的另一个自动配置类 `MybatisAutoConfiguration` 才是 MyBatis 整合 Spring Boot 的核心。MyBatis 内部支撑运行的所有组件都在这个配置类中创建。本节内容会完整讲解 `MybatisAutoConfiguration` 中的所有配置内容。

#### 1. SqlSessionFactory

熟悉 MyBatis 的读者一定清楚，MyBatis 的底层核心支撑是 `SqlSessionFactory`，有了 `SqlSessionFactory` 就可以创建 `SqlSession`，进而使用 `SqlSession` 进行 CRUD 操作。`MybatisAutoConfiguration` 中注册的最关键组件就是这个 `SqlSessionFactory`，代码清单 11-7 中列举了创建 `SqlSessionFactory` 的部分核心逻辑。

**代码清单 11-7　SqlSessionFactory 的创建**

```java
@Bean
@ConditionalOnMissingBean
public SqlSessionFactory sqlSessionFactory(DataSource dataSource) throws Exception {
    SqlSessionFactoryBean factory = new SqlSessionFactoryBean();
    // 数据源
    factory.setDataSource(dataSource);
    factory.setVfs(SpringBootVFS.class);
```

```
// 外部 MyBatis 原生配置文件
if (StringUtils.hasText(this.properties.getConfigLocation())) {
    factory.setConfigLocation(
        this.resourceLoader.getResource(this.properties.getConfigLocation()));
}
// 应用 properties 配置
applyConfiguration(factory);
if (this.properties.getConfigurationProperties() != null) {
    factory.setConfigurationProperties(this.properties.getConfigurationProperties());
}
// 设置插件
if (!ObjectUtils.isEmpty(this.interceptors)) {
    factory.setPlugins(this.interceptors);
}
// 更多 set 操作......

return factory.getObject();
}
```

总体看来，创建 `SqlSessionFactory` 的步骤只是把连接数据库的数据源、全局配置文件中提取出的配置对象、IOC 容器中注册的 MyBatis 拦截器等组件一一应用在 `SqlSessionFactoryBean` 中，而实际构建 `SqlSessionFactory` 的动作是最后一行的 `factory.getObject();`方法调用，这个方法中包含一个至关重要的方法 `afterPropertiesSet`，它会在经过一些前置判断后执行 `buildSqlSessionFactory` 方法，以构建实际的 `SqlSessionFactory` 对象，如代码清单 11-8 所示。

**代码清单 11-8　getObject 会触发 afterPropertiesSet 方法**

```
public SqlSessionFactory getObject() throws Exception {
    if (this.sqlSessionFactory == null){
        afterPropertiesSet();
    }
    return this.sqlSessionFactory;
}

public void afterPropertiesSet() throws Exception {
    // 判断......
    this.sqlSessionFactory=buildSqlSessionFactory();
}
```

正常情况下 `afterPropertiesSet` 是 `InitializingBean` 接口的方法，用于 Bean 初始化阶段的生命周期回调，而 `SqlSessionFactoryBean` 中调用 `getObject` 方法获取目标对象时主动回调的目的是，考虑到在 Spring Boot 的整合场景下 `SqlSessionFactoryBean` 本身不会注册到 IOC 容器中，因此需要手动调用 `afterPropertiesSet` 方法触发内置 `SqlSessionFactory` 对象的构建。

下面着重研究 `buildSqlSessionFactory` 方法。由于这个方法的源码篇幅非常长，为了便于读者更好地阅读和理解源码，这里将其拆分为多个片段讲解。

（1）处理 MyBatis 全局配置对象

`buildSqlSessionFactory`方法的第一部分内容的核心工作是预准备 MyBatis 的全局配

置对象 Configuration，并根据是否事先传入外部 configuration 对象或者传入全局配置文件路径来决定是否准备 XMLConfigBuilder，如代码清单 11-9 所示。如果确实需要 XMLConfigBuilder 的处理，在第 6 步会有配置文件解析。

**代码清单 11-9　buildSqlSessionFactory（1）**

```java
protected SqlSessionFactory buildSqlSessionFactory() throws Exception {
    final Configuration targetConfiguration;

    XMLConfigBuilder XMLConfigBuilder = null;
    // 如果构建 SqlSessionFactoryBean 时传入了外部 Configuration，则直接处理
    if (this.configuration != null) {
        targetConfiguration = this.configuration;
        if (targetConfiguration.getVariables() == null) {
            targetConfiguration.setVariables(this.configurationProperties);
        } else if (this.configurationProperties != null) {
            targetConfiguration.getVariables().putAll(this.configurationProperties);
        }
    } else if (this.configLocation != null) {
        // 如果传入全局配置文件路径，则封装 XMLConfigBuilder 对象以备加载
        xmlConfigBuilder = new XMLConfigBuilder(this.configLocation.getInputStream(), null,
                this.configurationProperties);
        targetConfiguration = xmlConfigBuilder.getConfiguration();
    } else {
        // 如果无外部 Configuration 对象，则执行默认策略
        targetConfiguration = new Configuration();
        Optional.ofNullable(this.configurationProperties)
                .ifPresent(targetConfiguration::setVariables);
    }
    // ......
```

（2）处理内置组件

紧接着的 3 个设置动作分别对应 MyBatis 中的三个组件，如代码清单 11-10 所示。这里简单解释一下。对象工厂 ObjectFactory 负责创建 MyBatis 查询包装结果集时创建结果对象（模型类对象、Map 等），对象包装工厂 ObjectWrapperFactory 负责对创建出的对象进行包装，虚拟文件系统 Vfs 用于加载项目中的资源文件。

**代码清单 11-10　buildSqlSessionFactory（2）**

```java
// ......
Optional.ofNullable(this.objectFactory).ifPresent(targetConfiguration::setObjectFactory);
Optional.ofNullable(this.objectWrapperFactory)
        .ifPresent(targetConfiguration::setObjectWrapperFactory);
Optional.ofNullable(this.vfs).ifPresent(targetConfiguration::setVfsImpl);
// ......
```

在这个环节中 Spring Boot 仅对虚拟文件系统部分进行了扩展，它额外定义了一个 SpringBootVFS 用于加载项目中的资源文件，除此之外没有任何多余的扩展。

（3）处理别名

代码清单 11-11 中的动作是别名的包扫描以及某些特定类的别名设置。MyBatis 允许在项目中为实体类定义别名，这种设计可以使得在映射文件 mapper.xml 中简化类型编写，使映射文件

更清爽简洁。在默认情况下包扫描注册的别名就是类名本身（首字母大写），如果类上标注有 @Alias 注解，则会取注解属性值。

**代码清单 11-11　buildSqlSessionFactory（3）**

```java
// ......
if (hasLength(this.typeAliasesPackage)) {
    scanClasses(this.typeAliasesPackage, this.typeAliasesSuperType).stream()
        .filter(clazz -> !clazz.isAnonymousClass()).filter(clazz -> !clazz.isInterface())
        .filter(clazz -> !clazz.isMemberClass())
            .forEach(targetConfiguration.getTypeAliasRegistry()::registerAlias);
}

if (!isEmpty(this.typeAliases)) {
    Stream.of(this.typeAliases).forEach(typeAlias -> {
        targetConfiguration.getTypeAliasRegistry().registerAlias(typeAlias);
        // logger ......
    });
}
// ......
```

**（4）处理插件、类型处理器**

代码清单 11-12 的逻辑是处理 MyBatis 插件以及 TypeHandler。注意，此处的 this.plugins 是在 MybatisAutoConfiguration 中利用 ObjectProvider 从 IOC 容器中获取到的。由此就可以得出一个简单结论：Spring Boot 整合 MyBatis 后可以直接向 IOC 容器中注册 MyBatis 的关键组件，底层的自动装配可以将这些组件应用到 MyBatis 中。

**代码清单 11-12　buildSqlSessionFactory（4）**

```java
// ......
if (!isEmpty(this.plugins)) {
    Stream.of(this.plugins).forEach(plugin -> {
        targetConfiguration.addInterceptor(plugin);
        // logger ......
    });
}

if (hasLength(this.typeHandlersPackage)) {
    scanClasses(this.typeHandlersPackage, TypeHandler.class).stream()
        .filter(clazz -> !clazz.isAnonymousClass())
        .filter(clazz -> !clazz.isInterface())
        .filter(clazz -> !Modifier.isAbstract(clazz.getModifiers()))
        .forEach(targetConfiguration.getTypeHandlerRegistry()::register);
}

if (!isEmpty(this.typeHandlers)) {
    Stream.of(this.typeHandlers).forEach(typeHandler -> {
        targetConfiguration.getTypeHandlerRegistry().register(typeHandler);
        // logger ......
    });
}

targetConfiguration.setDefaultEnumTypeHandler(defaultEnumTypeHandler);
// ......
```

（5）处理其他内部组件

除了内部核心的对象工厂、插件、类型处理器等组件，MyBatis 还有一些相对不太重要的内部组件，如脚本语言驱动器（支持多种映射文件格式的编写）、数据库厂商标识器（为数据库可移植性提供了可能）等，如代码清单 11-13 所示。在构建 `SqlSessionFactory` 的过程中，这部分组件也会被初始化。

**代码清单 11-13　buildSqlSessionFactory（5）**

```
// ......
if (!isEmpty(this.scriptingLanguageDrivers)) {
    Stream.of(this.scriptingLanguageDrivers).forEach(languageDriver -> {
        targetConfiguration.getLanguageRegistry().register(languageDriver);
        // logger ......
    });
}
Optional.ofNullable(this.defaultScriptingLanguageDriver)
    .ifPresent(targetConfiguration::setDefaultScriptingLanguage);

if (this.databaseIdProvider != null) {
    try {
        targetConfiguration.setDatabaseId(
                this.databaseIdProvider.getDatabaseId(this.dataSource));
    }// catch throw ex ......
}

    Optional.ofNullable(this.cache).ifPresent(targetConfiguration::addCache);
// ......
```

（6）解析 MyBatis 全局配置文件

在代码清单 11-9 中有一段配置文件的处理，如果项目配置中传入了 `configLocation`，则此处会使用 `XMLConfigBuilder` 解析 MyBatis 配置文件，并应用于 MyBatis 全局配置对象 `Configuration` 中。Spring Boot 已经将 MyBatis 中的配置尽可能地移植到 `application.properties` 中，所以代码清单 11-14 中的内容可以忽略。

**代码清单 11-14　buildSqlSessionFactory（6）**

```
// ......
if (XMLConfigBuilder != null) {
    try {
        xmlConfigBuilder.parse();
        // logger ......
    }// catch finally ......
}
// ......
```

（7）处理数据源和事务工厂

代码清单 11-15 中只有一行代码，但它完成了两个组件的初始化：**数据源与事务工厂**。默认情况下 MyBatis 与 Spring Boot 整合之后，底层使用的事务工厂是 `SpringManagedTransactionFactory`，了解 MyBatis 底层原理的读者可能会了解 `JdbcTransactionFactory`，它是 MyBatis 原生控制事务的事务工厂，`SpringManagedTransactionFactory`

与 `JdbcTransactionFactory` 的底层事务控制并无太大差别。

**代码清单 11-15　buildSqlSessionFactory（7）**

```
// ......
targetConfiguration.setEnvironment(new Environment(this.environment,
    this.transactionFactory == null
        ? new SpringManagedTransactionFactory()
        : this.transaction Factory,
this.dataSource));
// ......
```

（8）处理 Mapper

由于 `SqlSessionFactoryBean` 的构建中只能传入映射文件 mapper.xml 的路径，因此 `SqlSessionFactoryBean` 本身的逻辑中并无 Mapper 接口的扫描。从代码清单 11-16 可以发现，解析映射文件 mapper.xml 的逻辑是借助 XMLMapperBuilder 组件实现的，这个组件会逐个解析 mapper.xml，封装 `MappedStatement` 并注册到 MyBatis 的全局配置对象 Configuration 中。

**代码清单 11-16　buildSqlSessionFactory（8）**

```
// ......
if (this.mapperLocations != null) {
    if (this.mapperLocations.length == 0) {
        // logger ......
    } else {
        for (ResourcemapperLocation : this.mapperLocations) {
            if (mapperLocation == null) {
                continue;
            }
            try {
                XMLMapperBuilder xmlMapperBuilder = new XMLMapperBuilder(
                    mapperLocation.getInputStream(), targetConfiguration,
                    mapperLocation.toString(), targetConfiguration.getSqlFragments());
                xmlMapperBuilder.parse();
            } // catch finally ......
            // logger ......
        }
    }
} // else logger ......
return this.sqlSessionFactoryBuilder.build(targetConfiguration);
}
```

经过庞大的 `afterPropertiesSet` 方法处理后，`SqlSessionFactory` 被成功创建，MyBatis 的核心也就初始化完毕了。

#### 2. SqlSessionTemplate

相较于 `SqlSessionFactory` 的构建过程，`SqlSessionTemplate` 的构建逻辑非常简单，如代码清单 11-17 所示。`SqlSessionTemplate` 本身是一个实现了 `SqlSession` 接口的模板类，它可以非常简单地调用 MyBatis 的核心 CRUD 方法，而不必关心 `SqlSession` 的生命周期。在构建 `SqlSessionTemplate` 时，必传入的组件是 `SqlSessionFactory`，毕竟只有传入 `SqlSessionFactory` 之后，`SqlSessionTemplate` 才能获取到实际的 `SqlSession` 并调用其方法。除此之外，我们也可以关注一下构建方法的另一个参数，它的类型是一个枚举：`ExecutorType`。

**代码清单 11-17　SqlSessionTemplate 的创建**

```
@Bean
@ConditionalOnMissingBean
public SqlSessionTemplate sqlSessionTemplate(SqlSessionFactory sqlSessionFactory) {
    ExecutorType executorType = this.properties.getExecutorType();
    if (executorType != null) {
        return new SqlSessionTemplate(sqlSessionFactory, executorType);
    } else {
        return new SqlSessionTemplate(sqlSessionFactory);
    }
}
```

从类名上理解，`ExecutorType` 指代的是 SQL 语句的执行类型。MyBatis 内置的 `ExecutorType` 有三种，其中默认的模式是 `SIMPLE`，这种执行类型会在 `SqlSession` 执行具体的 CRUD 操作时为每条 SQL 语句创建一个预处理对象 `PreparedStatement` 并逐条执行；`REUSE` 模式会复用同一条 SQL 语句对应创建的 `PreparedStatement`，该模式会在一定程度上提高 MyBatis 的执行效率；`BATCH` 模式下不仅会复用 `PreparedStatement` 对象，还会执行批量操作，这使得 `BATCH` 模式的执行效率是最高的。但是使用 `BATCH` 模式有一个缺陷，即在执行 `insert` 语句时，如果插入的数据库表的主键是自增序列，则在事务提交之前无法从数据库获得实际的自增 ID 值，这种设计在某些业务场景下是不符合要求的。

### 3. AutoConfiguredMapperScannerRegistrar

由于 `SqlSessionFactory` 的构建中没有处理 Mapper 接口的扫描，因此 MyBatis 在整合 Spring Boot 时专门提供了一个适配 Spring Boot 项目模式的 Mapper 接口扫描注册器，如代码清单 11-18 所示，这个扫描器会在 Spring Boot 项目中未标注 `@MapperScan` 注解时生效。

**代码清单 11-18　@MapperScan 与 Mapper 接口扫描注册器**

```
@Configuration
@Import(AutoConfiguredMapperScannerRegistrar.class)
@ConditionalOnMissingBean({MapperFactoryBean.class, MapperScannerConfigurer.class})
public static class MapperScannerRegistrarNotFoundConfiguration implements InitializingBean {
    //......
}

@Import(MapperScannerRegistrar.class)
public @interface MapperScan
```

由于 `@MapperScan` 注解会向 IOC 容器中导入一个 `MapperScannerRegistrar` 组件，当项目中没有标注该注解时，默认的 `AutoConfiguredMapperScannerRegistrar` 就会被导入并生效（约定大于配置）。而从代码清单 11-19 中可以看出，默认注册的 `MapperScannerConfigurer` 扫描器中的扫描规则是扫描 Spring Boot 主启动类所在包及其子包下所有标注了 `@Mapper` 注解的接口。

### 代码清单 11-19 注册默认的 MapperScannerConfigurer

```java
public static class AutoConfiguredMapperScannerRegistrar
        implements BeanFactoryAware, ImportBeanDefinitionRegistrar {

    private BeanFactory beanFactory;

    @Override
    public void registerBeanDefinitions(AnnotationMetadata importingClassMetadata,
            BeanDefinitionRegistry registry) {
        // logger ......
        // 取出 SpringBoot 主启动类所在包
        List<String> packages = AutoConfigurationPackages.get(this.beanFactory);

        BeanDefinitionBuilder builder = BeanDefinitionBuilder
                .genericBeanDefinition(MapperScannerConfigurer.class);
        builder.addPropertyValue("processPropertyPlaceHolders", true);
        // 扫描标识@Mapper 接口的接口
        builder.addPropertyValue("annotationClass", Mapper.class);
        builder.addPropertyValue("basePackage",
                StringUtils.collectionToCommaDelimitedString(packages));
        BeanWrapper beanWrapper= new BeanWrapperImpl(MapperScannerConfigurer.class);
        Stream.of(beanWrapper.getPropertyDescriptors())
                .filter(x ->x.getName().equals("lazyInitialization")).findAny()
                .ifPresent(x -> builder.addPropertyValue(
                        "lazyInitialization","${mybatis.lazy-initialization: false}"));
        registry.registerBeanDefinition(MapperScannerConfigurer.class.getName(), builder.
getBeanDefinition());
    }
```

以此法注册 MapperScannerConfigurer 后，在项目开发中只需要编写 Mapper 接口并标注@Mapper 注解，即可被 MapperScannerConfigurer 自动扫描并注册到 IOC 容器中。

至此，MyBatis 整合 SpringBoot 的核心自动装配内容剖析完毕。

## 11.4 小结

本章主要研究了 Spring Boot 整合 MyBatis 持久层框架的自动装配核心，并简单了解了 MyBatis 中核心组件 SqlSessionFactory 的构建流程。MyBatis 的内部核心支撑是一个 SqlSessionFactory，在与 Spring Boot 框架整合时会借助 SqlSessionFactoryBean 的工厂 Bean 创建来完成 MyBatis 的框架初始化。此外，针对 Mapper 动态代理的开发，MyBatis 在整合 Spring Boot 时提供了默认的 Mapper 接口扫描器，以完成 Mapper 接口的代理装配。

# 第 12 章 Spring Boot 整合 WebMvc

本章主要内容：
◇ Spring Framework 与 Spring Boot 概述；
◇ Spring Boot 整合 WebMvc 的核心自动装配；
◇ WebMvc 核心组件的功能剖析；
◇ DispatcherServlet 的初始化原理与工作全流程解析。

除了与数据层的交互，Spring Boot 经常整合的另一个核心场景就是 Web 开发了。Spring Framework 5.x 中对于 Web 场景的开发提供了两套实现方案：WebMvc 与 WebFlux。本章先讲解读者所熟知的 WebMvc 部分。

## 12.1 整合 WebMvc 的核心自动装配

有关整合 WebMvc 的自动装配，我们在第 2 章已经大体了解，本节来回顾一下核心的装配内容。

- **WebMvcAutoConfiguration**

WebMvcAutoConfiguration 主配置类中核心注册的组件包括来自 WebMvcAutoConfigurationAdapter 的消息转换器 HttpMessageConverter、视图解析器 ContentNegotiatingViewResolver、国际化组件 LocaleResolver，以及来自 EnableWebMvcConfiguration 的 RequestMappingHandlerMapping、RequestMappingHandlerAdapter 和静态资源配置。

- **DispatcherServletAutoConfiguration**

DispatcherServletAutoConfiguration 的核心装配就是 DispatcherServlet 本身，另外它还注册了一个 ServletRegistrationBean 将 DispatcherServlet 注册到 IOC 容器中，以使 Servlet 生效。

> 提示：了解原生 SpringWebMvc 原理的读者可能知道，DispatcherServlet 的初始化逻辑中有一段默认的组件初始化逻辑，用于无配置的项目中初始化默认组件实现，这部分在 Spring Boot 中是否也会生效？这个疑虑是完全可以打消的，因为 Spring Boot 针对 WebMvc 场景已经在 WebMvcAutoConfiguration 中注册了默认的组件，所以不再需要 DispatcherServlet 主导创建。

# 第 12 章 Spring Boot 整合 WebMvc

- **ServletWebServerFactoryAutoConfiguration**

ServletWebServerFactoryAutoConfiguration 注册的内容是有关嵌入式 Web 容器的，它会根据当前项目中导入的项目式 Web 容器依赖决定选择装配何种容器实例，另外它还注册了 `BeanPostProcessorsRegistrar` 用于导入两个处理定制器的后置处理器、两个 `WebServerFactoryCustomizer` 用于将 `server.*` 系列配置应用到嵌入式 Web 容器中。

最后结合 Spring Boot 的官方文档，在 spring-boot-features 板块的 7.1.1 节有对 SpringWebMvc 自动装配的描述，这里截取出核心部分简单总结，文档原文如图 12-1 所示。

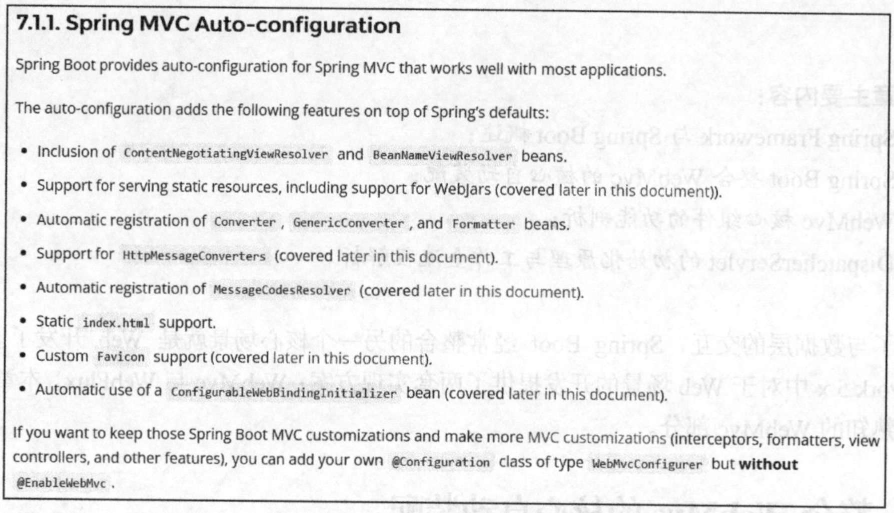

> **7.1.1. Spring MVC Auto-configuration**
>
> Spring Boot provides auto-configuration for Spring MVC that works well with most applications.
>
> The auto-configuration adds the following features on top of Spring's defaults:
>
> - Inclusion of `ContentNegotiatingViewResolver` and `BeanNameViewResolver` beans.
> - Support for serving static resources, including support for WebJars (covered later in this document)).
> - Automatic registration of `Converter`, `GenericConverter`, and `Formatter` beans.
> - Support for `HttpMessageConverters` (covered later in this document).
> - Automatic registration of `MessageCodesResolver` (covered later in this document).
> - Static `index.html` support.
> - Custom `Favicon` support (covered later in this document).
> - Automatic use of a `ConfigurableWebBindingInitializer` bean (covered later in this document).
>
> If you want to keep those Spring Boot MVC customizations and make more MVC customizations (interceptors, formatters, view controllers, and other features), you can add your own `@Configuration` class of type `WebMvcConfigurer` but **without** `@EnableWebMvc`.

图 12-1 Spring Boot 官方文档中描述的自动装配内容

官方文档列举的组件如下，列举的组件基本上都在官方文档中有相应的描述：

- 视图解析器；
- WebJars 的资源映射；
- 自动配置的转换器（`Converter`）、格式化器（`Formatter`）；
- HTTP 请求转换器（`HttpMessageConverter`）；
- 响应代码解析器；
- 静态主页映射；
- 网站图标映射；
- 可配置的 Web 初始化绑定器。

## 12.2 WebMvc 的核心组件

下面重点了解一些 WebMvc 中的核心组件的设计和作用，从 SpringWebMvc 的整体上了解 WebMvc 的架构设计以及整体的工作流程。通过学习本节的内容，可以帮助读者从宏观层面重新认识一下 WebMvc。

### 12.2.1 DispatcherServlet

DispatcherServlet 是读者最熟悉的 WebMvc 核心前端控制器，它统一接收客户端（浏览器)的所有请求,并根据请求的 URI 转发给项目中编写好的 Controller 中的方法。有一点请注意,有关匹配寻找和请求转发的工作,以及返回视图、响应 JSON 数据的处理,都不由 `DispatcherServlet` 完成,而是**委托**给其他组件,这些组件与 `DispatcherServlet` 共同协作完成整个 MVC 的工作。

这里需要读者认识到,WebMvc 中对于各组件作用的划分是很清晰的,每个组件各司其职,下面介绍核心组件的时候,读者可以试着体会一下。

### 12.2.2 Handler

首先解释 **Handler** 的概念。其实在项目开发中编写的 Controller 方法中,一个标注了 `@RequestMapping` 注解(或派生注解)的方法就是一个 **Handler**。很明显,Handler 中要做的事情就是处理客户端发送的请求,并声明响应视图/响应 JSON 数据。`DispatcherServlet` 在接收到请求后,只要能匹配到声明的那些标注了`@RequestMapping`注解的方法,最终就会将这些请求转发给编写好的 Handler。上述的处理逻辑如图 12-2 所示。

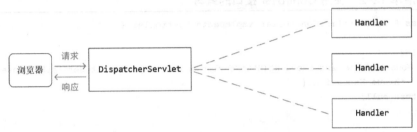

图 12-2 DispatcherServlet 与 Handler 的交互

### 12.2.3 HandlerMapping

上面已经解释过,WebMvc 的组件都是各司其职,根据`@RequestMapping` 去匹配 Handler 的工作,`DispatcherServlet` 不会自己去做,而是**委托给 HandlerMapping** 来负责。`HandlerMapping` 意为**处理器映射器**,它的作用是**根据 URI,去匹配查找能处理请求的 Handler**。注意,HandlerMapping 查找到可以处理请求的 Handler 之后,并不是立即将其返回,而是先组合了一个 HandlerExecutionChain,这样处理的原因是,如果一个请求要被拦截器处理的话,则在执行 Handler 之前需要先执行这些拦截器,之后再执行 Handler 本身,`HandlerMapping` 在封装的时候考虑到这一点,于是它把当前这次请求会涉及的拦截器和 Handler 一起封装起来组成一个 `HandlerExecutionChain` 的对象,交给 `DispatcherServlet`,具体逻辑如图 12-3 所示。

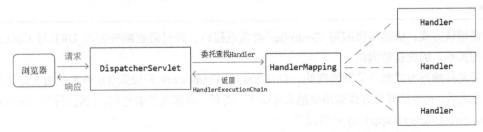

图 12-3 DispatcherServlet 委托 HandlerMapping 完成 Handler 匹配

从源码的角度看，HandlerMapping 的接口定义刚好符合以上描述，如代码清单 12-1 所示。

**代码清单 12-1　HandlerMapping 的接口定义**

```java
public interface HandlerMapping {
    HandlerExecutionChain getHandler(HttpServletRequest request) throws Exception;
}
```

HandlerMapping 接口有几个主要的落地实现，读者可以简单了解。

- **RequestMappingHandlerMapping**：支持 @RequestMapping 注解的处理器映射器【最常用】。这种 HandlerMapping 是项目开发中最常用的，无须多言。
- BeanNameUrlHandlerMapping：使用 bean 对象的名称作为 Handler 接收请求路径的处理器映射器。需要编写 Controller 类实现 Controller 接口（WebMvc 中有一个名为 Controller 的接口）。代码清单 12-2 中提供了一个简单示例。

**代码清单 12-2　使用 Controller 接口的编写**

```java
public class DepartmentListController implements Controller {

    @Override
    public ModelAndView handleRequest(HttpServletRequest request, HttpServletResponse response)
            throws Exception {
        return null;
    }
}
```

使用该方式编写的 Controller，一个类只能接收一个请求路径，目前已被淘汰，实际项目开发中不再使用。

- SimpleUrlHandlerMapping：集中配置请求路径与 Controller 的对应关系的处理器映射器。这种 HandlerMapping 可以配置请求路径与 Controller 的映射关系，例如代码清单 12-3 定义了一个简单的 SimpleUrlHandlerMapping 配置。

**代码清单 12-3　SimpleUrlHandlerMapping 的配置**

```xml
<bean class="org.springframework.web.servlet.handler.SimpleUrlHandlerMapping">
    <property name="mappings">
        <props>
            <prop key="/department/list">departmentListController</prop>
            <prop key="/department/save">departmentSaveController</prop>
        </props>
    </property>
</bean>
```

使用该方式，依然需要编写 Controller 类实现接口，并且需要额外配置 URI 与 Controller 的映射关系，因此也被淘汰。

后两种编写方式都是早期发现的，自从 Spring Framework 升级到 3.0 版本全面支持注解驱动开发之后，传统的开发方式就用得越来越少了，读者不必深入了解它们，只需要记住 RequestMappingHandlerMapping 就可以了。

## 12.2.4 HandlerAdapter

DispatcherServlet 获取到 HandlerExecutionChain 后，接下来要执行这些拦截器和 Handler，DispatcherServlet 依然选择交给 **HandlerAdapter** 去执行。HandlerAdapter 意为**处理器适配器**，它的作用是执行上面封装好的 HandlerExecutionChain。加入 HandlerAdapter 后的流程图如图 12-4 所示。

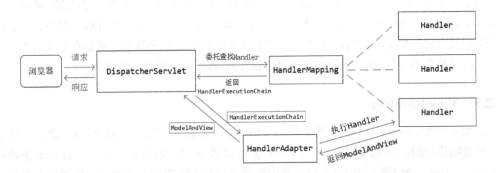

图 12-4　DispatcherServlet 委托 HandlerAdapter 执行 HandlerExecutionChain

注意 HandlerAdapter 与 Handler 的交互：执行 Handler 之后，虽然编写的返回值基本都是视图名称或者借助@ResponseBody 的响应 JSON 数据，但在 WebMvc 的框架内部，最终都是封装了一个 ModelAndView 对象返回给 HandlerAdapter。HandlerAdapter 再把这个 ModelAndView 对象交给 Dispatcher Servlet，这部分的工作就完成了。

从接口名上看，HandlerAdapter 蕴含着**适配器模式**，对于这个设计读者也需要了解一下。

### 1. 适配器模式

适配器模式可以简单地理解为，将一堆不兼容的产品通过一个中间"桥梁"整合好。举个例子：现实中的 U 盘和 TF 卡都是外置闪存设备，但是 U 盘可以直接插入 USB 口，TF 卡不可以，这个时候就需要一个读卡器作为中间"桥梁"，将 TF 卡插入读卡器里，读卡器就可以插入 USB 口了。

关于 HandlerAdapter 如何体现适配器模式，需要结合上面的 HandlerMapping 来看。Handler 的编写不仅有@Controller+@RequestMapping 的方式，还有实现 Controller 接口的方式（当然可能还有别的方式）。要实现用一个接口同时兼顾这两种方式，很明显适配器是一个不错的选择。将不同类型的 Handler 以 Object 的形式接收进来，再针对不同的 Handler 编写方式匹配对应的 HandlerAdapter 实现即可。

### 2. WebMvc 中的 HandlerAdapter 实现

WebMvc 中对于 HandlerAdapter 的核心实现主要有以下几种方式。

- **RequestMappingHandlerAdapter**：基于@RequestMapping 的处理器适配器，底层使用反射机制调用 Handler 的方法【最常见】。
- SimpleControllerHandlerAdapter：基于 Controller 接口的处理器适配器，底层会将 Handler 强转为 Controller，调用其 handleRequest 方法。
- SimpleServletHandlerAdapter：基于 Servlet 的处理器适配器，底层会将 Handler 强转为 Servlet，调用其 service 方法。

可以发现 WebMvc 中兼顾了多种编写 Handler 的方式，读者只需要重点关注 Request MappingHandlerAdapter。

**3. WebMvc 管理 Servlet**

注意，Servlet 对于 WebMvc 来讲也可以是一个 Handler，并且 WebMvc 为之提供了适配器，这就意味着在 WebMvc 的整合过程中，可以把 Servlet 直接注入 WebMvc 的 IOC 容器中，而不需要通过 web.xml 或者 @WebServlet 的方式注册到 ServletContext 中。

但是适配也有两面性，如果使用 Servlet 作为请求处理的载体，就只能利用 BeanNameUrlHandlerMapping 或者 SimpleUrlHandlerMapping 声明请求映射路径，这样处理的编码复杂度反而更高，所以这种写法在实际开发中也不会被采用，更多的方式是整合其他框架（将其他框架中的 Servlet 纳入 WebMvc 的 IOC 容器以实现统一管理）。

## 12.2.5 ViewResolver

DispatcherServlet 获取到 HandlerAdapter 返回的 ModelAndView 之后，接下来的工作是响应视图，DispatcherServlet 会将这部分工作委托给 ViewResolver 来处理。ViewResolver 会根据 ModelAndView 中存放的视图名称到预先配置好的位置去查找对应的视图文件（.jsp、.html 等），并进行实际的视图渲染。渲染完成后，将视图响应给 DispatcherServlet。具体的完整流程如图 12-5 所示。

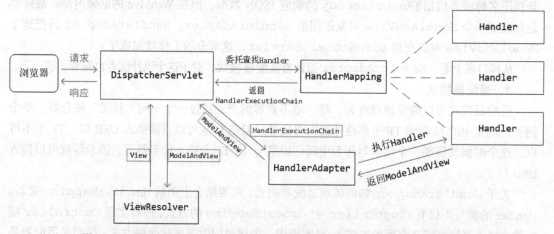

图 12-5　DispatcherServlet 委托 ViewResolver 处理视图和 JSON 数据响应

至此，DispatcherServlet 的核心工作流程已经处理完毕，这样看来 DispatcherServlet 其实没有做具体工作，而是扮演一个类似于"调度者"的角色，在不同的环节分发不同的工作给其他核心组件，由此读者也应该体会到，WebMvc 中的各组件是职责分明的。

回到本节的主题上，视图解析器 ViewResolver，顾名思义，它是解析和生成视图用的，默认情况下 WebMvc 只会初始化一个 ViewResolver，即在原生 SSM 框架组合开发中配置的 InternalResourceViewResolver，这个类继承自 UrlBasedViewResolver，它可以方便地声明页面路径的前后缀，以便于开发者在返回视图（JSP 页面）的时候编写视图名称。除了 InternalResourceViewResolver，WebMvc 还为一些模板引擎提供了支持类，例如 FreeMarker→FreeMarkerViewResolver，GroovyXML→GroovyMarkupViewResolver 等。

> 提示：可能有读者会产生疑惑，既然 Controller 不仅可以跳转视图，还可以响应 JSON 数据，那么响应 JSON 数据是否也由 ViewResolver 处理呢？答案是否定的，12.4 节会讲解该部分内容。

## 12.3 @Controller 控制器装配原理

当编写的 Controller 类中标注有 `@Controller` 或派生注解并声明 `@RequestMapping` 注解的方法时，即可装载到 WebMvc 中完成视图跳转/数据响应的功能。本节内容会深入初始化的流程中，探究这些 WebMvc 的 Handler 是如何被识别并装载到 WebMvc 中的。

### 12.3.1 初始化 RequestMapping 的入口

使用 `@RequestMapping` 的方式声明的 Handler，底层一定会与 RequestMappingHandlerMapping 有关联，而 WebMvcAutoConfiguration 中已经初始化了 RequestMappingHandlerMapping，下面会沿着这个思路深入分析。

借助 IDE 观察 RequestMappingHandlerMapping 的方法实现，可以发现有一个 `afterPropertiesSet` 方法，它来自 InitializingBean 接口，说明在 RequestMappingHandlerMapping 的对象初始化阶段有额外的扩展逻辑，而对应的 `afterPropertiesSet` 方法就是初始化 RequestMapping 的核心逻辑，通过一直向下调用，可以找到一个名为 `initHandlerMethods` 的方法，如代码清单 12-4 所示。

**代码清单 12-4　RequestMappingHandlerMapping 的初始化逻辑**

```java
public void afterPropertiesSet() {
    this.config = new RequestMappingInfo.BuilderConfiguration();
    this.config.setUrlPathHelper(getUrlPathHelper());
    // this.config.set ......

    super.afterPropertiesSet();
}

public void afterPropertiesSet() {
    initHandlerMethods();
}

private static final String SCOPED_TARGET_NAME_PREFIX = "scopedTarget.";
protected void initHandlerMethods() {
    // 此处会获取 IOC 容器中的所有 Bean
    for (String beanName : getCandidateBeanNames()) {
        if (!beanName.startsWith(SCOPED_TARGET_NAME_PREFIX)) {
            processCandidateBean(beanName);
        }
    }
    handlerMethodsInitialized(getHandlerMethods());
}
```

从 `initHandlerMethods` 的方法名就可以知道，该方法会初始化所有 HandlerMethod，即可以处理请求的 Controller 方法。从方法的实现上看，`initHandlerMethods` 方法会提取出 IOC 容器中所有名称前缀不是 `"scopedTarget."` 的 bean 对象，并逐一进行解析。

## 12.3.2 processCandidateBean

要判断一个 Bean 是否是一个 WebMvc 的 Controller，只需要判断类上是否标注了 @Controller 注解（或派生注解）或者 @RequestMapping 注解（或派生注解），只要二者中存在一个，就可判定当前 Bean 是一个前端控制器，如代码清单 12-5 所示。如果判断一个 Bean 的确是一个 Controller，则继续向下执行 detectHandlerMethods 方法，处理 Handler 方法。

代码清单 12-5　检查和处理候选的 bean 对象

```
protected void processCandidateBean(String beanName) {
    Class<?> beanType = null;
    try {
        beanType =obtainApplicationContext().getType(beanName);
    } // catch ......
    if (beanType != null && isHandler(beanType)) {
        detectHandlerMethods(beanName);
    }
}

protected boolean isHandler(Class<?> beanType) {
    return (AnnotatedElementUtils.hasAnnotation(beanType, Controller.class)
            || AnnotatedElementUtils.hasAnnotation(beanType, RequestMapping.class));
}
```

> 提示：可能有读者会产生疑惑，为什么一个类上只标注了 @RequestMapping 注解，也会判定为 Controller？如果这个类没有注册成为 IOC 容器的 Bean 怎么办？其实有这个想法是多虑了。注意观察代码清单 12-5 中的 try 块，beanType 是通过调用 ApplicationContext 的 getType 方法获取的，既然能从 ApplicationContext 中获取到 Bean 的类型，那么它必然是容器中的一个实际存在的 bean 对象。Spring Framework 在此处认为只要一个 bean 对象上标注了 @RequestMapping 注解，它就可以充当 Controller，而无须限定是否真的标注了 @Controller 注解，只是在实际项目开发中通常都是 @Controller 配合 @RequestMapping 同时出现而已。

## 12.3.3　detectHandlerMethods

继续向下执行 detectHandlerMethods 方法，这个方法整体分为两个步骤，首先会将一个 Controller 中的方法全部检查一遍，并提取出标注了 @RequestMapping 注解的方法，随后封装映射信息并注册方法映射，如代码清单 12-6 所示。

代码清单 12-6　解析 Controller 类型提取 Handler 方法

```
protected void detectHandlerMethods(Object handler) {
    Class<?> handlerType = (handler instanceof String
            ? obtainApplicationContext().getType((String) handler) : handler.getClass());
    if (handlerType != null) {
        Class<?> userType = ClassUtils.getUserClass(handlerType);
        // 解析筛选方法
```

```java
        Map<Method, T> methods = MethodIntrospector.selectMethods(userType,
                (MethodIntrospector.MetadataLookup<T>) method -> {
                    try {
                        return getMappingForMethod(method, userType);
                    } // catch ......
                });
    // logger ......
    // 注册方法映射
    methods.forEach((method, mapping) -> {
        Method invocableMethod = AopUtils.selectInvocableMethod(method, userType);
        registerHandlerMethod(handler, invocableMethod, mapping);
    });
    }
}
```

第一步检查筛选 Handler 方法的逻辑中利用了 MethodIntrospector 进行方法遍历，而 MethodIntrospector 的底层会利用反射机制逐个遍历一个 Class 中的所有方法，并执行 selectMethods 方法参数中的 lambda 表达式，而 lambda 表达式中调用的 getMappingForMethod 方法检查方法上是否标注了 @RequestMapping 注解（或派生注解），如果有标注则会封装一个 RequestMappingInfo 对象，如代码清单 12-7 所示。

**代码清单 12-7　检查方法并封装 RequestMappingInfo 对象**

```java
protected RequestMappingInfo getMappingForMethod(Method method, Class<?> handlerType) {
    // 创建方法级的 RequestMappingInfo
    RequestMappingInfo info = createRequestMappingInfo(method);
    if (info != null) {
        // 创建类级的 RequestMappingInfo
        RequestMappingInfo typeInfo = createRequestMappingInfo(handlerType);
        if (typeInfo != null) {
            // 拼接类级的 RequestMapping uri
            info = typeInfo.combine(info);
        }
        // 拼接路径前缀
        String prefix = getPathPrefix(handlerType);
        if (prefix != null) {
            info = RequestMappingInfo.paths(prefix).build().combine(info);
        }
    }
    return info;
}

private RequestMappingInfo createRequestMappingInfo(AnnotatedElement element) {
    RequestMapping requestMapping = AnnotatedElementUtils
            .findMergedAnnotation(element, RequestMapping.class);
    RequestCondition<?> condition =(element instanceof Class
            ? getCustomTypeCondition((Class<?>) element):getCustomMethodCondition((Method) element));
    return (requestMapping != null ? createRequestMappingInfo(requestMapping, condition) : null);
}
```

经过 getMappingForMethod 方法的处理后，就可以获得一个包含映射 URI、请求方式等信息的 RequestMappingInfo 对象。回到代码清单 12-6 的 detectHandlerMethods 方

法的下半段，它会将 Handler 方法逐个与 RequestMappingInfo 对象一一映射并注册到 MappingRegistry 中，如代码清单 12-8 所示。

**代码清单 12-8　注册 Handler 方法映射到 MappingRegistry**

```
protected void detectHandlerMethods(Object handler) {
    Class<?> handlerType = (handler instanceof String
            ? obtainApplicationContext().getType((String) handler): handler.getClass());
    if (handlerType != null) {
        // ......
        methods.forEach((method, mapping) -> {
            Method invocableMethod = AopUtils.selectInvocableMethod(method, userType);
            registerHandlerMethod(handler, invocableMethod, mapping);
        });
    }
}

protected void registerHandlerMethod(Object handler, Method method, T mapping) {
    this.mappingRegistry.register(mapping, handler, method);
}
```

MappingRegistry 中存放了 URI 与 HandlerMethod 的映射关系，在 DispatcherServlet 接收到客户端请求时，可以根据 URI 找到合适的 HandlerMethod，进而定位到可以处理请求的 Handler，将请求转发到实际的 Controller 中。

IOC 容器中的 bean 对象全部检查完成后，所有 Controller 中的 Handler 方法全部装载到 RequestMappingHandlerMapping 中，HandlerMapping 的初始化动作就完成了。

## 12.4　DispatcherServlet 的工作全流程解析

本节内容是本章的重点和难点，我们来探讨 WebMvc 在实际运行期间 DispatcherServlet 对于请求处理和响应的全流程执行原理。作为 WebMvc 的核心前端控制器，DispatcherServlet 默认会接收所有请求并分发处理。本节内容使用 springboot-01-quickstart 的示例项目，实际调试 DispatcherServlet 的核心工作全流程。

启动 springboot-01-quickstart 示例项目，并在 DispatcherServlet 的父类 FrameworkServlet 的 service 方法上打入断点，随后使用浏览器访问 http://localhost:8080/hello，待程序停在断点处，开始 Debug 调试。

### 12.4.1　DispatcherServlet#service

调试从 FrameworkServlet 开始，首先 service 方法会对 PATCH 类型的请求单独处理，PATCH 类型的请求本身是对 PUT 类型的补充，一般用于资源的局部更新。通常在项目开发中不会使用 PATCH 类型的请求。继续向下执行 else 块的 super.service 方法，而 FrameworkServlet 的父类 HttpServlet 会根据不同的请求类型将方法转发至 doxxx 方法中，所以最终执行的是 FrameworkServlet 中重写的 doGet、doPost 等方法，而这些方法最终调用的都是一个名为 processRequest 的方法，如代码清单 12-9 所示。

### 代码清单 12-9　FrameworkServlet#service

```java
@Override
protected void service(HttpServletRequest request, HttpServletResponse response)
        throws ServletException,IOException {
    HttpMethod httpMethod = HttpMethod.resolve(request.getMethod());
    if (httpMethod == HttpMethod.PATCH || httpMethod == null) {
        processRequest(request, response);
    } else {
        super.service(request, response);
    }
}

protected final void doGet(HttpServletRequest request, HttpServletResponse response)
        throws ServletException, IOException {
    processRequest(request, response);
}

protected final void doPost(HttpServletRequest request, HttpServletResponse response)
        throws ServletException,IOException {
    processRequest(request, response);
}
```

### 12.4.2　processRequest

processRequest 方法中会对请求进行一些前置处理,随后转调 doService 方法处理请求。请注意代码清单 12-10 中标有注释的部分，这里面有一个细节：**线程隔离**。

### 代码清单 12-10　processRequest 预处理请求

```java
protected final void processRequest(HttpServletRequest request, HttpServletResponse response)
        throws ServletException, IOException {
    // 记录接收请求的时间
    long startTime = System.currentTimeMillis();
    Throwable failureCause = null;
    // 获取当前线程的 LocaleContext，并创建当前线程的 LocaleContext
    LocaleContext previousLocaleContext = LocaleContextHolder.getLocaleContext();
    LocaleContext localeContext = buildLocaleContext(request);
    // 获取当前线程的 RequestAttributes，并创建当前线程对应的 ServletRequestAttributes
    RequestAttributes previousAttributes = RequestContextHolder.getRequestAttributes();
    ServletRequestAttributes requestAttributes =
            buildRequestAttributes(request, response, previousAttributes);

    WebAsyncManager asyncManager = WebAsyncUtils.getAsyncManager(request);
    asyncManager.registerCallableInterceptor(FrameworkServlet.class.getName(),
            new RequestBindingInterceptor());

    // 初始化 ContextHolder，传入新封装好的请求参数和上下文，目的是隔离线程
    initContextHolders(request, localeContext, requestAttributes);

    try {
        // 子类的模板方法
        doService(request, response);
    } // catch ......
```

```java
    finally {
        // 重新设置当前线程的 LocaleContext 和 RequestAttributes
        resetContextHolders(request, previousLocaleContext, previousAttributes);
        if (requestAttributes != null) {
            requestAttributes.requestCompleted();
        }
        logResult(request, response, failureCause, asyncManager);
        // 发布 ServletRequestHandledEvent 事件
        publishRequestHandledEvent(request, response, startTime, failureCause);
    }
}
protected abstract void doService(HttpServletRequest request, HttpServletResponse response)
    throws Exception;
```

注意,在 initContextHolders 方法执行之前,processRequest 方法中会获取当前线程的 LocaleContext 和 RequestAttributes 并暂存到方法中,随后使用当前请求的 REQUEST 与 RESPONSE 对象构造全新的 LocaleContext 与 RequestAttributes,并设置到当前线程上下文中(见代码清单 12-11),以此完成线程之间的隔离。

**代码清单 12-11　initContextHolders 设置当前请求对应的模型**

```java
private void initContextHolders(HttpServletRequest request,
        @Nullable LocaleContext localeContext, @Nullable RequestAttributes requestAttributes) {
    if (localeContext != null) {
        LocaleContextHolder.setLocaleContext(localeContext, this.threadContextInheritable);
    }
    if (requestAttributes != null) {
        RequestContextHolder.setRequestAttributes(requestAttributes,
                this.threadContextInheritable);
    }
}
```

### 12.4.3　doService

processRequest 方法的前置方法处理完成后,下一步的核心动作是 doService 方法,但是请注意,doService 方法仍然不是真正处理请求的方法,如代码清单 12-12 所示。

**代码清单 12-12　doService 预处理请求**

```java
protected void doService(HttpServletRequest request, HttpServletResponse response)
        throws Exception {
    logRequest(request);
    // 判断请求参数中是否存在 javax.servlet.include.request_uri
    Map<String,Object> attributesSnapshot = null;
    if (WebUtils.isIncludeRequest(request)) {
        attributesSnapshot = new HashMap<>();
        Enumeration<?> attrNames = request.getAttributeNames();
        while (attrNames.hasMoreElements()) {
            String attrName = (String) attrNames.nextElement();
            if (this.cleanupAfterInclude || attrName.startsWith(DEFAULT_STRATEGIES_PREFIX)) {
                attributesSnapshot.put(attrName, request.getAttribute(attrName));
            }
```

## 12.4 DispatcherServlet 的工作全流程解析

```
        }
    }
    // ......
    // flashMapManager
    if (this.flashMapManager != null) {
        FlashMap inputFlashMap = this.flashMapManager.retrieveAndUpdate(request, response);
        if (inputFlashMap != null) {
            request.setAttribute(INPUT_FLASH_MAP_ATTRIBUTE,
                    Collections.unmodifiableMap(inputFlashMap));
        }
        request.setAttribute(OUTPUT_FLASH_MAP_ATTRIBUTE, new FlashMap());
        request.setAttribute(FLASH_MAP_MANAGER_ATTRIBUTE, this.flashMapManager);
    }
    try {
        // 真正处理请求的方法
        doDispatch(request, response);
    } // finally ......
}
```

由代码清单 12-12 中可以大体了解 doService 执行的请求预处理逻辑，这里有两个动作，下面简单解释这两个动作对应的源码的作用。

#### 1. isIncludeRequest 的判断

在 doService 的第一个 if 判断结构中会调用 WebUtils.isInclude Request (request) 方法检查当前的 HttpServletRequest 是否是 "include" 请求，而方法的实现中会判断请求中是否包含名为 "javax.servlet.include.request_uri" 的属性，如代码清单 12-13 所示。

**代码清单 12-13 检查请求是否为 "include" 请求**

```
public static final String INCLUDE_REQUEST_URI_ATTRIBUTE = "javax.servlet.include.request_uri";

public static boolean isIncludeRequest(ServletRequest request) {
    return (request.getAttribute(INCLUDE_REQUEST_URI_ATTRIBUTE) != null);
}
```

有关 "javax.servlet.include.request_uri" 属性的含义，在 Servlet 3.0 规范中有这样一段描述：已经被另一个 Servlet 使用 RequestDispatcher 的 include 方法调用过的 Servlet，有权访问被调用过的 Servlet 的路径。这段描述中的 include 方法，其实指代的是在 JSP 中使用的 `<jsp:includepage="xxx.jsp"/>` 标签，使用该标签可以组合其他页面以完成页面共用部分的抽取。所以 isIncludeRequest 方法的作用是区别页面的加载是否由 `<jsp:include>` 标签而来。

#### 2. flashMapManager 的设计

有关 flashMapManager 的设计，需要读者回顾一个经典的业务场景：用户登录。在传统的前后端不分离的开发场景中，用户登录的场景通常是使用 POST 请求将用户名、密码等信息传入后端以供认证，当登录成功后使用重定向将客户端引导至系统主页。在这个前提下有一个特殊的场景：如果用户登录时提交的登录表单中有一些需要在跳转至主页时渲染的数据，则仅放入 request 域中无法解决问题。SpringWebMvc 在 3.1 版本后引入了 FlashMapManager 来解决该问题，可以在页面重定向发生跳转时将需要渲染的数据暂时放入 session 中，这样浏览器即便刷新也不会影响数据渲染。

该设计仅需了解，感兴趣的读者可以自行翻阅相关资料和文档加以了解。

### 12.4.4　doDispatch

doService 的预处理完成后，下一步是至关重要的核心处理请求和响应的方法：doDispatch。由于 doDispatch 方法的篇幅非常长，为了确保阅读体验和理解，下面会拆分为多个部分研究。

**1. 文件上传解析**

doDispatch 方法的第一段源码是处理带有文件上传的请求，如代码清单 12-14 所示。请注意，核心处理文件上传请求的 checkMultipart 方法传入一个 HttpServletRequest 对象，再返回一个 HttpServletRequest 对象，这一动作的目的是在保持接口类型不变的前提下改变具体的实现类。

**代码清单 12-14　doDispatch（1）**

```java
protected void doDispatch(HttpServletRequest request, HttpServletResponse response)
        throws Exception {
    HttpServletRequest processedRequest = request;
    HandlerExecutionChain mappedHandler = null;
    boolean multipartRequestParsed = false;

    WebAsyncManager asyncManager = WebAsyncUtils.getAsyncManager(request);
    try {
        ModelAndView mv = null;
        Exception dispatchException = null;
        try {
            // 此处会处理文件上传的情况
            processedRequest = checkMultipart(request);
            multipartRequestParsed = (processedRequest != request);
            // ......

protected HttpServletRequest checkMultipart(HttpServletRequest request)
        throws MultipartException {
    if (this.multipartResolver != null && this.multipartResolver.isMultipart(request)) {
        if (WebUtils.getNativeRequest(request, MultipartHttpServletRequest.class) != null) {
            if (request.getDispatcherType().equals(DispatcherType.REQUEST)) {
                // logger ......
            }
        } else if (hasMultipartException(request)) {
            // logger ......
        } else {
            try {
                return this.multipartResolver.resolveMultipart(request);
            } // catch ......
        }
    }
    return request;
}
```

## 12.4 DispatcherServlet 的工作全流程解析

通过代码清单 12-15 的下半部分可以看到，处理文件上传请求的核心组件是 `multipartResolver`，它可以将一个由 Servlet 容器处理的 `HttpServletRequest` 对象转换为可以访问请求中文件对象的 `MultipartHttpServletRequest` 子接口对象。而判断一个请求是否为 **multipart** 请求的依据是 `content-type` 是否以 `"multipart/"` 开头，由于当前正在 Debug 的请求很明显只是一个普通的 GET 请求，因此不会进入该部分。

**代码清单 12-15　doDispatch（2）**

```java
// ......
// Determine handler for the current request.
mappedHandler = getHandler(processedRequest);
if (mappedHandler == null) {
    noHandlerFound(processedRequest, response);
    return;
}
// ......

protected HandlerExecutionChain getHandler(HttpServletRequest request) throws Exception {
    if (this.handlerMappings != null) {
        for (HandlerMapping mapping : this.handlerMappings) {
            HandlerExecutionChain handler = mapping.getHandler(request);
            if (handler != null) {
                return handler;
            }
        }
    }
    return null;
}
```

感兴趣的读者可以将当前项目整合视图模板引擎（如 Thymeleaf 或 FreeMarker 等）后，编写带有文件上传的表单页面，配合后端 Controller 编写的方法进行处理和调试，本书不对此展开。

### 2. 获取第一个可用的 Handler

紧接着的部分是 DispatcherServlet 核心处理流程的第一步：**搜索 Handler**。代码清单 12-14 的 `getHandler` 方法的核心逻辑是从所有 `HandlerMapping` 中寻找可以返回非空 `HandlerExecutionChain` 对象的 `HandlerMapping`，逻辑与 12.2.3 节相呼应。

Debug 至 `getHandler` 方法的内部，可以发现 DispatcherServlet 中有 5 个 HandlerMapping。由于在示例项目中编写的 Handler 都是以 `@Controller` + `@RequestMapping` 注解实现的，因此对应的关键实现是在 `WebMvcAutoConfiguration` 中注册的 `RequestMappingHandlerMapping`，它位于 `handlerMappings` 的第 0 位，如图 12-6 所示。

```
∞ this.handlerMappings = {ArrayList@5274} size = 5
  > ≡ 0 = {RequestMappingHandlerMapping@5308}
  > ≡ 1 = {BeanNameUrlHandlerMapping@5309}
  > ≡ 2 = {RouterFunctionMapping@5310}
  > ≡ 3 = {SimpleUrlHandlerMapping@5311}
  > ≡ 4 = {WelcomePageHandlerMapping@5312}
```

图 12-6　Debug 可见有 5 个 HandlerMapping

## 第 12 章　Spring Boot 整合 WebMvc

将断点打在 `return handler;` 一行并放行程序，断点停在 `RequestMappingHandlerMapping` 的循环中，说明以 `@Controller` + `@RequestMapping` 注解编写的 `Handler` 底层由 `RequestMappingHandlerMapping` 负责处理。下面进入 `RequestMappingHandlerMapping` 的 `getHandler` 方法中。

> 提示：从此处开始，读者可以非常强烈地感受到 WebMvc 中源码的风格，父类提供流程的抽象定义，子类负责具体的功能实现。

进入 `getHandler` 方法后，发现程序实际进入的类是 `RequestMappingHandlerMapping` 的父类 `AbstractHandlerMapping`，而父类的方法中又定义了需要子类实现具体功能的 `getHandlerInternal` 方法，由此就体现出了上面提到的父类定义流程，子类负责实现，如代码清单 12-16 所示。

### 代码清单 12-16　AbstractHandlerMapping#getHandler

```java
public final HandlerExecutionChain getHandler(HttpServletRequest request) throws Exception{
    // 留给子类的模板方法
    Object handler = getHandlerInternal(request);
    // ......

    // 构建 HandlerExecutionChain
    HandlerExecutionChain executionChain = getHandlerExecutionChain(handler, request);
    //......
    return executionChain;
}
```

抽取 `getHandler` 方法的核心逻辑有两步：由子类具体获取 `Handler` 的具体对象；根据 `Handler` 对象构建 `HandlerExecutionChain` 对象。逐一来看。

（1）getHandlerInternal

Debug 进入 `getHandlerInternal` 方法，发现程序会先进入 `RequestMappingHandlerMapping` 的直接父类 `RequestMappingInfoHandlerMapping` 中，不过父类的方法实现是直接调用 `super.getHandlerInternal(request)`，而继续往上调用，会来到抽象父类 `AbstractHandlerMethodMapping` 中，如代码清单 12-17 所示。

### 代码清单 12-17　AbstractHandlerMethodMapping#getHandlerInternal

```java
protected HandlerMethod getHandlerInternal(HttpServletRequest request) throws Exception {
    String lookupPath = getUrlPathHelper().getLookupPathForRequest(request);
    request.setAttribute(LOOKUP_PATH, lookupPath);
    this.mappingRegistry.acquireReadLock();
    try {
        HandlerMethod handlerMethod =lookupHandlerMethod(lookupPath, request);
        return (handlerMethod != null ? handlerMethod.createWithResolvedBean() : null);
    } // finally ......
}
```

父类的实现有必要详细研究，第一行源码会借助 `UrlPathHelper` 获取到本次请求的 URI，即 "/hello"，接下来进入 **try-finally** 结构中执行 `lookupHandlerMethod` 方法，从本身的 `HandlerMethod` 集合中检查是否有可以处理当前 URI 请求的 `HandlerMethod` 对象。

经过 `lookupHandlerMethod` 方法检索后，可以成功获取到 `HelloController` 中的 `hello` 方法，如图 12-7 所示。

```
> ≡ lookupPath = "/hello"
∨   handlerMethod = {HandlerMethod@5273} ... toString()
  >   logger = {LogAdapter$Slf4jLocationAwareLog@5309}
  >   bean = "helloController"
  >   beanFactory = {DefaultListableBeanFactory@5311} ... toString()
  >   beanType = {Class@4492} ... Navigate
  >   method = {Method@5312} "public java.lang.String com.linkedbear.springboot.quickstart.controller.HelloController.hello()"
  >   bridgedMethod = {Method@5312} "public java.lang.String com.linkedbear.springboot.quickstart.controller.HelloController.hello()"
      parameters = {MethodParameter[0]@5313}
      responseStatus = null
      responseStatusReason = null
      resolvedFromHandlerMethod = null
      interfaceParameterAnnotations = null
  >   description = "com.linkedbear.springboot.quickstart.controller.HelloController#hello()"
```

图 12-7　依据 URI 可以检索到 HelloController 的 hello 方法

获取到 `HandlerMethod` 对象之后，下一步还要执行 `createWithResolvedBean` 方法创建一个全新的 `HandlerMethod` 对象。可能有读者不理解为什么还要再次创建，有这个疑惑的读者可以仔细观察图 12-7 中 `handlerMethod` 对象的 `bean` 属性，它是一个字符串而不是 `BeanFactory` 中真实存在的 `HelloController` 对象，而且其余的属性中也没有 `HelloController` 对象的持有，所以为了获取 `HelloController` 对象，需要执行代码清单 12-18 中的 `createWithResolvedBean` 方法，从 `BeanFactory` 中取出 `HelloController` 对象并重新封装。

**代码清单 12-18　createWithResolvedBean 从 BeanFactory 中取出 bean 对象**

```java
public HandlerMethod createWithResolvedBean() {
    Object handler = this.bean;
    if (this.bean instanceof String) {
        Assert.state(this.beanFactory != null, "Cannot resolve bean name without BeanFactory");
        String beanName = (String) this.bean;
        handler = this.beanFactory.getBean(beanName);
    }
    return new HandlerMethod(this, handler);
}
```

重新封装完成后，获取 Handler 的动作结束。

（2）构建 HandlerExecutionChain

检索并封装 `HandlerMethod` 之后，`getHandler` 方法并不会将其返回，而是会组合 WebMvc 中注册的拦截器，因为每个请求对应的所需执行的拦截器不同，需要在处理请求之前再匹配。构建 `HandlerExecutionChain` 的流程不太复杂，`getHandlerExecutionChain` 方法会根据 `HandlerInterceptor` 的类型分别处理，如代码清单 12-19 所示。

**代码清单 12-19　根据当前请求构建 HandlerExecutionChain**

```java
protected HandlerExecutionChain getHandlerExecutionChain(Object handler,
        HttpServletRequest request) {
    HandlerExecutionChain chain = (handler instanceof HandlerExecutionChain ?
            (HandlerExecutionChain) handler : new HandlerExecutionChain(handler));
```

```
        String lookupPath = this.urlPathHelper.getLookupPathForRequest(request, LOOKUP_PATH);
        for (HandlerInterceptor interceptor : this.adaptedInterceptors) {
            if (interceptor instanceof MappedInterceptor) {
                MappedInterceptor mappedInterceptor = (MappedInterceptor) interceptor;
                // 匹配路径的拦截器
                if (mappedInterceptor.matches(lookupPath, this.pathMatcher)) {
                    chain.addInterceptor(mappedInterceptor.getInterceptor());
                }
            } else {
                // 普通拦截器
                chain.addInterceptor(interceptor);
            }
        }
        return chain;
    }
```

匹配并组合拦截器后，HandlerMapping 的工作全部完成，接下来回到 DispatcherServlet 中。

### 3. 获取 HandlerAdapter

下面是 DispatcherServlet 核心处理流程的第二步，会根据 HandlerMapping 匹配到的 Handler 对象，寻找可以执行它的 HandlerAdapter，对应的流程参见 12.2.4 节。而 getHandlerAdapter 的方法实现与 getHandler 如出一辙，如代码清单 12-20 所示。

**代码清单 12-20　doDispatch（3）**

```
    // ......
    // 为当前请求确定 HandleraAdapter
    HandlerAdapter ha = getHandlerAdapter(mappedHandler.getHandler());
    // ......

protected HandlerAdapter getHandlerAdapter(Object handler) throws ServletException {
    if (this.handlerAdapters != null) {
        for (HandlerAdapter adapter : this.handlerAdapters) {
            if (adapter.supports(handler)) {
                return adapter;
            }
        }
    }
    // throw ex ......
}
```

Debug 至此，可以发现有 4 个 HandlerAdapter 的实现类对象，如图 12-8 所示。

```
∞ this.handlerAdapters = {ArrayList@5327} size = 4
> ≡ 0 = {RequestMappingHandlerAdapter@5427}
> ≡ 1 = {HandlerFunctionAdapter@5428}
> ≡ 2 = {HttpRequestHandlerAdapter@5429}
> ≡ 3 = {SimpleControllerHandlerAdapter@5430}
```

**图 12-8　Debug 可见有 4 个 HandlerAdapter**

实际负责匹配的实现对象显然是 RequestMappingHandlerAdapter，而它的匹配规则仅是判断 Handler 的类型是不是 HandlerMethod，如代码清单 12-21 所示。根据设计，RequestMappingHandlerMapping 与 RequestMappingHandlerAdapter 本身就是互相匹配的，所以这里的匹配结果必定返回 true。

**代码清单 12-21　RequestMappingHandlerAdapter 的匹配规则**

```java
public final boolean supports(Object handler) {
    return (handler instanceof HandlerMethod && supportsInternal((HandlerMethod) handler));
}

protected boolean supportsInternal(HandlerMethod handlerMethod) {
    return true;
}
```

#### 4．回调拦截器

经过 HandlerMapping 的检索和 HandlerAdapter 的适配后，下面的动作是执行 Handler 的逻辑。不过在执行 Handler 的逻辑之前，先要执行项目中注册的可以匹配当前请求的拦截器。根据 HandlerInterceptor 的设计规则，首先会执行 preHandle 方法的拦截，对应到代码清单 12-22 中可以发现，它的触发方式仅是简单的循环依次调用。

**代码清单 12-22　doDispatch（4）**

```java
        // ......
        if (!mappedHandler.applyPreHandle(processedRequest, response)) {
            return;
        }
        // ......
boolean applyPreHandle(HttpServletRequest request, HttpServletResponse response)
        throws Exception {
    HandlerInterceptor[] interceptors = getInterceptors();
    if (!ObjectUtils.isEmpty(interceptors)) {
        for (int i = 0; i < interceptors.length; i++) {
            HandlerInterceptor interceptor = interceptors[i];
            if (!interceptor.preHandle(request, response, this.handler)) {
                triggerAfterCompletion(request, response, null);
                return false;
            }
            this.interceptorIndex = i;
        }
    }
    return true;
}
```

另外请注意，从 applyPreHandle 方法的逻辑中可以看到，如果 preHandle 方法返回了 false，则内部的 if 结构会触发并返回 false。如果 mappedHandler.applyPreHandle 方法返回 false，则会直接返回，不会继续执行 doDispatch 方法的后续逻辑。

#### 5．执行 Handler+回调拦截器

接下来是执行 DispatcherServlet 核心处理流程的第三步，DispatcherServlet 会委托 HandlerAdapter 执行具体的 Handler 方法，而执行 Handler 方法的 HandlerAdapter

是 `RequestMappingHandlerAdapter` 的父类 `AbstractHandlerMethodAdapter`，其 `handle` 方法又会转调模板方法 `handleInternal` 方法，如代码清单 12-23 所示。

**代码清单 12-23　doDispatch（5）**

```
// ......
// 实际调用 Handler
mv = ha.handle(processedRequest, response, mappedHandler.getHandler());
// ......

// AbstractHandlerMethodAdapter
public final ModelAndView handle(HttpServletRequest request,
        HttpServletResponse response, Object handler) throws Exception {
    return handleInternal(request, response, (HandlerMethod) handler);
}
```

> 💡 提示：可能有读者会发现 WebMvc 中的源码风格与核心包不同，在前面有关 IOC 和 AOP 的章节中，父类调用子类的方法的命名规范通常是 **xxx** → do**xxx** 方法，但是在 WebMvc 中大多数情况是 **xxx** → **xxxInternal** 方法，读者在阅读源码时一定要多加注意。

（1）handleInternal

进入 `RequestMappingHandlerAdapter` 的 `handleInternal` 方法，如代码清单 12-24 所示。忽略与主线逻辑无关的源码后，可以提取出的最核心方法是 else 块中的 invokeHandlerMethod 方法，这个方法相当于将 Handler 方法的执行转移至 invokeHandlerMethod 方法内部，继续往下跟进。

**代码清单 12-24　RequestMappingHandlerAdapter 的 handleInternal 方法**

```
protected ModelAndView handleInternal(HttpServletRequest request,
        HttpServletResponse response, HandlerMethod handlerMethod) throws Exception {
    ModelAndView mav;
    checkRequest(request);
    // 同步 session 的配置，默认不同步，执行下面的 invokeHandlerMethod 方法
    if (this.synchronizeOnSession) {
        // ......
    } else {
        mav = invokeHandlerMethod(request, response, handlerMethod);
    }
    // Cache-Control 相关的处理 ......
    return mav;
}
```

（2）invokeHandlerMethod

接下来研究的 `invokeHandlerMethod` 方法篇幅较长，为便于读者更好地阅读和理解源码，下面将其拆分为多个片段讲解。

（a）参数绑定器初始化

第一个要重点研究的方法是 `getDataBinderFactory`，这个方法会初始化一个参数绑定器，用于将客户端请求中参数和请求体的数据映射到 Handler 的形参中，如代码清单 12-25 所示。WebMvc 中有一个注解@InitBinder，它可以单独声明在一个 Controller 中，执行当前 Controller

中的方法时,会先执行标注了@InitBinder 注解的方法,初始化一些参数绑定器的逻辑。另外,熟悉 WebMvc 的读者一定会了解注解@ControllerAdvice,这个注解可以配合@InitBinder 注解标注的方法,实现全局的参数绑定器预初始化。代码清单 12-26 展示了一个简单的@InitBinder 注解的使用。

**代码清单 12-25　初始化 WebDataBinderFactory**

```java
protected ModelAndView invokeHandlerMethod(HttpServletRequest request,
        HttpServletResponse response, HandlerMethod handlerMethod) throws Exception {

    ServletWebRequest webRequest = new ServletWebRequest(request, response);
    try {
        // 参数绑定器初始化
        WebDataBinderFactory binderFactory = getDataBinderFactory(handlerMethod);
        // ......
```

**代码清单 12-26　@ControllerAdvice + @InitBinder 注解的使用**

```java
@ControllerAdvice
public class ConversionBinderAdvice {

    @InitBinder
    public void addDateBinder(WebDataBinder dataBinder) {
        dataBinder.addCustomFormatter(new DateFormatter("yyyy年MM月dd日"));
    }
}
```

底层收集和执行@InitBinder 注解标注方法的逻辑就是这个 getDataBinderFactory 方法,该方法中会将当前请求的 Handler 所在 Controller 中标注了@InitBinder 注解的方法全部取出,之后组合所有标注了@ControllerAdvice 注解的 Controller 类中标注了@InitBinder 注解的方法,共同组成参数绑定器的初始化器,如代码清单 12-27 所示。

**代码清单 12-27　getDataBinderFactory 组合所有参数绑定器的初始化器**

```java
private WebDataBinderFactory getDataBinderFactory(HandlerMethod handlerMethod)
        throws Exception{
    Class<?> handlerType = handlerMethod.getBeanType();
    Set<Method> methods = this.initBinderCache.get(handlerType);
    if (methods == null) {
        // 筛选当前 Controller 中被@InitBinder 注解标注的方法
        methods = MethodIntrospector.selectMethods(handlerType, INIT_BINDER_METHODS);
        this.initBinderCache.put(handlerType, methods);
    }
    List<InvocableHandlerMethod> initBinderMethods = new ArrayList<>();
    // 组合全局的被@InitBinder 注解标注的方法
    this.initBinderAdviceCache.forEach((controllerAdviceBean, methodSet) -> {
        if (controllerAdviceBean.isApplicableToBeanType(handlerType)) {
            Object bean = controllerAdviceBean.resolveBean();
            for (Method method : methodSet) {
                initBinderMethods.add(createInitBinderMethod(bean, method));
            }
        }
    });
```

```java
    for (Method method : methods) {
        Object bean = handlerMethod.getBean();
        initBinderMethods.add(createInitBinderMethod(bean, method));
    }
    return createDataBinderFactory(initBinderMethods);
}
```

**（b）参数预绑定**

紧接着的初始化逻辑是参数预绑定部分，如代码清单 12-28 所示，预绑定这个概念不是很好理解，可能读者会更熟悉@ModelAttribute 注解，这个注解除了可以支持前后端不分离情况下的数据回显，还拥有公有数据暴露、请求数据处理的能力，它的收集规则与@InitBinder 注解大同小异，如代码清单 12-29 所示，不再赘述。

**┃代码清单 12-28   参数预绑定的初始化**

```java
// ......
// 参数预绑定
ModelFactory modelFactory = getModelFactory(handlerMethod, binderFactory);
// ......
```

**┃代码清单 12-29   getModelFactory 收集所有标注了@ModelAttribute 注解的方法**

```java
public static final MethodFilter MODEL_ATTRIBUTE_METHODS = method ->
        (!AnnotatedElementUtils.hasAnnotation(method, RequestMapping.class)
            && AnnotatedElementUtils.hasAnnotation(method, ModelAttribute.class));

private ModelFactory getModelFactory(HandlerMethod handlerMethod,
        WebDataBinderFactory binderFactory) {
    SessionAttributesHandler sessionAttrHandler =
            getSessionAttributesHandler(handlerMethod);
    Class<?> handlerType = handlerMethod.getBeanType();
    Set<Method> methods = this.modelAttributeCache.get(handlerType);
    if (methods == null) {
        // 筛选出被@ModelAttribute 标注且没有被@RequestMapping 标注的方法
        methods = MethodIntrospector.selectMethods(handlerType, MODEL_ATTRIBUTE_METHODS);
        this.modelAttributeCache.put(handlerType, methods);
    }
    // 组合全局的被@ModelAttribute 标注的方法 ......
    return new ModelFactory(attrMethods, binderFactory, sessionAttrHandler);
}
```

**（c）创建方法执行对象**

下一个动作是创建 ServletInvocableHandlerMethod 对象，源码中仅是对 HandlerMethod 进行二度封装，如代码清单 12-30 所示，封装后的模型会在下面执行 Controller 方法时用到。

**┃代码清单 12-30   创建 ServletInvocableHandlerMethod 对象**

```java
// ......
// 创建方法执行对象
ServletInvocableHandlerMethod invocableMethod = createInvocableHandlerMethod(handlerMethod);
// ......
```

```java
protected ServletInvocableHandlerMethod createInvocableHandlerMethod(HandlerMethod handlerMethod){
    return new ServletInvocableHandlerMethod(handlerMethod);
}
```

（d）ModelAndView、异步请求的处理

紧跟着的两段源码中完成的工作是创建 ModelAndView 的容器 ModelAndViewContainer 以及对异步请求的支持。由于这部分源码不是很重要，感兴趣的读者可以自行借助 IDE 翻阅和调试源码，这里不再展开。

（e）执行 Controller 的方法

接下来执行的核心逻辑就是上面创建的 ServletInvocableHandlerMethod 中的 invokeAndHandle 方法，如代码清单 12-31 所示。由于这个方法内部很长，因此下面单独讲解该方法。

**代码清单 12-31　执行 Controller 的方法**

```java
// ......
// 执行 Controller 的方法
invocableMethod.invokeAndHandle(webRequest, mavContainer);
if (asyncManager.isConcurrentHandlingStarted()) {
    return null;
}
// 包装 ModelAndView
return getModelAndView(mavContainer, modelFactory, webRequest);
} // finally ......
}
```

（3）invokeAndHandle（反射）

下一个步骤是利用反射机制调用 Handler 的方法，对应的 invokeAndHandle 方法（见代码清单 12-32）及后续涉及的环节篇幅较长，为了便于读者更好地阅读和理解源码，下面将其拆分为多个片段讲解。

**代码清单 12-32　invokeAndHandle**

```java
public void invokeAndHandle(ServletWebRequest webRequest, ModelAndViewContainer mavContainer,
        Object... providedArgs) throws Exception {
    Object returnValue = invokeForRequest(webRequest, mavContainer, providedArgs);
    // ......
}
```

（a）反射执行 Controller 方法

由代码清单 12-33 中可以明显看到，RequestMappingHandlerAdapter 执行 Controller 方法的核心是，在收集好执行 Handler 方法所需的参数列表后利用反射机制执行目标 Controller 的方法。doInvoke 方法执行完毕后，就意味着 Controller 的工作已经完成，下面的部分是对返回值的处理。

**代码清单 12-33　invokeForRequest 反射执行 Handler**

```java
public Object invokeForRequest(NativeWebRequest request,
        @Nullable ModelAndViewContainer mavContainer, Object... providedArgs) throws Exception {
    Object[] args = getMethodArgumentValues(request, mavContainer, providedArgs);
```

```java
        return doInvoke(args);
}

protected Object doInvoke(Object... args) throws Exception {
    ReflectionUtils.makeAccessible(getBridgedMethod());
    try {
        return getBridgedMethod().invoke(getBean(), args);
    }
    // catch ......
}
```

(b) 处理方法返回值

invokeAndHandle 方法最后的 try-catch 部分是解析并处理返回值的逻辑，如代码清单 12-34 所示。由于 returnValueHandlers 是一个复合对象，它的内部组合了一组 HandlerMethodReturnValueHandler，根据之前阅读源码的经验不难得知，它的内部会使用 for 循环去匹配可以处理当前 Controller 返回值的 HandlerMethodReturnValueHandler 并实际处理，如代码清单 12-35 所示。

**代码清单 12-34　invokeAndHandle 中处理返回值的部分**

```java
private HandlerMethodReturnValueHandlerComposite returnValueHandlers;

public void invokeAndHandle(ServletWebRequest webRequest, ModelAndViewContainer mavContainer,
        Object... providedArgs) throws Exception {
    Object returnValue = invokeForRequest(webRequest, mavContainer, providedArgs);
    // ......
    try {
        // 处理返回值
        this.returnValueHandlers.handleReturnValue(
                returnValue, getReturnValueType(returnValue), mavContainer, webRequest);
    } // catch ......
}
```

**代码清单 12-35　利用 HandlerMethodReturnValueHandler 处理返回值**

```java
public void handleReturnValue(@Nullable Object returnValue, MethodParameter returnType,
        ModelAndViewContainer mavContainer, NativeWebRequest webRequest) throws Exception {
    // 选择合适的返回值处理器
    HandlerMethodReturnValueHandler handler = selectHandler(returnValue, returnType);
    // 检查为空 ......
    handler.handleReturnValue(returnValue, returnType, mavContainer, webRequest);
}

private HandlerMethodReturnValueHandler selectHandler(@Nullable Object value,
        MethodParameter returnType) {
    boolean isAsyncValue = isAsyncReturnValue(value, returnType);
    for (HandlerMethodReturnValueHandler handler : this.returnValueHandlers) {
        // 前置检查 ......
        // 匹配支持处理返回值类型的 HandlerMethodReturnValueHandler
        if (handler.supportsReturnType(returnType)) {
```

```
            return handler;
        }
    }
    return null;
}
```

注意，对于返回视图和返回 JSON 数据，底层使用的 `HandlerMethodReturnValue Handler` 并不相同。下面分别就视图返回和 JSON 数据响应这两种常见场景展开讲解。

(c) 处理视图返回

如果一个 Controller 方法最后要跳转视图，则方法的返回值一定是一个字符串，并且方法和类上都没有标注 `@ResponseBody` 注解，这种情况下对应的视图返回值处理器是 `ViewNameMethodReturnValueHandler`，从它的类名上就可以直观地理解其作用，它处理视图名称的逻辑是，通过执行 `setViewName` 方法将 Handler 返回的逻辑视图名放入 `Model AndViewContainer` 中，如代码清单 12-36 的 `handleReturnValue` 方法所示。

**代码清单 12-36　ViewNameMethodReturnValueHandler 支撑视图返回**

```java
public boolean supportsReturnType(MethodParameter returnType) {
    Class<?> paramType = returnType.getParameterType();
    // 判断返回值是不是 CharSequence(String)
    return (void.class == paramType || CharSequence.class.isAssignableFrom(paramType));
}
public void handleReturnValue(@Nullable Object returnValue, MethodParameter returnType,
        ModelAndViewContainer mavContainer, NativeWebRequest webRequest) throws Exception {
    if (returnValue instanceof CharSequence) {
        String viewName = returnValue.toString();
        mavContainer.setViewName(viewName);
        if (isRedirectViewName(viewName)) {
            mavContainer.setRedirectModelScenario(true);
        }
    }
    // else if throw ex ......
}
```

请注意，这个操作非常类似于直接操作 `ModelAndView` 对象的 API，但又不完全相同，`ModelAndViewContainer` 的内部并没有直接组合 `ModelAndView` 对象，而是存储了一套与 `ModelAndView` 相同的内部结构（包括 `ModelMap` 集合以及 `VIEW` 视图对象）。此处执行的 `setViewName` 方法是将视图名称放入 `ModelAndViewContainer` 的内部，在下面即将看到的 `getModelAndView` 方法中会将 `ModelAndViewContainer` 转换为 `ModelAndView` 对象，此时视图名称会转移到 `ModelAndView` 对象中。

(d) 处理 JSON 数据响应

如果一个 Controller 方法需要响应 JSON 数据，则需要在方法或方法所在类上标注 `@ResponseBody` 注解。处理 JSON 数据响应的底层实现是 `RequestResponseBodyMethod Processor`，它在 `handleReturnValue` 方法中会执行 `writeWithMessageConverters` 方法，使用 JSON 序列化的方式将方法返回的数据（即 `returnValue`）转化为文本，并直接写入 `HttpServletResponse` 的输出流中，如代码清单 12-37 所示。

### 代码清单 12-37 RequestResponseBodyMethodProcessor 支撑 JSON 数据响应

```java
public boolean supportsReturnType(MethodParameter returnType) {
    return (AnnotatedElementUtils.hasAnnotation(returnType.getContainingClass(),
        ResponseBody.class) || returnType.hasMethodAnnotation(ResponseBody.class));
}
public void handleReturnValue(@Nullable Object returnValue, MethodParameterreturnType,
        ModelAndViewContainer mavContainer, NativeWebRequest webRequest)
        throws IOException, HttpMediaTypeNotAcceptableException,
            HttpMessageNotWritableException {

    mavContainer.setRequestHandled(true);
    ServletServerHttpRequest inputMessage = createInputMessage(webRequest);
    ServletServerHttpResponse outputMessage = createOutputMessage(webRequest);

    writeWithMessageConverters(returnValue, returnType, inputMessage, outputMessage);
}
```

有关核心的 JSON 写入方法 writeWithMessageConverters 的内部逻辑相对复杂，感兴趣的读者可以借助 IDE 自行深入探究，由于该部分内容相对不是重点，因此本书不再对此展开。

**（4）getModelAndView**

invokeAndHandle 方法执行完成后，ModelAndViewContainer 中已经封装了数据对象和视图响应，或是将需要响应的数据写入了 HttpServletResponse 中。回到 RequestMappingHandlerAdapter 的 invokeHandlerMethod 方法中，最后一步需要执行的方法是 getModelAndView，如代码清单 12-38 所示。getModelAndView 方法的核心动作是将 ModelAndViewContainer 中的 ModelMap 与 View 对象取出，以此构建一个全新的 ModelAndView 对象，并在一些简单的后置逻辑处理动作后返回。

### 代码清单 12-38 invokeHandlerMethod 的最后执行 getModelAndView 方法

```java
        // ......
        // 执行 Controller 的方法
        invocableMethod.invokeAndHandle(webRequest, mavContainer);
        // ......
        // 封装 ModelAndView
        return getModelAndView(mavContainer, modelFactory, webRequest);
    } // finally ......
}

private ModelAndView getModelAndView(ModelAndViewContainer mavContainer,
        ModelFactory modelFactory, NativeWebRequest webRequest) throws Exception {
    // 前置准备逻辑 ......
    // 注意此处将 ModelMap 取出
    ModelMap model = mavContainer.getModel();
    // 注意此处把视图名称取出，并组合生成 ModelAndView
    ModelAndView mav = new ModelAndView(mavContainer.getViewName(), model,
            mavContainer.getStatus());
    if (!mavContainer.isViewReference()) {
        mav.setView((View) mavContainer.getView());
```

```
    }
    // ......
    return mav;
}
```

简言之，getModelAndView 方法完成了 ModelAndViewContainer 到 ModelAndView 对象的转换。

至此，HandlerAdapter 的工作全部完成，下面回到 DispatcherServlet 中。

（5）回调拦截器

执行完 HandlerAdapter 的 handle 方法（见代码清单 12-39）后，下一步的核心动作是调用所有 HandlerInterceptor 的 postHandle 方法，进行请求的后置拦截。从代码清单 12-40 的方法实现上可以看出，它的调用机制与 applyPreHandle 几乎完全一致，唯一不同的是回调的顺序与 applyPreHandle 方法刚好相反。

**代码清单 12-39　doDispatch（6）**

```
// ......
// 实际调用 Handler
mv = ha.handle(processedRequest, response, mappedHandler.getHandler());
// ......
applyDefaultViewName(processedRequest, mv);
mappedHandler.applyPostHandle(processedRequest, response, mv);
}
// ......
```

**代码清单 12-40　倒序回调所有 HandlerInterceptor**

```
void applyPostHandle(HttpServletRequest request, HttpServletResponse response,
        @Nullable ModelAndView mv) throws Exception {
    HandlerInterceptor[] interceptors = getInterceptors();
    if (!ObjectUtils.isEmpty(interceptors)) {
        // 注意此处是倒序回调
        for (int i = interceptors.length - 1; i >= 0; i--) {
            HandlerInterceptor interceptor = interceptors[i];
            interceptor.postHandle(request, response, this.handler, mv);
        }
    }
}
```

**6. 处理视图、解析异常**

doDispatch 方法的核心 try-catch 块逻辑执行完毕后，进入 DispatcherServlet 核心工作流程的最后一个关键步骤 processDispatchResult 方法，如代码清单 12-41 所示。该方法会进行视图处理，以及解析整个请求处理中抛出的异常。processDispatchResult 方法分为三个步骤，为了便于读者更好地阅读和理解源码，下面将拆分每个步骤分别研究。

**代码清单 12-41　doDispatch（7）**

```
    // ......
    mappedHandler.applyPostHandle(processedRequest, response, mv);
}
// catch ......
```

```
        processDispatchResult(processedRequest, response, mappedHandler, mv, dispatchException);
    }
    // ......
```

(1) 处理异常

`DispatcherServlet` 在处理客户端发起的请求时，中间调用 Controller 或者 Service 等组件时抛出的异常几乎不可能是 `ModelAndViewDefiningException`（源码也没有任何构建 `ModelAndViewDefiningException` 的部分），所以代码清单 12-42 中判断异常类型的 if 结构可以忽略。在 else 块中的内部会调用 `processHandlerException` 方法，替换 Handler Adapter 返回的 `ModelAndView` 对象，可想而知这个 `processHandlerException` 方法就是处理异常的核心逻辑，并针对异常情况重新生成 `ModelAndView` 对象。

**代码清单 12-42　processDispatchResult（1）**

```
private void processDispatchResult(HttpServletRequest request, HttpServletResponse response,
        @Nullable HandlerExecutionChain mappedHandler, @Nullable ModelAndView mv,
        @Nullable Exception exception) throws Exception {
    boolean errorView = false;
    // 处理异常
    if (exception != null) {
        if (exception instanceof ModelAndViewDefiningException) {
            // ......
        } else {
            Object handler = (mappedHandler != null ? mappedHandler.getHandler() : null);
            mv = processHandlerException(request, response, handler, exception);
            errorView = (mv != null);
        }
    }
    // ......
```

`processHandlerException` 方法的内部实现逻辑与其他核心组件非常相似，主干逻辑都是从一组核心处理对象中筛选出一个可以处理当前逻辑的对象并返回，如代码清单 12-43 所示。不过在当前 Debug 调试的 `springboot-01-quickstart` 的示例项目中并没有定义异常处理器，Spring Boot 内置处理的异常处理器对于项目中出现的异常处理也只是简单抛出异常。如果在示例项目中添加一个标注有@ControllerAdvice 注解的 Controller，并声明标注有@ExceptionHandler 注解的方法，就可实现全局统一异常处理。

**代码清单 12-43　processHandlerException 处理异常**

```
protected ModelAndView processHandlerException(HttpServletRequest request,
        HttpServletResponse response, @Nullable Object handler, Exception ex) throws Exception {
    // ......
    ModelAndView exMv = null;
    if (this.handlerExceptionResolvers != null) {
        for (HandlerExceptionResolver resolver : this.handlerExceptionResolvers) {
            exMv = resolver.resolveException(request, response, handler, ex);
            if (exMv != null) {
                break;
            }
        }
    }
```

```
    // ......
    throw ex;
}
```

本部分的源码底层逻辑与 `HandlerAdapter` 相似，感兴趣的读者可以自行编写带有抛出异常的 `Controller` 方法，并配合统一异常处理完成异常处理的构建，在 `processHandlerException` 方法中自行调试源码，本书不再对此展开。

（2）渲染视图

无论是 `HandlerAdapter` 调用实际处理请求的 `Handler` 正常响应，还是抛出异常后被统一异常处理，最终都会生成一个 `ModelAndView` 对象。紧接着要处理的逻辑是视图的渲染，而渲染视图的核心方法是在代码清单 12-44 中 `if` 块的 `render` 方法。这个方法源码篇幅不长，读者只需要关注代码清单 12-45 中带有注释的源码，对于不重要的源码已进行了省略。

**代码清单 12-44　processDispatchResult（2）**

```
// ......
if (mv != null && !mv.wasCleared()) {
    render(mv, request, response);
    if (errorView) {
        WebUtils.clearErrorRequestAttributes(request);
    }
}
// ......
```

**代码清单 12-45　render 渲染视图**

```
protected void render(ModelAndView mv, HttpServletRequest request, HttpServletResponse response)
        throws Exception {
    // 国际化处理 ......

    View view;
    String viewName = mv.getViewName();
    if (viewName != null) {
        // 如果有视图名，则解析出视图
        view = resolveViewName(viewName, mv.getModelInternal(), locale, request);
    } else {
        // 否则，直接获取视图。如果还没有视图，则抛出异常
        view = mv.getView();
    }
    try {
        if (mv.getStatus() != null) {
            response.setStatus(mv.getStatus().value());
        }
        // 带入 Model，渲染视图
        view.render(mv.getModelInternal(), request, response);
    } // catch ......
}
```

通读整段源码可以发现，`render` 方法会从 `ModelAndView` 中获取逻辑视图的名称，如果可以成功获取到逻辑视图名称，则会借助 `ViewResolver` 去匹配视图（最常见的实现是拼接前后缀的 `InternalResourceViewResolver`），如果可以成功匹配到视图文件，则成功返回；

如果匹配不到则抛出异常。匹配生成 View 对象后，会在最下面的 try 块中执行 view.render 方法，该方法在整合不同的前端模板引擎（如 Thymeleaf、FreeMarker 等）时会有不同的实现，具体的实现不需要了解，读者只需要知道 View 对象的 render 方法可以实际渲染视图。

至此，ViewResolver 部分的工作执行完成。

（3）回调拦截器

视图渲染完成后，最后的收尾动作是回调拦截器的 afterCompletion 方法，如代码清单 12-46 所示，其底层实现与前置、后置拦截相似，这里不再展开。

**代码清单 12-46　processDispatchResult（3）**

```
// ......
// 回调拦截器的 afterCompletion
if (mappedHandler != null) {
    mappedHandler.triggerAfterCompletion(request, response, null);
}
}
```

processDispatchResult 方法执行完毕后，整个 doDispatch 方法的主干逻辑全部执行完毕，一次完整的 DispatcherServlet 请求处理与响应就完成了。

### 12.4.5　DispatcherServlet 工作全流程小结

简单总结一下 DispatcherServlet 的工作全流程，以下的全流程是针对 @RequestMapping 的场景总结。

1. 客户端向服务端发起请求，由 DispatcherServlet 接收请求。
2. DispatcherServlet 委托 HandlerMapping，根据本次请求的 URL 匹配合适的 Controller 方法。
3. HandlerMapping 找到合适的 Controller 方法后，组合可以应用于当前请求的拦截器，并封装为一个 Handler 对象返回给 DispatcherServlet。
4. DispatcherServlet 接收到 Handler 后委托 HandlerAdapter，将该请求转发给 HandlerMapping 选定的 Controller 中的 Handler。
5. Handler 接收到请求后，实际执行 Controller 中的方法。
6. Controller 方法执行完毕后返回 ModelAndView 对象。
7. HandlerAdapter 接收到 Handler 返回的 ModelAndView 后返回给 DispatcherServlet。
8. DispatcherServlet 获取 ModelAndView 后委托 ViewResolver，由 ViewResolver 负责渲染视图。
9. ViewResolver 渲染视图完成后返回给 DispatcherServlet，由 DispatcherServlet 负责响应视图。

## 12.5　小结

本章从 WebMvc 的自动装配开始，回顾 WebMvc 中的核心组件及用途。然后研究 WebMvc 中 Controller 中可以处理请求的 Handler 方法的收集逻辑。最后从一个示例出发，结合 Debug

研究 DispatcherServlet 的工作全流程。WebMvc 本质是以 DispatcherServlet 为核心组合几个关键功能组件，共同构成 WebMvc 的底层支撑。

在 Spring Framework 5.0 版本之后 WebMvc 多了一个孪生兄弟 WebFlux，它基于异步非阻塞场景的开发，第 13 章会从相关概念开始，逐步深入底层探究 WebFlux 的使用与原理。

# 第 13 章 Spring Boot 整合 WebFlux

本章主要内容：
◇ 响应式编程与 Reactive；
◇ Spring Boot 整合 WebFlux 的快速使用；
◇ Spring Boot 整合 WebFlux 的核心自动装配；
◇ DispatcherHandler 的工作全流程。

在第 12 章我们了解了 Spring Boot 整合 WebMvc 场景下的自动装配，WebMvc 的本质是基于 Servlet，无论设计得多强大，其本质都是阻塞的，每个连接都会占用一个线程。在 Servlet 容器处理客户端请求时，会为每一个请求分配一个工作线程进行处理。这样带来的后果是，基于 Servlet 的阻塞式 Web 框架在面对海量请求时性能上会捉襟见肘。为了解决该问题，在 Spring Framework 5.0 版本后引入了 WebMvc 的孪生兄弟 WebFlux，它是一个异步非阻塞式 Web 框架，且 Spring Framework 5.x 基于 JDK1.8，Java 底层就已经支持了函数式编程，这就为 WebFlux 提供了强有力的语言级支撑。本章就 WebFlux 的核心知识和 Spring Boot 整合的底层原理展开探讨和讲解。

## 13.1 快速了解响应式编程与 Reactor

WebFlux 的底层核心技术是响应式编程，这种编程思想和方式区别于命令式编程，是一种新的编程风格，下面就这两种编程风格进行简单对比。

### 13.1.1 命令式与响应式

在基于 WebMvc 的项目开发中，通过编写 Controller 前端控制器，注入 Service 业务逻辑类进行处理，Service 中包含与数据库的交互、与中间件的通信等，这种编码风格就是**命令式编程**。使用命令式编程的代码更像是一组前后联系紧密的任务，它们有明确的先后执行顺序，后面的任务通常需要依赖前面任务生成的结果才可以正确执行。命令式编程的特点是**串行**、**阻塞**。

响应式编程中不再将这些前后关联的任务看作一个整体，而是将其拆分为一个个可以**并行**执行的工作任务，这些工作任务之间互不干扰。每个工作任务都可以接收特定的数据，并在处理完成后传递给整体流程中的下一个任务，同时继续处理下一组数据。请注意，响应式编程中，每个工作任务不会主动获取数据，而是被动地等待数据提供方给它提供数据，这种设计的核心思想是"**被动接收**"而不是"**主动获取**"，它主张数据以订阅的方式推送，而不是以请求的方式拉取。

简单总结，响应式编程本身是一种代替命令式编程的**范式**，它并不能完美应对所有业务场景。通过合理选择编码方式，可以在实际的项目开发中实现最优效果。

### 13.1.2 概念和思想的回顾与引入

简单对比命令式与响应式编程之后，下面需要读者回顾一些概念和思想，它们会在后续了解响应式编程时起到引导作用。之后本节还会引入一些有关响应式编程的新概念，理解这些概念后，在学习响应式编程时会更加顺利。

#### 1. 异步非阻塞

有关对异步非阻塞概念的解释，有一个非常经典的示例，此处引用该示例并加以解释。

假设有一个老张烧水的场景，老张有两把烧水壶，分别是没有哨的普通水壶以及壶盖上带哨的响水壶。烧水的场景包含以下 4 种，逐一来看。

- **同步阻塞式**：使用普通水壶烧水，由于不清楚水烧开的时间，因此需要老张在水壶旁观察，等到水壶冒热气，壶里的水沸腾，老张将水壶离火，烧水结束。在该场景中，由于老张在烧水期间无法完成其他工作，只能等待水烧开，烧水占据了老张的注意力和时间，构成同步阻塞。
- **同步非阻塞**：经过上一次烧水后，老张发现烧水太浪费自己的时间，于是下一次烧水时老张选择同时打游戏，每隔一小段时间就去看一下水壶里的水是否烧开，如果水还没有烧开就继续打游戏，水烧开则将水壶离火，烧水结束。在该场景中，老张没有一直盯着水壶，但还是会间歇性消耗精力，只不过在整个烧水的过程中，老张没有一直被水壶占用全部精力和时间，构成同步非阻塞。
- **异步阻塞式**：间歇性观察水壶仍然不是最佳选择，老张选择使用响水壶烧水，但由于第一次使用响水壶烧水，老张不确定水壶上的哨是否好用，于是他像第一次烧水那样在水壶旁观察，等到水壶冒热气，同时哨声响起，老张将水壶离火，烧水结束。在该场景中，老张不再主动关心水壶的状态，但精力和时间仍然被水壶占用，构成异步阻塞。
- **异步非阻塞**：烧完水后老张发现自己很傻，因为哨声响起就意味着水已烧开，无须自己消耗精力和时间，于是后续烧水时，老张都是准备好后直接去打游戏，等到水壶哨声响起，再将水壶离火，烧水结束。在最终的场景中，老张不再主动关心水壶状态，也不需要间歇性检查水壶内水的状态，而只需要在水壶的哨声响起时处理水壶离火的任务，此场景就是异步非阻塞。

#### 2. 观察者模式

观察者模式是 GoF23 中非常经典的设计模式之一，它也被称为"发布-订阅模式"或"监听器模式"。观察者模式中关注的点是，**某一个对象被修改/做出某些反应/发布一个信息等时会自动通知依赖它的对象（订阅者）**。观察者模式的三大核心是**观察者、被观察主题和订阅者**。观察者（Observer）需要绑定要通知的订阅者（Subscriber），并且要观察指定的主题（Subject）。

#### 3. 前端框架的双向绑定

在 Vue、React、AngularJS 中有一个很基本的概念：**双向绑定**。修改输入框中的表单值时上方的 p 标签内容也会随之变化，如图 13-1 所示，这就是响应式的体现。

图 13-1  Vue 的双向绑定

仔细分析双向绑定的现象，可以从中提取出几个关键的信息。
- p 标签段落部分的内容更像是"观察"着下方表单输入框的内容，当输入框的内容发生变化时，段落的内容也会随之变化，这本身就是观察者模式的体现。由此可以引出响应式编程的第一个关键概念：**变化传递**。当表单输入框的内容变化时，输入框的值会传递至 p 标签段落中。
- 在实际测试中，每修改一次表单输入框的内容，p 标签段落中的内容就会随之改变，如果将这组变化的内容全部列举，可以形成一组表单输入框内容的变化事件记录。由此就可以引出响应式编程的第二个关键概念：**数据流**。事件源的每一次变化连起来就是一个事件流。
- 在 Vue.js 的实现方案中，双向绑定的示例代码如代码清单 13-1 所示。除了必要的页面元素和 Vue 对象的构建，没有任何多余绑定关系的代码，由此就可以引出响应式编程的第三个关键概念：**声明式**。不需要编写命令式代码，仅靠声明两者之间的关系就可以形成双向绑定。

**代码清单 13-1　Vue.js 中实现最简单的双向绑定**

```
<div id="app">
    <p>{{ message }}</p>
    <input v-model="message">
</div>
var app = new Vue({
    el: '#app',
    data: {
        message: 'SpringBoot good!'
    }
})
```

简单总结，响应式编程的三个关键点是**变化传递、数据流和声明式**。这里最关键的围绕着响应式编程的核心概念是**事件**，事件是观察者模式的核心，响应式也是观察者模式，自然响应式编程也需要依赖事件。

#### 4．响应式流

下面的两个概念是有关响应式编程的新概念。响应式流有别于 Java 8 中的 Stream，普通的 Stream 是同步阻塞的，在高并发场景下不能有效缓解压力大的问题，而**响应式流可以做到异步非阻塞**。另外，Stream 的一个关键特性是，一旦有了消费型方法，它就会将这个流中的所有方法处理完毕，如果这期间的数据量很大，Stream 就无法对海量数据进行妥善处理（相当于使用函数对集合进行迭代）；而响应式流可以通过背压对海量数据（甚至是无限量数据）进行**流量控制**，以确保数据的接收速度在处理能力之内。简单总结，响应式流的关键点是**异步非阻塞和数据流速控制**。

#### 5．背压

上面提到了背压是控制数据流速的关键手段，下面通过一个模拟场景来讲解背压。

假设你在一个知名手机生产大厂工作，你的职位是生产流水线上的一名普通工人，你的工作是负责流水线上的一个关键环节，该环节需要的加工时间比较长，而恰好近期与你共同负责相同工作的同事都请假了，剩下你单枪匹马仍然战斗在生产一线。

与此同时，负责你上游工作的同事似乎并不清楚你负责环节的生产现状，而且由于上司的激励政策，上游同事的生产效率非常高，导致你的待加工区积压了非常多半成品，但由于你负责的工序耗时长，积压的半成品过多无法及时处理，于是你不得不向上游同事反馈：**你们做慢点，我的工作吞吐量有限**。上游同事了解你的现状后改变了半成品处理策略，他们将**处理好的半成品不直接传递给你**，而暂时由上游同事保管，等你向他们反馈积压的半成品处理完毕后，再继续传递新的半成品。

> 💡 提示：由此可以体现出背压的第一个策略，即数据提供方将数据暂存，不传递给下游消费者。

一段时间之后，领导发现你的业绩非常好，于是你升职加薪，以经销商的身份销售该款手机。手机一上市就得到广大消费者的关注，你的店铺生意非常好。

正当你的生意做得风生水起时，这批手机在售卖后的一段时间后传出硬件问题，市面销量急剧下降，作为经销商，你自然也不想再销售该款手机，于是你向厂商反映：**请不要再提供该款手机**。厂商也非常无奈，手机还在正常生产，但经销商都不再提货，于是只好**将这部分成品废弃**。

> 💡 提示：由此可以体现出背压的第二个策略：数据提供方将数据丢弃。

简单总结，背压是下游消费者"倒逼"上游数据生产者的数据提供速率，以避免被海量数据压垮，达到两者之间的动态平衡。

### 13.1.3 快速体会 Reactor 框架

下面通过几个简单的示例体会响应式编程的具体落地使用。市面上流行的响应式编程框架包括 Reactor 与 ReactiveX（RxJava）。由于 WebFlux 底层使用 Reactor 提供响应式支撑，因此本书选择使用 Reactor 进行演示。

具体的项目搭建中，无须导入特定版本的 Reactor 框架，而是直接导入 `spring-boot-starter-webflux` 依赖即可，根据 Maven 的依赖传递原则，Reactor 框架会一并导入。

**1. 最简单的发布-订阅**

下面先编写一个最简单的发布-订阅实现，如代码清单 13-2 所示。

**代码清单 13-2　基于响应式的最简单发布-订阅模型**

```java
public class QuickDemo {
    public static void main(String[] args) {
        Flux<Integer> flux = Flux.just(1, 2, 3);
        flux.subscribe(System.out::println);
    }
}
```

示例代码非常简单，发布-订阅模型需要一个数据的生产者，即 `Publisher`，这里为了编码方便，代码清单 13-2 中选用 Reactor 中的实现类 `Flux` 作为数据生产者，数据接收者的类型是 Java 8 中的 `Consumer`，所以此处可以传入 Lambda 表达式或者方法引用。在这段代码中，生产者与接收者通过 `subscribe` 方法建立订阅关系（消费关系），一旦触发 `subscribe` 方法，接收者就可以接收生产者提供的数据并进行处理，直到生产者的数据全部处理完成或者出现异常终止。

## 2. 响应式的核心规范组件

上述的发布-订阅模型中涉及响应式编程的 4 个核心规范组件，逐一来看。

（1）Publisher

Publisher 作为数据生产者，它只有一个方法 subcribe，该方法会接收一个 Subscriber，构成"订阅"关系，如代码清单 13-3 所示。

**代码清单 13-3　Publisher 接口的方法**

```
public void subscribe(Subscriber<? super T> s);
```

请注意，subscribe 方法是一个类似于"工厂"的方法，它可以被多次调用，但是每次调用时都会创建一个新的订阅关系，且一个订阅关系只能关联一个接收者（即下面的 Subscriber）。

（2）Subscriber

Subscriber 作为数据接收者，它的接口方法有 4 个，如代码清单 13-4 所示。

**代码清单 13-4　Subscriber 接口的方法**

```
public void onSubscribe(Subscription s);
public void onNext(T t);
public void onError(Throwable t);
public void onComplete();
```

请注意，Subscriber 接口的方法都以 **on** 作为前缀，代表它属于事件形式（可以联想 JavaScript 中的 onclick 方法等），以此理解 Subscriber 接口的 4 个方法。

- onSubscribe：当触发订阅时触发。
- onNext：当接收到下一个数据时触发。
- onComplete：当生产者的数据都处理完成时触发。
- onError：当出现异常时触发。

（3）Subscription

Subscription 可以看作一个订阅"关系"，它归属于一个生产者和一个接收者（可以理解为关系型数据库的多对多中间表的一条数据）。Subscription 接口中有两个方法，如代码清单 13-5 所示。

**代码清单 13-5　Subscription 接口的方法**

```
public void request(long n);

public void cancel();
```

request 方法的作用是主动请求/拉取数据，cancel 方法的作用是放弃/停止拉取数据，它完成了数据接收者与生产者的交互，背压也是基于 Subscription 接口来实现的。

（4）Processor

Processor 从类名上可以翻译为处理器，这些处理器的特点是**有输入**，**有输出**，对应到 Reactor 的概念中则是生产者和接收者的合体，如代码清单 13-6 所示。

**代码清单 13-6　Processor 接口**

```
public interfac Processor<T, R> extends Subscriber<T>, Publisher<R> { }
```

从代码清单 13-6 中可以发现，`Processor` 直接继承了 `Publisher` 接口和 `Subscriber` 接口，它一般用于数据的中间环节处理（如数据转换 map、数据过滤 filter 等）。

### 3. Reactor 中的常用组件

上述 4 个响应式的核心规范组件对应到响应式编程的具体实现框架中，它们的实现类也需要简单了解。

（1）Flux

`Flux` 可以简单理解为 "非阻塞的 Stream"，它实现了 `Publisher` 接口，它的内部定义了很多 `subscribe` 方法，如图 13-2 所示。

```
subscribe(): Disposable
subscribe(Consumer<? super T>): Disposable
subscribe(Consumer<? super T>, Consumer<? super Throwable>): Disposable
subscribe(Consumer<? super T>, Consumer<? super Throwable>, Runnable): Disposable
subscribe(Consumer<? super T>, Consumer<? super Throwable>, Runnable, Consumer<? super Subscription>): Disposable
subscribe(CoreSubscriber<? super T>): void
subscribe(Subscriber<? super T>): void
subscribeOn(Scheduler): Flux<T>
subscribeOn(Scheduler, boolean): Flux<T>
```

图 13-2 Flux 中定义的 subscribe 方法

请注意，`Flux` 在实现 `Publisher` 原生的 `subscribe` 方法时，还扩展了几个方法，目的是简化操作，如代码清单 13-2 的简单示例中就使用了图 13-2 中的第二个重载的方法，只传入一个 `Consumer` 的实现对象（该方法只处理正常情况下的数据接收）。

另外需要读者了解的是，`Flux` 本身拥有非常多 Java 8 中 Stream 的操作，代码清单 13-7 中是一个简单示例。要学习多更具体的 API 操作，读者可以参照 javadoc 和官方文档，本书不再展开。

**代码清单 13-7 操作 Flux 的示例代码**

```java
public class FluxDemo {
    public static void main(String[] args) throws Exception {
        Flux<Integer> flux = Flux.just(1,2,3);
        flux.map(num -> num * 5) // 将所有数据扩大 5 倍
            .filter(num -> num > 10) // 只过滤出数值中超过 10 的数
            .map(String::valueOf) // 将数据转换为 String 类型
            .publishOn(Schedulers.elastic()) // 使用弹性线程池来处理数据
            .subscribe(System.out::println); // 消费数据
    }
}
```

（2）Mono

`Mono` 可以简单理解为 "非阻塞的 Optional"，它也实现了 `Publisher` 接口，具备生产数据的能力。`Mono` 在特征上与 Java 8 中的 `Optional` 类似，都是内部包含至多一个对象实例。`Mono` 的 API 具体操作与 `Flux` 相似，不再赘述。

```java
public abstract class Mono<T> implements Publisher<T> {...}
```

（3）Scheduler

`Scheduler` 可以简单理解为 "线程池"，从代码清单 13-7 的倒数第二行中可以看到，线

程池需要由 `Schedulers` 工具类产生（注意 `Scheduler` 不是 `public` 类型，只能由工具类产生）。响应式线程池有以下几种类型。

- **immediate**：与主线程一致。
- **single**：只有一个线程的线程池（可类比 `Executors.newSingleThreadExecutor()`）。
- **elastic**：弹性线程池，线程池中的线程数量原则上没有上限（底层创建线程池时指定了最大容量为 `Integer.MAX_VALUE`）。
- **parallel**：并行线程池，线程池中的线程数量等于 CPU 的数量（JDK 中的 `Runtime` 类可以调用 `availableProcessors` 方法来获取 CPU 数量）。

## 13.2　快速使用 WebFlux

通过 13.1 节的内容，想必读者已经对响应式编程和 Reactor 框架有了一个初步的了解。本节内容会搭建一个 Spring Boot 整合 WebFlux 的项目，以代替 WebMvc 进行实际的 Web 开发。由于 WebFlux 是与 WebMvc 地位等同的框架，Spring Framework 的作者为了避免开发者因 WebFlux 的使用门槛过高而放弃，在 WebFlux 的使用过程中允许完全采用 WebMvc 的开发风格，使用 `@Controller`+`@RequestMapping` 注解组合即可实现基于 WebFlux 的前端控制与响应。

为了让读者也能循序渐进地了解 WebFlux，本节内容会先从读者熟悉的 WebMvc 编码风格开始讲解。

### 13.2.1　WebMvc 的开发风格

首先创建新的项目。注意，此处导入的依赖不再是 Web，而是 WebFlux，如代码清单 13-8 所示。

**代码清单 13-8　导入 WebFlux 的依赖**

```xml
<dependencies>
    <dependency>
        <groupId>org.springframework.boot</groupId>
        <artifactId>spring-boot-starter-webflux</artifactId>
    </dependency>
</dependencies>
```

当导入完成后，编写主启动类 `SpringBootWebFluxApplication` 并启动应用，如代码清单 13-9 所示。注意观察控制台的输出，可以发现应用启动的嵌入式 Web 容器不再是 Tomcat，而是 Netty（本书在讲解有关嵌入式容器时没有刻意指 Servlet 容器，而是统称为 Web 容器，因为 Netty 不属于 Servlet 容器）。

**代码清单 13-9　WebFlux 的启动类**

```java
@SpringBootApplication
public class SpringBootWebFluxApplication {

    public static void main(String[] args) {
        SpringApplication.run(SpringBootWebFluxApplication.class, args);
    }
}
```

```
[main] c.l.s.SpringBootWebFluxApplication      : Starting SpringBootWebFluxApplication on Linked
Bear with PID 780...
[main] c.l.s.SpringBootWebFluxApplication      : No active profile set, falling back to default
profiles: default
[main] o.s.b.web.embedded.netty.NettyWebServer : Netty started on port(s): 8080
[main] c.l.s.SpringBootWebFluxApplication      : Started SpringBootWebFluxApplication in 2.306
seconds ...
```

接下来的内容与之前的 WebMvc 完全一致,首先简单编写一个 WebmvcStyleController,并声明两个方法,如代码清单 13-10 所示。

**代码清单 13-10　基于 WebMvc 风格的 Controller 编写**

```
@RestController
public class WebmvcStyleController {

    @GetMapping("/hello")
    public String hello() {
        return "Hello WebFlux";
    }

    @GetMapping("/list")
    public List<Integer>list(){
        return Arrays.asList(1,2,3);
    }
}
```

编写完成后,使用浏览器或 Postman 访问 `localhost:8080/hello`,客户端可以正常接收到服务端的"HelloWebFlux"字符串响应,说明 WebFlux 可以完美兼容 WebMvc 的编码方式。

### 13.2.2　逐步过渡到 WebFlux

从 13.1 节的内容中可以了解到,Reactor 中核心数据的封装模型是 Mono 和 Flux,下面根据该模型对 WebmvcStyleController 进行改造。当返回单个对象时,使用 Mono 封装;当返回一组数据时,使用 Flux 封装,如代码清单 13-11 所示。

**代码清单 13-11　使用 Reactor 的数据封装模型改造 Handler**

```
@RestController
public class WebmvcStyleController {

    @GetMapping("/hello2")
    public Mono<String> hello2() {
        return Mono.just("Hello WebFlux");
    }

    @GetMapping("/list2")
    public Flux<Integer> list2() {
        return Flux.just(1, 2, 3);
    }
}
```

> 提示：留意一个细节，目前的改造中仍然使用@RestController 注解中的@ResponseBody 响应数据。

重新启动 SpringBootWebFluxApplication，并访问 localhost:8080/hello2 和 /list2，客户端仍然可以接收到服务端响应的正常数据，说明利用 Reactor 中的数据模型作为响应主体完全可行。

### 13.2.3 WebFlux 的函数式开发

如果想要完全丢弃 WebMvc 的编码方式，转而使用 WebFlux 的风格，需要使用 WebFlux 提供的一套全新的函数式 API。在切换风格之前，先总结一下在 WebMvc 的开发中的关键点。

- Controller 类上要标注@Controller 注解或其派生注解。
- Controller 类中的 Handler 方法上要标注@RequestMapping 注解或其派生注解。

结合第 12 章 WebMvc 中的原理分析可以了解到，Controller 中标注了@RequestMapping 注解的方法在底层会封装为一个个 Handler，每个 Handler 中都封装有 **URL+执行方式**以及具体要反射执行的 **Method** 对象，这两个核心要素在 WebFlux 的编码风格中会转换为两个核心组件：**HandlerFunction** 和 **RouterFunction**。下面将 **WebmvcStyleController** 重构为 WebFlux 的风格代码。

#### 1. Controller 转 Handler

WebFlux 的编码风格中不再有 Controller 的概念，容器中的一切 Bean 都可以视为处理客户端请求的 Handler，所以在声明具体的前端控制器时，将不再使用@Controller 注解，而是使用原始的@Component 注解，且内部的方法不再需要多余的注解，只需要按照 WebFlux 的规则编写方法，如代码清单 13-12 所示。

**代码清单 13-12　HelloHandler 的编写**

```
@Component
public class HelloHandler {

    public Mono<ServerResponse> hello3(ServerRequest request) {
        return ServerResponse.ok().contentType(MediaType.TEXT_PLAIN)
                .body(Mono.just("Hello Handler"), String.class);
    }

    public Mono<ServerResponse> list3(ServerRequest request) {
        return ServerResponse.ok().contentType(MediaType.APPLICATION_JSON)
                .body(Flux.just(1, 2, 3), Integer.class);
    }
}
```

#### 2. RequestMapping 转 Router

注意对比 HelloHandler 与 WebmvcStyleController 的差别。由于 HelloHandler 中不再有@Controller 注解，方法上也不再使用@RequestMapping 注解封装 URL 信息，因此 Spring Framework 无法感知到 IOC 容器中的哪些 bean 对象具备 WebFlux 前端控制器的能力，此时就需要一个新的组件来定义 Bean 与具体路由的关系，这个组件就是 **RouterFunction**。

在编写具体的路由规则时，需要一个配置类来编程式创建 `RouterFunction` 对象，代码清单 13-13 展示了将 `HelloHandler` 应用到 WebFlux 的具体示例。

**代码清单 13-13　HelloRouterConfiguration 配置路由规则**

```java
import static org.springframework.web.reactive.function.server.RequestPredicates.*;

@Configuration(proxyBeanMethods = false)
public class HelloRouterConfiguration {

    @Autowired
    private HelloHandler helloHandler;

    @Bean
    public RouterFunction<ServerResponse> helloRouter() {
        return RouterFunctions
                .route(GET("/hello3")
                .and(accept(MediaType.TEXT_PLAIN)), helloHandler::hello3)
                .andRoute(GET("/list3")
                .and(accept(MediaType.APPLICATION_JSON)), helloHandler::list3);
    }
}
```

可能有部分读者不理解上述的配置方式，注意代码清单 13-13 中的第一行，它静态导入了 `RequestPredicates` 中的所有静态成员，所以下面的路由配置构造中会简洁许多。简单归纳路由配置类中的核心要素：路由配置类也是一个配置类，其中包含 `@Bean` 注解标注的、返回值类型是 `RouterFunction` 的方法；`RouterFunction` 的构造过程中，配置的核心要素也是 Handler 方法 +URL 路径和请求方式。

至此，一个基于 WebFlux 编码风格的示例项目搭建完成。

### 13.2.4　WebMvc 与 WebFlux 的对比

示例项目搭建完毕后，下面简单对比一下 WebFlux 与传统的 WebMvc 的异同点。在 Spring Framework 的官方文档中有一张图，清楚直观地描述了 WebMvc 与 WebFlux 的共性和特性，如图 13-3 所示。

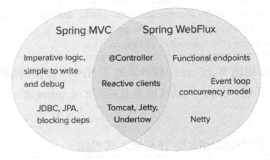

图 13-3　WebMvc 与 WebFlux 的共性和个性

简单总结一下图 13-3 中的内容。

- WebMvc 基于原生 Servlet，它是命令式编程 + 声明式映射，编码简单、便于调试；Servlet

可以是阻塞的,它更适合与传统的关系型数据库等阻塞 I/O 的组件进行交互。
- WebFlux 基于 Reactor,它是异步非阻塞的,使用函数式编程,相较于命令式编程和声明式映射更灵活,而且它可以运行在 Netty 等纯异步非阻塞的 Web 容器,以及同时支持同步阻塞和异步非阻塞的基于 Servlet 3.1 及以上规范的 Servlet 容器中(如高版本的 Tomcat、Undertow 等)。
- WebMvc 和 WebFlux 都可以使用声明式映射注解编程,配置控制器和映射路径。

在实际的项目技术选型中,需要综合考虑项目中使用的技术栈、用户群规模、开发团队能力等多方面因素决定使用 WebMvc 还是 WebFlux。

## 13.3 WebFlux 的自动装配

下面探讨 Spring Boot 整合 WebFlux 的底层原理。根据前面讲过的内容以及参考 WebMvc 的自动装配,不难得出 WebFlux 的核心自动装配类似于 WebMvc,对应的类名是 `WebFluxAutoConfiguration`。与 WebMvc 的自动装配类似,`WebFluxAutoConfiguration` 的装配也有几个前置的装配(如嵌入式容器、`DispatcherHandler` 等),下面对其中涉及的核心配置类逐一展开讲解。

### 13.3.1 ReactiveWebServerFactoryAutoConfiguration

与 WebMvc 中的 `ServletWebServerFactoryAutoConfiguration` 类似,它导入的核心配置类也是一个 `BeanPostProcessorsRegistrar` 和几个嵌入式 Web 容器的内部类,如代码清单 13-14 所示。

**代码清单 13-14　ReactiveWebServerFactoryAutoConfiguration**

```
@AutoConfigureOrder(Ordered.HIGHEST_PRECEDENCE)
@Configuration(proxyBeanMethods = false)
@ConditionalOnClass(ReactiveHttpInputMessage.class)
@ConditionalOnWebApplication(type = ConditionalOnWebApplication.Type.REACTIVE)
@EnableConfigurationProperties(ServerProperties.class)
@Import({ ReactiveWebServerFactoryAutoConfiguration.BeanPostProcessorsRegistrar.class,
        ReactiveWebServerFactoryConfiguration.EmbeddedTomcat.class,
        ReactiveWebServerFactoryConfiguration.EmbeddedJetty.class,
        ReactiveWebServerFactoryConfiguration.EmbeddedUndertow.class,
        ReactiveWebServerFactoryConfiguration.EmbeddedNetty.class})
public class ReactiveWebServerFactoryAutoConfiguration
```

请注意,与 WebMvc 相比 WebFlux 中额外添加了 Netty 的内部容器支持,且 WebFlux 默认使用的嵌入式 Web 容器就是 Netty,所以这里重点展开 EmbeddedNetty 的源码研究,如代码清单 13-15 所示。

**代码清单 13-15　EmbeddedNetty 注册嵌入式 Netty**

```
@Configuration(proxyBeanMethods = false)
@ConditionalOnMissingBean(ReactiveWebServerFactory.class)
@ConditionalOnClass(HttpServer.class)
static class EmbeddedNetty {
```

```java
@Bean
@ConditionalOnMissingBean
ReactorResourceFactory reactorServerResourceFactory() {
    return new ReactorResourceFactory();
}

@Bean
NettyReactiveWebServerFactory nettyReactiveWebServerFactory(
        ReactorResourceFactory resourceFactory, ObjectProvider<NettyRouteProvider> routes,
        ObjectProvider<NettyServerCustomizer> serverCustomizers) {
    NettyReactiveWebServerFactory serverFactory = new NettyReactiveWebServerFactory();
    serverFactory.setResourceFactory(resourceFactory);
    routes.orderedStream().forEach(serverFactory::addRouteProviders);
    serverFactory.getServerCustomizers().addAll(
            serverCustomizers.orderedStream().collect(Collectors.toList()));
    return serverFactory;
}
```

由 EmbeddedNetty 中注册的 Bean 来看，NettyReactiveWebServerFactory 显然类比于 WebMvc 中注册的 TomcatServletWebServerFactory，它会在 IOC 容器的初始化阶段创建嵌入式 Netty 服务器。除此之外，另一个注册的 Bean 类型是 ReactorResourceFactory，它是一个可以管理 Reactor Netty 资源的工厂，这个设计类似于线程池。在 ReactorResourceFactory 的内部组合了一个 ConnectionProvider，它会初始化一个弹性连接提供者，而这个提供者的落地实现类是 PooledConnectionProvider，如代码清单 13-16 所示。从类名上可以直观地体现出"池"的概念，由此可以简单理解 ReactorResourceFactory 就是一个 Reactor Netty 的连接池。

**代码清单 13-16　ReactorResourceFactory 中组合的核心成员**

```java
public class ReactorResourceFactory implements InitializingBean, DisposableBean {
    // ......
    private Supplier<ConnectionProvider> connectionProviderSupplier =
            () -> ConnectionProvider.elastic("webflux");
    // ......
}

// elastic 静态方法
static ConnectionProvider elastic(String name) {
    return new PooledConnectionProvider(name,
            (bootstrap, handler, checker) -> new SimpleChannelPool(bootstrap,
                    handler, checker, true, false));
}
```

此外，在 ReactorResourceFactory 中还有 select 与 worker 的概念，这些概念都是 Netty 的核心。有关 Netty 的相关知识，读者可以自行借助搜索引擎，查询相关资料进行学习，本书不对此展开讲解。

### 13.3.2　WebFluxAutoConfiguration

下面是自动配置类核心 WebFluxAutoConfiguration，如代码清单 13-17 所示。从它的类头可以看出，它生效的前提是当前项目的 Web 类型为 REACTIVE，以及需要当前项目类路径

下存在 `WebFluxConfigurer` 类，这与 WebMvc 的判断条件相似（WebMvc 中判断的项目 Web 类型为 SERVLET，而且检查的类是 `Servlet`、`DispatcherServlet` 等）。

**代码清单 13-17　WebFluxAutoConfiguration**

```
@Configuration(proxyBeanMethods = false)
@ConditionalOnWebApplication(type = ConditionalOnWebApplication.Type.REACTIVE)
@ConditionalOnClass(WebFluxConfigurer.class)
@ConditionalOnMissingBean({ WebFluxConfigurationSupport.class })
@AutoConfigureAfter({ ReactiveWebServerFactoryAutoConfiguration.class,
        CodecsAutoConfiguration.class, ValidationAutoConfiguration.class })
@AutoConfigureOrder(Ordered.HIGHEST_PRECEDENCE+10)
public class WebFluxAutoConfiguration
```

在 `WebFluxAutoConfiguration` 的内部没有任何 Bean 的注册，而是由几个静态内部类根据不同功能和场景分别配置对应的组件，下面就其中重要的静态内部类展开讲解。

### 13.3.3　WebFluxConfig

`WebFluxConfig` 是 WebFlux 的第一个核心配置类，它本身导入了 `EnableWebFlux Configuration`，如代码清单 13-18 所示。本节先关注 `WebFluxConfig` 的配置，13.3.4 节会对 `EnableWebFluxConfiguration` 展开研究。

**代码清单 13-18　WebFluxConfig**

```
@Configuration
@EnableConfigurationProperties({ ResourceProperties.class, WebFluxProperties.class })
@Import({ EnableWebFluxConfiguration.class })
public static class WebFluxConfig implements WebFluxConfigurer { ... }
```

`WebFluxConfig` 本身实现了 `WebFluxConfigurer` 接口，它具备配置 WebFlux 的能力。这个接口与 WebMvc 中的 `WebMvcConfigurer` 类似，在编程式定制 WebMvc 中编写的配置类就是实现 `WebMvcConfigurer` 接口，配置 WebMvc 中的静态资源映射、拦截器等。`WebFlux Config` 中配置了部分 WebFlux 的内容，下面就重点部分逐一来看。

#### 1. 静态资源映射

第一个重点关注的是静态资源映射。WebFlux 本身只是 WebMvc 的异步非阻塞替代品，它同样可以支持前后端不分离的 Web 项目开发，所以对于静态资源映射，WebFlux 同样需要配置。而可以支持的静态资源路径，除 webjars 之外，代码清单 13-19 的最后一个 if 结构中从 `resource Properties` 提取出的静态路径与 WebMvc 完全一致，如代码清单 13-19 所示。

**代码清单 13-19　addResourceHandlers 配置静态资源映射**

```
public void addResourceHandlers(ResourceHandlerRegistry registry) {
    // 前置检查 ......
    if (!registry.hasMappingForPattern("/webjars/**")) {
        ResourceHandlerRegistration registration =
                registry.addResourceHandler("/webjars/**").
                addResourceLocations("classpath:/META-INF/resources/webjars/");
    // ......
    }
```

```java
        String staticPathPattern = this.webFluxProperties.getStaticPathPattern();
        if (!registry.hasMappingForPattern(staticPathPattern)) {
            ResourceHandlerRegistration registration = registry
                    .addResourceHandler(staticPathPattern)
                    .addResourceLocations(this.resourceProperties.getStaticLocations());
            // ......
        }
    }
}

public class ResourceProperties {
    private static final String[] CLASSPATH_RESOURCE_LOCATIONS = {
            "classpath:/META-INF/resources/", "classpath:/resources/",
            "classpath:/static/", "classpath:/public/"};
    private String[] staticLocations = CLASSPATH_RESOURCE_LOCATIONS;
```

#### 2. 视图解析器

与 WebMvc 相同，WebFlux 默认也可以支持视图跳转，所以底层也有视图解析器的配置，如代码清单 13-20 所示。源码逻辑相对简单，不再展开。

**代码清单 13-20　视图解析器的配置**

```java
public void configureViewResolvers(ViewResolverRegistry registry) {
    this.viewResolvers.orderedStream().forEach(registry::viewResolver);
}
```

#### 3. 类型转换器和格式转换器

**代码清单 13-21　Converter 和 Formatter**

```java
public void addFormatters(FormatterRegistry registry) {
    for (Converter<?, ?> converter : getBeansOfType(Converter.class)) {
        registry.addConverter(converter);
    }
    for (GenericConverter converter : getBeansOfType(GenericConverter.class)) {
        registry.addConverter(converter);
    }
    for (Formatter<?> formatter : getBeansOfType(Formatter.class)) {
        registry.addFormatter(formatter);
    }
}
```

WebFlux 中的类型转换器和格式转换器与 WebMvc 部分的配置完全一致，不过对于这部分组件的功能读者可以不用过多关注，只需要体会 WebFlux 的大多数配置在本质上与 WebMvc 并无区别。

### 13.3.4　EnableWebFluxConfiguration

观察 EnableWebFluxConfiguration 的类名，它与 WebMvc 中的 EnableWebMvcConfiguration 非常相近，事实上它们继承的父类也非常类似，如代码清单 13-22 所示。

**代码清单 13-22　EnableWebFluxConfiguration 与 EnableWebMvcConfiguration**

```java
@Configuration(proxyBeanMethods = false)
public static class EnableWebFluxConfiguration extends DelegatingWebFluxConfiguration

@Configuration(proxyBeanMethods = false)
```

```
public static class EnableWebMvcConfiguration extends DelegatingWebMvcConfiguration
        implements ResourceLoaderAware
```

它们继承的父类 `Delegating***Configuration` 的作用在 2.6 节中讲过，它们是对应 `@EnableWebMvc` 或者 `@EnableWebFlux` 注解导入的配置类。如果 Spring Boot 的项目中显式标注了 `@EnableWebFlux` 注解，则 `WebFluxAutoConfiguration` 不会生效，改由项目自身接管 WebFlux，具体原因和底层机制可参照 2.6 节的提示，此处不再赘述。

下面简单列举 `EnableWebFluxConfiguration` 中注册的核心组件。由于源码中注册组件的逻辑相对简单，故本节中不再展示具体源码，感兴趣的读者可以借助 IDE 自行查阅。

- `FormattingConversionService`：参数类型转换器，用于数据的类型转换，如日期与字符串之间的互相转换（在 WebMvc 中同样有注册）。
- `Validator`：JSR-303 参数校验器（在 WebMvc 中同样有注册）。
- `HandlerMapping` & `HandlerAdapter`：覆盖父类。
- `WebFluxConfigurationSupport` 的注册逻辑（可以忽略不看）。

### 13.3.5 WebFluxConfigurationSupport

`EnableWebFluxConfiguration` 继承自 `WebFluxConfigurationSupport`，这个父类中也有一些核心组件的注册，下面列举其中重要的几个组件。

#### 1. DispatcherHandler

WebFlux 中的核心前端控制器是 `DispatcherHandler`，对应到 WebMvc 中的组件是 `DispatcherServlet`，不过 `DispatcherHandler` 的注册逻辑比 `DispatcherServlet` 简单，如代码清单 13-23 所示。

**代码清单 13-23　DispatcherHandler 的注册**

```
@Bean
public DispatcherHandler webHandler() {
    return new DispatcherHandler();
}
```

#### 2. WebExceptionHandler

WebFlux 的异常状态响应处理器用于处理异常情况下的 HTTP 状态码响应，如代码清单 13-24 所示。

**代码清单 13-24　WebFluxResponseStatusExceptionHandler**

```
@Bean
@Order(0)
public WebExceptionHandler responseStatusExceptionHandler() {
    return new WebFluxResponseStatusExceptionHandler();
}
```

#### 3. RequestMappingHandlerMapping&RequestMappingHandlerAdapter

因为 WebFlux 可以完美支持 WebMvc 中使用标准 `@RequestMapping` 注解的方式定义 Handler，而支持的底层是与 WebMvc 中相同的 `RequestMappingHandlerMapping` 和 `RequestMappingHandlerAdapter`，如代码清单 13-25 所示。

**代码清单 13-25　RequestMappingHandlerMapping 的注册**

```
@Bean
public RequestMappingHandlerMapping requestMappingHandlerMapping() {
    RequestMappingHandlerMapping mapping = createRequestMappingHandlerMapping();
    mapping.setOrder(0);
    // ......
    return mapping;
}

@Bean
public RequestMappingHandlerAdapter requestMappingHandlerAdapter() {
    RequestMappingHandlerAdapter adapter = createRequestMappingHandlerAdapter();
    // ......
    return adapter;
}
```

### 4. RouterFunctionMapping

`RouterFunctionMapping` 是基于函数式端点路由编程的 `Mapping` 处理器，如代码清单 13-26 所示。它的优先级高于 `RequestMappingHandlerMapping`，这意味着 WebFlux 倾向于开发中使用函数式端点的 Web 开发，而不是传统的 `@RequestMapping` 注解式开发。有关 `RouterFunctionMapping` 的作用，在 13.5 节中会讲解。

**代码清单 13-26　RouterFunctionMapping 的注册**

```
@Bean
public RouterFunctionMapping routerFunctionMapping() {
    RouterFunctionMapping mapping = createRouterFunctionMapping();
    mapping.setOrder(-1); // 注意此处设置的优先级高于 RequestMappingHandlerMapping
    mapping.setMessageReaders(serverCodecConfigurer().getReaders());
    mapping.setCorsConfigurations(getCorsConfigurations());
    return mapping;
}
```

### 5. HandlerFunctionAdapter

函数式端点路由的处理器在底层也需要由具体的 `HandlerAdapter` 负责调用，对应的支撑组件是 `HandlerFunctionAdapter`，如代码清单 13-27 所示。它可以直接提取出 `HandlerFunction` 中的 `Handler` 方法进行调用，具体的调用逻辑统一放在 13.5 节中讲解。

**代码清单 13-27　HandlerFunctionAdapter 的注册**

```
@Bean
public HandlerFunctionAdapter handlerFunctionAdapter() {
    return new HandlerFunctionAdapter();
}
```

### 6. ResultHandler

WebFlux 中同样需要有对返回值进行处理的组件，在 WebMvc 中的类型是 `HandlerMethodReturnValueHandler`，对应到 WebFlux 中则是 `ResultHandler`。默认情况下

WebFlux 会注册 4 种不同的 `ResultHandler` 实现类。
- `ResponseEntityResultHandler`：处理 `HttpEntity` 和 `ResponseEntity`。
- `ResponseBodyResultHandler`：处理 @`RequestMapping` 的标注了 @`Response` `Body` 注解的 `Handler`。
- `ViewResolutionResultHandler`：处理逻辑视图返回值。
- `ServerResponseResultHandler`：处理返回值类型为 `ServerResponse` 的（WebFlux 中可以直接返回 `ServerResponse`，如代码清单 13-12 所示）。

本节中没有提到的组件相对而言重要程度不高，感兴趣的读者可以借助 IDE 自行翻阅没有提到的组件注册。

## 13.4　DispatcherHandler 的传统方式工作原理

`DispatcherHandler` 作为 WebFlux 的核心前端控制器，它的作用必然与 `Dispatcher` `Servlet` 相同，都是负责统一接收客户端请求并处理，然后将结果响应给客户端。由于 WebFlux 可以完美兼容 @`RequestMapping` 注解式开发，本节内容先研究传统的开发方式下 `DispatcherHandler` 的工作原理。

以本章的示例项目 `springboot-13-webflux` 为调试基准，Debug 启动项目并将断点打在 `DispatcherHandler` 的 `handle` 方法中，随后用浏览器访问 http://localhost:8080/hello，待程序停在断点处时开始 Debug 调试。

### 13.4.1　handle 方法概览

进入 `DispatcherHandler` 的 `handle` 方法中，最直观的体现是源码篇幅更短，相较于 `DispatcherServlet` 的 `doDispatch` 方法实现更精炼，如代码清单 13-28 所示。

**代码清单 13-28　DispatcherHandler#handle**

```java
public Mono<Void> handle(ServerWebExchange exchange) {
    if (this.handlerMappings == null) {
        return createNotFoundError();
    }
    return Flux.fromIterable(this.handlerMappings)
            .concatMap(mapping -> mapping.getHandler(exchange))
            .next()
            .switchIfEmpty(createNotFoundError())
            .flatMap(handler -> invokeHandler(exchange, handler))
            .flatMap(result -> handleResult(exchange, result));
}
```

注意 `handle` 方法的入参是一个 `ServerWebExchange` 对象，联想 `DispatcherServlet` 中传入的是 `HttpServletRequest` 和 `HttpServletResponse` 对象，不难推测 `ServerWeb` `Exchange` 就是 `request` 和 `response` 的组合体，如代码清单 13-29 所示。

**代码清单 13-29　ServerWebExchange 接口的方法定义**

```java
public interface ServerWebExchange {
```

```
    ServerHttpRequest getRequest();
    ServerHttpResponse getResponse();
    //......
}
```

总观 handle 方法的实现，首先 if 分支中会检查 DispatcherHandler 中是否注册有 HandlerMapping，由于上面的自动配置类导入的配置类已经注册了必要的 HandlerMapping，所以不会进入该分支。后面的 return 结构是一串链式调用，源码本身的可读性比较高，可以结合第 12 章中 DispatcherServlet 的工作原理对应地分析 DispatcherHandler 中的三个关键步骤。

> 提示：由于函数式编程和响应式的设计问题，会导致实际调试程序时 Debug 难度非常大，在 Debug 的过程中随时可能出现来回跳转的情况，读者在结合 IDE 调试 WebFlux 相关源码时，一定要格外注意方法的执行轨迹。

### 13.4.2 筛选 HandlerMapping

DispatcherHandler 的 handle 方法的第一个动作与 DispatcherServlet 相同，都是寻找可以匹配当前请求的 HandlerMapping 对象，在 handle 方法中负责筛选 HandlerMapping 的步骤如代码清单 13-30 所示。首先会将 DispatcherHandler 中保存的所有 HandlerMapping 封装为一个 Flux 对象，之后使用 concatMap 方法使所有 HandlerMapping 都尝试匹配当前的请求，并将结果收集合并，最后调用 next 方法提取出第一个匹配成功的 HandlerMapping 解析后的对象。

**代码清单 13-30　DispatcherHandler 筛选 HandlerMapping**

```
return Flux.fromIterable(this.handlerMappings)
        .concatMap(mapping -> mapping.getHandler(exchange))
        .next() //......
```

对于源码中的第一步和最后一步都不难理解，中间的 concatMap(mapping -> mapping.getHandler(exchange)) 需要重点理解。

#### 1. concatMap

concatMap 是 Flux 中的一个动作，执行 concatMap 方法可以将 Flux 管道中的对象转换为另一种类型的对象，该操作非常类似于 Stream 中的 map 方法，但又不完全相同。Stream 中的 map 方法执行完成后，整个管道中的元素个数不变，而 Flux 的 concatMap 方法执行完成后，管道中的元素个数可能会发生改变。concatMap 的逻辑示意如图 13-4 所示。

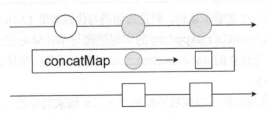

图 13-4　concatMap 方法示意

DispatcherHandler 之所以选择使用 concatMap 方法转换 HandlerMapping，是希望通过一个循环后将仅支持当前请求的 HandlerMapping 筛选出来，忽略不支持的 HandlerMapping。

#### 2. mapping.getHandler

筛选 HandlerMapping 的实际动作需要借助 HandlerMapping 的 getHandler 方法，如果 getHandler 方法返回的 Mono 对象中有具体值则认为匹配，反之不匹配。而 getHandler 方法定义在所有 HandlerMapping 实现类的基础父类 AbstractHandlerMapping 中，方法内部又会调用模板方法 getHandlerInternal，源码的设计与 WebMvc 中完全一致，如代码清单 13-31 所示。

**代码清单 13-31　getHandler 调用 getHandlerInternal**

```
public Mono<Object> getHandler(ServerWebExchange exchange) {
    return getHandlerInternal(exchange).map(handler -> {
        // logger ......
        // 跨域相关处理 ......
        return handler;
    });
}
```

基于 @RequestMapping 注解式开发的解析，底层由 RequestMappingHandlerMapping 负责匹配，而 getHandlerInternal 方法的实现逻辑在父类 AbstractHandlerMethodMapping 中，如代码清单 13-32 所示。

**代码清单 13-32　AbstractHandlerMethodMapping#getHandlerInternal**

```
public Mono<HandlerMethod> getHandlerInternal(ServerWebExchange exchange) {
    this.mappingRegistry.acquireReadLock();
    try {
        HandlerMethod handlerMethod;
        try {
            // 搜索处理器方法（真正处理请求的 RequestMapping）
            handlerMethod = lookupHandlerMethod(exchange);
        } // catch ......
        // 将方法分离出来，单独形成一个 Bean
        if (handlerMethod != null) {
            handlerMethod = handlerMethod.createWithResolvedBean();
        }
        return Mono.justOrEmpty(handlerMethod);
    } // finally ......
}
```

仔细观察代码清单 13-32 的获取逻辑，将这段源码与代码清单 12-16 对比，可以发现 WebFlux 中的 RequestMappingHandlerMapping 的实现逻辑与 WebMvc 完全一致，同样都是先搜索 Controller 方法，后封装为独立的 HandlerMethod 对象并返回。整体逻辑可以完全参照 12.4.4 节的内容，本节不再赘述。

通过 Debug，可以发现此处可以成功匹配到 /hello 请求对应的 Controller 方法，如图 13-5 所示。

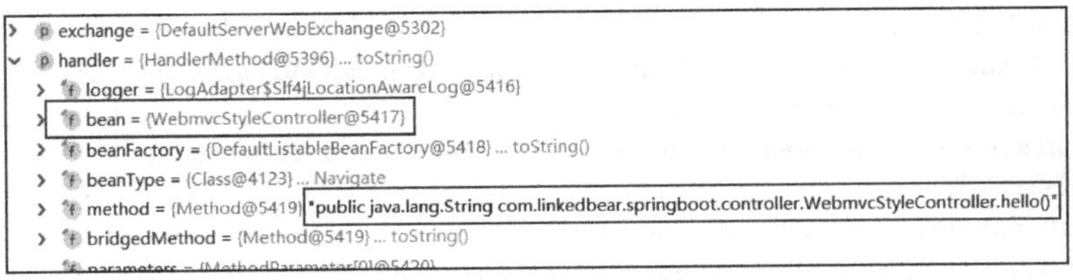

图 13-5 通过 RequestMapping 匹配到的 HandlerMethod

### 13.4.3 搜寻 HandlerAdapter 并执行

`HandlerMapping` 获取完毕后，`DispatcherHandler` 的下一个步骤也是搜寻合适的 `HandlerAdapter`，用于执行目标 `Handler`。该步骤对应 `handle` 方法中的 `flatMap(handler -> invokeHandler(exchange, handler))` 步骤。

#### 1. 匹配合适的 HandlerAdapter

进入 `invokeHandler` 方法中，可以发现源码中会逐个检查 `DispatcherHandler` 中的 `HandlerAdapter` 是否支持执行当前的 `Handler`，如果支持，则会直接调用 `HandlerAdapter` 的 `handle` 方法执行目标 `Handler`，如代码清单 13-33 所示。

**代码清单 13-33 invokeHandler**

```java
private Mono<HandlerResult> invokeHandler(ServerWebExchange exchange, Object handler) {
    if (this.handlerAdapters != null) {
        for (HandlerAdapter handlerAdapter : this.handlerAdapters) {
            if (handlerAdapter.supports(handler)) {
                return handlerAdapter.handle(exchange, handler);
            }
        }
    }
    return Mono.error(new IllegalStateException("No HandlerAdapter: " + handler));
}
```

整个逻辑与 WebMvc 中的逻辑有一点区别，`DispatcherServlet` 会逐个检查 `HandlerAdapter`，在找到合适的 `HandlerAdapter` 后会返回，之后再执行 `Handler`；而 `DispatcherHandler` 中会遍历检查，检查到 `HandlerAdapter` 支持执行当前 `Handler` 时会直接执行，不再添加额外的动作。

#### 2. 执行 Handler

基于 @RequestMapping 注解式开发的 Handler，底层使用 `RequestMappingHandlerAdapter` 执行，进入其 `handle` 方法中可以发现，在 WebMvc 的 `RequestMappingHandlerAdapter` 中处理的逻辑对应到 WebFlux 中基本上都会执行，只是执行顺序和方式不同。核心 Handler 执行的动作是在封装好 `InvocableHandlerMethod` 之后在 `return` 部分的链式调用中执行 `invocableMethod.invoke(exchange,bindingContext)` 方法，利用反射机制执行目标 Controller 方法，如代码清单 13-34 所示。

**代码清单 13-34 RequestMappingHandlerAdapter#handle**

```java
public Mono<HandlerResult>handle(ServerWebExchange exchange, Object handler) {
    HandlerMethod handlerMethod = (HandlerMethod) handler;
    // 检查 ......
```

```
// 初始化参数绑定上下文
InitBinderBindingContext bindingContext = new InitBinderBindingContext(
        getWebBindingInitializer(),
        this.methodResolver.getInitBinderMethods(handlerMethod));
// 创建方法执行对象
InvocableHandlerMethod invocableMethod =
        this.methodResolver.getRequestMappingMethod(handlerMethod);
// 异常处理器的准备
Function<Throwable, Mono<HandlerResult>> exceptionHandler =
        ex -> handleException(ex, handlerMethod, bindingContext, exchange);
// 执行目标方法,处理返回值和异常
return this.modelInitializer
        .initModel(handlerMethod, bindingContext, exchange)
        .then(Mono.defer(() -> invocableMethod.invoke(exchange, bindingContext)))
        .doOnNext(result -> result.setExceptionHandler(exceptionHandler))
        .doOnNext(result -> bindingContext.saveModel())
        .onErrorResume(exceptionHandler);
}
```

源码中的其余部分基本与 WebMvc 中的对应,部分 API 甚至直接复制了 WebMvc 中的源码,感兴趣的读者可以借助 IDE 自行翻阅和调试,本节不再过多展开。

### 13.4.4 返回值处理

InvocableHandlerMethod 的 invoke 方法执行完成后会返回一个 HandlerResult 对象,其中封装了执行 Controller 方法后返回的真实值。DispatcherHandler 的最后一个关键步骤是对该返回值进行处理,对应源码的动作是 flatMap(result -> handleResult(exchange, result)),如代码清单 13-35 所示。

**代码清单 13-35　handleResult 处理返回值**

```
private Mono<Void> handleResult(ServerWebExchange exchange, HandlerResult result) {
    return getResultHandler(result).handleResult(exchange, result)
            .onErrorResume(ex -> result.applyExceptionHandler(ex).flatMap (exceptionResult ->
                    getResultHandler(exceptionResult).handleResult(exchange, exceptionResult)));
}
```

handleResult 本身定义在 DispatcherHandler 中,它的筛选逻辑与返回值处理逻辑 HandlerMapping 和 HandlerAdapter 不太相同。HandlerMapping 和 HandlerAdapter 的筛选逻辑都是边检查边执行,而返回值处理逻辑是获取和处理分开为两步执行。

进入获取 ResultHandler 的方法 getResultHandler 中,如代码清单 13-36 所示。通过 Debug 观察到 DispatcherHandler 中默认组合了 4 种不同的 ResultHandler,刚好与 13.3.5 中看到的部分一样,如图 13-6 所示。

**代码清单 13-36　getResultHandler 匹配 ResultHandler**

```
private HandlerResultHandler getResultHandler(HandlerResult handlerResult) {
    if (this.resultHandlers != null) {
        for (HandlerResultHandler resultHandler : this.resultHandlers) {
            if (resultHandler.supports(handlerResult)) {
                return resultHandler;
```

                }
            }
        }
        throw new IllegalStateException("No HandlerResultHandler for " + handlerResult.get
ReturnValue());
    }

```
∞ this.resultHandlers = {ArrayList@5737} size = 4
    > ≡ 0 = {ResponseEntityResultHandler@7247}
    > ≡ 1 = {ServerResponseResultHandler@7248}
    > ≡ 2 = {ResponseBodyResultHandler@7242}
    > ≡ 3 = {ViewResolutionResultHandler@7249}
```

图 13-6  DispatcherHandler 中组合的 ResultHandler

由于 `WebmvcStyleController` 类中标注了 `@RestController` 注解，因此此处匹配的返回值处理器一定是 `ResponseBodyResultHandler`，而它的 `handleResult` 方法中会将 Handler 的返回值取出，并执行 `writeBody` 方法将返回值结果写入响应流，如代码清单 13-37 所示。

**代码清单 13-37　ResponseBodyResultHandler#handleResult**

```java
public Mono<Void> handleResult(ServerWebExchange exchange, HandlerResult result) {
    Object body = result.getReturnValue();
    MethodParameter bodyTypeParameter = result.getReturnTypeSource();
    return writeBody(body, bodyTypeParameter, exchange);
}
```

`ResultHandler` 的工作处理完成后，`handle` 方法的链式调用执行完毕，一次完整的请求处理结束。

## 13.4.5　工作流程小结

本节的最后，以流程图的形式总结上述 `DispatcherHandler` 处理的全流程，如图 13-7 所示。

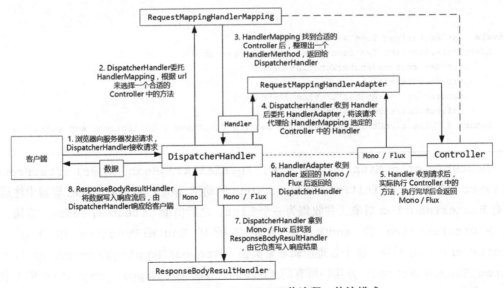

图 13-7　DispatcherHandler 工作流程：传统模式

## 13.5 DispatcherHandler 的函数式端点工作原理

使用传统的 `@RequestMapping` 注解式开发与使用函数式端点开发的最大区别在于，底层支持的 `HandlerMapping` 与 `HandlerAdapter` 的实现类不同。对于整体的请求处理流程而言，依然是遵循 `DispatcherHandler` 的 `handle` 方法。本节中着重研究基于函数式端点的开发中底层 `HandlerMapping` 与 `HandlerAdapter` 的工作原理。

为测试函数式端点的底层执行流程，下面使用浏览器访问 http://localhost:8080/hello3，待程序停在 `DispatcherHandler` 的 `handle` 方法中标注的断点处时开始 Debug 调试。

### 13.5.1 HandlerMapping 的不同

对于使用函数式端点开发的 Handler，底层使用的 `HandlerMapping` 不再是 `RequestMappingHandlerMapping`，而是在 13.3.5 节中提到的 `RouterFunctionMapping`，这个类的内部会收集所有注册的函数式端点，并组合为一个 `RouterFunction` 对象。下面先简单了解 `RouterFunctionMapping` 的初始化逻辑，相关源码如代码清单 13-38 所示。

**代码清单 13-38　RouterFunctionMapping 的初始化动作**

```java
public void afterPropertiesSet() throws Exception {
    // ......
    if (this.routerFunction == null) {
        initRouterFunctions();
    }
}

protected void initRouterFunctions() {
    List<RouterFunction<?>> routerFunctions = routerFunctions();
    this.routerFunction = routerFunctions.stream().reduce(RouterFunction::andOther)
            .orElse(null);
    logRouterFunctions(routerFunctions);
}

private List<RouterFunction<?>> routerFunctions() {
    List<RouterFunction<?>> functions = obtainApplicationContext()
        .getBeanProvider(RouterFunction.class)
        .orderedStream()
        .map(router -> (RouterFunction<?>) router)
        .collect(Collectors.toList());
    return (!CollectionUtils.isEmpty(functions) ? functions : Collections.emptyList());
}
```

`RouterFunctionMapping` 本身实现了 `InitializingBean`，对应的 `afterPropertiesSet` 方法中有一个 `initRouterFunctions` 的动作，该动作会提取出 IOC 容器中注册的所有 `RouterFunction` 对象，并收集为一个 `List`，之后借助 `Stream` 的 `reduce` 方法，调用 `RouterFunction` 的 `andOther` 方法将所有 `RouterFunction` 组合为一个 `RouterFunction` 对象。这个处理逻辑非常类似于 `RequestMappingHandlerMapping` 的 `detectHandlerMethods` 方法将所有的 Controller 方法存入 `MappingRegistry` 中（参见 12.3.3 节）。

通过实际 Debug，当 DispatcherHandler 中的 getHandler 方法执行完毕后，获取的 Handler 已经可以定位到注册函数式端点的配置类 HelloRouterConfiguration 中，如图 13-8 所示。

```
> ▸ exchange = {DefaultServerWebExchange@5335}
∨ ▸ handler = {HelloRouterConfiguration$lambda@5399}
    > ▸ arg$1 = {HelloHandler@5409}
```

图 13-8　RouterFunctionMapping 可以找到具体的 Handler 方法

### 13.5.2　HandlerAdapter 的不同

HandlerMapping 的工作处理完成后，下面是 HandlerAdapter 执行 Handler 的步骤。函数式端点开发中，底层配合完成 Handler 调用的 HandlerAdapter 也发生了变化，具体的实现类是 HandlerFunctionAdapter，从它的 handle 方法实现来看，核心动作是获取 HandlerFunction 并调用其 handle 方法，如代码清单 13-39 所示。

**代码清单 13-39　HandlerFunctionAdapter#handle**

```java
public boolean supports(Object handler) {
    return handler instanceof HandlerFunction;
}

public Mono<HandlerResult> handle(ServerWebExchange exchange, Object handler) {
    HandlerFunction<?> handlerFunction = (HandlerFunction<?>) handler;
    ServerRequest request = exchange.getRequiredAttribute(RouterFunctions.REQUEST_ATTRIBUTE);
    return handlerFunction.handle(request).map(response ->
        new HandlerResult(handlerFunction, response, HANDLER_FUNCTION_RETURN_TYPE));
}
```

到这里可能有读者会产生疑问，为什么进入 handle 方法的 handler 参数类型是 HandlerFunction？除了 HandlerFunctionAdapter 是否可以处理当前 Handler 的检查（supports 方法），更重要的其实是在注册函数式端点的配置类中。注意观察代码清单 13-40 中注册 RouterFunction 时传入的方法引用 helloHandler::hello3，它的本质是一个简化版的 Lambda 表达式，或者从原始的角度说，它的本质是一个匿名内部类，而该匿名内部类的类型就是 HandlerFunction。

**代码清单 13-40　HelloRouterConfiguration 中注册函数式端点路由**

```java
@Bean
public RouterFunction<ServerResponse> helloRouter() {
    return RouterFunctions
        // 注意下面
        .route(GET("/hello3").and(accept(MediaType.TEXT_PLAIN)), helloHandler::hello3)
        .andRoute(GET("/list3").and(accept(MediaType.APPLICATION_JSON)), helloHandler::list3);
}

// RouterFunctions#route
public static <T extends ServerResponse> RouterFunction<T> route(
        RequestPredicate predicate, HandlerFunction<T> handlerFunction) {
    return new DefaultRouterFunction<>(predicate, handlerFunction);
```

```
@FunctionalInterface
public interface HandlerFunction<T extends ServerResponse> {
    Mono<T> handle(ServerRequest request);
}
```

通过观察 RouterFunctions 的静态方法 route 也能获取到该信息，该方法的第二个参数会传入一个 HandlerFunction 类型的对象，而 HandlerFunction 本身是一个函数式接口，所以可以使用 Lambda 表达式或方法引用进行简化。如果不进行简化，则注册 RouterFunction 的代码如代码清单 13-41 所示。

**代码清单 13-41　不使用方法引用注册 RouterFunction**

```java
@Bean
public RouterFunction<ServerResponse> helloRouter() {
    return RouterFunctions.route(GET("/hello3").and(accept(MediaType.TEXT_PLAIN)),
            new HandlerFunction<ServerResponse>() {
                @Override
                public Mono<ServerResponse> handle(ServerRequest request) {
                    return helloHandler.hello3(request);
                }
            });
}
```

由以上分析，回到 HandlerFunctionAdapter 中，显然 HandlerFunctionAdapter 的 handle 方法所做的工作就是直接调用项目中编写的具体的 Controller 方法，相较于 RequestMappingHandlerAdapter 的处理方式更简单直接。

### 13.5.3　返回值处理的不同

对于基于函数式端点的开发方式，因为 Handler 最终返回的对象是 Mono<ServerResponse>，所以负责返回值处理的 ResultHandler 不再是 ResponseBodyResultHandler，而是 ServerResponseHandlerResult，而它处理返回值的方式仅是把方法返回的 ServerResponse 对象写入 ServerWebExchange 中，如代码清单 13-42 所示（由于 ServerResponse 的构建中一般会向响应体中写入数据，因此该步骤相当于将响应体的数据写入 ServerWebExchange 中）。

**代码清单 13-42　handleResult 处理 Handler 的返回值**

```java
public Mono<Void> handleResult(ServerWebExchange exchange, HandlerResult result) {
    ServerResponse response = (ServerResponse) result.getReturnValue();
    Assert.state(response != null, "No ServerResponse");
    return response.writeTo(exchange, new ServerResponse.Context() {
        @Override
        public List<HttpMessageWriter<?>> messageWriters() {
            return messageWriters;
        }
        @Override
        public List<ViewResolver> viewResolvers() {
            return viewResolvers;
```

        }
    });
}
```

返回值处理完成后,一次完整的请求处理流程结束。

### 13.5.4 工作流程小结

同样,对于基于函数式端点的工作全流程,本节的最后也以流程图的形式总结,如图 13-9 所示。

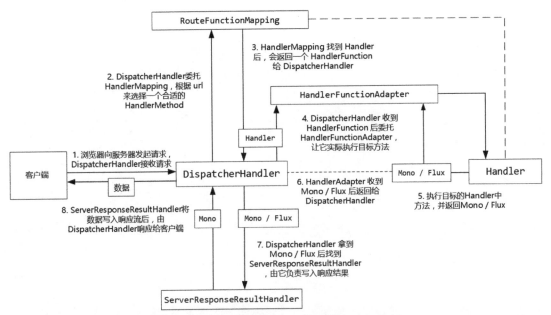

图 13-9 DispatcherHandler 工作流程:函数式端点

## 13.6 小结

本章从 WebFlux 的底层思想响应式编程以及底层框架 Reactor 着手,通过简单示例体会 WebFlux 的几种开发方式,之后从 WebFlux 的自动装配切入,研究 WebFlux 中注册的核心组件,以及与 WebMvc 中注册组件的对比,最后以两种不同的开发方式,结合两个示例 Debug 分析和研究 WebFlux 的核心前端控制器 DispatcherHandler 的工作全流程。WebFlux 的核心设计与 WebMvc 并无太大差别,本质上还是通过一个核心前端控制器 + 周边核心组件的方式共同完成与客户端之间的请求响应处理。
```

# 第 4 部分
# Spring Boot 应用的运行

▶ 第 14 章　运行 Spring Boot 应用

# 第 14 章 运行 Spring Boot 应用

**本章主要内容：**
- Spring Boot 应用的项目打包方式；
- 基于 jar 包独立运行的核心底层解析；
- 基于 war 包运行的引导机制解析；
- Spring Boot 的优雅停机。

通过前面 3 个部分的学习，有关 Spring Boot 以及 Spring Framework 相关的设计、开发、核心底层的研究已基本结束。在一个项目开发完成后，最终需要将应用部署到服务器使其正常运行，以提供功能服务使用。

部署运行 Spring Boot 的方式一般采用打包部署为主。下面先回顾一下 Spring Boot 应用的项目打包方式。

## 14.1 部署打包的两种方式

大多数情况下，通常会选择将 Spring Boot 打包为一个可运行的 jar 包，或者去掉内置的嵌入式 Web 容器，以 war 包形式部署到外置的容器中，这取决于开发者最终要部署的目标环境。

### 14.1.1 以可独立运行 jar 包的方式

Spring Boot 默认以独立可运行的 jar 包方式运行，当使用 Spring Initializer 创建 Spring Boot 应用时，可以在 `pom.xml` 文件中找到 `spring-boot-maven-plugin` 插件，如代码清单 14-1 所示。

**代码清单 14-1　项目中默认会引入一个 Spring Boot 集成 Maven 的插件**

```
<build>
    <plugins>
        <plugin>
            <groupId>org.springframework.boot</groupId>
            <artifactId>spring-boot-maven-plugin</artifactId>
        </plugin>
    </plugins>
</build>
```

在这种情况下，执行 `mvn package` 命令，就可以在项目根目录的 `target` 下获得一个 jar 包，如图 14-1 所示。

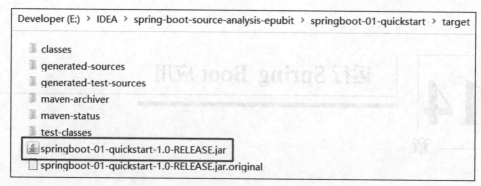

图 14-1 打包完成后生成的可独立运行 jar 包

在该目录下，可以直接执行 `java -jar` 命令启动该项目。关于具体操作读者可以自行完成实践。

### 14.1.2 以 war 包的方式

如果部署的目标环境是一个外置的 Web 容器，就需要以 war 包的方式打包项目，在这种情况下，需要修改 pom.xml 文件添加一些配置，如代码清单 14-2 所示。

**代码清单 14-2　Spring Boot 以 war 打包时需要修改 pom.xml 文件的部分配置**

```xml
<packaging>war</packaging>

<!-- 在 dependencies 中添加 -->
<dependency>
    <groupId>org.springframework.boot</groupId>
    <artifactId>spring-boot-starter-tomcat</artifactId>
    <scope>provided</scope>
</dependency>
```

除此之外，还需要修改主启动类或者添加新的类，使其继承 `SpringBootServletInitializer` 类，并重写 `configure` 方法指定配置源为 Spring Boot 主启动类，如代码清单 14-3 所示。

**代码清单 14-3　以 war 打包时需要修改主启动类添加新的继承**

```java
@SpringBootApplication
public class SpringBootQuickstartApplication extends SpringBootServletInitializer {

    public static void main(String[] args) {
        SpringApplication.run(SpringBootQuickstartApplication.class, args);
    }

    @Override
    protected SpringApplicationBuilder configure(SpringApplicationBuilder builder) {
        return builder.sources(SpringBootQuickstartApplication.class);
    }
}
```

以上修改完成后，重新执行 `mvn package` 命令，就生成一个可以部署到外置 Web 容器的

war 包了。接下来，将该 war 包部署至外置的 Web 容器（如 Tomcat Server）中并启动，就可以运行 Spring Boot 应用了。

简单回顾两种最常见的打包方式后，下面分别解读这两种方式的运行机制和底层原理。

## 14.2 基于 jar 包的独立运行机制

想要搞明白基于 jar 包的独立运行机制，需要先了解可运行 jar 包的一些前置知识。

### 14.2.1 可运行 jar 包的前置知识

从 Oracle 的官网上可以找到有关 jar 文件规范的文档，文档中提到了一个核心的目录 `META-INF`，这个目录中会存放当前 jar 包的一些扩展和配置数据，其中有一个核心配置文件 `MANIFEST.MF`，它以 properties 的配置形式保存了 jar 包的一些核心元信息。MANIFEST.MF 文件中的核心配置项主要包含以下几个核心内容，如表 14-1 所示（配置项比较多，本书只挑选几个下面会提到的配置，有关完整的配置读者可以参照规范文档自行了解）。

表 14-1　　　　　　　　　　MANIFEST.MF 中的核心配置项（节选）

| 配置项 | 配置含义 | 配置值示例 |
| --- | --- | --- |
| Manifest-Version | 定义 MANIFEST.MF 文件的版本 | 1.0（通常） |
| Class-Path | 指定当前 jar 包所依赖的 jar 包路径（一般是相对路径） | servlet.jar、config/ |
| Main-Class | 对于可运行 jar 包中引导的主启动类的全限定类名 | org.springframework.boot.loader.JarLauncher |

重点关注最后一个配置项 `Main-Class`，它需要指定一个可以在 jar 包的顶层结构中可以直接找到的、带有 main 方法的启动类的全限定类名。注意，这里所谓的顶层结构是指 jar 包中可以直接在目录中找到的，不需要再解压/探寻 jar 包内部（换句话说，被 `Main-Class` 配置项引用的类必须同它所属的包一起放在可运行 jar 包的顶层）。这里又出现了一个新的概念：jar 包中可能会嵌套 jar 包。这种类型的 jar 包被称为 Fat Jar，可以解决第三方库不在 classpath 下的加载失败问题。Spring Boot 生成的可运行 jar 包本身就是一种 Fat Jar，下面通过一个示例来具体研究。

### 14.2.2 Spring Boot 的可运行 jar 包结构

对于 Spring Boot 通过 Maven / Gradle 插件打包的可运行 jar 包，它的内部由 3 个目录构成，如图 14-2 所示。

- `BOOT-INF`：存放项目编写且编译好的字节码文件、静态资源文件、配置文件，以及依赖的 jar 包。
- `META-INF`：存放 `MANIFEST.MF` 等配置元文件。
- `org.springframework.boot.loader`：`spring-boot-loader` 的核心引导类。

其中 `META-INF` 中的 `MANIFEST.MF` 文件如代码清单 14-4 所示。它的内容比较多，而且其中不乏有一些熟悉的身影。

# 第 14 章 运行 Spring Boot 应用

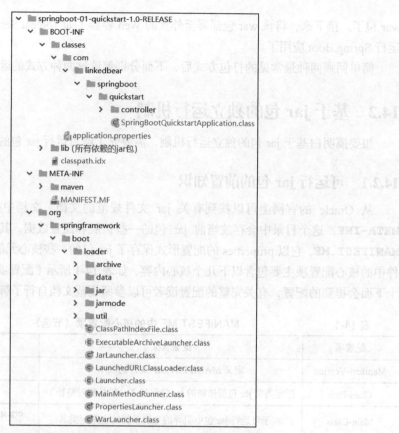

图 14-2 Spring Boot 打包好的 jar 包解压结构

**代码清单 14-4    MANIFEST.MF 文件的内容**

```
Manifest-Version: 1.0
Spring-Boot-Classpath-Index: BOOT-INF/classpath.idx
Implementation-Title: springboot-01-quickstart
Implementation-Version: 1.0-RELEASE
Start-Class: com.linkedbear.springboot.quickstart.SpringBootQuickstartApplication
Spring-Boot-Classes: BOOT-INF/classes/
Spring-Boot-Lib: BOOT-INF/lib/
Build-Jdk-Spec: 1.8
Spring-Boot-Version: 2.3.11.RELEASE
Created-By: Maven Jar Plugin 3.2.0
Main-Class: org.springframework.boot.loader.JarLauncher
```

注意其中的两个配置项：Main-Class:org.springframework.boot.loader.JarLauncher 以及 Start-Class:com.linkedbear.springboot.quickstart.SpringBootQuickstartApplication。这两个配置分别定义了两个类，其中 SpringBootQuickstartApplication 是我们在第 1 章中编写的示例项目的主启动类，这里它用了一个特殊的配置项 **Start-Class** 来引用，这个配置项本身并不是 MANIFEST.MF 文件标准规范中的配置项，而是 Spring Boot 自行定义的，因此我们有理由认为，如果直接用 SpringBootQuickstartApplication 来引导这个可运行 jar 包，是无法启动项目的。

将 MANIFEST.MF 文件中 Main-Class 属性的值改为 SpringBootQuickstart

Application,并执行java-jar命令,发现直接使用SpringBootQuickstartApplication根本无法引导启动 Spring Boot 项目，如图 14-3 所示。

```
Main-Class: com.linkedbear.springboot.quickstart.SpringBootQuickstartApplication
```

```
E:\>java -jar springboot-01-quickstart-1.0-RELEASE.jar
错误: 找不到或无法加载主类 com.linkedbear.springboot.quickstart.SpringBootQuickstartApplication

E:\>
```

图 14-3　Spring BootQuickstartApplication 无法引导 jar 包启动

无法启动的原因是引导的 `SpringBootQuickstartApplication` 并没有完全放在 jar 包的顶层目录下，而是放在了 `BOOT-INF/classes/` 目录下，中间隔了两层包，所以无法引导启动。

如果用默认打包好的可运行 jar 包中的 `JarLauncher`，就可以正常引导 Spring Boot 项目启动，这说明这个 `JarLauncher` 是引导的核心，需要重点研究它的设计。

### 14.2.3　JarLauncher 的设计及工作原理

`JarLauncher` 本身是来自 `spring-boot-loader` 依赖中一个普通的带 `main` 方法的类，只不过 Spring Boot 需要它来引导可运行 jar 包启动，它才以一种特殊的方式"降临"到 jar 包中。可以发现连同 `JarLauncher` 在内的非常多字节码文件直接存在于可运行 jar 包的顶层目录之下，这是因为可运行 jar 包的规范要求如此，所以 Spring Boot 在打包时不得不将这些 .class 文件全部复制到 jar 包中。

**1. JarLauncher 的继承结构**

借助 IDEA 可以生成包括 `JarLauncher` 在内的类继承关系图，如图 14-4 所示。

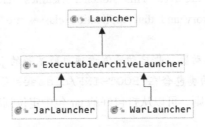

图 14-4　JarLauncher 的继承关系

从图 14-4 中可以看到，Spring Boot 项目的启动器是通过两个 `Launcher` 的落地子类实现的，它们分别处理 jar 包启动和 war 包启动，而这两个落地子类又同时继承自 `ExecutableArchiveLauncher`。下面先简单了解一下这几个类的设计。

（1）Launcher

`Launcher` 是所有启动 Spring Boot 项目的顶层引导类，它的内部定义了一个非常关键的 `launch` 方法，该方法就是用于启动 Spring Boot 应用的核心方法。

（2）ExecutableArchiveLauncher

`ExecutableArchiveLauncher` 从类名上可以理解为"可执行归档文件的启动器"，这里涉及一个概念"归档文件"，虽然陌生，但理解起来相对容易。读者可以简单地理解为，一个

Spring Boot 的独立可运行 jar 包就是一个归档文件，可以放在外置的 Web 容器中运行的 war 包也是一个归档文件。`ExecutableArchiveLauncher` 类中额外拥有的能力主要是可以从归档文件中检索到 Spring Boot 的主启动类，并提供给父类 `Launcher` 以完成主启动类的引导。

（3）JarLauncher

JarLauncher 是基于 Spring Boot 独立可运行 jar 包的启动引导器，它的内部定义了几个核心的常量，它们与打包好的 jar 包中 `BOOT-INF` 文件夹中的内容一一对应，从图 14-5 中可以看出它们的对应关系。

```
public class JarLauncher extends ExecutableArchiveLauncher {
    private static final String DEFAULT_CLASSPATH_INDEX_LOCATION = "BOOT-INF/classpath.idx";
    static final EntryFilter NESTED_ARCHIVE_ENTRY_FILTER = (entry) -> {
        if (entry.isDirectory()) {
            return entry.getName().equals("BOOT-INF/classes/");
        }
        return entry.getName().startsWith("BOOT-INF/lib/");
    };
```

图 14-5　JarLauncher 的常量与 BOOT-INF 中的成员一一对应

请注意，`classpath.idx` 是 Spring Boot 2.3.0 版本之后新添加的成员，在此版本之前的 JarLauncher 中的成员更简单清晰，如代码清单 14-5 所示。

**代码清单 14-5　Spring Boot 2.3.0 版本之前的 JarLauncher 定义的常量**

```
public class JarLauncher extends ExecutableArchiveLauncher {
    static final String BOOT_INF_CLASSES = "BOOT-INF/classes/";
    static final String BOOT_INF_LIB = "BOOT-INF/lib/";
```

由此可见，无论 Spring Boot 版本的高低，这两个指定 `BOOT-INF` 目录的路径都是关键。在 javadoc 中已经清楚地解释了这两个路径的作用：

> Launcher for JAR based archives. This launcher assumes that dependency jars are included inside a /BOOT-INF/lib directory and that application classes are included inside a /BOOT-INF/classes directory.
>
> 基于 jar 包归档文件的启动器。此启动器假设项目所依赖的 jar 包包含在 /BOOT-INF/lib 目录中，并且项目所中定义的类包含在 /BOOT-INF/classes 目录中。

依照 javadoc 中描述的规则，可以从打包好的 jar 包中定位到项目中编写的类、配置文件，以及所依赖的所有 jar 包。

（4）WarLauncher

WarLauncher 是基于 Spring Boot 以外置 Web 容器运行时打包 war 包的启动类引导器，请注意它本身也是一个启动器，可以将打包好的 war 包使用 `java-jar` 命令引导启动 Spring Boot 应用，如图 14-6 所示。

图 14-6　打包好的 war 包也可以独立运行

与 jar 包不同，war 包对于所依赖的 jar 包和项目中的 class 有一定限制，对于一个标准 war 包而言，项目中定义的所有类应当放在 `WEB-INF/classes` 目录下，所依赖的 jar 包则放在 `WEB-INF/lib` 下。除此之外，Spring Boot 为了使 war 包也能独立运行，会将所有作用范围为 provided 的依赖统一放在 `WEB-INF/lib-provided` 中，这部分在独立运行时可以与 `WEB-INF/lib` 下的依赖同时被加载，而当 war 包放置于 Web 容器时，由于 Web 容器不会读取 `lib-provided` 目录，因此这部分不会被加载，这样就同时兼容了两种启动方式。

通过外置 Web 容器引导的机制会在 14.3 节中讲解，而独立 war 包运行的引导机制与 JarLauncher 基本一致。下面会分别讲解 JarLauncher 与 WarLauncher 的引导原理。

**2．JarLauncher 的引导原理**

在 JarLauncher 内部定义了一个 main 方法，它就是整个可运行 jar 包在运行时的入口。JarLauncher 的构造方法中没有任何动作，也没有调用父类的构造方法，所以所有动作都在后面的 launch 方法中，如代码清单 14-6 所示。注意，launch 方法没有在 JarLauncher 中定义，而是在顶层父类 Launcher 中定义。

**│代码清单 14-6　JarLauncher 的引导入口**

```java
public JarLauncher() { }

public static void main(String[] args) throws Exception {
    new JarLauncher().launch(args);
}

// 父类 Launcher
protected void launch(String[] args) throws Exception {
    // 注册 URL 协议并清除应用缓存
    if (!isExploded()) {
        JarFile.registerUrlProtocolHandler();
    }
    // 创建类加载器
    ClassLoader classLoader = createClassLoader(getClassPathArchivesIterator());
    String jarMode = System.getProperty("jarmode");
    // 获取主启动类的类名
    String launchClass =(jarMode != null && !jarMode.isEmpty())
            ? JAR_MODE_LAUNCHER : getMainClass();
    // 执行主启动类的 main 方法
    launch(args, launchClass, classLoader);
}
```

launch 方法的文档注释中有一句话：本方法是一个入口点，且应该被一个 `public static void main(String[] args)`（即 main 方法）调用。这句话刚好与代码清单 14-7 上方的 main 方法相呼应（WarLauncher 中也是如此）。整个 launch 方法中的核心步骤可以拆分为 3 步，下面分步解释其原理。

**（1）创建 ClassLoader**

由于可独立运行的 jar 包中使用常规的类加载器无法加载内部 jar 包中的类，因此 Spring Boot 需要对其进行特殊处理。createClassLoader 方法中会创建一个特殊的类加载器 LaunchedURLClassLoader，它可以加载内部嵌套 jar 包中的类。

具体分析源码，需要进入 createClassLoader 方法中，但是在此之前需要读者先关注

createClassLoader 方法中传入的参数：getClassPathArchivesIterator()。很明显这是将一个方法执行完毕后的返回值传入了 createClassLoader 方法中，所以需要先进入获取参数的 getClassPathArchivesIterator 方法中，如代码清单 14-7 所示。

**代码清单 14-7　ExecutableArchiveLauncher#getClassPathArchivesIterator**

```
protected Iterator<Archive> getClassPathArchivesIterator() throws Exception {
    Archive.EntryFilter searchFilter = this::isSearchCandidate;
    Iterator<Archive> archives = this.archive.getNestedArchives(searchFilter,
            (entry) -> isNestedArchive(entry) && !isEntryIndexed(entry));
    if (isPostProcessingClassPathArchives()) {
        archives = applyClassPathArchivePostProcessing(archives);
    }
    return archives;
}
```

从代码清单 14-7 中来看，getClassPathArchivesIterator 方法的可读性并不强，大概总结下来可以理解为，通过 getClassPathArchivesIterator 方法，可以将当前 Spring Boot 应用中依赖的嵌套 jar 包和字节码文件都获取到，并且以迭代器的形式返回。获取的关键动作是中间的 this.archive.getNestedArchives 方法，它需要传入两个过滤器，一个是搜索范围，另一个是过滤条件。对于可独立运行的 jar 文件来讲，搜索范围是所有路径以 BOOT-INF 开头的文件，而过滤条件是筛选所有 BOOT-INF/lib 目录下的文件以及 BOOT-INF/classes 目录下的所有文件夹，相关源码如代码清单 14-8 所示。

**代码清单 14-8　从 BOOT-INF 下搜索 jar 包和字节码文件**

```
// 搜索范围
protected boolean isSearchCandidate(Archive.Entry entry) {
    return entry.getName().startsWith("BOOT-INF/");
}

// 过滤条件
protected boolean isNestedArchive(Archive.Entry entry) {
    return NESTED_ARCHIVE_ENTRY_FILTER.matches(entry);
}
static final EntryFilter NESTED_ARCHIVE_ENTRY_FILTER = (entry) -> {
    // 收集 BOOT-INF/classes 下的文件夹
    if (entry.isDirectory()) {
        return entry.getName().equals("BOOT-INF/classes/");
    }
    // 收集 BOOT-INF/lib 下的所有文件
    return entry.getName().startsWith("BOOT-INF/lib/");
};
```

通过观察源码，读者可以很明确地理解 Spring Boot 打包好的 jar 包中每个文件夹的含义：**BOOT-INF/classes** 目录存放的是当前 Spring Boot 项目中编写的所有类经过编译之后的字节码文件，**BOOT-INF/lib** 目录存放的是当前 Spring Boot 项目所依赖的所有 jar 包。

加载上述文件之后，就进入了创建 ClassLoader 的方法中，如代码清单 14-9 所示。在创建类加载器之前，createClassLoader 方法会将上一步获取到的 Archive 对象转换为一个个 URL 对象，每个 URL 对象对应一个 jar 包或者字节码文件目录的路径。转换完成后，最终要

创建的 `ClassLoader` 实现类是 `LaunchedURLClassLoader`,虽然它的构造方法中传入的参数比较多,但读者只需要关心 `urls` 数组。

**代码清单 14-9　createClassLoader 创建特殊的类加载器**

```java
protected ClassLoader createClassLoader(Iterator<Archive> archives) throws Exception {
    List<URL> urls = new ArrayList<>(guessClassPathSize());
    while (archives.hasNext()) {
        urls.add(archives.next().getUrl());
    }
    if (this.classPathIndex != null) {
        urls.addAll(this.classPathIndex.getUrls());
    }
    return createClassLoader(urls.toArray(new URL[0]));
}

protected ClassLoader createClassLoader(URL[] urls) throws Exception {
    return new LaunchedURLClassLoader(isExploded(), getArchive(), urls,
        getClass().getClassLoader());
}
```

`LaunchedURLClassLoader` 创建完成后,`launch` 的第一步就执行完毕了。

（2）获取主启动类名

`launch` 方法的倒数第二行会执行 `getMainClass` 方法,定位当前 Spring Boot 应用的主启动类,它的实现要向下找到 `Launcher` 的子类 `ExecutableArchiveLauncher`,如代码清单 14-10 所示。

**代码清单 14-10　ExecutableArchiveLauncher#getMainClass**

```java
private static final String START_CLASS_ATTRIBUTE = "Start-Class";

protected String getMainClass() throws Exception {
    Manifest manifest = this.archive.getManifest();
    String mainClass = null;
    if (manifest != null) {
        mainClass = manifest.getMainAttributes().getValue(START_CLASS_ATTRIBUTE);
    }
    if (mainClass == null) {
        // throw ex ...
    }
    return mainClass;
}
```

从源码中可以清晰地看出,解析 Spring Boot 主启动类的方式就是提取 MANIFEST.MF 文件中"Start-Class"属性对应的值,而在代码清单 14-4 中读者已经看到过 MANIFEST.MF 文件的内容,`Start-Class` 刚好就是定义了 Spring Boot 应用主启动类的全限定类名。

（3）执行主启动类的 main 方法

特殊的类加载器 `LaunchedURLClassLoader` 以及主启动类都获取到之后,最后一步是真正启动 Spring Boot 应用。进入重载的 `launch` 方法中,可以发现触发主启动类 `main` 方法的机制是借助 `MainMethodRunner`,利用反射机制调用 Spring Boot 主启动类的 `main` 方法,如代码清单 14-11 所示。

**代码清单 14-11　launch 借助反射机制启动 Spring Boot 主启动类**

```java
protected void launch(String[] args, String launchClass, ClassLoader classLoader)
        throws Exception {
    Thread.currentThread().setContextClassLoader(classLoader);
    // 创建 MainMethodRunner，并调用 run 方法
    createMainMethodRunner(launchClass, args, classLoader).run();
}

protected MainMethodRunner createMainMethodRunner(String mainClass, String[] args,
        ClassLoader classLoader) {
    return new MainMethodRunner(mainClass, args);
}

public void run() throws Exception {
    Class<?> mainClass = Class.forName(this.mainClassName, false,
            Thread.currentThread().getContextClassLoader());
    Method mainMethod = mainClass.getDeclaredMethod("main", String[].class);
    mainMethod.setAccessible(true);
    mainMethod.invoke(null, new Object[] {this.args});
}
```

当 Spring Boot 主启动类的 `main` 方法被反射调用成功后，Spring Boot 应用即可顺利启动，基于 `JarLauncher` 的启动引导完成。

### 3. WarLauncher 的引导原理

使用 `WarLauncher` 的引导原理在本质上与 `JarLauncher` 并无太大区别，只是在定位 jar 包和字节码文件时搜索的目录不同，如代码清单 14-12 所示。

**代码清单 14-12　WarLauncher 中搜索 jar 包和字节码文件的规则**

```java
protected boolean isSearchCandidate(Entry entry) {
    return entry.getName().startsWith("WEB-INF/");
}

public boolean isNestedArchive(Archive.Entry entry) {
    if (entry.isDirectory()) {
        return entry.getName().equals("WEB-INF/classes/");
    }
    return entry.getName().startsWith("WEB-INF/lib/")
            || entry.getName().startsWith("WEB-INF/lib-provided/");
}
```

由于标准 war 包中项目编译后的字节码文件和 jar 包会分别放入 `WEB-INF/classes` 和 `WEB-INF/lib` 中，而部署到外置 Servlet 容器中解压后的 Web 项目也是如此，因此 Spring Boot 为了同时兼容这两种情况，在搜索 Archive 归档文件时做出了一些调整。

除此之外，`WarLauncher` 的启动引导原理与 `JarLauncher` 并无区别，不再赘述。

## 14.3　基于 war 包的外部 Web 容器运行机制

基于 war 包的外置容器运行需要借助 Servlet 3.0 规范的一个引导机制，这个机制是引导 Spring Boot 应用启动的核心，我们需要先对它有所了解。

### 14.3.1 Servlet 3.0 规范中引导应用启动的说明

在 Servlet 3.0 规范文档的 8.2.4 节中有对运行时插件的描述，以下内容节选自该小节。

> An instance of the ServletContainerInitializer is looked up via the jar services API by the container at container / application startup time. The framework providing an implementation of the ServletContainerInitializer MUST bundle in the META-INF/services directory of the jar file a file called javax.servlet.ServletContainerInitializer, as per the jar services API, that points to the implementation class of the ServletContainerInitializer.
>
> 在容器/应用程序启动时，容器通过 SPI 机制查找 ServletContainerInitializer 的实例。提供 ServletContainerInitializer 实现的框架必须在 jar 包的 META-INF/services 目录中定义一个名为 javax.servlet.ServletContainerInitializer 的文件，根据 SPI 机制，找到对应的 ServletContainerInitializer 接口的实现类。

由该段描述可以得知，Servlet 容器启动应用时会扫描项目及依赖 jar 包中 ServletContainerInitializer 接口的实现类，如果项目依赖的框架需要在启动时初始化，就必须在 jar 包的 META-INF/services 目录中提供一个名为 javax.servlet.ServletContainerInitializer 的文件，文件内容要标明 ServletContainerInitializer 接口实现类的全限定名。从代码清单 14-13 中可以发现，ServletContainerInitializer 本身是一个接口，它仅有一个 onStartUp 方法，不难推测出 Servlet 容器启动时会回调 onStartUp 方法以完成应用的初始化逻辑。

**代码清单 14-13　ServletContainerInitializer**

```
public interface ServletContainerInitializer {
    void onStartup(Set<Class<?>> c, ServletContext ctx) throws ServletException;
}
```

此外，实现了 ServletContainerInitializer 接口的实现类可以在类上标注 @HandlesTypes 注解，并指定一些感兴趣的类（或接口类型），Servlet 容器初始化时会将这些感兴趣的类（或接口的实现类）传入 onStartUp 方法的第一个参数中，以此可以完成一些更高级的处理。

Spring Boot 为了适配外置 Servlet 容器启动的方式，提供了一个特殊的 ServletContainerInitializer 实现类 **SpringServletContainerInitializer**，这个类会使用上述 Servlet 规范中的特性。

### 14.3.2　Spring BootServletInitializer 的作用和原理

在研究 SpringBootServletInitializer 的作用机制之前，先请读者回想一下 Spring Boot 打包 war 包时的必要步骤。除了修改 pom.xml 文件中的打包方式、修改嵌入式 Web 容器的作用域，还需要编写一个 SpringBootServletInitializer 的子类，指定 Spring Boot 主启动类作为启动源，如代码清单 14-14 所示。

## 第 14 章 运行 Spring Boot 应用

**代码清单 14-14  SpringBootServletInitializer 的子类用于配置主启动类**

```
public class ServletInitializer extends SpringBootServletInitializer {
    @Override
    protected SpringApplicationBuilder configure(SpringApplicationBuilder application) {
        return application.sources(SpringBootLaunchApplication.class);
    }
}
```

如此编写的目的是在当前的 Spring Boot 项目中提供一个 `SpringBootServletInitializer` 的子类，从而让外置 Servlet 容器在启动时可以加载该子类，从而初始化和启动 Spring Boot 应用。下面结合源码分析外置 Servlet 容器如何引导和启动 Spring Boot 应用。

### 1. ServletContainerInitializer 的加载

当外置 Servlet 容器启动时，默认会加载 webapp 目录下的 war 包，此时被打包成 war 包的 Spring Boot 项目被解压，Servlet 容器会从当前项目及项目所依赖的 jar 包中搜索一个全路径名为 `META-INF/services/javax.servlet.ServletContainerInitializer` 的文件（该特性基于 JDK SPI 机制）。如果可以成功搜索到该文件，则会加载文件中定义的全限定类名对应的类。从 spring-web 依赖中可以找到该文件，对应的类是 `SpringServletContainerInitializer`，这个类上标注了 `@HandlesTypes` 注解，它感兴趣的类型是 `WebApplicationInitializer`，这意味着 `SpringServletContainerInitializer` 的 onStartUp 方法会获取当前项目中所有实现了 `WebApplicationInitializer` 接口的最终落地实现类，如代码清单 14-15 所示。

**代码清单 14-15  SpringServletContainerInitializer**

```
@HandlesTypes(WebApplicationInitializer.class)
public class SpringServletContainerInitializer implements ServletContainerInitializer {
    @Override
    public void onStartup(Set<Class<?>> webAppInitializerClasses,
            ServletContext servletContext) throws ServletException {
        // 加载、实例化 WebApplicationInitializer 对象 ......
        for (WebApplicationInitializer initializer : initializers) {
            initializer.onStartup(servletContext);
        }
    }
}
```

### 2. SpringBootServletInitializer 的加载

请注意，上面提到的 `ServletInitializer` 本身就是一个 `WebApplicationInitializer`，如代码清单 14-16 所示。所以在 `SpringServletContainerInitializer` 的 onStartUp 方法中实际上获取的就是当前项目中定义的 `ServletInitializer` 类，并在实例化对象后调用其 onStartup 方法。

**代码清单 14-16  ServletInitializer 本身是一个 WebApplicationInitializer**

```
public class ServletInitializer extends SpringBootServletInitializer
public abstract class SpringBootServletInitializer implements WebApplicationInitializer
```

由于 `onStartUp` 方法定义在 `ServletInitializer` 的父类 `SpringBootServletInitializer` 中，下面的研究重点放在父类的源码上。

#### 3. SpringApplication 的构建与启动

`SpringBootServletInitializer` 的 `onStartUp` 方法的核心动作是创建一个 `WebApplicationContext`，而创建的过程需要构建 `SpringApplication` 并启动，具体的逻辑如代码清单 14-17 所示（难度不大，读者通读一遍即可）。注意 `createRootApplicationContext` 方法的中间有 `configure` 方法的调用动作，该方法就是子类 `ServletInitializer` 中用来编程式指定 Spring Boot 主启动类的关键步骤。

**代码清单 14-17　SpringBootServletInitializer#onStartUp**

```java
public void onStartup(ServletContext servletContext) throws ServletException {
    // ......
    WebApplicationContext rootApplicationContext = createRootApplicationContext(servletContext);
    // ......
}

protected WebApplicationContext createRootApplicationContext(ServletContext servletContext) {
    SpringApplicationBuilder builder = createSpringApplicationBuilder();
    builder.main(getClass());
    ApplicationContext parent = getExistingRootWebApplicationContext(servletContext);
    // 存在父容器的处理
    builder.initializers(new ServletContextApplicationContextInitializer(servletContext));
    builder.contextClass(AnnotationConfigServletWebServerApplicationContext.class);
    // 【关键】注意此处会跳转至自定义的 ServletInitializer 子类
    builder = configure(builder);
    builder.listeners(new WebEnvironmentPropertySourceInitializer(servletContext));
    // 构建 SpringApplication
    SpringApplication application = builder.build();
    // ......
    // 基于外置 Servlet 容器启动不需要注册回调钩子
    application.setRegisterShutdownHook(false);
    // 启动 SpringApplication
    return run(application);
}

protected WebApplicationContext run(SpringApplication application) {
    return (WebApplicationContext) application.run();
}
```

经过 `SpringApplicationBuilder` 的构建并调用 `SpringApplication` 的 `run` 方法，Spring Boot 项目即可成功启动。

## 14.4　Spring Boot 2.3 新特性：优雅停机

在 Spring Boot 的应用运行需要停机时，如果直接关闭应用，会导致部分正在处理中的请求被强制中断，在某些特殊的业务场景中会产生脏数据。为了解决这一问题，Spring Boot 2.3 中引入了"优雅停机"的新特性，在 Spring Boot 应用被关闭时（注意此处的关闭可以是 `kill -2`，但不能是 `kill -9`），Spring Boot 会预留一小段时间，使应用内部的业务线程执行完毕，此时嵌入式 Web 容器不允许客户端有新的请求进入，以此达到优雅停机的效果。

### 14.4.1 测试优雅停机场景

下面通过一个简单的测试场景来体会优雅停机的作用效果。想要模拟优雅停机的效果，只需要编写一个处理时间很长的接口，如代码清单 14-18 所示，Handler 中使线程休眠 10s，模拟业务逻辑的处理，在线程休眠的前后添加时间戳的打印，以便于测试观察。

**代码清单 14-18　GracefulTestController 用于测试长时间响应接口**

```java
@RestController
public class GracefulTestController {

    @GetMapping("/test")
    public String test() throws Exception {
        System.out.println(System.currentTimeMillis());
        TimeUnit.SECONDS.sleep(10);
        System.out.println(System.currentTimeMillis());
        return "success";
    }
}
```

默认情况下 Spring Boot 的停机方式是立即停机（immediate），若想启用优雅停机，需要在 `application.properties` 中配置 `server.shutdown=graceful`，配置完毕即代表开启默认策略的优雅停机。此外还可以通过配置 `spring.lifecycle.timeout-per-shutdown-phase` 属性，自定义缓冲时间（默认为 30s）。

请注意，如果读者在学习时使用 IDE 测试优雅停机，需要修改主启动类的内容（见代码清单 14-19），或者引入 actuator 依赖实现优雅停机，而不是借助 IDE 的停止应用。如果直接单击 IDE 中的停止按钮，会直接关闭 JVM（类似于 `kill -9`），从而导致无法触发优雅停机的回调机制。

**代码清单 14-19　使用 IDE 测试优雅停机的必要修改**

```java
@SpringBootApplication
public class SpringBootLaunchApplication {

    public static void main(String[] args) {
        ApplicationContext ctx = SpringApplication.run(SpringBootLaunchApplication.class, args);

        // 借助 Scanner 实现控制台软退出
        Scanner scanner = new Scanner(System.in);
        while (true) {
            String input = scanner.nextLine();
            if ("exit".equals(input)) {
                break;
            }
        }

        SpringApplication.exit(ctx);
    }
}
```

准备工作完成后启动项目,使用浏览器访问 `localhost:8080/test`,此时若不加干预,浏览器在等待 10s 后会接收到"success"的字符串响应;若访问请求用手动停止 Spring Boot 项目,可以发现浏览器仍然在等待响应,并在 10s 后接收到来自服务端的"success"响应。

```
DispatcherServlet : Completed initialization in 4 ms
1640966400000
// 此行为手动输入
exit
// 输入之后控制台会打印 GracefulShutdown 的缓冲等待
GracefulShutdown : Commencing graceful shutdown. Waiting for active requests to complete
1640966410000
GracefulShutdown : Graceful shutdown complete
ThreadPoolTaskExecutor : Shutting down ExecutorService 'applicationTaskExecutor'
```

不同的嵌入式 Web 容器在优雅停机期间应对客户端新请求的响应策略不同:嵌入式 Tomcat 和 Netty 不会接收请求,客户端会响应超时;嵌入式 Undertow 则会直接响应 503 错误。

### 14.4.2 优雅停机的实现原理

简单了解优雅停机的使用与现象后,下面解释优雅停机的实现原理。以嵌入式 Tomcat 为例,在 `TomcatServletWebServerFactory` 创建 `TomcatWebServer` 时,传入的 `Shutdown` 枚举即代表停机策略,如代码清单 14-20 所示。注意,此处的参数 `shutdown` 就是 Spring Boot 全局配置文件中 `server.shutdown` 的值。

**代码清单 14-20　创建 TomcatWebServer 时指定停机策略**

```java
protected TomcatWebServer getTomcatWebServer(Tomcat tomcat) {
    return new TomcatWebServer(tomcat,getPort() >= 0, getShutdown());
}

public TomcatWebServer(Tomcat tomcat, boolean autoStart, Shutdown shutdown) {
    Assert.notNull(tomcat, "Tomcat Server must not be null");
    this.tomcat = tomcat;
    this.autoStart = autoStart;
    // 此处会初始化优雅停机的回调钩子
    this.gracefulShutdown=(shutdown == Shutdown.GRACEFUL) ? new GracefulShutdown(tomcat) : null;
    initialize();
}
```

当 Spring Boot 项目关闭时,根据第 8 章中了解的回调钩子,`WebServerGracefulShutdownLifecycle` 会被调用,触发停机逻辑判断,如代码清单 14-21 所示。默认情况下停机的模式是 `IMMEDIATE`,对应的 if 结构会立即关闭;而如果在嵌入式 Web 容器初始化时设置了优雅停机,则会执行 if 结构下面的最后一行代码,即调用 `GracefulShutdown` 的 `shutDownGracefully` 方法。

**代码清单 14-21　WebServerGracefulShutdownLifecycle 的关闭回调**

```java
public void stop(Runnable callback) {
    this.running = false;
```

## 第 14 章 运行 Spring Boot 应用

```java
        this.webServer.shutDownGracefully((result) -> callback.run());
    }
    public void shutDownGracefully(GracefulShutdownCallback callback) {
        if (this.gracefulShutdown == null) {
            callback.shutdownComplete(GracefulShutdownResult.IMMEDIATE);
            return;
        }
        // 此处会执行优雅停机
        this.gracefulShutdown.shutDownGracefully(callback);
    }
```

shutDownGracefully 方法完成的工作是延迟关闭嵌入式 Web 容器,它会在内部启动一个新的线程,执行 doShutdown 方法,如代码清单 14-22 所示。doShutdown 方法中首先会关闭 Connector,由此 Tomcat 就失去了接收客户端新请求的能力;随后该方法中会提取出嵌入式 Tomcat 中所有 Engine 中的所有 Context,每隔 50ms 检查其是否停止,当所有 Context 中的线程全部执行完毕,即 Context 全部停止时,优雅停机流程执行完毕。

**代码清单 14-22　doShutdown 方法实现优雅停机**

```java
void shutDownGracefully(GracefulShutdownCallback callback) {
    logger.info("Commencing graceful shutdown. Waiting for active requests to complete");
    new Thread(() -> doShutdown(callback), "tomcat-shutdown").start();
}

private void doShutdown(GracefulShutdownCallback callback){
    // 关闭 Connector, 失去接收客户端新请求的能力
    List<Connector> connectors = getConnectors();
    connectors.forEach(this::close);
    try {
        for (Container host : this.tomcat.getEngine().findChildren()) {
            for (Container context : host.findChildren()) {
                // 每隔 50ms 检查一次 Container 是否停止
                while (isActive(context)) {
                    if (this.aborted) {
                        // logger ......
                        callback.shutdownComplete(GracefulShutdownResult.REQUESTS_ACTIVE);
                        return;
                    }
                    Thread.sleep(50);
                }
            }
        }
    } // catch ......
    logger.info("Graceful shutdown complete");
    callback.shutdownComplete(GracefulShutdownResult.IDLE);
}
```

由代码清单 14-22 中可以看到上述测试中打印的两行日志,证明优雅停机在底层的确生效了。

## 14.5 小结

本章从 Spring Boot 的部署运行出发，分别分析了可独立运行的 jar 包和借助外置 Web 容器的 war 包运行的底层机制，以及其中的特殊设计。Spring Boot 的强大特性之一就是可独立运行，它通过定制的 jar/war 包目录规则，配合特殊的类加载器，可以实现项目的可独立运行。此外 Spring Boot 2.3 新提供的优雅停机特性，可以使项目更可靠地关闭。

## 14.5 小结

本章从 Spring Boot 的框架着述于出发，介绍基于它构建立行的 Java 应用的跟著开发 Web 项或例如 war 包或者使用部署的包。以及其他的特殊用法。Spring Boot 的最大特点之一就是可以独立运行，它通过在面的 jar/war，包自行交流的的，结合秘器服务类型和部署，可以实现项目的独立运行，此处 Spring Boot 2.5 基础中的其容易进行部，可以使项目更可容实现。